Advances in Biochemical Engineering/ Biotechnology

Volume 196

Aims and Scope

This book series reviews current trends in modern biotechnology and biochemical engineering. Its aim is to cover all aspects of these interdisciplinary disciplines, where knowledge, methods and expertise are required from chemistry, biochemistry, microbiology, molecular biology, chemical engineering and computer science. This series supports and advances the UN's sustainable development goals, mainly SDG7 (Affordable and Clean Energy), SDG3 (Good Health and Well Being), and SDG13 (Climate Change).

Volumes are organized topically and provide a comprehensive discussion of developments in the field over the past 3–5 years. The series also discusses new discoveries and applications. Special volumes are dedicated to selected topics which focus on new biotechnological products and new processes for their synthesis and purification.

In general, volumes are edited by well-known guest editors. The series editor and publisher will, however, always be pleased to receive suggestions and supplementary information. Manuscripts are accepted in English.

In references, Advances in Biochemical Engineering/Biotechnology is abbreviated as *Adv. Biochem. Engin./Biotechnol.* and cited as a journal.

Xin-Qing Zhao • Yue-Qin Tang • Feng-Wu Bai
Editors

Biotechnology for Utilization of Agricultural Residues

With contributions by

A. E. K. Afedzi · F.-W. Bai · T. Bamba · L. Bingfeng · M. Cai · Y. Chang · Y. Chen · C. Cheng · Y. Cheng · C. Chi · N. Chiarasumran · X. Chuang · G. De Bhowmick · A. Debnath · J.-X. Feng · S. Ghosh · C. Guangli · X. Guojun · T. Hasunuma · X. Hu · S. Jiazheng · M. Jin · W. Jiwen · Y. Kawahara · I. Kazunori · A. Kondo · K. Li · K. Lueangwattanapong · R. Mitra · R. Nasuno · P. Parakulsuksatid · M. Saisriyoot · A. S. Sen · R. Sen · G. Shen · F. Shengqiang · P. Srinophakun · R.-Q. Tang · Y.-Q. Tang · M.-R. Tao · A. Thanapimmetha · N. Thongchul · W.-B. Wang · Z. Wenshuai · L. Xiaoxing · C.-Y. Xie · Q. Yang · S. Yang · X. Yang · T. Yuan · X. Yuan · G. Yujing · D.-W. Zeng · R. Zhai · S. Zhao · X.-Q. Zhao · Z. Zhao · S. Zhaoyong · Y. Zhong · D. Zhu · L. Ziyu

Springer

Editors
Xin-Qing Zhao
Shanghai Jiao Tong University
Shanghai, China

Yue-Qin Tang
Sichuan University
Sichuan, China

Feng-Wu Bai
Shanghai Jiao Tong University
Shanghai, China

ISSN 0724-6145 ISSN 1616-8542 (electronic)
Advances in Biochemical Engineering/Biotechnology
ISBN 978-3-032-38277-1 ISBN 978-3-032-38278-8 (eBook)
https://doi.org/10.1007/978-3-032-38278-8

This Springer imprint is published by the registered company Springer Nature Switzerland AG
The registered company address is: Gewerbestrasse 11, 6330 Cham, Switzerland

Preface

Agricultural residues, predominately with lignocellulosic biomass, are renewable, environmentally friendly, and abundantly available with less geographical limitations for sustainable socioeconomic development to address challenges with fossil resources such as crude oil. As a result, growing interests have been focused on producing biofuels, biobased chemicals, and biomaterials through the biorefinery. However, achieving cost-effective bioconversion and industrial production for those biobased products is still challenging, making it essential to stay informed about the latest technological advances and current developments in this regard.

The bioconversion of agricultural residues into valuable products is inherently a multidisciplinary endeavor, demanding seamless integration across diverse scientific and engineering domains. The plant cell wall is a recalcitrant matrix of cellulose, hemicellulose, and lignin that is highly resistant to microbial and enzymatic attack. Overcoming this "recalcitrance barrier" is the central challenge that unites the diverse topics for biorefinery of the biomass. From pretreatment technologies that dismantle this barrier to cellulosic enzyme systems that hydrolyze the resulting polysaccharides and to fermentation strategies that convert mixed sugars into fuels and biochemicals, each step in the biorefinery chain demands innovation, integration, and a deep understanding of the underlying mechanisms.

This volume systematically addresses topics in the bioconversion of agricultural residues. It begins with a survey of biomass resources and their conversion pathways, followed by an examination of pretreatment technologies and recent advances. The discussion then moves to enzymatic saccharification, emphasizing the need for cost-effective cellulase production to be achieved through both high-performance enzyme producers (e.g., *Trichoderma reesei*) and optimized fermentation processes. Subsequent sections cover metabolic engineering strategies for efficient co-utilization of C5 and C6 sugars to produce biofuels and biobased chemicals including butanol, butanediol, and succinic acid from lignocellulosic hydrolysates. In addition, producing soil conditioners, animal feed, biogas, and biohydrogen is also highlighted. Finally, environmental implications are considered, with the dedicated life cycle assessment of agricultural residues and their derived products.

With a comprehensive and up-to-date overview on the latest knowledge in this rapidly evolving field, we hope that this book volume serves as a valuable resource for researchers, engineers, industry professionals, and policymakers to work together and collectively for a truly low-carbon society and sustainable bioeconomy.

At the end, we express our sincere thanks to all authors and reviewers for their valuable contributions and comments and the production editor, Ms. Alamelu Damodharan, as well for her kind coordination of the publication work.

Shanghai, China — Xin-Qing Zhao
Chengdu, China — Yue-Qin Tang
Shanghai, China — Feng-Wu Bai

Contents

Adv Biochem Eng Biotechnol (2026) 196: 1–32
https://doi.org/10.1007/10_2026_316

Published online: 7 May 2026

Lignocellulosic Biomass Resources for Sustainable Socioeconomic Development

Pramuk Parakulsuksatid and Alfred Elikem Kwami Afedzi

Contents

P. Parakulsuksatid (✉) and A. E. K. Afedzi
Department of Biotechnology, Faculty of Agro-Industry, Kasetsart University, Bangkok, Thailand
e-mail: pramuk.p@ku.ac.th; alfredelikemkwami.a@ku.th

Abstract Agricultural residues including corn stover, rice straw, sugarcane bagasse, cereal straws, oil palm biomass, and many others are often burned or left to decompose, releasing avoidable emissions and losing nutrients. Redirected to modern biorefineries, they become a renewable carbon feedstock for biofuels, biogas, chemicals, and biomaterials. This chapter quantifies regional residue supplies and finds that Asia, the Americas, Europe, and Africa together generate more than 3 billion dry tons of residues yearly. Life cycle assessments report up to 90% greenhouse gas savings compared with fossil routes, particularly when open-field burning is eliminated. Technical advances in pretreatment, engineered enzymes, co-fermentation, and thermochemical upgrading now enable high-yield multiproduct conversion. Policy instruments such as feed-in tariffs, biogas purchase guarantees, carbon credits, and sustainability certifications lower investment risk and accelerate deployment. Cooperative models for collection, densification, transport, and processing create employment for farmers, women, and rural youth. Upgraded roads, storage, and off-grid energy systems follow. By integrating job creation, income diversification, and infrastructure development with rigorous soil and carbon safeguards, agricultural residues can transform from low-value waste into the foundation of a circular, low-carbon bioeconomy. Future studies should integrate advanced process innovations with equity strategies to secure resilient rural livelihoods.

Graphical Abstract

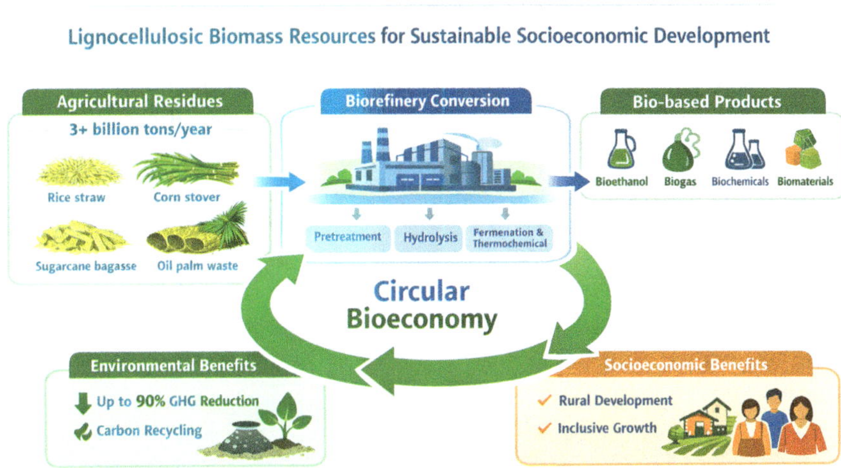

Keywords Agricultural residues, Biorefinery processes, Carbon footprint, Circular bioeconomy, Lignocellulosic biomass, Rural development

1 Introduction

1.1 Background and Significance of Lignocellulosic Biomass

Plants assimilate atmospheric CO_2 via photosynthesis, producing lignocellulosic biomass (LCB) that serves as a renewable carbon reservoir across croplands, forests, and dedicated energy crop systems. LCB is one of the most abundant renewable feedstocks on Earth. Global reliance on fossil fuels poses pressing challenges. These include finite resource availability, volatile energy prices, and severe environmental impacts [1]. Unlike fossil resources, when lignocellulosic feedstocks are managed well, they can deliver low-carbon fuels, power, and chemicals that support climate goals [2, 3]. Yet, the same evolutionary traits that make cell walls strong also hinder some industrial use. Cellulose microfibrils, hemicellulose, and lignin intertwine to form a lignin-carbohydrate complex (LCC) that resists chemical and enzymatic attack. Breaking that recalcitrance without raising costs or emissions remains a central technical and economic challenge. In addition, lignocellulosic feedstocks possess low bulk density and dispersed supply chains, raising logistics costs [4].

Annual global production of LCB exceeds 181.5 billion tons, yet only about 8.2 billion tons are currently harnessed [5], primarily from forest resources, agricultural

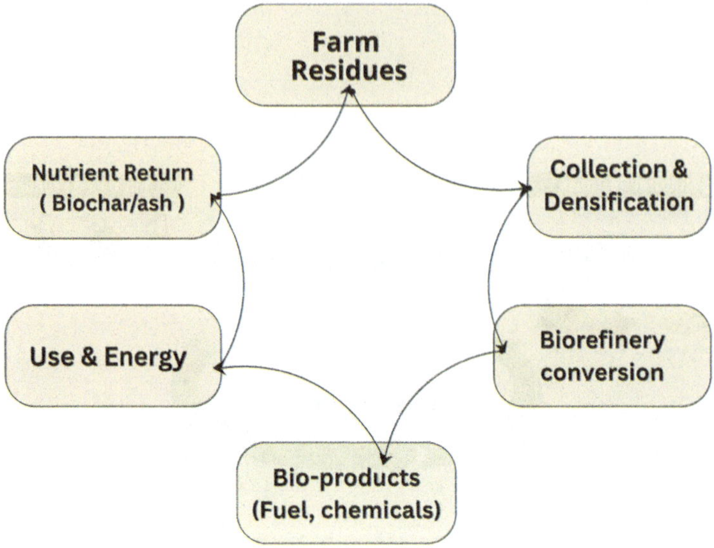

Fig. 1 Circular bioeconomy loop for agriculture residues

residues, grasses, and industrial waste streams [6]. Among all lignocellulosic options, agricultural residues – corn stover, rice straw, wheat straw, sugarcane bagasse, etc. – stand out. They are already harvested, typically undervalued, and regionally concentrated near farming communities. Redirecting them from open-field burning or landfilling into biorefineries offers dual benefits: lower carbon emissions and new rural income streams [7]. Additionally, advancements in biomass processing and bioconversion technologies have opened new pathways for its utilization in the sustainable production of biofuels, platform chemicals, and biodegradable materials. Hence, it aligns with strategies that aim to address global warming, resource depletion, and socioeconomic disparities. Governments, industries, and scientific communities now view lignocellulosic biomass as a cornerstone of the emerging bioeconomy [8].

1.2 The Role of Lignocellulosic Biomass in Climate Resilience, Circular Bioeconomy, and Rural Development

Circular bioeconomy principles aim to recirculate carbon, nutrients, and materials. Valorizing crop residues supports this vision by turning waste into biofuels, biogas, biochemicals, and biomaterials [9]. This is especially true when feedstock collection and conversion include energy recovery from byproducts. Many studies have demonstrated the potential to upcycle residues, such as sugarcane bagasse, oil palm biomass, or rice husks, into valuable commodities [10–13]. Such innovations reduce reliance on virgin fossil resources and help sustain natural ecosystems. They also

Table 1 Chemical composition of major agriculture residues

Agriculture residue	Cellulose (%)	Hemicellulose (%)	Lignin (%)	Other components (%)	References
Rice straw	29–44	16–34	8–20	Extractives: 4–16, Ash: 14–15	[20–24]
Wheat straw	32–46	28–31	8–22	Ash: 4–12	[25–28]
Corn stover	24–38	25–51	13–20	Extractives: 3–7, Ash: 5–11	[29–32]
Sugarcane bagasse and trash	34–50	23–32	21–34	Extractives: 7–14, Ash: 2–3	[12, 33–36]
Oil palm trunk, bunch, shell	21–42	23–36	22–51	Extractives: 4–9, Ash: 2–16	[13, 37–40]

keep carbon in circulation for a longer period, lowering net emissions [14]. Rural communities stand to benefit significantly. Agricultural residues are often overlooked in traditional supply chains. Through local biorefineries or decentralized conversion facilities, these communities can process residues close to their source [15]. This decreases transportation costs and improves operational efficiencies. It also creates jobs in feedstock collection, handling, and processing.

In turn, these new industries support improved livelihoods, financial inclusion, and local infrastructure development. Rural regions can thus become key players in a greener economy. Figure 1 illustrates the circular bioeconomy loop for crop residues, tracing carbon, nutrients, and value from the farm through the biorefinery and back to the soil as biochar and fertilizer ash. Residue utilization is not a linear disposal route but a regenerative cycle that locks biogenic carbon into repeated use.

1.3 Scope and Objectives

This chapter provides a concise overview of the composition, availability, distribution, environmental footprint, and socioeconomic value of lignocellulosic biomass, with a special focus on agricultural residues. Subsequent sections explore composition, availability, technologies, and policy frameworks. By the end, readers should grasp the fundamental role of lignocellulosic biomass in a sustainable bioeconomy.

2 Overview of Lignocellulosic Biomass

2.1 Composition and Structural Characteristics

Typical chemical composition of major agricultural residues includes cellulose, hemicellulose, and lignin (as shown in Table 1). Cellulose is a linear β-1,4-glucan that crystallizes into microfibrils and can reach 35–50 wt% in agricultural residues.

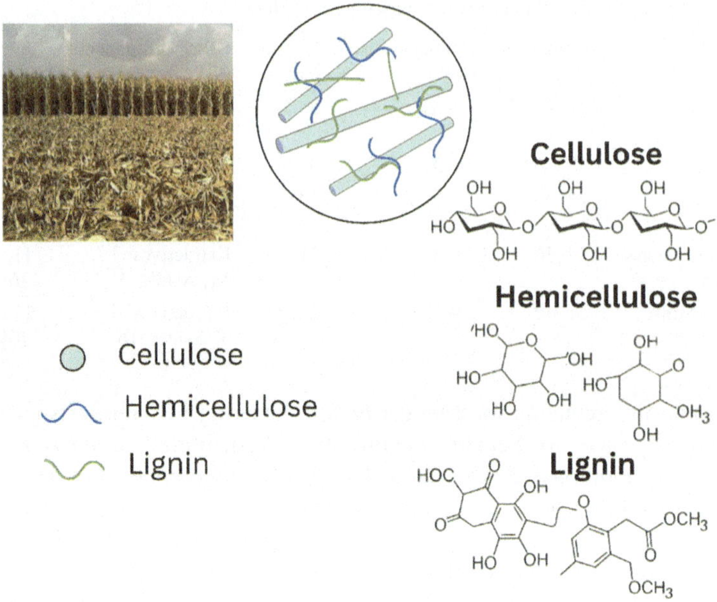

Fig. 2 Structural composition of lignocellulosic biomass (agricultural residues)

Hemicellulose is amorphous and branched; it carries pentoses (mainly xylose) and hexoses, making up 20–50 wt% depending on species. It holds the cellulose fibrils together [16]. Lignin is an irregular phenyl-propanoid network that locks fibers in place and confers hydrophobicity. Its content ranges from 10 wt% in cereal straw to >25 wt% in softwoods [17]. The physical and chemical arrangement of cellulose, hemicellulose, and lignin within agricultural residues highlights the complex structural interactions that confer biomass recalcitrance (as shown in Fig. 2). The relative proportions of cellulose, hemicellulose, and lignin vary among feedstocks. For example, agricultural residues often have moderate lignin content and high cellulose, whereas woody biomass contains higher lignin contents [18, 19]. These differences influence pretreatment severity and conversion efficiency.

Minor constituents such as extractives, proteins, ash, and minerals [41, 42] affect processing and product quality, ash melting behavior, and catalyst poisoning in thermochemical routes [43]. Compositional analysis, typically using protocols such as those from the National Renewable Energy Laboratory (NREL) and the Technical Association of the Pulp and Paper Industry (TAPPI), helps researchers design efficient bioconversion methods [44]. Understanding the chemical makeup is crucial for optimizing valorization routes in both laboratory and industrial settings.

2.2 *Classification of Biomass Resources*

2.2.1 Agricultural Residues

Agricultural residues are byproducts of crop harvest and postharvest processing. Common examples include corn stover, wheat straw, rice husks, oil palm biomass, and sugarcane bagasse. These residues are often abundant. Many farmers either burn these residues in the field or leave them to decompose, causing air pollution and carbon emissions [45, 46]. Valorization through biofuel or biochemical production can reduce waste and air pollution while offering economic returns.

2.2.2 Energy Crops

Energy crops are dedicated plant species cultivated explicitly for biomass production. They include perennial grass like switchgrass (*Panicum virgatum*) and miscanthus (*Miscanthus* × *giganteus*), as well as short-rotation woody crops such as poplar (*Populus* spp.) and willow (*Salix* spp.). These crops are bred for high biomass yield, efficient water use, and adaptability to marginal lands [47]. Because they can be harvested repeatedly over many years, they offer consistent feedstock supply for biorefinery operations. Such crops often exhibit favorable energy balances and reduced net greenhouse gas emissions when managed with sustainable practices [48, 49].

2.2.3 Forest Residues

Forest residues come from logging operations, forest thinning, and sawmill activities. These include branches, bark, sawdust, and offcuts. Woody biomass is rich in lignin [50]. This makes it more durable but harder to break down. Advanced pretreatment methods, such as steam explosion or organosolv, can help unlock its value [51, 52]. Collecting and using these residues also lower fire hazards by reducing excess biomass on forest floors.

2.2.4 Industrial Byproducts and Waste Streams

Industrial byproducts include pulp and paper sludge, brewer's spent grain, and fruit pomace. These streams can have high moisture content and variable composition [53, 54]. While they may pose disposal challenges, they also hold unique opportunities. Biochemical platforms can convert these materials into enzymes, bioethanol, or biogas [55, 56]. The circular bioeconomy gains momentum when industries recycle their own byproducts instead of discarding them.

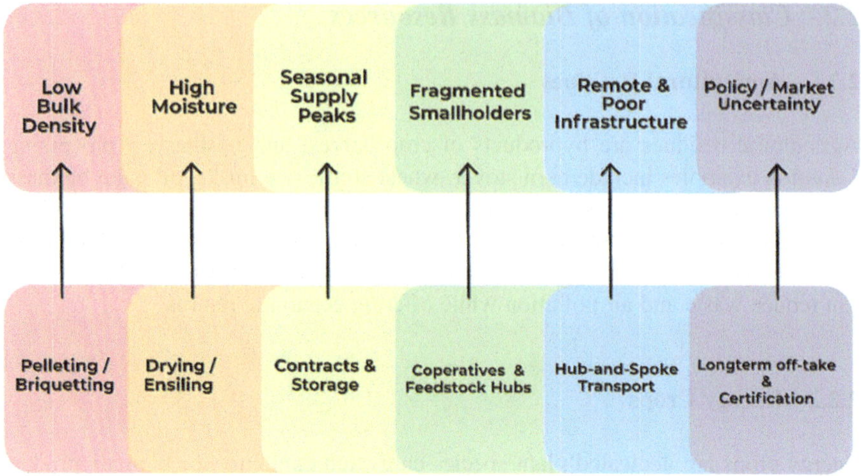

Fig. 3 Logistics bottlenecks and practical mitigations for agriculture residues

2.3 Key Factors Influencing Lignocellulosic Biomass Valorization

1. High lignin content hinders enzymatic access, while high cellulose boosts sugar yield. Yet, isolated lignin as a co-product is a valuable precursor for resins, carbon fibers, and bio-aromatics. Tailored pretreatment is needed to balance cost and conversion [57].
2. Wet biomass ferments well but hinders combustion and pyrolysis. Ensiling or hydrothermal storage stabilizes sugars for biochemical routes, while torrefaction upgrades drier streams for thermochemical uses [58].
3. Supply-chain logistics presents another challenge. To enhance operational efficiency and reduce transportation expenses, mechanized systems for rice straw collection have become essential. Baling equipment enables the aggregation and compression of straw into various shapes and densities, facilitating handling, storage, and logistics [59].
4. The seasonal availability of agricultural residues creates storage and transport issues [60]. Forest residues may be accessible year-round, but they are often located in remote areas with limited infrastructure.
5. Policy frameworks and market dynamics: Stable policy instruments such as long-term biomass procurement agreements, carbon credit systems, and sustainability certification schemes (e.g., RED II, ISO 13065) play a pivotal role in de-risking investments and ensuring market uptake. Conversely, policy uncertainty undermines investor confidence and disrupts supply-chain development [61, 62].

Critical logistical bottlenecks such as low bulk density, high moisture content, fragmented feedstock ownership, seasonal availability, and long transportation distances significantly affect the economic feasibility of agricultural residue valorization (as shown in Fig. 3). The proposed targeted solutions include pelleting, drying or ensiling, cooperative aggregation, contract farming, and hub-and-spoke logistics. This effectively mitigates these constraints and enhances feedstock reliability for biorefineries.

3 Global Availability and Geographic Distribution

3.1 Overview of Major Biomass-Producing Regions

The global production and distribution of lignocellulosic biomass vary widely due to climatic differences, agricultural practices, and land availability [63]. Effective utilization strategies require region-specific analyses that consider biomass types, production volumes, and local socioeconomic contexts.

3.1.1 North America

The 2024 Billion-Ton Report [64] estimates that the USA can supply 1.1–1.5 billion tons (Bt) of biomass annually in a mature market. This includes agricultural residues, energy crops, forest biomass, and waste materials. North America, particularly the USA and Canada, is a major producer of corn and wheat. As a result, the USA generates approximately 159 million tons (Mt) of corn stover and 18 million tons of wheat straw each year. Among energy crops, switchgrass shows the highest potential, with a projected yield of 230 million tons annually. These resources offer a sustainable pathway for producing renewable fuels, especially for sectors less suited to electrification while supporting food systems and environmental goals. Forest residues from Canada and the northern USA contribute significantly to the regional biomass supply [65].

3.1.2 South America

In South America, the theoretical biomass potential is estimated at 1,214 Mt and is projected to rise to 1,371 Mt by 2050. The technical potential available for energy applications ranges from 796 Mt in 2021 to 916 Mt by 2050. Of this, approximately 66% is derived from agricultural biomass, 10% from agro-industrial residues, 17% from forestry sources, and 7% from municipal solid waste [66]. However, this projected increase is not uniform across all countries. Brazil and Colombia collectively account for 97% of the region's direct waste biomass, primarily from large-scale sugarcane cultivation. Meanwhile, Argentina contributes about 45% of the region's indirect biomass, followed by Brazil, Peru, Chile, and Paraguay [67]. Brazil leads in sugarcane cultivation, producing significant bagasse for cogeneration, bioethanol, and biohydrogen [68, 69]. Other countries like Argentina and Colombia also contribute through residues from maize and other crops.

3.1.3 Asia

Asia produces a significant portion of the world's biomass, accounting for approximately 40% of the global domestic supply of biomass in 2020. The primary sources of biomass in Asia are solid biomass (wood, agricultural residues, industrial waste), which constitutes 86% of the global supply [70]. Countries like China, India, Indonesia, Malaysia, and Thailand also produce large volumes of wheat straw, oil palm residues, and corn stover. Vietnam produces an estimated 39 Mt of paddy residue annually. While a portion is utilized for purposes such as animal feed and soil mulching, open-field burning remains a widespread disposal method. In China, substantial quantities of crop residues, primarily from wheat, rice, and maize, are harvested across six key agricultural regions [71].

3.1.4 Africa

Africa has abundant agricultural land and diverse cropping systems. Across Africa, fresh agricultural residues are estimated to total around 102.26 gigatons (Gt), equivalent to approximately 62.12 Gt on a dry-weight basis. Cocoa pods constitute the predominant fraction, contributing nearly 19% of the total residue. Significant contributions also come from pineapple (11%), maize (10%), sorghum (7%), sugarcane (5%), and rice (4%), while the remainder originates from a wide variety of other crop sources [72]. These residues are primarily utilized to enhance soil fertility and structure, contributing to improved soil health through organic matter enrichment and nutrient cycling [73].

3.1.5 Europe

The agricultural sector within the European Union utilizes a combination of primary crops and residual biomass to support the production of biofuels and a range of bio-based materials. It is projected that approximately 83.3 Mt of agricultural residues are available for processing, encompassing residues from key crops such as barley, maize, oats, olives, rapeseed, paddy rice, rye, soybeans, sunflower, triticale, and wheat [74]. Within the EU, Finland, Sweden, Austria, and Denmark are among the leading contributors to agricultural residue generation [75].

3.2 *Quantifying Biomass Potential at Local and Global Scales*

Accurately quantifying agricultural residue availability is crucial for strategic planning and sustainable bioeconomy development. Globally, agricultural residues alone exceed 3 Bt annually, yet their practical availability for utilization depends significantly on sustainable extraction rates that maintain ecological balance and soil health [76]. Modern approaches employ geographic information systems (GIS) and remote sensing technologies to provide accurate biomass resource mapping. These technologies facilitate detailed assessments of residue production, accessibility, and sustainable collection rates at both regional and global scales [77, 78]. Incorporating agronomic data, seasonal variability, and ecological constraints, these methodologies help optimize the location and size of biorefineries, thus enhancing both economic viability and environmental sustainability of biomass resource utilization [79, 80].

4 Environmental Impacts and Carbon Footprint of Agricultural Residue Utilization

The use of agricultural residues as feedstocks for bioenergy and bio-based products offers substantial environmental advantages over fossil-based systems. By redirecting residues from traditional disposal pathways toward value-added conversion, it is possible to reduce greenhouse gas emissions, improve carbon efficiency, and mitigate air, soil, and water pollution [81]. However, the environmental benefits depend strongly on feedstock management, processing routes, and regional practices.

4.1 *Life Cycle Assessment and Carbon Footprint*

Life cycle assessment (LCA) is a widely used tool for evaluating the environmental performance of biomass-based systems. It encompasses all stages of the product chain – from residue collection and transportation to conversion and end use, which allows for a holistic comparison with fossil-based alternatives [82]. Studies consistently show that biofuels derived from agricultural residues exhibit significantly lower net greenhouse gas (GHG) emissions than gasoline or diesel [83, 84]. Bioethanol from lignocellulosic feedstocks offers substantial climate benefits. When accounting for avoided burning and residue removal, corn stover can reduce GHG emissions by 90–103%. Ethanol from switchgrass achieves a 77–97% reduction. Miscanthus-based ethanol shows the highest savings, ranging from 101% to 115% compared to fossil fuels [85]. The key carbon-saving factors include the noncompetitive nature of residues (no land-use change), biogenic carbon neutrality during combustion, and displacement of high-emission fossil resources.

Additionally, the use of renewable electricity or biogas in processing stages (e.g., enzymatic hydrolysis, fermentation) can further reduce the carbon footprint. Conversely, long-distance transport, inefficient pretreatment, or overharvesting residues beyond sustainable limits may offset these gains, emphasizing the need for regionally optimized systems [86].

4.2 Role in Carbon Sequestration and Emission Reduction

While agricultural residues themselves are not long-term carbon sinks, their responsible management contributes indirectly to carbon sequestration. Residue removal from the field when balanced against soil health needs prevents methane (CH_4) and nitrous oxide (N_2O) emissions associated with anaerobic decomposition or burning [87, 88]. In tropical regions, where biomass burning is common, redirecting residues to biorefineries offers one of the fastest routes to reduce air pollution and GHG emissions.

Furthermore, biochar, a byproduct of thermochemical processes like pyrolysis, can be returned to the soil. This not only stabilizes carbon over long periods but also improves soil fertility and water retention [89]. As part of an integrated biomass management strategy, biochar application transforms agricultural residues into tools for climate adaptation and mitigation, aligning well with carbon farming and regenerative agriculture initiatives.

4.3 Resource Management Strategies for Minimal Ecological Footprint

To ensure environmental sustainability, agricultural residue utilization must follow ecologically sound harvesting and land management practices. Over-removal of residues can degrade soil organic matter, disrupt microbial ecosystems, and reduce long-term crop productivity [90, 91]. Therefore, to maintain healthy soil and ecosystem services, it is crucial to enforce sustainable extraction thresholds for biomass, typically limiting the removal of aboveground residue to 30–50%. This ensures that enough organic matter remains in the soil to support nutrient cycling, water infiltration, and soil structure, ultimately enhancing soil health and ecosystem functions [92, 93].

Integrated assessment tools are essential to balance the trade-offs between residue valorization and agroecological integrity. Decision support systems now incorporate

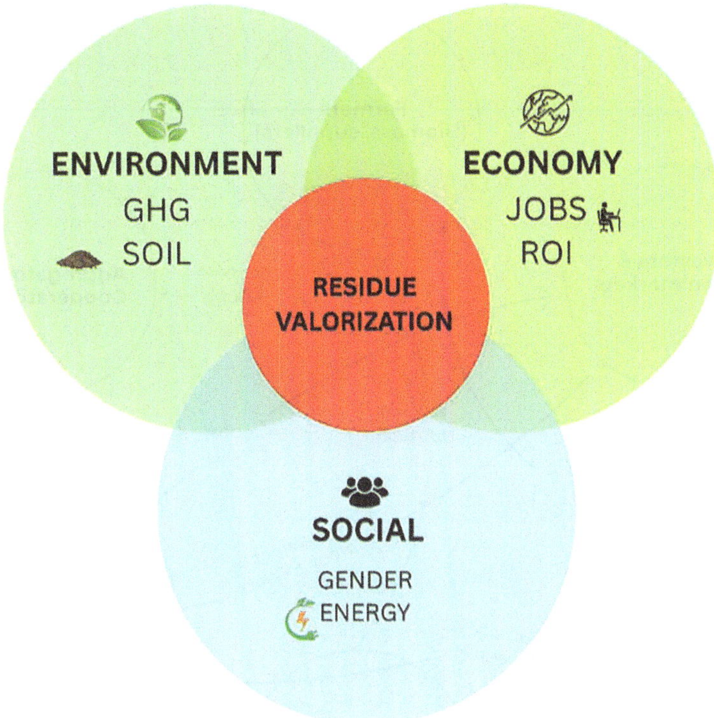

Fig. 4 Sustainability nexus of agriculture residues

LCA outputs, soil modeling, and remote sensing to guide residue removal limits, location of processing hubs, and optimal valorization routes [94, 95]. Policy instruments, such as sustainability certifications, carbon offset schemes, and residue management guidelines, further promote best practices across the value chain.

5 Contribution of Agricultural Residues to Rural and Socioeconomic Development

The utilization of agricultural residues extends beyond environmental benefits. It serves as a catalyst for rural transformation, economic diversification, and inclusive development. As agriculture remains the primary livelihood in many low- and middle-income countries, leveraging biomass byproducts like crop residues presents an opportunity to unlock local value chains, stimulate job creation, and build community resilience [96–98]. This section explores the multifaceted ways in which agricultural residue valorization contributes to sustainable socioeconomic development. The multi-ring "sustainability nexus" shown in Fig. 4 integrates the environmental, economic, and social gains of residue valorization into a single

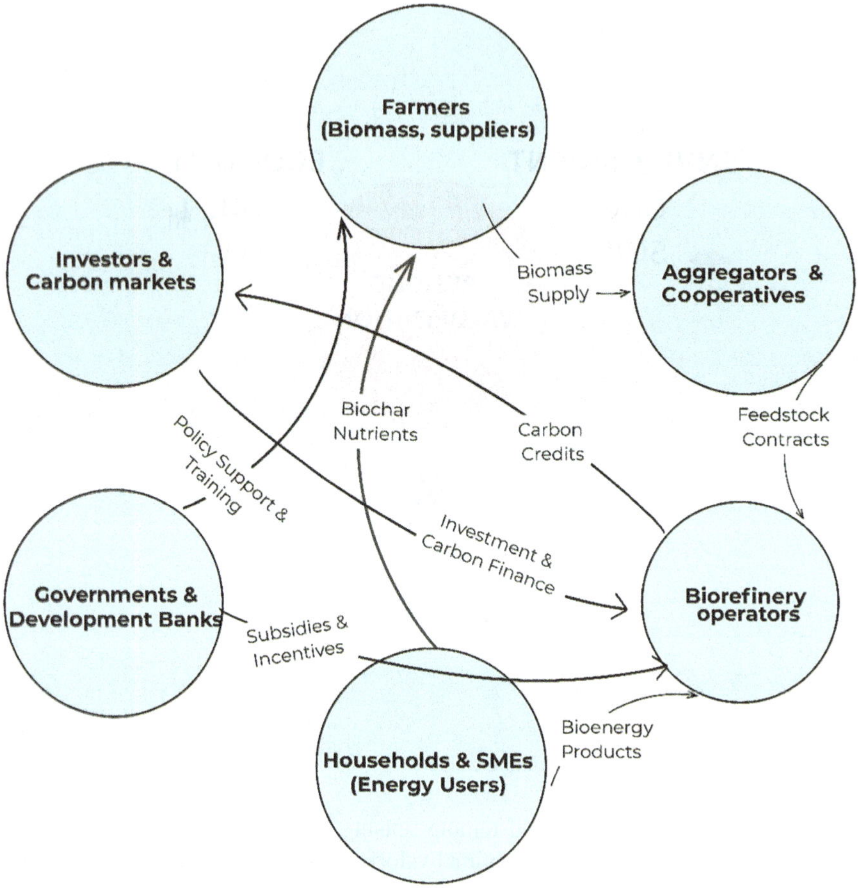

Fig. 5 Circular investment and value flow in the rural bioeconomy

graphic, illustrating how these three domains overlap to support a circular bioeconomy.

5.1 Job Creation and Income Diversification

Agricultural residue-based bioeconomy activities offer employment across multiple segments of the value chain: residue collection and aggregation, preprocessing (e.g., baling, drying), transport logistics, and biomass conversion through fermentation, gasification, or composting [99, 100]. These operations create both direct and indirect jobs, especially in rural areas where economic opportunities are often limited.

In South Asia, biomass power plants located near rice mills ensure a steady supply of husks and improve plant utilization, lowering electricity costs. This setup provides reliable energy, reduces reliance on diesel, and supports rural economic development [101]. Similarly, in sub-Saharan Africa, integrating corn stover and cassava peels into small-scale biogas systems has created alternative income sources for farming households [102, 103]. Women and youth, who are often excluded from formal employment, particularly benefit when inclusive models are applied. Furthermore, the development of rural biorefineries stimulates ancillary services such as machinery leasing, maintenance, feedstock brokerage, and cooperative marketing. This supports economic diversification and enhances rural resilience against agricultural market shocks. To effectively sustain job creation and income diversification, investments should follow a circular flow, connecting biomass producers, aggregators, biorefinery operators, energy consumers, policymakers, and investors in mutually reinforcing economic exchanges (as shown in Fig. 5).

5.2 *Strengthening Community Resilience Through Bio-Based Ventures*

The ability of rural communities to adapt to climate and economic shocks improves when they are integrated into local bio-based economies [104, 105]. Agricultural residues, when valorized at the point of origin, provide a buffer against volatile fuel prices and erratic electricity supply. For example, locating a multi-feedstock biorefinery near avocado and plantain residue sources and distribution centers improved economic, technical, and environmental performance. This proximity enabled efficient feedstock use, reduced transport needs, and supported the production of high-value commercial products [106]. In remote regions, converting residues to biogas, briquettes, or ethanol enables clean cooking, off-grid electrification, and access to productive energy for small enterprises.

Community-owned biomass projects, such as cooperatives and producer groups, can enhance social capital, enabling collective bargaining, better market access, and stronger local governance [107–110]. In such settings, profits from biomass sales or energy savings are reinvested into community infrastructure, education, or health services, reinforcing long-term resilience.

5.3 *Policy Initiatives and Funding Mechanisms*

National and regional governments play a critical role in mainstreaming the use of agricultural residues for rural development. Supportive policy instruments include feed-in tariffs for biomass electricity, subsidies for briquetting and pelletizing units, and tax incentives for bio-based product manufacturing [111–113]. In Thailand, competitive feed-in tariffs through bidding schemes can incentivize rice husk-based power generation. This approach supports grid integration while reducing the need for subsidies on renewable electricity from rice husk biomass [114]. Development agencies and climate funds such as the Global Environment Facility (GEF) and Green Climate Fund (GCF) also provide financial support for pilot projects, technology transfer, and capacity building in biomass utilization.

In the EU, the Common Agricultural Policy (CAP) now supports sustainable residue management as part of broader environmental and climate objectives. Likewise, India's Sustainable Alternative Towards Affordable Transportation ("SATAT") initiative promotes compressed biogas (CBG) from agricultural and agro-industrial waste, offering price guarantees and assured offtake for rural entrepreneurs [115, 116]. Strategic funding and institutional frameworks lower the entry barriers for rural participation in biomass ventures, enabling long-term market integration.

5.4 Gender Empowerment and Inclusive Growth

Women play a central role in agricultural systems, particularly in crop residue management, livestock care, and household energy use. Yet, they often face restricted access to land, finance, and decision-making power. Agricultural residue-based bioeconomy models, especially those built on decentralized, community-managed systems, offer an entry point for women's empowerment [117, 118].

In projects involving residue collection, composting, or bioenergy production, women have been successfully trained as operators, managers, and entrepreneurs [119, 120]. Participatory design of these systems ensures that gender norms and practical constraints such as time, poverty, and mobility are addressed from the outset. Moreover, access to clean bioenergy from residues reduces reliance on firewood, saving time and improving indoor air quality for women and children [121, 122]. Inclusive models that integrate gender equality not only enhance the social sustainability of biomass systems but also improve productivity and innovation.

5.5 Infrastructure Development and Energy Access

The establishment of agricultural residue-based enterprises necessitates and often accelerates rural infrastructure development. Storage facilities, feeder roads, biomass collection hubs, and microgrids are typically co-developed with biomass ventures, producing co-benefits for the broader community [120, 123].

Rural energy access is another major outcome. Agricultural residues can be converted to electricity, heat, or clean cooking fuels, displacing kerosene, diesel, or traditional biomass. In regions where grid connectivity is poor or unreliable, localized biomass energy solutions provide a critical lifeline for households, clinics, and small businesses [124, 125]. These infrastructure gains have long-term development impacts, facilitating agricultural mechanization, improving education through better lighting, and enabling digital inclusion via powered devices.

6 Technological Advances and Biotechnological Applications for Agricultural Residues

The efficient conversion of agricultural residues into fuels, chemicals, and materials requires tailored technological pathways that overcome the structural complexity and recalcitrance of lignocellulosic biomass. Recent decades have witnessed considerable progress in biorefinery development, particularly in pretreatment technologies, enzyme engineering, microbial fermentation, and product diversification. This section offers a concise overview of key technological advances underpinning the sustainable utilization of agricultural residues and sets the stage for the more detailed discussions presented in the subsequent chapters, including *Challenging Pretreatment Technologies of Lignocellulosic Biomass, Robust Production of Cellulases to Develop the Sugar Platform, Engineered Microbial Strains for Co-Fermentation of C5 and C6 Sugars to Produce Cellulosic Ethanol, and Producing Advanced Biofuel Butanol through Acetone-Butanol-Ethanol Fermentation.*

6.1 Pretreatment, Conversion, and Process Optimization Strategies

Pretreatment is a critical step in lignocellulosic biomass valorization. It aims to disrupt the lignin-carbohydrate complex, increase surface area, reduce cellulose crystallinity, and enhance enzyme accessibility [126]. Agricultural residues such as rice straw, wheat straw, and corn stover generally require milder pretreatment conditions than woody biomass due to their lower lignin content [127, 128]. Prominent pretreatment methods include dilute acid hydrolysis, alkaline treatment (e.g., sodium hydroxide or ammonia fiber explosion), steam explosion, and organosolv processes. More recent advances involve the use of deep eutectic solvents (DESs) and ionic liquids, which offer higher selectivity and lower environmental impact [128–130]. Effective pretreatment improves saccharification yields and reduces the formation of inhibitory byproducts.

Following pretreatment, enzymatic hydrolysis converts cellulose and hemicellulose into fermentable sugars. The process depends heavily on cellulase and hemicellulase enzyme systems, which must be both robust and cost-effective.

Advances in enzyme production, especially through genetically engineered fungi and bacteria (e.g., *Trichoderma reesei, Aspergillus niger*), have significantly improved hydrolysis efficiency and reduced costs [131, 132]. Enzyme cocktails are now optimized to match specific feedstock compositions and pretreatment histories, enhancing process flexibility.

6.2 Value-Added Products: Biofuels, Biochemicals, and Biomaterials

Agricultural residues can be transformed into a wide range of bio-based products, broadly classified into three categories:

Biofuels: The most advanced technologies produce bioethanol, biobutanol, and biogas. Bioethanol from corn stover and sugarcane bagasse is already commercially produced in countries such as the USA and Brazil. Biogas, generated through anaerobic digestion of residues like cassava peels or rice straw, is widely adopted in decentralized systems [133–135].

Biochemicals: Fermentation and catalytic routes convert lignocellulosic sugars into organic acids (e.g., succinic, lactic, acetic), solvents (e.g., ethanol, butanol), and biopolymers (e.g., polyhydroxyalkanoates) [12, 136, 137]. These chemicals serve as renewable alternatives to petroleum-based building blocks in industries such as packaging, textiles, and pharmaceuticals.

Biomaterials: Agricultural residues are increasingly used in producing composite materials, biodegradable plastics, and carbon-rich materials like biochar. Lignin, often a byproduct of biorefining, is gaining attention as a precursor for adhesives, resins, and nanomaterials [138–140]. These applications not only diversify revenue streams but also add circularity to the biorefinery model.

6.3 Innovations in Biotechnology for Enhanced Sustainability

Biotechnological advances continue to expand the possibilities for agricultural residue valorization. Engineered microbial strains are now capable of co-fermenting both hexose (C6) and pentose (C5) sugars such as glucose and xylose derived from cellulose and hemicellulose, respectively [141]. This capability is essential for improving yields and reducing waste in lignocellulosic fermentations, as hemicellulose-derived sugars make up a significant portion of most agricultural residues. CRISPR/Cas and synthetic biology tools have enabled precision editing of microbial pathways, enhancing tolerance to inhibitors, improving sugar uptake rates, and enabling the production of new metabolites. Consolidated bioprocessing (CBP), which integrates enzyme production, hydrolysis, and fermentation in a single step, is also under development and may drastically reduce capital and operating costs in future biorefineries [142, 143]. On the enzyme front, protein engineering and directed evolution have produced thermostable and pH-tolerant enzymes that maintain high activity under industrial conditions [144, 145]. These innovations not only improve efficiency but also enable greater compatibility with different pretreatment and fermentation environments.

7 Sustainable Management and Policy Frameworks for Agricultural Residue Utilization

The sustainable and large-scale utilization of agricultural residues depends not only on technological feasibility but also on enabling policy, governance, and institutional frameworks. Effective policies can guide biomass extraction rates, ensure ecological balance, incentivize rural participation, and attract investment into biomass-based industries. This section highlights key components of sustainable biomass governance, international policy alignment, and practical challenges associated with implementation.

7.1 Biomass Governance and Land-Use Policies

Sustainable biomass governance begins with clear guidelines on land use, residue removal limits, and environmental safeguards. Agricultural residues are technically "waste," but their extraction must be balanced with ecological functions such as soil organic matter retention, erosion prevention, and nutrient cycling [146, 147]. National frameworks must define what proportion of crop residues can be safely removed without compromising long-term soil health and agricultural productivity. These thresholds vary by region, crop type, and agronomic practice but often fall within 30–50% of total residue yield [148–150].

Countries with advanced biomass sectors have integrated residue management into broader land-use planning systems. For example, in the USA, the Department of Energy's *Billion-Ton Study* incorporates sustainable yield criteria based on erosion control and nutrient management [151, 152]. In the European Union, Common Agricultural Policy (CAP) instruments support environmentally sound residue collection through cross-compliance mechanisms tied to soil conservation [153]. In many low- and middle-income countries, however, biomass governance remains fragmented. Lack of enforcement, weak land tenure systems, and absence of formal residue markets hinder progress. Strengthening institutional capacity at the local level is crucial for operationalizing sustainable management principles and for integrating smallholders into the biomass value chain.

7.2 Incentives, Subsidies, and International Frameworks

Well-designed incentives can accelerate the adoption of biomass technologies and reduce investment risks. These include capital subsidies for equipment (e.g., briquetting machines, digesters), preferential tariffs for biomass electricity, tax exemptions for bio-based products, and guaranteed procurement schemes for agricultural residues [154–156]. India's Sustainable Alternative Towards Affordable Transportation (SATAT) scheme offers a notable example. It promotes the conversion of agricultural residues into compressed biogas (CBG) with guaranteed buy-back by oil marketing companies [157, 158]. Similarly, the European Renewable Energy Directive (RED II) mandates minimum shares of advanced biofuels, including those from agricultural residues, in the transport energy mix [159].

Effective biomass utilization depends significantly on supportive and progressively structured policies. These typically begin with foundational regulatory measures, advance through targeted economic incentives such as subsidies and feed-in tariffs, and culminate in robust, long-term market development frameworks incorporating carbon credits and sustainability certifications (as shown in Fig. 6).

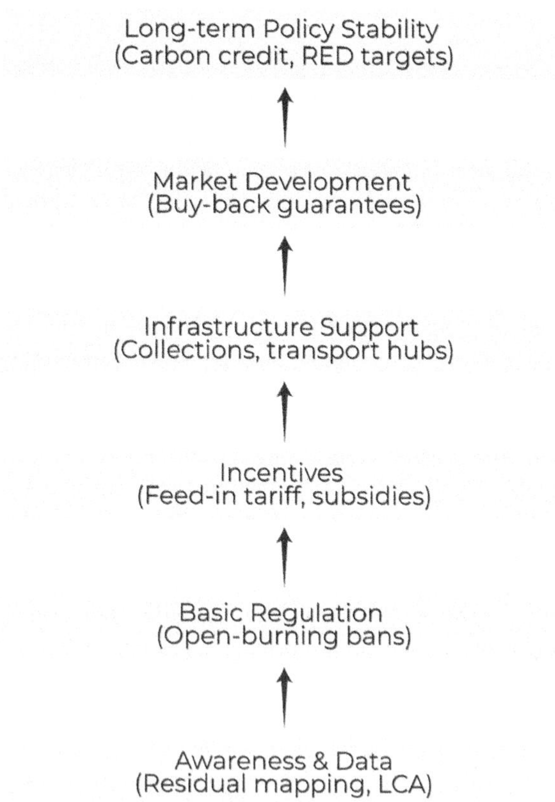

Fig. 6 Biomass policy ladder for residue utilization. *RED* Renewable Energy Directive

At the international level, frameworks such as the Paris Agreement, the Sustainable Development Goals (SDGs), and the Global Bioenergy Partnership (GBEP) have provided guiding principles and benchmarks for sustainable biomass deployment. Specifically, SDG 7 (Affordable and Clean Energy), SDG 12 (Responsible Consumption and Production), and SDG 13 (Climate Action) are directly advanced by residue-based bioenergy systems. Donor agencies and climate finance mechanisms (e.g., the Green Climate Fund) increasingly support biomass projects with dual environmental and development objectives. These funding mechanisms are critical for de-risking early-stage investments in regions where capital availability is limited.

7.3 Challenges in Implementation: Land Tenure, Logistics, and Environmental Trade-Offs

Despite promising policies and frameworks, implementation at the ground level faces several barriers.

1. **Land tenure insecurity** is a persistent issue, especially in sub-Saharan Africa and parts of Asia. Where farmers lack formal rights to land or associated residues, they are excluded from participating in biomass markets or accessing credit and incentives [160, 161]. Strengthening land governance and recognizing customary rights are essential steps toward inclusive residue valorization.
2. **Logistical complexity** also constrains residue collection. Agricultural residues are dispersed, bulky, and seasonal, requiring coordination among farmers, aggregators, and processors. Inadequate transport and storage infrastructure raise costs and reduce feedstock quality. Innovative models such as mobile preprocessing units and farmer cooperatives have shown promise in overcoming these challenges [162, 163].
3. **Environmental trade-offs** must be carefully managed. Overextraction of residues can degrade soil fertility, while poorly planned biomass combustion facilities may increase local air pollution. LCA-based decision tools, combined with sustainability certification schemes (e.g., RSB, ISCC), are valuable for assessing and mitigating such risks.

8 Future Directions and Research Priorities

Rapid strides in biomass science have moved agricultural residue valorization from concept to early commercialization. Yet, large gaps remain in logistics, technology cost, digital integration, and social acceptance. This section maps key research and innovation pathways that can unlock the next wave of sustainable growth.

8.1 Innovations in Biomass Logistics and Preprocessing

Residues are bulky and seasonally abundant. New densification techniques such as high-moisture pelleting, flash torrefaction, and ammonia fiber expansion can raise bulk density four- to sixfold, cutting transport costs and spoilage losses [164, 165]. Field-side mobile preprocessors, powered by tractor PTO or small gasifiers, now fractionate straw into clean cellulose fiber and lignin-rich fines, enabling higher-value markets and reducing haulage mass [166]. Drone-based yield mapping and blockchain traceability can be piloted to match feedstock supply with plant demand and to certify sustainability in real time.

8.2 Research Needs and Emerging Technologies

Pretreatment remains the largest cost and energy sink in biochemical routes. Deep eutectic solvents, low-melt salt hydrates, and aqueous ammonia oxidation show promise for selective lignin removal at milder conditions, but solvent recovery and scale-up require further study. Electrobiocatalytic systems that pair enzymatic depolymerization with in situ electrochemical cofactor regeneration could slash enzyme use [167]. Consolidated bioprocessing microorganisms capable of secreting robust cellulases while fermenting both C5 and C6 sugars need accelerated metabolic engineering efforts and adaptive laboratory evolution to improve inhibitor tolerance.

8.3 Digital Technologies for Resource Assessment and Plant Optimization

High-resolution satellite imagery, LiDAR, and machine-learning algorithms now quantify residue yields with subfield accuracy, guiding optimal removal rates and logistics routes [168, 169]. Digital twins, real-time virtual replicas of biorefineries, enable dynamic process control, predictive maintenance, and rapid scenario testing for feedstock variability. Edge-based IoT sensors track moisture, temperature, and microbial load in storage piles, preventing fires and quality loss [170, 171] Open-source platforms that integrate agronomic, economic, and emissions data can inform policymakers and investors, fostering transparent and adaptive supply chains.

8.4 Multidisciplinary Approaches to Maximize Rural Impact

Technical breakthroughs alone will not secure sustainable adoption. Integrated techno-economic analysis (TEA) and life cycle sustainability assessment (LCSA) must be co-developed with community co-design, gender analysis, and land tenure studies. Agronomists, soil scientists, and extension services need to refine residue removal guidelines that protect long-term soil health while supplying biorefineries. Participatory business models such as farmer cooperatives with equity stakes in processing hubs can align incentives across the chain. Finally, harmonized policy benchmarks and sustainability certification (e.g., RSB, ISCC) should be expanded to include smallholders, ensuring equitable market access.

9 Conclusion

Agricultural residues including corn stover, rice straw, sugarcane bagasse, and other crop byproducts form a vast renewable carbon reserve capable of slashing life cycle greenhouse-gas emissions by up to 90% when converted into fuels, chemicals, and materials under sustainable harvest limits. Recent advances in mild pretreatments, tailored enzymes, engineered microbes, and thermochemical upgrading now make high-yield, multiproduct biorefineries technically feasible. Forward-thinking policies such as the EU's RED II and India's SATAT demonstrate how price guarantees, sustainability certification, and inclusive business models can mobilize investment and empower rural communities. Remaining challenges such as seasonal logistics, first-plant capital costs, land tenure insecurity, and soil health safeguards require integrated research that couples digital supply-chain tools with social impact metrics. If these hurdles are addressed, crop residues will shift from waste to cornerstone of a circular, low-carbon bioeconomy that delivers climate mitigation and rural prosperity together.

References

1. IPCC (2019) Climate change and land: an IPCC Special Report on climate change, desertification, land degradation, sustainable land management, food security, and greenhouse gas fluxes in terrestrial ecosystems, p 41
2. USAD (2024) Building a resilient biomass supply: a plan to enable the bioeconomy in America. Mary Ann Liebert, Inc.
3. IEA (2021) Zero by 2050: a roadmap for the global energy sector. International Energy Agency, p 224
4. Ilic D, Williams K, Farnish R et al (2018) On the challenges facing the handling of solid biomass feedstocks. Biofuels Bioprod Biorefin 12:187–202. https://doi.org/10.1002/bbb.1851
5. Ashokkumar V, Venkatkarthick R, Jayashree S et al (2022) Recent advances in lignocellulosic biomass for biofuels and value-added bioproducts – a critical review. Bioresour Technol 344: 126195. https://doi.org/10.1016/j.biortech.2021.126195
6. Zabed H, Sahu JN, Suely A et al (2017) Bioethanol production from renewable sources: current perspectives and technological progress. Renew Sustain Energy Rev 71:475–501. https://doi.org/10.1016/j.rser.2016.12.076

7. Afedzi AEK, Afrakomah GS, Gyan K et al (2025) Enhancing economic and environmental sustainability in lignocellulosic bioethanol production: key factors, innovative technologies, policy frameworks, and social considerations. Sustainability 17:499. https://doi.org/10.3390/su17020499
8. Ogwu MC, Osawaru ME, Kosoe EA, Izah SC (2025) Processes associated with bioeconomy in the global South. In: Sustainable bioeconomy development in the global south: volume I status and perspectives. Springer, pp 209–238
9. Wong A, Zhang H, Kumar A (2016) Life cycle assessment of renewable diesel production from lignocellulosic biomass. Int J Life Cycle Assess 21:1404–1424. https://doi.org/10.1007/s11367-016-1107-8
10. Maluin FN (2024) Enhancing chitosan nanofilm with agricultural waste fillers for sustainable and safe functional food packaging. ACS Agric Sci Technol 4:1136–1162. https://doi.org/10.1021/acsagscitech.4c00398
11. Afedzi AEK, Poolsawat T, Wanikorn B, Parakulsuksatid P (2025) Effect of subcritical water extraction and precipitation methods on the functional properties, antioxidant activity, and ACE1 inhibition of palm kernel cake protein. Biocatal Agric Biotechnol 66:103569. https://doi.org/10.1016/j.bcab.2025.103569
12. Pakchamni P, Afedzi AEK, Parakulsuksatid P (2022) Optimization of alkaline-assisted organosolv pretreatment of sugarcane trash for the production of succinic acid using response surface methodology. Biocatal Agric Biotechnol 43:102374. https://doi.org/10.1016/j.bcab.2022.102374
13. Afedzi AEK, Rattanaporn K, Sultan IN et al (2023) Comparison of marine and rushton impeller in high solid loading of two-step pretreated oil palm trunk using simultaneous saccharification and fermentation for ethanol production. Results Eng 19:101273. https://doi.org/10.1016/j.rineng.2023.101273
14. Gupta N, Wijenayake WPT, Roy D et al (2025) Smart recycling and sustainable lignocellulosic waste management. In: Value addition and utilization of lignocellulosic biomass: through novel technological interventions. Springer, pp 221–250
15. Panoutsou C, Singh A (2020) A value chain approach to improve biomass policy formation. GCB Bioenergy 12:464–475. https://doi.org/10.1111/gcbb.12685
16. Chen H, Chen H (2014) Chemical composition and structure of natural lignocellulose. In: Biotechnology of lignocellulose: theory and practice, pp 25–71. https://doi.org/10.1007/978-94-007-6898-7_2
17. Yadav S, Chattopadhyay D (2023) Lignin: the building block of defense responses to stress in plants. J Plant Growth Regul 42:6652–6666. https://doi.org/10.1007/s00344-023-10926-z
18. Pisanó I, Gottumukkala L, Hayes DJ, Leahy JJ (2021) Characterisation of Italian and Dutch forestry and agricultural residues for the applicability in the bio-based sector. Ind Crop Prod 171:113857. https://doi.org/10.1016/j.indcrop.2021.113857
19. García A, González Alriols M, Labidi J (2014) Evaluation of different lignocellulosic raw materials as potential alternative feedstocks in biorefinery processes. Ind Crop Prod 53:102–110. https://doi.org/10.1016/j.indcrop.2013.12.019
20. Chen X, Yu J, Zhang Z, Lu C (2011) Study on structure and thermal stability properties of cellulose fibers from rice straw. Carbohydr Polym 85:245–250
21. Bhattacharyya P, Bhaduri D, Adak T et al (2020) Characterization of rice straw from major cultivars for best alternative industrial uses to cutoff the menace of straw burning. Ind Crop Prod 143:111919. https://doi.org/10.1016/j.indcrop.2019.111919
22. Harun S, Geok SK (2016) Effect of sodium hydroxide pretreatment on rice straw composition. Indian J Sci Technol 9:1–9
23. El-Sayed MA, El-Samni TM (2006) Physical and chemical properties of rice straw ash and its effect on the cement paste produced from different cement types. J King Saud Univ Sci 19:21–29

24. Gu F, Wang W, Jing L, Jin Y (2013) Effects of green liquor pretreatment on the chemical composition and enzymatic digestibility of rice straw. Bioresour Technol 149:375–382. https://doi.org/10.1016/j.biortech.2013.09.064
25. Wu Z, Hao H, Zahoor et al (2014) Diverse cell wall composition and varied biomass digestibility in wheat straw for bioenergy feedstock. Biomass Bioenergy 70:347–355. https://doi.org/10.1016/j.biombioe.2014.08.025
26. Liu X, Zicari SM, Liu G et al (2015) Improving the bioenergy production from wheat straw with alkaline pretreatment. Biosyst Eng 140:59–66. https://doi.org/10.1016/j.biosystemseng.2015.09.006
27. Paniagua S, García-Pérez AI, Calvo LF (2019) Biofuel consisting of wheat straw–poplar wood blends: thermogravimetric studies and combustion characteristic indexes estimation. Biomass Convers Biorefin 9:433–443
28. Tutt M, Kikas T, Olt J (2012) Influence of different pretreatment methods on bioethanol production from wheat straw. Agron Res 10:269–276
29. Zheng P, Fang L, Xu Y et al (2010) Succinic acid production from corn stover by simultaneous saccharification and fermentation using Actinobacillus succinogenes. Bioresour Technol 101: 7889–7894. https://doi.org/10.1016/j.biortech.2010.05.016
30. Singh HK, Patil T, Vineeth SK et al (2020) Isolation of microcrystalline cellulose from corn stover with emphasis on its constituents: corn cover and corn cob. Mater Today Proc 27:589–594
31. Shankar TJ (2015) Comparison of chemical composition and energy property of torrefied switchgrass and corn stover. Front Energy Res 3:1–11
32. Li Z, Zhai H, Zhang Y, Yu L (2012) Cell morphology and chemical characteristics of corn stover fractions. Ind Crop Prod 37:130–136
33. Ávila PF, Forte MBS, Goldbeck R (2018) Evaluation of the chemical composition of a mixture of sugarcane bagasse and straw after different pretreatments and their effects on commercial enzyme combinations for the production of fermentable sugars. Biomass Bioenergy 116:180–188. https://doi.org/10.1016/j.biombioe.2018.06.015
34. Bu J, Yan X, Wang Y-T et al (2019) Co-production of high-gravity bioethanol and succinic acid from potassium peroxymonosulfate and deacetylation sequentially pretreated sugarcane bagasse by simultaneous saccharification and co-fermentation. Energy Convers Manag 186: 131–139. https://doi.org/10.1016/j.enconman.2019.02.038
35. Chen J, Yang S, Alam MA et al (2021) Novel biorefining method for succinic acid processed from sugarcane bagasse. Bioresour Technol 324:124615. https://doi.org/10.1016/j.biortech.2020.124615
36. Qin Y, Li Y, Liang G et al (2024) Efficient production of succinic acid from sugarcane bagasse hydrolysate by Actinobacillus succinogenes GXAS137. Fermentation 10:22
37. Akhtar J, Idris A (2017) Oil palm empty fruit bunches a promising substrate for succinic acid production via simultaneous saccharification and fermentation. Renew Energy 114:917–923. https://doi.org/10.1016/j.renene.2017.07.113
38. Bukhari NA, Jahim JM, Loh SK et al (2020) Organic acid pretreatment of oil palm trunk biomass for succinic acid production. Waste Biomass Valoriz 11:5549–5559. https://doi.org/10.1007/s12649-020-00953-2
39. Asadieraghi M, Daud WMAW (2015) In-depth investigation on thermochemical characteristics of palm oil biomasses as potential biofuel sources. J Anal Appl Pyrolysis 115:379–391
40. Abnisa F, Daud WMAW, Husin WNW, Sahu JN (2011) Utilization possibilities of palm shell as a source of biomass energy in Malaysia by producing bio-oil in pyrolysis process. Biomass Bioenergy 35:1863–1872. https://doi.org/10.1016/j.biombioe.2011.01.033
41. Praspaliauskas M, Pedišius N, Čepauskienė D, Valantinavičius M (2020) Study of chemical composition of agricultural residues from various agro-mass types. Biomass Convers Biorefin 10:937–948. https://doi.org/10.1007/s13399-019-00457-7

42. Rambo MKD, Schmidt FL, Ferreira MMC (2015) Analysis of the lignocellulosic components of biomass residues for biorefinery opportunities. Talanta 144:696–703. https://doi.org/10.1016/j.talanta.2015.06.045
43. Puri L, Hu Y, Naterer G (2024) Critical review of the role of ash content and composition in biomass pyrolysis. Front Fuels 2:1378361. https://doi.org/10.3389/ffuel.2024.1378361
44. Sluiter JB, Ruiz RO, Scarlata CJ et al (2010) Compositional analysis of lignocellulosic feedstocks. 1. Review and description of methods. J Agric Food Chem 58:9043–9053. https://doi.org/10.1021/jf1008023
45. Ravindra K, Singh T, Mor S (2019) Emissions of air pollutants from primary crop residue burning in India and their mitigation strategies for cleaner emissions. J Clean Prod 208:261–273. https://doi.org/10.1016/j.jclepro.2018.10.031
46. Bhattacharyya P, Barman D (2018) Chapter 21 – Crop residue management and greenhouse gases emissions in tropical rice lands. In: Muñoz MÁ, Zornoza R (eds) Soil management and climate change: effects on organic carbon, nitrogen dynamics, and greenhouse gas emissions. Academic Press, pp 323–335
47. Blanco-Canqui H (2016) Growing dedicated energy crops on marginal lands and ecosystem services. Soil Sci Soc Am J 80:845–858. https://doi.org/10.2136/sssaj2016.03.0080
48. Tuomisto HL, Hodge ID, Riordan P, Macdonald DW (2012) Comparing energy balances, greenhouse gas balances and biodiversity impacts of contrasting farming systems with alternative land uses. Agric Syst 108:42–49. https://doi.org/10.1016/j.agsy.2012.01.004
49. Del Grosso S, Smith P, Galdos M et al (2014) Sustainable energy crop production. Curr Opin Environ Sustain 9:20–25. https://doi.org/10.1016/j.cosust.2014.07.007
50. Yang H, Zhang X, Luo H et al (2019) Overcoming cellulose recalcitrance in woody biomass for the lignin-first biorefinery. Biotechnol Biofuels 12:1–18. https://doi.org/10.1186/s13068-019-1503-y
51. Matsakas L, Nitsos C, Raghavendran V et al (2018) A novel hybrid organosolv: steam explosion method for the efficient fractionation and pretreatment of birch biomass. Biotechnol Biofuels 11:1–14. https://doi.org/10.1186/s13068-018-1163-3
52. Meng X, Bhagia S, Wang Y et al (2020) Effects of the advanced organosolv pretreatment strategies on structural properties of woody biomass. Ind Crop Prod 146:112144. https://doi.org/10.1016/j.indcrop.2020.112144
53. Romaní A, Michelin M, Domingues L, Teixeira JA (2018) Valorization of wastes from agrofood and pulp and paper industries within the biorefinery concept: southwestern Europe scenario. In: Bhaskar T, Pandey A, Mohan SV et al (eds) Waste biorefinery. Elsevier, pp 487–504
54. dos Santos Mathias TR, Alexandre VMF, Cammarota MC et al (2015) Characterization and determination of brewer's solid wastes composition. J Inst Brew 121:400–404. https://doi.org/10.1002/jib.229
55. Khaswal A, Mishra SK, Chaturvedi N et al (2024) Microbial enzyme production: unlocking the potential of agricultural and food waste through solid-state fermentation. Bioresour Technol Rep:101880. https://doi.org/10.1016/j.biteb.2024.101880
56. Nutrizio M, Dukić J, Sabljak I et al (2024) Upcycling of food by-products and waste: nonthermal green extractions and life cycle assessment approach. Sustainability 16:9143. https://doi.org/10.3390/su16219143
57. Renders T, Van den Bossche G, Vangeel T et al (2019) Reductive catalytic fractionation: state of the art of the lignin-first biorefinery. Curr Opin Biotechnol 56:193–201. https://doi.org/10.1016/j.copbio.2018.12.005
58. Angenent LT, Usack JG, Xu J et al (2018) Integrating electrochemical, biological, physical, and thermochemical process units to expand the applicability of anaerobic digestion. Bioresour Technol 247:1085–1094. https://doi.org/10.1016/j.biortech.2017.09.104
59. Balingbing C, Van Hung N, Nghi NT et al (2020) Mechanized collection and densification of rice straw. In: Gummert M, Van Hung N, Chivenge P, Douthwaite B (eds) Sustainable rice straw management. Springer International Publishing, Cham, pp 15–32

60. Nunes LJR, Silva S (2023) Optimization of the residual biomass supply chain: process characterization and cost analysis. Logistics 7:48. https://doi.org/10.3390/logistics7030048
61. Junginger HM, Mai-Moulin T, Daioglou V et al (2019) The future of biomass and bioenergy deployment and trade: a synthesis of 15 years IEA Bioenergy Task 40 on sustainable bioenergy trade. Biofuels Bioprod Biorefin 13:247–266. https://doi.org/10.1002/bbb.1993
62. Monti A, Polugodina M (2021) Biofuel and biogas policies: economic, regulatory, and sustainability challenges. In: Affordable and clean energy. Springer, pp 87–104
63. Honorato-Salazar JA, Sadhukhan J (2020) Annual biomass variation of agriculture crops and forestry residues, and seasonality of crop residues for energy production in Mexico. Food Bioprod Process 119:1–19. https://doi.org/10.1016/j.fbp.2019.10.005
64. USDOE (2024) 2023 Billion-ton report: an assessment of U.S. renewable carbon resources
65. IEA (2022) Development of techno-economic model for assessment of bio-hubs in Canada
66. Magne A, Khatiwada D, Cardozo E (2024) Assessing the bioenergy potential in South America: projections for 2050. Energy Sustain Dev 82:101535. https://doi.org/10.1016/j.esd.2024.101535
67. Sampaolesi S, Briand LE, Saparrat MC, Toledo MV (2023) Potentials of biomass waste valorization: case of South America. Sustainability 15:8343
68. Coelho Junior LM, Santos Júnior EP, da Silva CFF et al (2024) Supply of bioelectricity from sugarcane bagasse in Brazil: a space–time analysis. Sustain Environ Res 34:17. https://doi.org/10.1186/s42834-024-00223-z
69. Romeu da Silva GH, Nascimento A, Nascimento D et al (2025) Analysis of the sugarcane biomass use to produce green hydrogen: Brazilian case study. Appl Sci 15:1675
70. WBA (2023) Global bioenergy statistics report
71. Rao MM, Botsa SM, Rao TP et al (2024) A comprehensive review on agricultural waste production and onsite management with circular economy opportunities. Discov Sustain 5:288. https://doi.org/10.1007/s43621-024-00492-z
72. Same NN, Yakub AO, Chaulagain D et al (2024) The future of clean energy: agricultural residues as a bioethanol source and its ecological impacts in Africa. Renew Energy 237:121612. https://doi.org/10.1016/j.renene.2024.121612
73. AU (2024) Africa Fertilizer and Soil Health (AFSH) summit 2024. African Union, Nairobi
74. Jane O, Baldino C (2024) Availability of biomass feedstocks in the European Union to meet the 2035 ReFuelEU Aviation SAF target. Washington, DC
75. Pelkmans L (2024) IEA Bioenergy: countries' report – update 2024: implementation of bioenergy in the IEA Bioenergy member countries. IEA Bioenergy
76. Sarkar S, Skalicky M, Hossain A et al (2020) Management of crop residues for improving input use efficiency and agricultural sustainability. Sustainability 12:9808. https://doi.org/10.3390/su12239808
77. Chakraborty A, Biswal A, Pandey V et al (2022) Developing a spatial information system of biomass potential from crop residues over India: a decision support for planning and establishment of biofuel/biomass power plant. Renew Sustain Energy Rev 165:112575. https://doi.org/10.1016/j.rser.2022.112575
78. Heck R, Rudi A, Lauth D, Schultmann F (2024) An estimation of biomass potential and location optimization for integrated biorefineries in Germany: a combined approach of GIS and mathematical modeling. Sustainability 16:6781. https://doi.org/10.3390/su16166781
79. Larnaudie V, Ferrari MD, Lareo C (2022) Switchgrass as an alternative biomass for ethanol production in a biorefinery: perspectives on technology, economics and environmental sustainability. Renew Sustain Energy Rev 158:112115. https://doi.org/10.1016/j.rser.2022.112115
80. Giuliano A, Poletto M, Barletta D (2016) Process optimization of a multi-product biorefinery: the effect of biomass seasonality. Chem Eng Res Des 107:236–252. https://doi.org/10.1016/j.cherd.2015.12.011
81. Gueboudji Z (2024) Waste valorization techniques. In: Generation of energy from municipal solid waste: circular economy and sustainability. Springer, pp 29–52

82. Black MJ, Whittaker C, Hosseini SA et al (2011) Life Cycle Assessment and sustainability methodologies for assessing industrial crops, processes and end products. Ind Crop Prod 34: 1332–1339. https://doi.org/10.1016/j.indcrop.2010.12.002
83. Wasilewski J, Zając G, Szyszlak-Bargłowicz J, Kuranc A (2022) Evaluation of greenhouse gas emission levels during the combustion of selected types of agricultural biomass. Energies 15:1
84. Akkoli KM, Gangavati PB, Banapurmath NR, Yaliwal VS (2020) Comparative study of various biofuel combinations derived from agricultural residues on the performance and emissions of CI engine. Int J Sustain Eng 13:140–150. https://doi.org/10.1080/19397038.2019.1634774
85. Wang M, Han J, Dunn JB et al (2012) Well-to-wheels energy use and greenhouse gas emissions of ethanol from corn, sugarcane and cellulosic biomass for US use. Environ Res Lett 7:45905
86. Ostovari H, Sternberg A, Bardow A (2020) Rock 'n' use of CO2: carbon footprint of carbon capture and utilization by mineralization. Sustain Energy Fuels 4:4482–4496. https://doi.org/10.1039/d0se00190b
87. Verma B, Porwal M, Jha AK, Sharma T (2023) Agriculture practices to reduce in-field greenhouse gas emissions. In: Climate smart agriculture: principles and practices. Kripa Drishti Publication
88. De Melo RO, De Barros NF, da Silva Teixeira R et al (2023) CO2, N2O and CH4 emissions and C storage in eucalyptus forests with different management practices of harvest residues. Bioenergy Res 16:398–415. https://doi.org/10.1007/s12155-022-10468-4
89. Lehmann J, Joseph S (2024) Biochar for environmental management: science, technology and implementation. Taylor & Francis
90. Dutta A, Bhattacharyya R, Chaudhary VP et al (2022) Impact of long-term residue burning versus retention on soil organic carbon sequestration under a rice-wheat cropping system. Soil Tillage Res 221:105421. https://doi.org/10.1016/j.still.2022.105421
91. Sena KN, Boni TS, Maltoni KL et al (2023) Post-harvest eucalyptus residue removal reduces soil aggregation and biological activities in Central-West Brazil. Sustainability 15:8790
92. Van den Born GJ, Van Minnen JG, Olivier JGJ, Ros JPM (2014) Integrated analysis of global biomass flows in search of the sustainable potential for bioenergy production. PBL 1509, p 24
93. Ansari MA, Babu S, Choudhary J et al (2022) Soil quality restoration and yield stabilization in acidic soils of northeastern Himalayas: five years impact of green manuring and crop residue management. Front Environ Sci 10:940349. https://doi.org/10.3389/fenvs.2022.940349
94. Khan MA, Khan R, Al-Zghoul TM et al (2024) Optimizing municipal solid waste management in urban Peshawar: a linear mathematical modeling and GIS approach for efficiency and sustainability. Case Stud Chem Environ Eng 9:100704. https://doi.org/10.1016/j.cscee.2024.100704
95. Longato D, Gaglio M, Boschetti M, Gissi E (2019) Bioenergy and ecosystem services trade-offs and synergies in marginal agricultural lands: a remote-sensing-based assessment method. J Clean Prod 237:117672. https://doi.org/10.1016/j.jclepro.2019.117672
96. Gonocruz RA, Kudoh Y, Anbumozhi V (2024) Analysis of the water-energy-food nexus for sustainable biomass utilisation for fuel, fibre, and food in selected EAS countries-phase I (2023–2024) report
97. Salleh SF, Mohd Roslan ME, Abd Rahman A et al (2020) Transitioning to a sustainable development framework for bioenergy in Malaysia: policy suggestions to catalyse the utilisation of palm oil mill residues. Energy Sustain Soc 10:1–20. https://doi.org/10.1186/s13705-020-00269-y
98. Anokye K (2024) From waste to wealth: exploring biochar's potential in energy generation and waste mitigation. Clean Circ Bioecon:100101. https://doi.org/10.1016/j.clcb.2024.100101
99. ENRD (2021) European rural bioeconomy: policy and tools. Conclusions from the ENRD: Thematic Group on 'Mainstreaming the bioeconomy' – Part 1
100. IRENA (2023) Agricultural residue-based bioenergy: regional potential and scale-up strategies. Abu Dhabi

101. Bhattacharyya SC (2014) Viability of off-grid electricity supply using rice husk: a case study from South Asia. Biomass Bioenergy 68:44–54. https://doi.org/10.1016/j.biombioe.2014.06.002
102. Jindo K, Ghaffari G, Lamichhane M et al (2025) Assessment of trade-off balance of maize stover use for bioenergy and soil erosion mitigation in Western Kenya. Front Sustain Food Syst 9:1409457. https://doi.org/10.3389/fsufs.2025.1409457
103. Feleke S, Cole SM, Sekabira H et al (2021) Circular bioeconomy research for development in sub-Saharan Africa: innovations, gaps, and actions. Sustainability 13:1926
104. Lopolito A, Prosperi M, Sisto R, De Meo E (2015) Translating local stakeholders' perception in rural development strategies under uncertainty conditions: an application to the case of the bio-based economy in the area of Foggia (South Italy). J Rural Stud 37:61–74. https://doi.org/10.1016/j.jrurstud.2014.12.005
105. Bennich T, Belyazid S, Kopainsky B, Diemer A (2018) Understanding the transition to a bio-based economy: exploring dynamics linked to the agricultural sector in Sweden. Sustainability 10:1504. https://doi.org/10.3390/su10051504
106. Piedrahita-Rodríguez S, Cardona Urrea S, Escobar García DA et al (2023) Life cycle assessment and potential geolocation of a multi-feedstock biorefinery: integration of the avocado and plantain value chains in rural zones. Bioresour Technol Rep 21:101318. https://doi.org/10.1016/j.biteb.2022.101318
107. Pagot G, Andrighetto N (2024) Fuel for collective action: a SWOT analysis to identify social barriers and drivers for a local woody biomass supply chain in an Italian alpine valley. Heliyon 10. https://doi.org/10.1016/j.heliyon.2024.e38170
108. Koltunov M, Möller J, Szántó Z (2019) The impact of renewable energy cooperatives on the welfare of local communities. Corvinus University of Budapest, Budapest
109. Oteman M, Wiering M, Helderman J-K (2014) The institutional space of community initiatives for renewable energy: a comparative case study of the Netherlands, Germany and Denmark. Energy Sustain Soc 4:1–17. https://doi.org/10.1186/2192-0567-4-11
110. Siamanta ZC (2021) Conceptualizing alternatives to contemporary renewable energy development: Community Renewable Energy Ecologies (CREE). J Polit Ecol 28:47–69
111. Bangalore M, Hochman G, Zilberman D (2016) Policy incentives and adoption of agricultural anaerobic digestion: a survey of Europe and the United States. Renew Energy 97:559–571. https://doi.org/10.1016/j.renene.2016.05.062
112. Li Y, Lin J, Qian Y, Li D (2023) Feed-in tariff policy for biomass power generation: incorporating the feedstock acquisition process. Eur J Oper Res 304:1113–1132. https://doi.org/10.1016/j.ejor.2022.05.011
113. Li Y, Sohail MT, Zhang Y, Ullah S (2024) Bioenergy for sustainable rural development: elevating government governance with environmental policy in China. Land 13:2147
114. Peerapong P, Limmeechokchai B (2020) A prospective feed-in tariff of rice husk power plants in Thailand under the power development plan 2018. In: 2020 International conference and utility exhibition on energy, environment and climate change (ICUE). IEEE, pp 1–9
115. Pe'er G, Bonn A, Bruelheide H et al (2020) Action needed for the EU Common Agricultural Policy to address sustainability challenges. People Nat 2:305–316. https://doi.org/10.1002/pan3.10080
116. Singh P, Kalamdhad AS (2023) A comprehensive assessment of state-wise biogas potential and its utilization in India. Biomass Convers Biorefin 13:12557–12579. https://doi.org/10.1007/s13399-021-02001-y
117. Singh N (2024) Gender, environment degradation, and bioenergy trajectories: eco-feminist perspective. In: SDGs in the Asia and Pacific region. Springer, pp 1387–1406
118. Gannan I, Kubaji H, Siwale W et al (2023) Streamlined social footprint analysis of the nascent bio-pellet sub-sector in Zambia. Sustainability 15:5492. https://doi.org/10.3390/su15065492
119. Chineme A, Assefa G, Herremans IM et al (2022) African indigenous female entrepreneurs (IFÉs): a closed-looped social circular economy waste management model. Sustainability 14:11628. https://doi.org/10.3390/su141811628

120. Singh SK (2013) Recycling of within farm renewable resources: an entrepreneurial opportunity for woman folk. Lecture delivered during Model Train Course, 24 Jan 2013, ICAR Research Complex for EAS 82
121. Carvalho RL, Yadav P, García-López N et al (2020) Environmental sustainability of bioenergy strategies in western Kenya to address household air pollution. Energies 13:719. https://doi.org/10.3390/en13030719
122. Kyayesimira J, Florence M (2021) Health concerns and use of biomass energy in households: voices of women from rural communities in Western Uganda. Energy Sustain Soc 11:1–13. https://doi.org/10.1186/s13705-021-00316-2
123. MPC (2023) Action plan: national biomass action plan 2023–2030. Putrajaya, Malaysia
124. Abdulkarim A, Faruk N, Alozie E et al (2024) Advances in the design of renewable energy power supply for rural health clinics, case studies, and future directions. Clean Technol 6:921–953. https://doi.org/10.3390/cleantechnol6030047
125. Vallecha H, Bhola P (2019) Sustainability and replicability framework: actor network theory based critical case analysis of renewable community energy projects in India. Renew Sustain Energy Rev 108:194–208. https://doi.org/10.1016/j.rser.2019.03.053
126. Shafiei M, Kumar R, Karimi K (2015) Pretreatment of lignocellulosic biomass. In: Lignocellulose-based bioproducts, pp 85–154. https://doi.org/10.1007/978-3-319-14033-9_3
127. Margellou AG, Pappa CP, Psochia EA et al (2023) Mild isolation and characterization of surface lignin from hydrothermally pretreated lignocellulosic forestry and agro-industrial waste biomass. Sustain Chem Pharm 33:101056. https://doi.org/10.1016/j.scp.2023.101056
128. Carvalheiro F, Duarte LC, Pires F et al (2022) Effective mild ethanol-based organosolv pre-treatment for the selective valorization of polysaccharides and lignin from agricultural and forestry residues. Energies 15:5654
129. Tran TVH, Nguyen TAH, Nguyen MV (2024) Pretreatment techniques for derivation of value-added products from agro-waste biomass. In: Valorization of biomass wastes for environmental sustainability: green practices for the rural circular economy. Springer, pp 207–221
130. El Achkar T, Greige-Gerges H, Fourmentin S (2021) Basics and properties of deep eutectic solvents: a review. Environ Chem Lett 19:3397–3408. https://doi.org/10.1007/s10311-021-01225-8
131. Zhao C, Deng L, Fang H (2018) Mixed culture of recombinant Trichoderma reesei and Aspergillus niger for cellulase production to increase the cellulose degrading capability. Biomass Bioenergy 112:93–98. https://doi.org/10.1016/j.biombioe.2018.03.001
132. Raheja Y, Gaur P, Islam T et al (2025) Advancement in lignocellulolytic enzyme production: tailored strategies to overcome challenges in biomass hydrolysis. Syst Microbiol Biomanuf 5: 948–968. https://doi.org/10.1007/s43393-025-00342-7
133. Bezerra PXO, De Farias Silva CE, Soletti JI, de Carvalho SHV (2021) Cellulosic ethanol from sugarcane straw: a discussion based on industrial experience in the Northeast of Brazil. Bioenergy Res 14:761–773. https://doi.org/10.1007/s12155-020-10169-w
134. Ismail N, Fauzi NF, Salehabadi A et al (2022) A study on biogas production from cassava peel and stem in anaerobic digestion reactor. Int J Environ Sci Technol 19:1695–1704. https://doi.org/10.1007/s13762-021-03222-4
135. Hoang T-D, Nghiem N (2021) Recent developments and current status of commercial production of fuel ethanol. Fermentation 7:314
136. Tran PHN, Jung JH, Ko JK et al (2023) Co-production of ethanol and polyhydroxybutyrate from lignocellulosic biomass using an engineered Saccharomyces cerevisiae. Renew Energy 212:601–611. https://doi.org/10.1016/j.renene.2023.05.080
137. de Cárdenas LZ, de Cárdenas BZ (2020) Production of organic acids via fermentation of sugars generated from lignocellulosic biomass. In: Lignocellulosic biorefining technologies. Wiley, pp 203–246. https://doi.org/10.1002/9781119568858.ch10
138. Aup-Ngoen K, Noipitak M (2020) Effect of carbon-rich biochar on mechanical properties of PLA-biochar composites. Sustain Chem Pharm 15:100204. https://doi.org/10.1016/j.scp.2019.100204

139. Iwuozor KO, Abdulkareem SA, Amoloye MA et al (2024) Eco-friendly composite materials: enhancing sustainability with sugarcane bagasse biochar and polystyrene resin. Sugar Tech 26: 408–421. https://doi.org/10.1007/s12355-023-01350-4
140. Mostafa NA, Farag AA, Abo-dief HM, Tayeb AM (2018) Production of biodegradable plastic from agricultural wastes. Arab J Chem 11:546–553
141. Yan N, Luan T, Yin M et al (2023) Co-fermentation of glucose–xylose–cellobiose–XOS mixtures using a synthetic consortium of recombinant Saccharomyces cerevisiae strains. Fermentation 9:775. https://doi.org/10.3390/fermentation9080775
142. Dempfle D, Kröcher O, Studer MH-P (2021) Techno-economic assessment of bioethanol production from lignocellulose by consortium-based consolidated bioprocessing at industrial scale. New Biotechnol 65:53–60. https://doi.org/10.1016/j.nbt.2021.07.005
143. Maitra S, Singh V (2022) A consolidated bioprocess design to produce multiple high-value platform chemicals from lignocellulosic biomass and its technoeconomic feasibility. J Clean Prod 377:134383. https://doi.org/10.1016/j.jclepro.2022.134383
144. Ogunyewo OA, Okereke OE, Kumar S, Yazdani SS (2022) Characterization of a GH5 endoxylanase from Penicillium funiculosum and its synergism with GH16 endo-1,3(4)-glucanase in saccharification of sugarcane bagasse. Sci Rep 12:17219. https://doi.org/10.1038/s41598-022-21529-1
145. Mushtaq Q, Ishtiaq U, Joly N et al (2024) Harnessing Bacillus subtilis QY5 PP784163 for bioethanol production from potato peel waste and nutrient recovery for animal feed: maximizing resource efficiency. Fermentation 10:523
146. Gregg JS, Izaurralde RC (2010) Effect of crop residue harvest on long-term crop yield, soil erosion and nutrient balance: trade-offs for a sustainable bioenergy feedstock. Biofuels 1:69–83. https://doi.org/10.4155/bfs.09.8
147. Lal R (2005) World crop residues production and implications of its use as a biofuel. Environ Int 31:575–584. https://doi.org/10.1016/j.envint.2004.09.005
148. Derpsch R, Kassam A, Reicosky D et al (2024) Nature's laws of declining soil productivity and conservation agriculture. Soil Secur 14:100127. https://doi.org/10.1016/j.soisec.2024.100127
149. Salinas-Garcia JR, Báez-González AD, Tiscareño-López M, Rosales-Robles E (2001) Residue removal and tillage interaction effects on soil properties under rain-fed corn production in Central Mexico. Soil Tillage Res 59:67–79. https://doi.org/10.1016/S0167-1987(00)00187-2
150. Andrade Díaz C, Clivot H, Albers A et al (2023) The crop residue conundrum: maintaining long-term soil organic carbon stocks while reinforcing the bioeconomy, compatible endeavors? Appl Energy 329:120192. https://doi.org/10.1016/j.apenergy.2022.120192
151. Gan J, Stupak I, Smith CT (2019) Integrating policy, market, and technology for sustainability governance of agriculture-based biofuel and bioeconomic development in the US. Energy Sustain Soc 9:1–25. https://doi.org/10.1186/s13705-019-0223-2
152. Field J, Kline KL, Langholtz M, Singh N (2023) Sustainably sourcing biomass feedstocks for bioenergy with carbon capture and storage in the United States. Energy Futures Initiative
153. Alliance Environnement (2020) Evaluation support study on the impact of the CAP on sustainable management of the soil. Final report
154. Kaoma M, Gheewala SH (2021) Techno-economic assessment of bioenergy options using crop and forest residues for non-electrified rural growth centres in Zambia. Biomass Bioenergy 145:105944. https://doi.org/10.1016/j.biombioe.2020.105944
155. Phomsoda K, Puttanapong N, Piantanakulchai M (2021) Economic impacts of Thailand's biofuel subsidy reallocation using a dynamic computable general equilibrium (CGE) model. Energies 14:2272
156. Diakosavvas D, Frezal C (2019) Bio-economy and the sustainability of the agriculture and food system: opportunities and policy challenges. OECD Food, Agriculture and Fisheries Papers

157. Yadav K, Sircar A (2022) Fundamentals and developments of compressed biogas in city gas distribution network in India: a review. Petrol Res 7:409–418. https://doi.org/10.1016/j.ptlrs.2021.12.003
158. Sawale S, Patil D, Joshi C et al (2020) Biogas commercialization: commercial players, key business drivers, potential market, and fostering investment. In: Biogas production: from anaerobic digestion to a sustainable bioenergy industry, pp 343–387. https://doi.org/10.1007/978-3-030-58827-4_16
159. Prussi M, Panoutsou C, Chiaramonti D (2022) Assessment of the feedstock availability for covering EU alternative fuels demand. Appl Sci 12:740. https://doi.org/10.3390/app12020740
160. Hoque MM, Artz GM, Jarboe DH, Martens BJ (2015) Producer participation in biomass markets: farm factors, market factors, and correlated choices. J Agric Appl Econ 47:317–344. https://doi.org/10.1017/aae.2015.9
161. Tagwi A, Chipfupa U (2023) Participation of smallholder farmers in modern bioenergy value chains in Africa: opportunities and constraints. Bioenergy Res 16:248–262. https://doi.org/10.1007/s12155-022-10451-z
162. Peres IF, Slating TA, Kesan JP (2014) Case for vertical integration in the developing bioenergy industry. William Mary Environ Law Policy Rev 39:575
163. Zhu X, Wang G (2024) Impact of agricultural cooperatives on farmers' collective action: a study based on the socio-ecological system framework. Agriculture 14:96
164. Chen C, Yang R, Wang X et al (2022) Effect of in-situ torrefaction and densification on the properties of pellets from rice husk and rice straw. Chemosphere 289:133009. https://doi.org/10.1016/j.chemosphere.2021.133009
165. Sundaram V, Muthukumarappan K, Kamireddy SR (2015) Effect of ammonia fiber expansion (AFEX™) pretreatment on compression behavior of corn stover, prairie cord grass and switchgrass. Ind Crop Prod 74:45–54. https://doi.org/10.1016/j.indcrop.2015.04.027
166. Mehta CR, Badegaonkar UR, Singh PL, Singh KK (2023) Integrated straw management in India. South and South-West Asia Development Papers
167. Ma C, Wang Y, Guo K et al (2025) Enzymatic electrosynthesis system based on multi-enzyme catalysis or coupled with microbial transformation. Catal Sci Technol. https://doi.org/10.1039/D4CY01381F
168. Udali A, Talbot B, Ackerman S et al (2024) Enhancing precision in quantification and spatial distribution of logging residues in plantation stands. Eur J For Res 144:817–830. https://doi.org/10.1007/s10342-024-01699-5
169. Wang S, Guan K, Zhang C et al (2023) Cross-scale sensing of field-level crop residue cover: integrating field photos, airborne hyperspectral imaging, and satellite data. Remote Sens Environ 285:113366. https://doi.org/10.1016/j.rse.2022.113366
170. Kar E, Maity S, Kar A, Sen S (2024) Agricultural waste rice husk/poly (vinylidene fluoride) composite: a wearable triboelectric energy harvester for real-time smart IoT applications. Adv Compos Hybrid Mater 7:87. https://doi.org/10.1007/s42114-024-00896-5
171. Sharma A, Singh PK (2023) Applicability of UAVs in detecting and monitoring burning residue of paddy crops with IoT integration: a step towards greener environment. Comput Ind Eng 184:109524. https://doi.org/10.1016/j.cie.2023.109524

Adv Biochem Eng Biotechnol (2026) 196: 33–78
https://doi.org/10.1007/10_2026_334

Published online: 8 July 2026

Processes and Products Derived from Lignocellulosic Feedstock: A Biorefinery Approach

Romit Mitra, Aritro Sanjib Sen, Supratit Ghosh, Goldy De Bhowmick, Ankita Debnath, and Ramkrishna Sen

Contents

Romit Mitra, Aritro Sanjib Sen, Supratit Ghosh and Goldy De Bhowmick contributed equally with all other contributors.

R. Mitra
P K Sinha Centre for Bioenergy and Renewables, Indian Institute of Technology Kharagpur, Kharagpur, West Bengal, India
e-mail: romit.be18@iitkgp.ac.in

A. S. Sen, S. Ghosh, A. Debnath, and R. Sen (✉)
Department of Bioscience and Biotechnology, Indian Institute of Technology Kharagpur, Kharagpur, West Bengal, India
e-mail: aritrosen2001@kgpian.iitkgp.ac.in; supratit.g@kgpian.iitkgp.ac.in; ankitadebnath@iitkgp.ac.in; rksen@bt.iitkgp.ac.in

G. De Bhowmick
School of Chemical Sciences, The University of Auckland, Auckland, New Zealand
e-mail: gdeb449@aucklanduni.ac.nz

Abstract The global imperative to transition from fossil-dependent industries toward sustainable, circular production models has positioned lignocellulosic biomass as a cornerstone feedstock for the modern bioeconomy. This chapter provides a critical synthesis of lignocellulosic biorefineries, evaluating the integrated processes and diverse value chains required to transform complex plant matter into high-value commodities. While the tripartite structure of cellulose, hemicellulose, and lignin offers immense potential for the production of biofuels, bioenergy, and platform chemicals, its inherent recalcitrance poses a formidable barrier to commercial viability. To address this, the chapter systematically examines various pretreatment approaches, including physical, chemical, physicochemical, and biological methods, highlighting their mechanisms, advantages, and limitations. Emerging green and sustainable technologies, such as deep eutectic solvents, ionic liquids, and supercritical fluid-based processes, are also critically discussed in the context of improving biomass fractionation and process sustainability. Subsequently, key conversion pathways such as thermal, thermochemical, and biochemical are explored for the valorization of lignocellulosic biomass into fuels, platform chemicals, and other value-added products. It highlights recent breakthroughs in enzymatic saccharification and microbial fermentation, with a focus on how metabolic engineering and process optimization are currently overcoming traditional yield limitations. The chapter further addresses integrated and hybrid biorefinery concepts, such as consolidated bioprocessing and process intensification, which aim to maximize resource utilization and economic feasibility. Overall, this chapter provides critical insights into current advancements, challenges, and future prospects of lignocellulosic biorefineries in advancing a sustainable and carbon-neutral bioeconomy.

Graphical Abstract

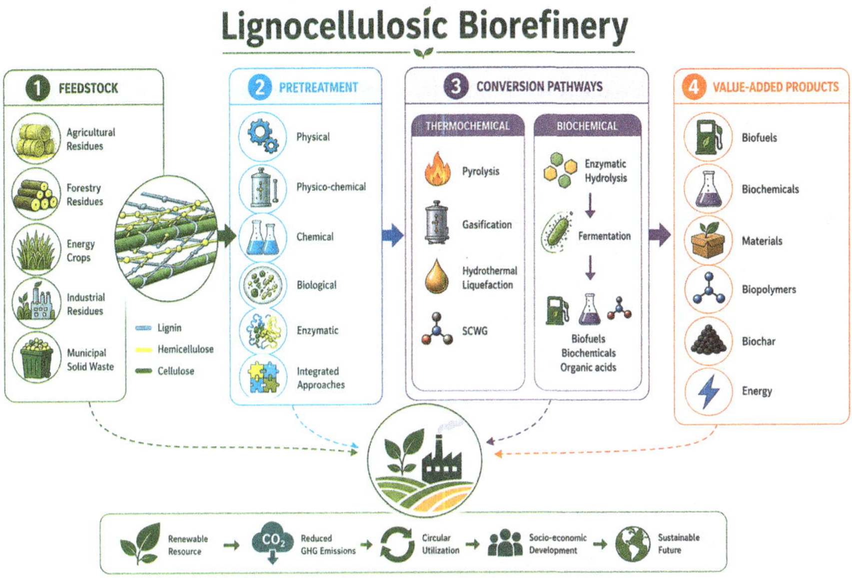

Keywords Biomass conversion, Biorefinery, Circular bioeconomy, Lignocellulosic biomass, Pretreatment technologies

1 Introduction

Petrochemical refineries have served as the primary source of fuels and chemical commodities for decades to meet the growing global energy demand [1]. Currently, more than 85% of the world's energy consumption is achieved by using fossil fuels [2]. Apparently, intensive usage of fossil fuels resulted in increased greenhouse gas (GHG) emissions and air pollution. Moreover, limited availability coupled with high production cost of fossil fuel posed significant challenges to the global energy security and dismantling economic stability [3]. In response to these challenges, many countries have introduced policies and initiatives aimed at reducing their dependence on fossil fuel-based energy resources while transitioning to renewable and sustainable alternatives [4].

Alternative to fossilized resources, lignocellulosic biomass is the most abundant and underutilized bio-resource for renewable energy generation [5, 6]. Lignocellulosic materials such as crop residues, agro-industrial wastes, and woody biomass through bioprocessing or bio-refinement can be effectively converted into a wide range of fuels and high-value bioproducts [1]. Possible replacement of fossil fuel-based chemicals with renewable biomass resources looks promising for the advancement of global bioeconomy while meeting the global agenda to become carbon negative by 2030 [7].

In a lignocellulosic biorefinery, the key components, such as cellulose and hemicellulose, are broken down into monomeric sugars. These sugars can be subsequently converted into liquid biofuels and bioproducts via fermentation or anaerobic digestion techniques [8]. However, the economic viability largely depends on the efficiency of cellulose saccharification into fermentable sugars [5], as these C_6 and C_5 sugars could serve as critical intermediates in various biological and chemical conversion pathways [9, 10]. Therefore, utilizing lignocellulosic biomass as a feedstock in a circular economy framework would not only resolve the food-versus-fuel conflict but also eliminate additional land requirements for energy crop cultivation. Currently, a large portion of lignocellulosic waste is either burned, thrown away as landfill, or dumped into water bodies. Thus, utilizing such waste biomass for energy production would significantly improve environmental conditions by lowering the GHG emissions and water pollution [11]. In spite of certain advantages of using lignocellulosic biomass, the inherent recalcitrance nature of the lignocellulosic biomass poses significant challenges, hindering enzymatic hydrolysis of cellulose, thereby reducing yield [12]. To improve the digestibility of lignocellulosic materials, various pretreatment methods have been investigated previously. However, these processes often demand high energy inputs involving intense mechanical force and elevated temperature and pressure to disrupt the complex lignocellulosic network [13]. Therefore, the ability to minimize energy consumption while maximizing product yield is crucial to achieving a positive energy balance in lignocellulosic biorefineries. In recent years, advances in industrial biotechnology have significantly accelerated the development and global expansion of lignocellulosic biorefineries [10]. As a result, the conversion of lignocellulosic biomass into value-added products has been extensively studied at the laboratory level, scaled up in pilot facilities, and, in some instances, successfully implemented at the commercial scale [6, 14, 15]. Nonetheless, a thorough evaluation of capital and operational expenditures (CAPEX and OPEX) associated with product development is essential for ensuring the economic viability of lignocellulose-based bioprocesses.

2 Role of Lignocellulosic Biorefineries in Bio-Based Manufacturing and Bioeconomy

Extensive usage of fuels and chemicals produced from fossil fuels has accelerated the rate of global warming by releasing greenhouse gases into the atmosphere. Alarming environmental issues necessitate transitioning of fossil-based resources to alternative and sustainable bio-resources. In the recent past, LC biorefinery concept has gained immense interest due to its potential to manufacture a range of bioproducts starting from biofuels and bioenergy alternatives to various high-value chemicals such as xylitol, polyols, furfural, and aromatics [4, 16]. The National Renewable Energy Laboratory (NREL) describes biorefinery approach as an integration of biomass conversion processes and equipment to produce power, chemical, and fuels. However, according to the International Energy Agency (IEA), a biorefinery concept means processing of biomass into a spectrum of marketable products (food, feed, materials, chemicals) and energy (fuels, power, heat) [17].

LC biorefinery focuses upon different phases [17]. In phase I, unitary or single conversion mode is followed that incorporates a single feedstock and is converted into a single product. Phase I of biorefinery often limits the complete utilization of different parts of lignocellulosic biomass to its full capacity. In phase II, conversion of a single type of feedstock to multiple products following multiple conversion pathways is achieved. However, phase II is also limited due to usage of a single type of feedstock for operation. A wide array of lignocellulosic biomass derived from the agricultural and forestry remnants to different energy crops provides innovative options to simultaneously utilize multiple feedstocks efficiently. To achieve this goal, the phase III concept in the LC biorefinery was further introduced. In phase III, multiple products like chemicals, biofuels, and bioenergy were produced through multiple conversion routes from a variety of feedstock. Nonetheless, this multi-feedstock multi-product kind of biorefinery was well established as it not only ensures the constant feedstock availability through diversified feedstock usage but also provides an evolved and promising model by flexibly responding toward market demand for product diversity.

In case of bio-manufacturing, LC biorefineries comply with modern industrial goals by enabling the conversion of high-value products like nutraceuticals, food additives, cosmetics, and platform chemicals (like furfural, levulinic acid, lactic acid) along with commercially viable biofuels and biopolymers [18]. LC biorefineries efficiently integrate with the existing infrastructure facilities like pulp-paper mills or ethanol plants and turn their waste streams into an alternative revenue source. Moreover, LC biorefineries advance the concept of efficient waste valorization by utilizing nonfood biomass and agricultural residues. Additionally, replacing fossil fuels and chemicals with biomass-derived substitutes not only lowers net greenhouse gas emissions but also advances toward a carbon-negative system [17, 18].

Furthermore, LC biorefineries catalyze inclusive bioeconomy by promoting rural and regional development through decentralized biomass collection and processing, creating multiple job opportunities. It also facilitates industrial symbiosis wherein a by-product, waste, or energy stream from one industrial process could serve as an input stream for another pathway integrated in the biorefinery approach. Thus, lignocellulosic biorefineries are rapidly becoming a central player in the global manufacturing process, shifting the paradigm.

3 Pretreatment Strategies for Lignocellulosic Feedstock

Lignocellulosic biomass is composed of lignin, hemicellulose, and cellulose. The cellulosic components of lignocellulosic feedstock serve as the primary source of fermentable sugars. Subsequently, these sugars can be extracted from carbohydrate polymers by enzymatic hydrolysis of delignified feedstock. Nevertheless, accessing the cellulosic fractions of the lignocellulosic biomass is challenging due to the presence of lignin. Consequently, it is essential to eliminate the lignin fraction of the biomass and segregate the cellulosic components of the feedstock from the lignin. It is vital to depolymerize and delignify the biomass while lowering the crystallinity of cellulose via biological, physical, chemical, or physicochemical methods. The major goal of the pretreatment process is to remove the lignin from the lignocellulosic biomass and break down the cellulosic fractions into smaller and simpler fermentable sugar fragments. Pretreatment therefore becomes a vital step in the LC biorefinery process.

3.1 Physical Methods

Physical pretreatment usually involves conventional mechanical and thermal procedures that aid in delignification. Physical methods are energy intensive that can increase the overall expenses of the bio-manufacturing process. Additionally, in order to effectively extract the fermentable sugars, these methods must be combined with other chemical and enzymatic pretreatment procedures. An overview of different physical pretreatment methods is shown in Table 1.

3.2 Chemical Methods

In contrast to physical techniques like milling, steam explosion, or ultrasound, which increase surface area and partially break down the biomass matrix, chemical pretreatment uses hydrolysis, delignification, or solubilization targeting particular structural components such as lignin and hemicellulose. The use of oxidizing agents, solvents, acids, or alkalis enables the deeper penetration and structural modification of cellulose, thereby improving the accessibility of intracellular compounds. Chemical pretreatment improves the efficiency of enzymatic hydrolysis or microbial conversion by enhancing the overall disruption of biomass structure. However, when combined with physical approaches, chemical pretreatment can dramatically reduce recalcitrance, eliminate enzyme inhibition, and boost the yield of fermentable sugars or extractable biomolecules.

Table 1 Overview of physical pretreatment strategies for lignocellulosic biomass

Technique	Principle	Effect on biomass	Advantages	Disadvantages	References
Milling/grinding	Mechanical size reduction via ball mill, hammer mill, etc.	Reduces particle size, increases surface area, and slightly disrupts structure	• Mass transfer improvements • More available space for hydrolysis • Higher surface-to-volume ratio	• Energy intensive • Limited delignification • Nonuniform particle sizes	[9]
Steam explosion	High-pressure steam exposure followed by sudden decompression	Disrupts lignocellulosic structure, hemicellulose solubilization, and lignin removal	• Further chemical pretreatments not needed • Efficient lignin degradation • Versatile process	• Release of inhibitory compounds from lignin depolymerization • Energy intensive • Expensive equipment	[19]
Autoclaving	Saturated steam at ~121°C, 15 psi for 15–60 min	Partial cell wall disruption, hemicellulose softening, sterilization	• Simple and inexpensive process • Results in sterilization • Effective for wet biomass	• Multiple cycles needed • Not effective for high-scale applications • Prolonged and extensive process	[20]
Microwave	Dielectric heating causes rapid internal heating	Breaks cell walls, releases sugars/lipids, disrupts structure	• Energy efficient • Quick and uniform disruption • Limited solvent requirement	• Formation of inhibitory compounds • Not ideal for hard biomass • Difficult to scale up	[21]
Ultrasound	Acoustic cavitation from high-frequency sound waves	Causes microbubble collapse, mechanical disruption of cells	• Energy efficient • Enhances enzymatic saccharification and lipid extraction • Limited solvent requirement	• Not ideal for hard biomass • Difficult to scale up • Energy intensive	[22]

3.2.1 Acid Pretreatment

Acid pretreatment is primarily utilized to eliminate hemicellulose fraction of the LC biomass. Generally, dilute acids such as sulfuric or hydrochloric acid (0.5–2%) are used to hydrolyze hemicellulose into monomeric sugars, thereby improving cellulose accessibility for subsequent enzymatic hydrolysis. Although concentrated acids can hydrolyze cellulose directly, they are poisonous and corrosive, and they need to be recovered, rendering the pretreatment costly. Further, acid pretreatment is more effective when integrated with physical techniques like steam explosion or microwave treatment, resulting in improved acid infiltration and reaction kinetics.

Numerous ways have been devised to enhance the yield of reducing sugar using dilute acid hydrolysis. For instance, Gao et al. utilized acid mine drainage as the source of dilute acid for maize cob hydrolysis. In this study, yield of reducing sugar was approximately 93% greater using acid mine drainage than dilute sulfuric acid hydrolysis alone keeping other pretreatment conditions (121°C for 60 min) [23]. In another study, Gutierrez et al. used dilute sulfuric acid along with the biomass in a bioreactor resulting in 77% xylan removal, with the hydrolysate yielding 19.9 g/L of total reducing sugars. It was anticipated that the improvement in the mass transfer and uniform distribution of the acid in the biomass certainly increased the effectiveness of the pretreatment process [24]. Dilute acid pretreatment, however, may lead to the generation of inhibitory chemicals, including furfurals, vanillin, and HMF [25].

Dilute acid pretreatment operates or functions under elevated temperatures leading to higher energy consumption. Alternative options such as organic acids have gained considerable attention recently, due to their sustainability and environmentally friendly approach. Organic acid pretreatment facilitates selective extraction of hemicellulose and lignin, resulting in efficient biomass fractionation. Organic acid (a weak acid) usage also results in the release of lower concentration of inhibitory products from delignification [26]. It was observed that the generation of degradation products and the release of sugar were influenced by the acid's dissociation constant (K_a). The level of degradation products increased with the K_a value. Consequently, compared to mineral acids, organic acid pretreatment produced less breakdown products. For instance, Hermsdorff et al. during enzymatic hydrolysis used oxalic acid as catalyst for the hydrolysis of sugarcane bagasse and obtained 86% delignification with minimal concentrations of fermentation inhibitors like HMF and furfural [27].

3.2.2 Alkaline Pretreatment

Alkali pretreatment is an efficient chemical process that is mostly used to remove lignin while breaking down the structure of lignocellulosic biomass. It entails mild-to-moderate treatment of biomass with alkaline agents like ammonia, potassium hydroxide (KOH), sodium hydroxide (NaOH), or calcium hydroxide (lime). By cleaving lignin-carbohydrate bonds and saponifying ester bonds, the alkali causes lignin solubilization, cellulose fiber swelling, and partial removal of hemicellulose. Eventually, enzymes can more easily access cellulose as a result of the increased porosity and surface area. Alkali pretreatment produces less fermentation inhibitors than acid pretreatment and works particularly well for biomass rich in lignin, including woody biomass and agricultural leftovers. They could, however, necessitate more thorough cleaning procedures, lengthier reaction times, and cautious management of chemical waste. Alkali pretreatment can improve biomass conversion efficiency in biofuel and bioproduct processes when paired with physical pretreatment like milling or ultrasonication. However, alkaline pretreatment can also lead to the formation of insoluble salts which get stuck with the biomass and are difficult to recover.

Sun et al. optimized the alkaline pretreatment conditions for lignocellulosic yard waste, where they were able to recover 80% lignin while pretreating 150–200 mg/g feedstock with NaOH at 105°C for 30 min [28]. Tharunkumar et al. focused on optimizing the alkaline pretreatment for the delignification of paddy straw and sugarcane bagasse [29]. In this study, 2–10% of NaOH concentrations were tested for the pretreatment, and amongst the tested range, 2% NaOH resulted in the highest amount of reducing sugar yield of ~8.37 g/L for paddy straw and 7.635 g/L for sugarcane bagasse.

3.2.3 Organosolv Pretreatment

Using organic solvents (such as ethanol, methanol, acetone, or mixes of these) in conjunction with catalysts (such as diluted acid or alkali) at high temperatures and pressures, organosolv pretreatment is proved to be an effective method for removal of lignin. By dissolving lignin and hemicellulose into the solvent phase, this process breaks down the lignin-carbohydrate matrix leaving behind a cellulose-rich solid mass that is comparatively pure and easily digested. Usually, this pretreatment procedure runs for 30–90 min at temperatures between 120 and 200°C. Organosolv pretreatment results in high-quality yields and sulfur-free lignin as a useful by-product that can be utilized in synthesizing other materials or biochemicals. In addition, the process makes subsequent enzymatic hydrolysis simpler and produces less fermentation inhibitors than acid pretreatment. However, effective solvent recovery and recycling are necessary for economic viability. Chen et al. reported improved organosolv pretreatment conditions by employing Taguchi experimental design, wherein the lignin affinity was strongest at a volume ratio of 5:11:4 for acetone, phenoxyethanol, and water for a combined lignocellulosic biomass. Further, 95.6% lignin removal was achieved under ideal conditions of 130°C, 70 min, 0.15 M sulfuric acid, and a liquid-to-solid ratio of 20 [30].

3.2.4 Deep Eutectic Solvents (DESs)

Lower toxicity, biodegradability, and adjustable characteristics of deep eutectic solvents (DESs) have classified them as a green solvent for biomass pretreatment. In order to create a eutectic mixture with a melting point much lower than the melting points of the constituent parts, DESs are created by combining a hydrogen bond donor (HBD) and a hydrogen bond acceptor (HBA), such as choline chloride (HBA) with urea, glycerol, or lactic acid (HBD). DESs effectively delignify, solubilize hemicellulose, and increase cellulose accessibility by breaking down lignocellulosic structures in biomass pretreatment, mostly by hydrogen bonding and mild solvation. DES pretreatment can be used alone or in conjunction with light physical or enzymatic treatments, and it is frequently carried out at moderate temperatures (80–120°C). DESs' low environmental impact and recyclability are two of its main advantages, and they fit well under sustainable biorefinery portfolio. However, their industrial application is currently hampered by issues like high viscosity, delayed mass transfer, and a lack of large-scale investigations. Zhang et al. established a unique DES ferric chloride-based pretreatment method for corn stover consisting of *N*-(2-hydroxyethyl)ethylenediamine as the HBD for the ChCl-based DES. Using a 60% DES and 0.5 wt.% $FeCl_3$ mixture, they achieved 87% lignin removal and 98.6% enzymatic saccharification in the resultant hydrolysate at optimal conditions of 110°C for 50 min [31]. In another study, Wu et al. devised a DES with benzyltriethylammonium chloride (TEBAC) and glycerol (Gly) with *p*-toluenesulfonic acid (*p*-TsOH) as a catalyst. Here, under optimized conditions of 90°C for 160 min and 1.6 wt.% *p*-TsOH, around 92% delignification and 88.2% hemicellulose removal were achieved [32].

3.2.5 Ionic Liquids (ILs)

Ionic liquids (ILs) are a family of low-melting-point organic salts that have demonstrated considerable potential for biomass pretreatment due to their unique physicochemical features, including minimal vapor pressure, thermal stability, and high solvation capabilities. ILs, particularly those based on imidazolium, pyridinium, or ammonium cations with anions like chloride (Cl^-), acetate (CH_3COO^-), or phosphate, can efficiently break hydrogen bonding networks in cellulose and solubilize lignin and hemicellulose. This results in a substantially more accessible cellulose fraction for subsequent enzymatic hydrolysis. For example, $[C_2mim][OAc]$ (1-ethyl-3-methylimidazolium acetate) is one of the most investigated ILs due to its great capacity to dissolve cellulose and destroy lignin structures. IL occurs at 80–140°C and can be modified for selectivity by changing the IL's chemical structure. Despite their usefulness, the high cost of ILs, difficulty in solvent recovery, and possible toxicity remain challenging. Sunar et al. optimized the conditions for two different ILs *N,N*-diisopropylethylammonium hydrogen sulfate (IL-DHS) and *N,N*-diisopropylethylammonium perchlorate (IL-DP) for the pretreatment of sugarcane bagasse. Reducing sugar concentrations of 53.4% and 55.9% with 3.9- and 4.7-fold increase in enzymatic saccharification efficiency for IL-DHS and IL-DP was achieved successfully [33]. In another study, Zhu et al. pretreated corn stalk with [BHEM]mesy-ethylene glycol co-solvent system and obtained 93% delignification [34].

3.2.6 Supercritical Fluids

Supercritical fluid (SCF) pretreatment employs fluids at temperatures and pressures beyond their critical point, where they display unusual properties, combining gas-like diffusivity with liquid-like solvating power to efficiently disturb the structure of lignocellulosic biomass. Amongst the most widely utilized supercritical fluids, supercritical carbon dioxide ($scCO_2$) and supercritical water (scH_2O) are typically used. Briefly, $scCO_2$ employed with co-solvents like water or ethanol enters biomass pores, produces swelling, and promotes enzymatic hydrolysis without creating inhibiting by-products. Supercritical water with its capacity to hydrolyze hemicellulose, while solubilizing lignin under severe circumstances (>374°C and >22.1 MPa), effectively depolymerizes biomass, but it requires careful supervision to minimize sugar degradation. SCF pretreatment is considered eco-friendly, solvent-free (in the case of $scCO_2$), and ideal for integrated biorefinery systems. However, the process needs high-pressure equipment, leveraging energy demands, and is currently constrained by scale-up issues combined with higher capital expenditure. When maximized, SCF pretreatment offers a clean and selective process for biomass breakdown with low chemical waste. Bertran-Llorens et al. reported that during alkaline and organosolv pretreatment, wheat straw was exposed to supercritical CO_2 and the lignin solubilization yields rose from 49% to 79% and 74% to 91%, respectively [35].

3.2.7 Ammonia Fiber Explosion

Ammonia fiber explosion is a physicochemical pretreatment process that utilizes liquid ammonia at moderate pressure (usually 1–2 MPa) and extreme temperatures (60–100°C) to break down the lignocellulosic structure of biomass. In this method, biomass is exposed to liquid ammonia and then rapidly depressurized, generating a physical explosion that breaks the cell wall matrix. The abrupt pressure release leads to fiber expansion, increased surface area, and partial decrystallization of cellulose while successfully eliminating hemicellulose-lignin cross-links and enhancing enzyme accessibility. Unlike acid-based approaches, AFEX creates low fermentation inhibitors and maintains most of the carbohydrates, making it particularly useful for herbaceous biomass such as maize stover, wheat straw, and switchgrass. However, this approach is less successful on woody biomass due to its high lignin concentration. The ammonia recovery and recycling system is crucial to achieve cost-effectiveness and environmental compliance. Despite the necessity for high-pressure equipment, AFEX remains a viable and scalable technology for biorefineries looking for high sugar yields with low degradation. Zhang et al. used ammonia fiber explosion pretreatment on herbaceous wood, hardwood, and softwood in combination with alkaline pretreatment and obtained 84.2%, 59.7%, and 36.7% lignin removal, respectively [36].

3.3 *Physicochemical Methods*

Physicochemical pretreatments use both chemical action (solvent penetration, acid/base reactions) and physical disruption (pressure, temperature, particle breakage) to break down the lignocellulosic biomass. These procedures are aimed to promote cellulose accessibility, partly eliminate hemicellulose and lignin, and minimize crystallinity. An overview of different physicochemical pretreatment methods for lignocellulosic biomass is shown in Table 2.

3.4 *Biological Methods*

Biological pretreatment improves accessibility for enzymatic saccharification by using microbes or enzymes to break down lignin, hemicellulose, and occasionally cellulose. Though usually slower than other pretreatment methods, this approach is environmentally safe and affordable and runs under mild conditions.

3.4.1 Fungal Pretreatment

Fungal pretreatment is a biological technique that utilizes lignin-degrading fungus to boost the enzymatic accessibility of lignocellulosic biomass. White-rot fungi, including *Phanerochaete chrysosporium*, *Trametes versicolor*, and *Pleurotus ostreatus*, are the most efficient strains of the main fungal groups. These strains produce oxidative enzymes like laccase, manganese peroxidase (MnP), and lignin peroxidase (LiP), which selectively break down lignin with minimal cellulose loss. Brown-rot fungi, such as *Postia placenta* and *Gloeophyllum trabeum*, primarily decompose cellulose and hemicellulose via the Fenton reaction, resulting in a lignin-rich residue that renders them less appropriate for delignification-oriented treatments. Soft-rot fungi, such as *Chaetomium globosum* and *Kretzschmaria deusta*, generally flourish in humid conditions and chemically alter lignin instead of significantly depolymerizing it, rendering them less efficient yet valuable for specific niche uses. Overall, fungal pretreatment is ecologically benign and energy efficient and operates under moderate temperatures, but it suffers from sluggish reaction rates, necessity for regulated growing conditions, and probable cellulose damage depending on fungal specificity. The growth of various white-rot fungal species,

Table 2 Overview of physicochemical methods for pretreatment of lignocellulosic biomass

Technique	Principle	Effect on biomass	Advantages	Disadvantages	Reference
Ammonia fiber explosion	Uses explosive depressurization after applying pressure on liquid ammonia	Breaks down lignin-carbohydrate complexes and increases the accessibility of enzymes	High sugar yield, minimal inhibitors	Ammonia recovery required, less effective on woody biomass	[36]
Liquid hot water	Hemicellulose is broken down by water under high pressure and temperature	Solubilizes hemicellulose, preserves cellulose and lignin	No chemicals needed, low inhibitor formation	High water and energy demand	[37]
CO_2 explosion	High-pressure CO_2 penetrates the biomass and is suddenly released	Mechanically alters cell structure, particularly hemicellulose	Nontoxic, environmentally friendly	High-pressure requirement, less effective on lignin	[35]
Wet oxidation	High-temperature oxidative breakdown in water utilizing oxygen or air	Increases the digestibility of enzymes by eliminating lignin and hemicellulose	Effective delignification	Requires high pressure, risk of carbohydrate degradation	[38]
Ozonolysis	Lignin reacts with ozone	Selectively cleaves lignin aromatic rings, preserves polysaccharides	Effective at room temperature, minimal inhibitor formation	Ozone cost, long reaction time	[39]

including *Ceriporiopsis subvermispora*, *Lentinula edodes*, and *Pleurotus eryngii*, on rapeseed straw for biogas production, was investigated by Zheng et al. [40]. In this study, the degradation of rapeseed straw by *Lentinula edodes* showed the highest improvement in in vitro gas production (86%). Summary of fungi-based pretreatment methods for lignocellulosic biomass is shown in Table 3.

3.4.2 Bacterial Pretreatment

Numerous extracellular enzymes secreted by bacteria like *Bacillus*, *Cellulomonas*, *Clostridium*, *Pseudomonas*, and *Streptomyces* can break down cellulose, hemicellulose, and occasionally lignin. These microbes operate under mild environmental conditions such as neutral pH and moderate temperature, requiring low energy. Cellulases, xylanases, lignin peroxidases, and laccases (particularly in actinobacteria) belong to their enzymatic repertory. Several bacteria form synergistic consortia to increase the effectiveness of their enzymatic reaction capacity. Although typically being slower than physical or chemical approaches, bacterial pretreatment has other benefits such as reduced inhibitor formation, selective breakdown, and compatibility with downstream enzymatic saccharification. Longer residence time, sensitivity to environmental changes, and lower delignification capability are some of the drawbacks. *Paenibacillus alvei* and *Bacillus haynesii* are two lignin-degrading bacterial strains that were isolated from agricultural feedstock and were used for pretreatment in a study by Rabi Prasad et al. [44]. In this study, *Bacillus haynesii* performed better under mild conditions requiring a shorter incubation period, which made it appropriate for pretreatment of rice straw and sugarcane bagasse. However, *Paenibacillus alvei* required a longer incubation period, making it suitable for pretreatment of maize stalks.

3.4.3 Enzymatic Pretreatment

Enzymatic pretreatment is a moderate, eco-friendly approach that involves usage of specialized enzymes to selectively break down lignocellulosic biomass into cellulose, hemicellulose, and lignin. Enzymatic pretreatment method often follows physical or chemical pretreatment processes to minimize biomass recalcitrance and promote enzymatic access to structural polysaccharides. Laccases and lignin peroxidases are important enzymes that break down cellulose into monomeric sugars like xylose. Likewise, cellulases hydrolyze cellulose into glucose, hemicellulases hydrolyze cellulose into glucose, and β-glucosidases convert cellobiose into glucose. Enzymatic pretreatment due to its substrate specificity and mild operation conditions (moderate temperature and pH) reduces the generation of inhibitory by-products. The high cost of enzymes, decreased effectiveness on untreated biomass, sensitivity to substrate composition, and buildup of by-products are some of the drawbacks of the enzymatic method. For instance, Parveen and researchers investigated enzymatic activity of cellulase, pectinase, and xylanase isolated from 56 different cellulolytic

Table 3 Summary of fungi-based pretreatment methods for lignocellulosic biomass

Fungal type	Target	Enzymes	Mechanism	Species	Remarks	Reference
White rot	Lignin	Lignin peroxidase, manganese peroxidase, laccase	Enzymatic oxidative cleavage of lignin		Best for removing lignin; enhances enzymatic saccharification	[41]
Brown rot	Cellulose, hemicellulose	Cellulases, hemicellulases	Fenton reaction	*Postia placenta*, *Gloeophyllum trabeum*	Polysaccharides break down quickly, although lignin mostly stays intact	[42]
Soft rot	Cellulose	Cellulases, xylanases	Hydrolytic degradation of carbohydrates	*Chaetomium globosum*, *Kretzschmaria deusta*	Active in humid environments; slight but not significant lignin modification	[43]

microorganisms. They observed that *Aspergillus fischeri*, *Talaromyces radicus*, and *Trichoderma reesei* had the highest enzymatic activities. They prepared an enzyme cocktail of cellulase, pectinase, and xylanase mixed at a ratio of 1:1:1 according to their activities and compared the reducing sugar concentrations with other pretreatment methods (acid, alkali, and DES pretreatments). They observed that the enzymatic mixture yielded 790 μg/mL of reducing sugar following enzymatic hydrolysis of sugarcane bagasse, which was comparatively better than the other pretreatment techniques [45].

3.5 In Vivo Pretreatment

Under in vivo pretreatment, direct application of live cellulolytic and ligninolytic microorganisms results in the breakdown of lignocellulosic biomass by secretion of oxidative and hydrolytic enzymes. Natural metabolic activity of microorganisms like fungi (like *Phanerochaete chrysosporium*, *Trametes versicolor*), actinomycetes (like *Streptomyces* spp.), and bacteria (like *Clostridium thermocellum*, *Bacillus subtilis*) is used to deconstruct biomass under controlled aerobic or anaerobic conditions unlike enzymatic pretreatment using purified enzymes. Microbes involved generate a wide variety of enzymes that actively break down the complex matrix. Enzymes such as lignocellulose, including cellulases, hemicellulases, lignin peroxidases, manganese peroxidases, and laccases, are usually produced. In vivo pretreatment offers benefits including low energy input and lesser chemical usage. Solid-state or submerged fermentation systems can be used during this pretreatment method. Nevertheless, the process is typically slower than traditional pretreatment methods. To achieve efficient biomass breakdown, it is necessary to precisely regulate the growth parameters (pH, temperature, and oxygen levels) and extend residence time. In integrated biorefineries, this approach is especially useful when looking for sustainable and affordable biomass valorization. Seesatat et al. examined the ligninolytic and cellulolytic capabilities of 106 bacterial isolates on rice straw. The isolate that performed the best had laccase activity of 0.06 U/mL and cellulase activity of 1.1 U/mL. The highest yields of biogas and methane were determined to be 318.15 and 177.93 mL/g_{TS}, respectively [46].

3.6 Comparative Analysis of All Pretreatment Methods

Depending on the biomass type and intended result, each pretreatment technique has unique advantages and disadvantages. Physical pretreatment like milling is easy to use, increases surface area, and does not significantly change the chemical structure of biomass but the process is energy intensive. Physical methods require expensive equipment that is often challenging to scale up. In this regard, microwave and ultrasonic procedures prove to be useful to loosen the rigid cell wall structure while increasing the porosity of the cell wall. Effective hemicellulose removal and partial lignin disruption are achieved by physicochemical techniques such as steam explosion and autoclaving. These methods result in uneven biomass treatment, necessitating high-pressure setups and producing fermentation inhibitors. Ammonia fiber explosion (AFEX) effectively eliminates lignin while maintaining sugars and meticulously recovering ammonia. Although supercritical fluids allow for clean and efficient pretreatment, regular usage is however restricted because of high operation pressures and cost associated with the method. Although acid pretreatment (diluted sulfuric or hydrochloric acid) looks promising to effectively remove hemicellulose, consequently it can also cause corrosion and result in production of inhibitory compounds. On the other hand, alkali pretreatment uses NaOH or $Ca(OH)_2$ enhancing cellulose accessibility and effectively solubilizing lignin, while it requires longer treatment time and leads to difficult recovery. Organic acids such as formic or acetic acid offer a gentler substitute with reduced toxicity but are less effective in comparison to other methods. Although organosolv pretreatment is successful in producing high-purity lignin, the solvents used are costly and need targeted recovery. Additionally, lignocellulose can be successfully dissolved by ionic liquids (ILs); these highly adjustable solvents have drawbacks in terms of cost, toxicity, and recyclability. Despite their lower efficiency, deep eutectic solvents (DESs) look promising due to its biodegradability and affordability as an environment-friendly substitute for ILs. Furthermore, biological pretreatment method utilizing fungi or bacteria can selectively break down lignin or cellulose. This method is a low-energy alternative but slower and extremely susceptible to environmental factors. Brown-rot and soft-rot fungi concentrate mostly on cellulose by modifying surfaces, but white-rot fungi effectively break down lignin. Although bacterial pretreatment method is appealing for commercial scale-up opportunities, it is far from reality due to the requirements of regulated environments with extended residence periods. Commercial enzymes are costly and less effective on untreated biomass; however, enzymatic pretreatment particularly with cellulases, hemicellulases, or laccases can provide targeted approach improving the quality of pretreatment resulting in little to no inhibitors during the reaction. In vivo microbial pretreatment using the entire lignocellulolytic microbial communities can be a more sustainable and affordable option in comparison to other methods discussed. Raina et al. evaluated the effects of organosolv and alkaline pretreatment techniques on sugarcane biomass, wherein alkaline pretreatment with 10% NaOH resulted in 11% more cellulose production than organosolv pretreatment, while 10% KOH yielded maximum lignin removal [47]. In a different study, Kaur et al. compared alkaline, steam explosion, and organosolv pretreatment, wherein alkaline pretreatment resulted in similar % of cellulose (48.6%) in comparison to organosolv ~46.5% and steam explosion ~43.7%, respectively. Additionally, this method also resulted in hydrolysate containing about 10% less lignin, and by performing an extra bleaching step, the yield of cellulose in the alkaline hydrolysate was further increased by 16% [48] (Table 4).

Table 4 Comparative analysis of different pretreatment strategies for lignocellulosic biomass

Category	Method	Advantages	Disadvantages
Physical	Milling	Increases surface area; simple and scalable	High energy consumption; no chemical modification
	Microwave	Rapid, uniform heating; enhances other treatments	High equipment cost; limited penetration and scalability
	Ultrasound	Disrupts biomass structure; increases porosity	Low effectiveness alone; equipment cost
Physicochemical	Steam explosion	Effective hemicellulose removal; low chemical use	Inhibitor formation; uneven treatment
	Autoclave	Simple setup; enhances digestibility	Less efficient delignification; energy intensive
	AFEX	Preserves sugars; efficient for herbaceous biomass	Requires ammonia recovery; limited for woody biomass
	Supercritical fluids	High penetration; clean and effective	High cost and pressure; specialized equipment required
Chemical	Acid	Rapid hemicellulose removal; widely studied	Corrosive; inhibitor formation
	Alkaline	Effective lignin solubilization; enhances cellulose digestibility	Long treatment time; chemical recovery needed
	Organic acids	Milder and biodegradable; less toxic	Lower efficiency than mineral acids
	Organosolv	High lignin recovery; produces pure cellulose and hemicellulose	Costly solvents; solvent recovery required
	Ionic liquids	Dissolves lignocellulose completely; tunable properties	Toxicity and cost; difficult recyclability
	Deep eutectic solvents	Green and biodegradable; cost-effective	Moderate efficiency; optimization still ongoing
Biological	White-rot fungi	Selective lignin degradation; mild and eco-friendly	Very slow; sensitive to conditions
	Brown-rot fungi	Attacks cellulose; effective on carbohydrate fraction	Little lignin removal
	Soft-rot fungi	Works in moist environments; surface-level modification	Limited efficiency compared to white rot
	Bacteria	Mild and environmentally friendly	Long processing time; growth control needed
Enzymatic	Enzymatic pretreatment	Specific to targets; minimal inhibitor production	Expensive enzymes; less effective on untreated biomass
Whole cell (in vivo)	Microbial whole-cell systems	Integrated degradation and conversion; low energy and cost	Very slow; sensitive to environmental parameters

4 Conversion Pathways of Lignocellulosic Feedstock

The potential conversion of lignocellulosic biomass (LCB) predominantly consisting of cellulose, hemicellulose, and lignin to various end products including green energy, fuels, alcohols, polyols, phenolic compounds, and a range of other platform chemicals has increasingly popularized the use of LCB as an environment-friendly alternative to petrochemical biorefinery [49].

There are a number of conversion methods through which the LCB can be valorized into different high-value, low-volume as well as low-value, high-volume end products. Primarily, the conversion pathways are categorized into four major groups such as thermal, hydrothermal, thermochemical, and biological/biochemical conversion methods including certain engineered, hybrid, and integrated technologies.

4.1 Thermal Conversion

Thermal conversion is one of the oldest conversion pathways that requires no pretreatment, wherein the feedstock containing less than 50% moisture is directly subjected to combustion in the presence of oxygen at 800–1,600°C [50]. However, in some cases, pretreatment can be beneficial to specifically separate lignin from cellulose/hemicellulose. Moreover, cellulose and hemicellulose are used to produce fuels and other platform chemicals through various other conversion methods, while lignin undergoes combustion to generate thermal energy. For example, separating lignin from LCB from pulp and paper industry liberates significant proportion of heat during the extraction process; this heat can then be alternatively used to operate other processes like power generation, powering lime kiln [51]. Indeed, one of the major drawbacks of this thermal conversion process is generation of flue gas primarily through the combustion and could potentially lead to greenhouse gas (GHG) emission [52]. LCB feedstock includes energy crops, grasses/herbaceous crops, softwoods, and hardwoods.

4.2 Hydrothermal and Thermochemical (HT/TC) Conversion

Amongst various conversion strategies, hydrothermal and thermochemical processes offer robust and versatile approaches for depolymerizing and valorizing complex LCB into different value-added products under elevated temperature and pressure. Thermochemical conversion is defined as a chemical process of reformation, which converts long-chain organic compounds from biomass into short-chain oxygenated hydrocarbons [53, 54]. In general, pyrolysis, gasification, and hydrothermal liquefaction are few of the thermochemical conversion techniques used for conversion of LCB feedstock [55].

4.2.1 Pyrolysis

Pyrolysis is defined as the process of decomposition of organic compounds present in the biomass at elevated temperature in the absence of oxygen [56]. During the pyrolysis process based on the temperature range 300–1,000°C, three main products are produced such as bio-oil, syngas, and biochar. Pyrolysis depends on few important factors such as temperature range, heating rate, and residence time of reaction. Depending on temperature, pyrolysis is further categorized into slow or conventional pyrolysis, fast pyrolysis, and flash pyrolysis [57]. Slow or conventional pyrolysis occurs at temperature ranging between 300 and 700°C at a heating rate of ~0.1–1°C/s, and residence time varying between 10 and 100 min. Fast pyrolysis occurs at comparatively higher temperature ranging from 400 to 800°C with a heating rate of 10–200°C/s and residence time of 0.5–5 s [58]. Likewise, flash pyrolysis typically occurs at significantly higher temperature ranging between 800 and 1,000°C, at very low residence time typically <0.5–1 s, and at a heating rate of ~1,000°C/s [59]. Amongst other factors, particle size of the feedstock plays an integral part determining the rate of formation of either bio-oil via fast pyrolysis or biochar via slow pyrolysis. In case of slow pyrolysis, optimum particle size ranges between 5 and 50 mm, whereas for fast and flash pyrolysis, the particle size is usually less than 1 and 0.2 mm, respectively [60]. Lower particle size results in effective conversion by lowering the residence time and increasing the surface area aggravating the conversion rate [61]. Slow pyrolysis is the most effective method for biochar production [62]. In recent times, bio-oil and oxygenated hydrocarbons have emerged as efficient alternatives to fossil fuels, produced primarily through fast pyrolysis [63]. An advantage of pyrolysis is its ability to process both wet and dry lignocellulosic biomass (LCB), although a key limitation is the additional energy required to evaporate moisture, which may increase energy consumption.

4.2.2 Gasification

Gasification is a thermochemical process that encompasses thermal decomposition of carbonaceous biomass into combustible gas mixtures called syngas (synthesis gas) at high temperatures around 600–1,000°C in a low-oxygen atmosphere to avoid combustion. Syngas is primarily composed of carbon monoxide (CO), hydrogen (H_2), carbon dioxide (CO_2), methane (CH_4), and other trace gases [64]. LCB consisting of cellulose, hemicellulose, and lignin presents an abundant and renewable feedstock for gasification. The advantage of gasification is that it is not limited to one particular type of lignocellulosic feedstock; however, the disadvantages include high energy requirement to operate in high reaction temperatures and failure to exploit the unique structure of lignin.

The entire gasification process of LCB typically involves four stages such as (i) drying (Eq. 1), (ii) pyrolysis, (iii) combustion/oxidation (Eq. 2), and (iv) reduction

(Eqs. 3–6) [54]. Drying removes moisture from the biomass at temperatures up to 200°C:

$$H_2O\ (l) \rightarrow H_2O\ (g) \quad [\text{Drying}] \tag{1}$$

After drying, the biomass is subjected to pyrolysis, wherein the LCB decomposes into volatile gases, tar (the liquid fraction), and char (the solid fraction). During the combustion/oxidation step, gases produced react with biomass in the absence of O_2 at higher temperatures of ~800–900°C:

$$C_mH_n + (m + n/4)\ O_2 \ {}^{\Delta}\!\rightarrow mCO_2 + n/2H_2O \quad [\text{Combustion}] \tag{2}$$

Moreover, in the reduction phase, the hot gases released from the combustion step are converted to synthesize gas (syngas) by the following endothermic reactions between char and gases.

[Reduction]:

$$C + CO_2 \rightarrow 2CO \tag{3}$$

$$C + H_2O \rightarrow CO + H_2 \tag{4}$$

$$CO + H_2O \rightarrow CO_2 + H_2 \tag{5}$$

$$C + 2H_2 \rightarrow CH_4 \tag{6}$$

Syngas produced via gasification can be upgraded into higher-value biofuels and biochemicals majorly through two conversion pathways, namely syngas fermentation and Fischer-Tropsch (FT) catalysis.

Syngas Fermentation

Raw syngas obtained from the gasification of LC biomass can be used directly for ethanol production through a biosynthesis route in the presence of acetogens. Acetogens are anaerobic bacteria that mostly use CO_2 as a carbon source and H_2 as an energy source to generate acetate via acetogenesis. Several mesophilic bacteria including *Acetobacterium woodii*, *Clostridium carboxidivorans*, *Peptostreptococcus* sp., *Clostridium aceticum*, *Clostridium ljungdahlii*, *Clostridium carboxidivorans*, *Clostridium ragsdalei*, *Clostridium autoethanogenum*, etc. are involved in fermentation of syngas through the acetyl-CoA pathway to liquid fuels like acetic acid, ethanol, and butanol more effectively than other conventional processes [65].

Fischer-Tropsch (FT) Catalysis

Conversion of syngas to liquid fuels in the presence of various metal/nonmetal catalysts is known as FT catalysis [66]. The purified syngas produced through LCB gasification is adjusted for their H_2/CO ratio between ~1 and 4 via water-gas shift (WGS) reaction (Eq. 7):

$$\text{WGS reaction}: \quad CO + H_2O \rightarrow CO_2 + H_2 \tag{7}$$

Firstly, CO reacts with H_2 under higher pressure (10–200 atm) and elevated temperature (~200–400°C) to produce linear and branched hydrocarbons as follows in the reaction (Eq. 8) [66]:

$$n\mathrm{CO} + (n + m/2)\mathrm{H}_2 \rightarrow \mathrm{C}_n\mathrm{H}_m + n\mathrm{H}_2\mathrm{O} \tag{8}$$

Moreover, the length of carbon chain (n) depends on specific operating conditions; specifically, temperature plays an integral part. In case of low-temperature FT, the temperature range revolves around 200–250°C in the presence of metal catalysts like Co or Fe yielding long-chain hydrocarbons. Conversely, the high-temperature FT operates under a temperature range of 320–375°C in the presence of Fe-based catalysts yielding short-chain hydrocarbon molecules such as gasoline and light olefins [67].

4.2.3 Supercritical Water Gasification (SCWG)

SCWG is an advanced thermochemical conversion process which utilizes the unique properties of water above its critical temperature (374°C) and pressure (22.1 MPa). Supercritical water converts LCB into valuable gaseous products like H_2 and CH_4. The nonpolar properties of supercritical water (SCW) accelerate its ability to dissolve nonpolar compounds homogeneously [68]. The SCW functions as a reactant as well as a medium to convert LCB into gaseous fuels. LCB upon conventional gasification produces hydrocarbon-rich syngas along with a considerable amount of biochar and tar components. The major issues related to conventional gasification are biochar formation and requirement of low-moisture-content biomass. Both of these can be avoided by using SCW gasification, which functions on the conversion of wet LC biomass into hydrocarbon-rich gaseous products in the presence of catalysts at high temperatures (400–450°C) and pressures ($\leq$35 MPa).

4.2.4 Hydrothermal Liquefaction (HTL)

The process of liquefaction involves thermal breakdown of LCB into its monomeric units, which further repolymerize to produce liquid oils and undesirable solid chars. Liquefaction is also known as hydrothermal liquefaction (HTL), which is a fragmentation technique that works for both wet and dry LCB. Liquefaction requires a catalyst which reduces the reaction temperature, a solvent which slows down the solid-state reactions and reduces the undesirable condensation reactions for char formation by improving the reaction kinetics. It operates under 5–20 MPa pressure within a temperature range of 250–450°C, which is lower than the temperature needed for pyrolysis or gasification. Lower temperature does not break down cellulose, while higher temperatures may restrict gas formation. Methanol, ethanol, isopropanol, or a specific combination of these solvents, or water, can be used to conduct the reaction. Basic alkali metal catalysts like Na_2CO_3, K_2CO_3, KOH, and Ca $(OH)_2$ are primarily favored for increasing the yield of bio-oil by facilitating the overall liquefaction process [54, 69]. Bio-oil, biochar, and gases are produced during the liquefaction process, and these products can be further upgraded to liquid fuels, hydrogen, and methane.

There are several complex reactions that occur during the liquefaction process, as follows: (a) cracking and reduction of LCB components cellulose, hemicellulose, and lignin; (b) hydrolysis of cellulose/hemicellulose to glucose/fructose and other monomeric sugars; (c) hydrogenolysis in the presence of hydrogen, etc. The efficiency of HTL reaction is determined by the composition, structure, and morphology of LCB.

4.3 *Biological/Biochemical Conversion*

Effective conversion of LC biomass into fermentable sugars requires a pretreatment step to break the lignin seal and reduce the crystallinity of the cellulose. Pretreatment methods are classified as physical (milling, irradiation), chemical (acidic, alkaline, pyrolysis), physicochemical (steam explosion, fiber explosion, ammonia), and biological (using microorganisms that degrade lignin, such as white-rot fungi). Each method has a specific impact on the composition and structure of the biomass, and the choice of method must be optimized according to the type of feedstock, the desired end product, and the compatibility with the waste stream.

4.3.1 Saccharification or Enzymatic Hydrolysis

Enzymatic hydrolysis of LCB is an important bioconversion process used for the conversion of renewable plant-based feedstock into fermentable sugars, which serve as precursors for the synthesis of biofuel, platform chemicals, and other value-added bioproducts. LCB is structurally complex consisting of three interconnected biopolymers – cellulose (35–50%), hemicellulose (20–35%), and lignin (10–25%), the exact proportions of which vary depending upon the source of the LC biomass [70]. These macromolecules are intricately woven into a compact, heterogeneous matrix, which gives rise to considerable resistance to enzymatic deformation. Lignin in particular forms a hydrophobic, cross-linked network which binds with the carbohydrate components, severely limiting its availability to bind with the enzymes. The pretreatment strategies vary from dilute acid hydrolysis, steam blast, and alkaline treatment to ionic liquids that are designed to break down lignin-carbonate complexes, reduce the crystallinity of cellulose, and solubilize hemicellulose fractions. These physicochemical changes significantly increase the surface area, enzymatic accessibility of cellulose, and residual hemicellulose for subsequent hydrolysis. The enzymatic hydrolysis process involves a highly coordinated enzymatic cascade, catalyzed by a complex consortium of glycoside hydrolases (GHs), carbohydrate-binding modules (CBMs), and auxiliary activity (AA) enzymes. Cellulolytic enzymes are secreted by filamentous fungi such as *Trichoderma reesei*, genetically engineered microbial hosts expressing endo-α-β-1,4-glucanases (EGs; EC 3.2.1.91 and 3.2.1.4), which randomly cleave the amorphous regions of cellulose fibrils to glucose and thereby generate new chain ends; exo-β-1,4-cellobiohydrolases (CBHs; EC 3.2.1.91 and 3.2.1.176), which cleave cellobiose units from either the reducing or the nonreducing ends of crystalline cellulose fibrils; and β-glucosidases (BGLs; EC 3.2.1.21), which catalyze the hydrolysis of cellobiose and soluble cello-oligosaccharides to glucose [71, 72]. At the same time, hemicellulose is degraded by a variety of enzymes, including α-arabinofuranosidases, α-glucuronidases, and feruloyl esterases, which together deplete the branched and substituted backbone of hemicellulose (e.g., arabinofuranosidases, β-1,4-xylosidases) [73]. The limitation of enzymatic saccharification is the removal of the noble species enhancing the activity of the backbone-cleaving enzymes, thereby accelerating the rate and extent of hydrolysis. This leads to a decrease in the effective concentration of the enzyme in the solvent as well as a decrease in the potency of the solvent. In addition, hydrolysis is inhibited by the accumulation of monosaccharide and disaccharide compounds, particularly glucose and cellobiose, which compete for the active site of the cellulase.

4.3.2 Fermentation

The fermentation of lignocellulosic biomass (LC) post-hydrolysis involves the microbial transformation of simple sugars (mainly glucose, xylose, and arabinose) into biofuels or organic acids or other chemicals. The process starts with the hydrolysis of complex polyols of cellulose and hemicellulose into fermentable sugars by acidic or enzymatic means. During fermentation, these sugars are metabolized by specific microorganisms such as *Saccharomyces cerevisiae* (for glucose) or engineered strains of *Escherichia coli* or *Zymomonas mobilis* (for pentoses) [74]. The sugars are broken down into pyruvate, which is then converted into the desired final product via alcoholic fermentation or mixed-acid fermentation. This biochemical transformation occurs under anaerobic or microaerobic conditions, depending on the agent used, and is affected by factors such as pH, temperature, and presence of inhibitors. Effective fermentation requires overcoming problems such as sugar co-utilization and inhibitor decomposition to maximize the yield of the product.

The fermentable hydrolysate obtained by enzymatic saccharification is a complex combination of monosaccharides, which comprises predominantly glucose from cellulose, and various pentoses such as xylose and arabinose from hemicellulose. Microbial fermentation of this sugar-rich hydrolysate is a critical step in the bioconversion of lignocellulosic biomass into value-added products such as ethanol, butanol, and various organic acids. The diversity of the sugar composition is a major challenge, as many commercially important microorganisms mostly metabolize hexose that is either inefficient or completely unable to utilize pentose. Effective fermentation therefore requires microorganisms with a reliable co-fermentation capacity that is capable of fermenting both hexoses and pentoses under anaerobic or microaerobic conditions.

Amongst the fermenting microorganisms, *Saccharomyces cerevisiae* continues to be a favorite strain for the production of ethanol due to its exceptional tolerance to ethanol and well-established genetic structure. However, native *S. cerevisiae* lacks endogenous metabolic pathways for the use of pentose sugars such as D-xylose and L-arabinose, which may account for up to 30–40% of the total sugars in lignocellulosic hydrolysis. Extensive genetic engineering has been carried out to extend the range of substrate species of *S. cerevisiae*. This usually involves heterologous expression of the xylose pathway, namely either the oxidoreductive or the isomerase pathway. The oxidoreductive pathway involves the concerted use of xylose reductase (XR) and xylitol dehydrogenase (XDH), often derived from *Pichia stipitis*, to sequentially convert xylose to xylulose with the aid of xylitol as an intermediate. These enzymes require a strict balance of redox cofactors (NADPH for XR, NAD^+ for XDH), and imbalances may lead to the accumulation of xylitol and reduced ethanol yield [75]. Alternatively, the xylose isomerase pathway, which is obtained from bacteria such as *Clostridium phytofermentans* and *Thermus thermophilus*, allows for the direct isomerization of xylose to xylulose, which is then phosphorylated by xylulokinase and fed to the non-oxidative pentose phosphate pathway [76]. Engineering of downstream enzymes such as transketolase and transaldolase along with transporter proteins for pentose uptake is crucial for optimizing flux toward glycolytic intermediates.

4.4 *Emerging Engineered, Hybrid, and Integrated Technology*

The development of different hybrid and engineered technologies for LCB conversion is one of the rapidly advancing approaches, which aim to overcome the limitations arising from the conventional single-pathway conversion approaches to improve the overall efficiency and yields of various value-added products and platform chemicals. The conventional approaches including thermal, thermochemical, and biochemical pathways often result in incomplete utilization of LC biomass due to its structural complexity arising from cellulose, hemicellulose, and lignin. Therefore, the combination of different thermochemical and biochemical methods

Table 5 Engineered and integrated approaches for effective conversion of lignocellulosic biomass

Approaches	Aim	Strategy	Reference
Process intensification by increment of solid loading capacity	Enhanced biomass production and cost reduction	The ratio of insoluble solid to liquid of ~15% with no free water	[80]
Simultaneous saccharification and fermentation process (SSFP)	Alleviating product inhibition during enzymatic hydrolysis of cellulose	Using different thermophilic and/or engineered strains to overcome temperature variation in enzymatic hydrolysis and fermentation process	[54]
Consolidated bioprocessing (CBP) by using engineered microorganisms	Increasing the cost-effectiveness	Combining fermentation process with cellulolytic enzymes and incorporating within a single reactor system	[77, 78]
Consolidated bioprocessing (CBP) employing microbial co-cultures and communities	Increasing effectiveness and targeted bioproduct formation	Using microbial consortia for enhanced efficiency	[79]

allows utilization of LCB and valorization of all biomass components while minimizing GHG emissions, waste generation, raw material inputs, and overall production cost.

Various approaches like process intensification, simultaneous saccharification and fermentation process (SSFP), and consolidated bioprocessing (CBP) by employing engineered microorganisms and/or microbial consortia were successfully implemented for enhanced effectivity (see the table). The engineered strains used in CBP process should possess some desirable trait – (**A**) robust strain able to grow under C_5 and C_6 sugar, (**B**) efficient cellulase enzyme production capacity, (**C**) rapid and high tolerance to pretreatment inhibitor and process fluctuations, and (**D**) high yielding. Recombinant strategy involves engineered organisms for efficient cellulose and hemicellulose utilization. For instance, *Clostridium thermocellum*, an anaerobic thermophilic cellulose-forming bacterium, hydrolyzes cellulose and hemicellulose at a much faster rate following the CBP pathway. However, for enhancing the effectiveness of *C. thermocellum*, the organism is engineered by knocking out organic acid-producing genes. The engineered *C. thermocellum* variant resulted in 4.2-fold higher ethanol production compared to the wild type [77]. Likewise, the lactate dehydrogenase and acetaldehyde/alcohol dehydrogenase gene deletion in *C. thermocellum* and in *Caldicellulosiruptor* yielded around 70% of ethanol during fermentation [78]. Moreover, the co-culture of two different microbes allows the generation of a defined system which can be employed for the targeted bioproduct formation. The co-cultivation of fungi *T. reesei* and yeast *S. cerevisiae* was employed for ethanol production in a specially designed membrane-aerated biofilm reactor [79]. Summary of different engineered and integrated approaches for effective conversion of lignocellulosic biomass is shown in Table 5.

5 Value-Added Products from Lignocellulosic Feedstock

5.1 Biohydrogen

Biohydrogen, a molecular hydrogen, is produced from microorganisms via different microbial metabolisms under several biological processes including dark fermentation, photofermentation, and biophotolysis [81]. Recently, biohydrogen produced from lignocellulosic biomass (LB) has drawn attention due to its concomitant ability to convert organic waste into energy carrier. The sugars liberated during the pretreatment process from LB provide electron donors to the hydrogen-generating metabolizing bacteria. During the metabolic process along with biohydrogen, various other by-products, such as organic acids, ethanol, and carbon dioxide, can be produced. Previous studies have shown that generation and usage of biohydrogen from LB do not add to the net greenhouse gas emission pool [82]. Thus, LB can be a sustainable and environmentally friendly substitute for hydrogen production. Additionally, the clean combustion rate and greater energy density (122 kJ/g) of the bio-hydrogen make it a favorable bio-resource than fossil-based hydrogen.

Dark fermentation utilizing pretreated lignocellulosic materials is one of the most commonly studied methods for hydrogen production. The wide range of applications of dark fermentation starting from organic wastes to biomass-derived feedstocks, such as grass, food-processing effluent, agricultural leftovers, and simple sugars like glucose, makes it an attractive technique for biohydrogen production. In this process, a diverse array of obligatory and facultative anaerobic bacteria, primarily from the genera *Clostridium* and *Enterobacter*, could be used, thereby making the process more robust. For instance, anaerobic microbes produce molecular hydrogen by decomposing organic substrates while transferring electrons to protons through intracellular carriers. In addition, organic acids and alcohols are concurrently produced as metabolic by-products but depend on the microbial strain and fermentation process. Further, the type of microorganism employed significantly influences the efficacy of hydrogen production. Facultative anaerobes, such as *Enterobacter* species, often produce up to 2 mol H_2 per mol of glucose due to their reliance on reductive metabolic pathways, whereas obligatory anaerobes, such as *Clostridium* spp., can theoretically release up to 4 mol H_2 per mol of glucose. During glycolysis, glucose gets converted to pyruvate, a crucial step in hydrogen production, wherein pyruvate ferredoxin oxidoreductase enzymatically transforms pyruvate into acetyl-CoA, yielding butyric acid and hydrogen, alongside acetic acid [83] (Eq. 9):

$$C_6H_{12}O_6 \rightarrow 2CH_3COCOO^-(\text{pyruvate}) + 2ATP + 2NADH + 2H^+ \quad (9)$$

Such activity is distinctive of the butyrate-type fermentation pathway mostly prevalent in *Clostridium* sp. generating carbon dioxide, volatile acids, and up to 2 mol of hydrogen per mol of glucose (Eq. 10):

$$C_6H_{12}O_6 \rightarrow CH_3CH_2CH_2COOH\text{ (butyric acid)} + 2CO_2 + 2H_2 \quad (10)$$

The maximum possible production of 4 mol H_2 per mol of glucose occurs during acetate-type fermentation (Eq. 11):

$$C_6H_{12} + 2H_2O \rightarrow 2CH_3COOH + 2CO_2 + 4H_2 \quad (11)$$

Nonetheless, not all metabolic pathways derived from pyruvate facilitate hydrogen production. The conversion of pyruvate to ethanol or lactic acid may diminish overall yields (Eqs. 12 and 13). Additionally, ethanol buildup can disrupt microbial metabolism and damage cell membrane integrity by feedback inhibition:

$$C_6H_{12}O_6 \rightarrow 2C_2H_5OH + 2CO_2 \quad (12)$$

$$C_6H_{12}O_6 \rightarrow 2CH_3CH(OH)COOH\text{ (lactic acid)} \quad (13)$$

Creation of propionic acid in the metabolic process is detrimental as it vigorously depletes hydrogen. Also, glucose on hydrogenation with 2 mol of H_2 yields 2 mol of propionic acid (CH_3CH_2COH) and water resulting in net hydrogen consumption (Eq. 14):

$$\text{Glucose} + 2H_2 \rightarrow 2CH_2CH_2COOH\text{ (propionic acid)} + 2H_2O \quad (14)$$

Additionally, hydrogen is generated during the formate metabolism by facultative anaerobes, wherein pyruvate formate lyase (PFL) catalyzes the cleavage of pyruvate to yield formate, which is subsequently transformed into hydrogen and carbon dioxide by formate hydrogen lyase (FHL) (Eq. 15):

$$CHOO^- + H^+ \rightarrow H_2 + CO_2 \quad (15)$$

During this process, operational parameters must be meticulously tuned to promote the synthesis of acetic acid, butyric acid, and formic acid while inhibiting the creation of ethanol, lactic acid, and propionic acid, hence augmenting biohydrogen yields during dark fermentation. Strategies like pH regulation, substrate loading rate, and inoculum selection are crucial for directing metabolic fluxes toward hydrogen production. The progression of dark fermentative hydrogen production from lignocellulosic biomass relies on a comprehensive understanding of microbial biochemistry and fermentation dynamics.

5.2 *Biobutanol*

Biobutanol, a four-carbon alcohol (C_4H_9OH) produced through microbial fermentation, has garnered attention as a prospective next-generation biofuel due to its beneficial physicochemical properties. Biobutanol exhibits greater compatibility with existing fuel infrastructure compared to ethanol, demonstrating higher energy density (~29.2 MJ/L) and lower volatility and hygroscopy [84]. Lower corrosiveness of biobutanol upon blending with gasoline at elevated ratios makes it highly compatible for industrial usage without compromising on engine performance.

Acetone-butanol-ethanol (ABE) fermentation was initially developed using *Clostridium acetobutylicum* and *Clostridium beijerinckii*, a well-known process. The two-stage fermentation process commences with acidogenesis, producing acetic and butyric acids (Eq. 16) [85]:

$$C_6H_{12}O_6 \rightarrow CH_3CH_2COOH + CH_3COOH + 2CO_2 + 2H_2 \tag{16}$$

These acids are subsequently reabsorbed and converted into solvents, including ethanol, butanol, and acetone, during the solventogenesis phase (Eqs. 17–19):

$$CH_3CH_2COOH + H_2 \rightarrow \text{Butanol } (C_4H_{10}O) \tag{17}$$

$$CH_3COOH \rightarrow CH_3COCH_3 \text{ (acetone)} \tag{18}$$

$$CH_3COOH + NADH \rightarrow C_2H_5OH \text{ (ethanol)} + CO_2 \tag{19}$$

Due to butanol's greater molecular weight and increased ATP requirement during biosynthesis, its potential yield is inferior to ethanol [86]. Therefore, recently, metabolic engineering approaches have been employed to redirect carbon flux toward butanol synthesis while inhibiting undesired by-product formation. However, there are some disadvantages of ABE fermentation such as high recovery expenses with lower butanol concentrations (<2% w/v) and additional end-product inhibition. Alternatively, robust microbial strains, in situ product recovery, continuous fermentation, and feedstocks such as glycerol, lignocellulosic hydrolysates, and food industry waste can improve the sustainability of the process. Apparently, biobutanol is considered as a viable product in integrated biorefineries due to its dual usage as an industrial solvent and biofuel.

5.3 Biogas

Biogas production from lignocellulosic biomass holds significant potential for sustainable energy generation but faces several technical challenges. Primarily, the inherent recalcitrant nature of lignocellulosic material posed by tightly bound cellulose, hemicellulose, and lignin impedes enzymatic hydrolysis, thereby limiting the conversion of biomass into fermentable sugars and ultimately reducing the biogas yield during anaerobic digestion [87]. A significant proportion of the biomass, especially cellulose and lignin, often stays undigested as a solid residue, thereby making the removal process more difficult and cumbersome. Additionally, the anaerobic digestion process usually necessitates lengthy retention periods, between

15 and 40 days, which restricts operating efficiency and throughput, especially in large-scale or continuous systems [88].

Two approaches have been extensively studied to increase the efficiency of biogas production, such as incorporating pretreatment techniques and increasing robust microbial strain with higher culture density. Accelerated hydrolysis and methane production primarily depend on maintaining an ideal ratio of methanogenic bacteria to lignocellulose. Firstly, the lignin-carbohydrate matrix is broken down during the pretreatment methods, which include mechanical disruption, chemical treatment, and biological processing. These procedures not only reduce digestion time but also increase the surface area enhancing substrate digestibility and enzyme accessibility. However, the structural and compositional variations of biomass determine the effectiveness of the pretreatment method used. For instance, pretreated cardboard has demonstrated a 140% increase in biogas yield as opposed to untreated cardboard, whereas pretreated rice straw has only demonstrated a 40% increase as opposed to untreated rice straw. Here, the difference in the composition and percentage of cellulose, hemicellulose, and lignin present in the material was responsible for a varied outcome. Furthermore, an increase in the availability of fermentable carbohydrates is the main cause for such variation. A systems-level strategy incorporating microbial optimization, customized pretreatment techniques, and strategic process integration is required to overcome the difficulties for biogas production. Shifting industrial and agricultural waste into clean renewable energy can lead to a low-carbon-based circular bioeconomy.

5.4 Bioethanol

Global fuel ethanol production has increased from 25.7 billion gal in 2015 to approximately 26.1 billion gal in 2020. About 80% of ethanol supply originates from the United States and Brazil, where corn and sugarcane serve as the primary feedstock for bioethanol production. In contrast, developing countries often utilize nonfood biomass such as cassava and sweet sorghum for bioethanol production, helping to avoid competition with food resources.

It has been well established that ethanol can be produced from natural and agricultural materials rich in starch and sugar [89]. Microorganisms are used to ferment carbohydrate-based feedstock, converting them into ethanol. In this context, lignocellulosic biomass (LCB) serves as a second-generation feedstock for bioethanol production. LCB is widely available in large quantities worldwide in the form of fruit and vegetable peels, leaves, branches, and agricultural residues. Hence, LCB could be a promising alternative for bioethanol production, especially for the countries generating greater agricultural and forestry wastes. The process of bioethanol production from LCB involves two main techniques such as thermochemical conversion and microbial fermentation. Thermochemical process includes pyrolysis at temperatures of 500–600°C and gasification producing synthesis gas ($CO + H_2$). The syngas is produced in a catalytic procedure generating ethanol and other liquid co-products. At temperatures above 700°C, biomass gasification occurs resulting in syngas production that can be converted again to bioethanol by either metallic catalysts (aluminum, cobalt, etc.) or Fischer-Tropsch synthesis. However, solubility of syngas and gas-to-liquid mass transfer are the major challenges faced by the industry for the commercialization of syngas to ethanol production techniques. On the other hand, physicochemical and enzymatic pretreatment methods release fermentable sugars from lignocellulosic biomass, which can then be converted to ethanol through microbial fermentation.

5.5 Biokerosene

Biokerosene is a renewable aviation fuel derived from vegetable oils and lignocellulosic biomass, via diverse thermochemical and biochemical conversion methods [90]. The lignocellulosic biomass, consisting of cellulose, hemicellulose, and lignin, initially undergoes pretreatment to deconstruct its intricate structure. Subsequently, it undergoes gasification at elevated temperatures with restricted oxygen, transforming it into syngas, abundant in carbon monoxide and hydrogen. The syngas is purified and catalytically transformed into liquid hydrocarbons using Fischer-Tropsch (FT) synthesis. The resultant hydrocarbon mixture is then catalytically upgraded, hydrotreated, and distilled to extract the kerosene portion, yielding renewable, drop-in biokerosene. It possesses chemical similarities to traditional petroleum-based kerosene and can be blended with or entirely substitute fossil-derived aviation fuels in current jet engines without necessitating modifications. Moreover, biokerosene presents a diminished carbon footprint and decreased greenhouse gas emissions throughout its life cycle, rendering it a viable solution for decarbonizing the aviation industry.

5.6 Other Value-Added Products

Lignocellulosic biomass also serves as a promising feedstock for the production of several high-value platform chemicals, including furfural, levulinic acid, and xylitol. These compounds are typically derived through selective conversion of the carbohydrate fractions, primarily hemicellulose and cellulose via hydrolysis and catalytic processes.

5.6.1 Furfural Production

Furfural is a significant bio-based chemical mostly derived from hemicellulose fraction abundantly found in pentosans and xylose. Firstly, under acidic environment, hemicellulose is hydrolyzed to yield pentose sugars. These sugars are further dehydrated to produce furfural. The process often involves acid hydrolysis, which depolymerizes hemicellulose into monomeric xylose, followed by the dehydration of xylose to furfural at elevated temperatures (150–200°C) and in the presence of mineral acids. Furfural is extensively utilized as a solvent in resins and as a precursor for many furan-derived compounds.

5.6.2 Levulinic Acid Production

Levulinic acid is a significant platform chemical derived from the cellulose and hexose sugars present in lignocellulosic biomass. The procedure occurs in three stages. Firstly, hydrolysis of cellulose into glucose takes place by acid catalysis followed by dehydration of glucose to 5-hydroxymethylfurfural (HMF). Subsequently, under enhanced hydration, the ring of HMF finally opens to produce levulinic acid and formic acid. Levulinic acid functions as a multifaceted intermediate for biofuels, plasticizers, medicines, and synthetic rubber production.

5.6.3 Xylitol Production

Xylitol, a low-calorie sugar alcohol, is synthesized from xylose. Xylitol production from the hemicellulose fraction of LCB occurs in three stages. Firstly, xylose is liberated via acidic or enzymatic hydrolysis of hemicellulose. Secondly, xylose obtained from the hydrolysate undergoes a rigorous purification step followed by catalytic hydrogenation to produce xylitol under elevated pressure utilizing a nickel-based catalyst. Xylitol is widely utilized as a sweetener in food items, dental care goods, and medications.

5.7 Biomaterials and Biopolymers from LCB

Wastes like sugarcane bagasse have been mostly utilized in biotechnology for biofuel production [91]. Recently, the focus of bioethanol industry mostly shifted toward 2G biofuels that are primarily produced from lignocellulosic biomass (LCB) such as rice husk, straw, sugarcane bagasse, wood, and rice straw among others. Reports on fermentation of other value-added microbial products such as bioplastics and biopolymers like bacterial extracellular polymeric substances (EPSs) from LCB are scarce [92]. Although some reports on PHA production from LCB can be found, EPS production via fermentation of LCB is particularly limited. Generally, bioplastics like polyhydroxyalkanoates are produced from the sugar-enriched hydrolysates of LCB usually obtained after acid or alkaline pretreatment [93]. It is anticipated that these biopolymers produced from LCB would be able to replace petroleum-derived polymers soon. Thus, to build a circular bioeconomy, utilizing biopolymers from greener renewable resources like LCBs could be a more sustainable approach. In order to achieve this goal, the energy-intensive chemical pretreatment method should be substituted by biological pretreatment techniques

Table 6 Simultaneous saccharification and fermentation of LCB for EPS production by different microorganisms

EPS	Microorganism	Substrate	Fermentation	Product yield	Reference
EPS	*Aspergillus* sp.	Wheat bran	Solid state	5.36 mg/g	[98]
EPS	*Bacillus megaterium* RB05	Jute	Solid state	0.297 g/g of substrate	[99]
Schizophyllan	*Schizophyllum commune* ISTL04	Leucaena leucocephala wood	Submerged	4.2 g/L	[100]
EPS	*Lasiodiplodia theobromae* CCT 3966	Sugarcane straw hydrolysate	Submerged	0.047 g/g	[101]
Pullulan	*Aureobasidium pullulans*	De-oiled rice bran	Submerged	5.48 g/L	[102]

[94]. Moreover, the cost of chemical and enzymatic pretreatment combined is significantly higher than their standalone counterparts. Furthermore, the major challenge remains in the delignification step, which requires chemical intervention. Using delignifying enzymes like laccases and lignin peroxidases to degrade lignin present in the LCB, followed by other enzymes such as cellulases and xylanases to hydrolyze the polysaccharides present in the cellulose and hemicellulose fraction, could aid in the mitigation process. However, in vitro enzymatic pretreatment techniques are quite expensive due to the high enzyme purification and downstream processing cost involved. Possibly, using cell-free extracts during the pretreatment process or whole cells as a biocatalyst transforming recalcitrant LCB into highly concentrated pool of reducing sugars seems to be a value proposition.

Biopolymers like pullulan, curdlan, hyaluron, and other EPSs have shown significant commercial relevance with wide applications. To avoid extra pretreatment steps and costs associated with the technique, certain classes of microorganisms which have intrinsic delignifying and cellulolytic potential can be used. For instance, *Cellulomonas flavigena* (CF), a common cellulolytic bacterium [95], has been used to ferment LCB in a single step. Additionally, few reports are available on water-insoluble EPS production from *Cellulomonas* via fermentation utilizing hydrocarbons and dextrose or commercially available cellulose powder [96, 97]. Apart from these, the usage of robust microbes having delignifying and cellulolytic potential is currently limited. Thus, identification of such microbial strains using high-throughput sequencing and mass spectrometry approaches is essential to utilize LCBs in a sustainable manner. Additionally, cutting-edge genetic engineering strategies using robust microbes like *E. coli* could be implemented at a larger scale (Table 6).

PHAs are intracellular polyesters synthesized by certain bacteria for carbon and energy storage in nutrient-limited, carbon-excess conditions. Amongst other intracellular polyesters, polyhydroxybutyrate (PHB) is the most extensively studied PHA. PHB is synthesized through the condensation of acetyl-CoA intermediates in the sugar metabolism pathway. Briefly, condensation of two acetyl-CoA molecules occurs to form acetoacetyl-CoA catalyzed by the enzyme β-ketothiolase (PhaA), wherein NADPH acts as the reducing agent. Concomitantly, acetoacetyl-CoA is further reduced to (*R*)-3-hydroxybutyryl-CoA by the enzyme acetoacetyl-CoA reductase (PhaB). Upon release of CoA, the enzyme PHB synthase (PhaC) facilitates the polymerization of multiple units of 3-hydroxybutyryl-CoA to synthesize PHB [103] (Eqs. 20–22):

$$2\text{Acetyl} - \text{CoA} \xrightarrow{\text{PhaA}} \text{Acetoacetyl} - \text{CoA} \tag{20}$$

$$\text{Acetoacetyl} - \text{CoA} + \text{NADPH} \xrightarrow{\text{PhaB}} 3 - \text{Hydroxybutyryl} - \text{CoA} + \text{NADP}^{+} \tag{21}$$

$$n3 - \text{Hydroxybutyryl} - \text{CoA} \xrightarrow{\text{PhaC}} (\text{PHB})_n + n\text{CoA} \tag{22}$$

Feedstocks for PHB synthesis by *Cupriavidus necator*, *Bacillus megaterium*, and *Pseudomonas putida* originate from lignocellulosic biomass-based hydrolysates. Substrates enriched in propionate or valate derived from hemicellulose degradation can be utilized to synthesize copolymers such as polyhydroxybutyrate-co-valerate (PHBV).

Pullulan is an extracellular α-D-glucan composed of maltotriose units linked by α-1,6 glycosidic bonds, generated by *Aureobasidium pullulans*. Ideal for edible films and pharmaceutical capsules, it possesses excellent film-forming and oxygen barrier properties. Pullulan biosynthesis begins with the absorption and digestion of simple sugars such as glucose, which is phosphorylated and converted through glycolysis into glucose-6-phosphate. Subsequently, glucose-6-phosphate is converted into UDP-glucose, the active sugar nucleotide utilized as a precursor for polymerization. Pullulan synthase (enzyme) reconstructs maltotriose subunits by catalyzing the sequential transfer of UDP-glucose units and linked via α-1,6 bonds [104] (Eq. 23):

$$\text{Glucose} \xrightarrow{\text{UTP}} \text{UDP} - \text{Glucose} \xrightarrow{\text{Pullulan synthase}} (C_6H_{10}O_5)_n + n\text{UDP} \tag{23}$$

Curdlan, a water-insoluble β-1,3-glucan produced by *Agrobacterium* and *Alcaligenes* species, is utilized in the food and pharmaceutical industries for its gelling and thickening properties. Similar to pullulan, it is synthesized from UDP-glucose but polymerized by β-linkages. The production of curdlan commences with the intracellular conversion of glucose to glucose-6-phosphate by hexokinase. This route subsequently integrates with UDP-glucose biosynthesis. Upon isomerization and activation, glucose-6-phosphate transforms into UDP-glucose, serving as the immediate sugar source for curdlan polymerization. The primary enzyme curdlan synthase catalyzes the polymerization of UDP-glucose monomers via β-1,3-glycosidic bonds to generate a high-molecular-weight, unbranched polymer [105] (Eq. 24). The polymer, secreted extracellularly, precipitates as an insoluble gel with thermoreversible gas properties:

$$n\text{UDP} - \text{Glucose} \xrightarrow{\text{Curdlan synthase}} (-\text{Glc} - \beta(1 \rightarrow 3) -)_n + n\text{UDP} \quad (24)$$

Through suitable medium optimization, glucose-rich hydrolysates derived from wheat straw and sugarcane bagasse can facilitate and improve curdlan synthesis.

Xanthan, an anionic heteropolysaccharide produced by *Xanthomonas campestris*, is widely employed as a food stabilizer and rheology modifier. It is released during aerobic fermentation and possesses a β-1,4-glucan backbone with trisaccharide side chains (mannose–glucuronic acid–mannose). Xanthan biosynthesis commences with the absorption of sugars, which are then transformed into sugar nucleotides, including UDP-glucose, GDP-mannose, and UDP-glucuronic acid, acting as precursors for polysaccharide formation. The xanthan biosynthetic gene cluster (gum genes) produces a sequence of glycosyltransferases that successively construct the repeating unit on a lipid carrier molecule attached in the inner membrane. Upon completion of the pentasaccharide repeat, it undergoes polymerization by xanthan polymerase and is subsequently translocated across the membrane, ultimately being secreted into the extracellular medium [106] (Eq. 25):

$$\text{Sugars}(C_6H_{12}O_6)_n \rightarrow \text{Xanthan}(C_{35}H_{49}O_{29})_n + CO_2 + H_2O \quad (25)$$

While glucose is the favored substrate, mutant strains can degrade xylose and arabinose derived from hemicellulosic hydrolysates. Hyaluronic acid (HA) is a high-molecular-weight glycosaminoglycan composed of repeating disaccharide units of D-glucuronic acid and *N*-acetyl-D-glucosamine. It is utilized in regenerative medicine, drug delivery, and cosmetics. HA biosynthesis commences with the transformation of glucose into two essential nucleotide sugar precursors: UDP-glucuronic acid (UDP-GlcA) and UDP-*N*-acetylglucosamine (UDP-GlcNAc). The enzyme hyaluronan synthase (HAS), which is membrane associated, employs these activated sugar donors to catalyze the sequential addition of GlcA and GlcNAc to the elongating HA chain (Eq. 26). The polymer is produced at the inner membrane and simultaneously expelled into the extracellular area [107]. Hyaluronic acid is highly esteemed for its viscoelastic, moisturizing, and biocompatible characteristics, rendering it an essential element in medicines, cosmetics, and tissue engineering applications. Genetically modified strains of *E. coli*, *Bacillus subtilis*, and *Streptococcus zooepidemicus* biosynthesize hyaluronic acid:

$$\text{UDP} - \text{GlcA} + \text{UDP} - \text{GlcNAc} \xrightarrow{\text{HA synthase}} (-\text{GlcA} - \beta(1 \longrightarrow 3) - \text{GlcNAc} - \beta(1 \longrightarrow 4) -)_n \quad (26)$$

Particularly with synthetic microbial platforms capable of co-factor regeneration, glucose and glucosamine precursors derived from lignocellulosic biomass hydrolysates can serve as substrates for hyaluronic acid synthesis.

Bacterial cellulose (BC) is produced in its pure form by *Komagataeibacter xylinus* and related strains. Cellulose production in microbial cultures commences

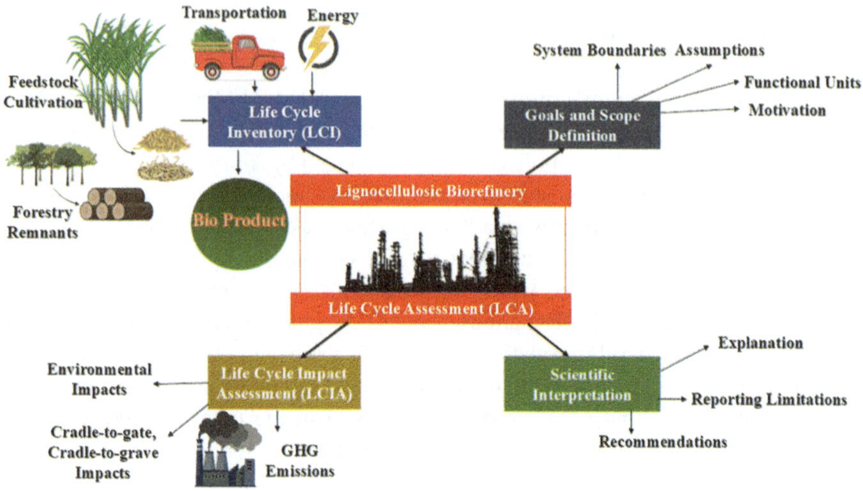

Fig. 1 Scheme for life cycle assessment (LCA) of lignocellulosic biorefineries. (Created using BioRender)

with the conversion of glucose to glucose-6-phosphate, which is subsequently isomerized to glucose-1-phosphate and activated by UDP-glucose pyrophosphorylase. A membrane-bound glycosyltransferase, cellulose synthase catalyzes the polymerization of UDP-glucose into β-1,4-glucan chains. These chains, subsequently secreted extracellularly, align by hydrogen bonding to form highly crystalline microfibrils, which impart structural rigidity [108] (Eq. 27). Cellulose frequently forms biofilms or pellicles at the air-liquid interface, enhancing bacterial environmental resilience and surface adherence. The applications of bacterial cellulose in wound healing, tissue scaffolding, and high-performance materials arise from its biocompatibility, exceptional tensile strength, and biodegradability:

$$n\text{UDP} - \text{glucose} \xrightarrow{\text{Cellulose synthase}} (-\text{Glc} - \beta(1 \rightarrow 4) -)_n + n\text{UDP} \tag{27}$$

While acetic acid and furfural residues must be minimized for optimal cell proliferation and polymer production, glucose-rich lignocellulosic biomass hydrolysates are advantageous for bacterial cellulose production.

6 Life Cycle Assessment and Environmental Impacts of the Lignocellulosic Feedstock-Based Biorefinery

Life cycle assessment (LCA) of lignocellulosic feedstock-based biorefineries offers a methodical and scientific evaluation of the environmental effects linked with each stage of the value chain of the biorefinery. According to ISO 14040, LCA conventionally consists of four principal phases, including purpose and scope definition, inventory analysis, impact assessment, and scientific interpretation [109] (Fig. 1). The scope for the LCA lies in delineating the system boundaries, functional unit (e.g., 1 kg of bioplastic), assumptions, limitations, allocation methods, and category-specific data quality requirements. In lignocellulosic biorefineries, inventory analysis or life cycle inventory (LCI) encompasses the aggregation of all raw material inputs, process activities, and resultant output and emissions. Upstream processes such as feedstock production, harvesting and collection, transportation to processing hub, and land use are some of the important boundaries. Likewise, in the midstream processing phase, pretreatment and conversion of feedstock to intended product, as well as downstream processes including product recovery and by-product management, also play an important part during the boundary selection. Impact assessment or life cycle impact assessment (LCIA) phase illustrates the collection of raw data from the LCI including CO_2 emissions, water, and energy inputs and transforms them into easily understandable environmental impacts. The last phase includes analysis, explanation, reporting limitations, uncertainties, and making recommendations associated with technologies used for running any LC biorefinery [110].

The primary benchmark of environmental sustainability is the carbon footprint (equivalent to CO_2) representing the total greenhouse gas (GHG) production throughout the life cycle analysis of the biorefinery. In contrast to traditional first-generation biofuels, lignocellulosic feedstock does not compete with food production while helping in sequestering CO_2. Nonetheless, emissions may still occur from activities such as fertilizer usage, biomass harvesting, and transportation using fossil fuel. Such activity may also leverage the carbon footprint, especially when the biomass is of low density or requires long-distance transit. LCA of lignocellulosic biofuels and biochemicals often indicates a reduction of 50–90% in GHG emissions compared to fossil fuel equivalents, provided that the raw materials, processing configurations, and energy sources are not directly incorporated [111]. Facilities utilizing agricultural leftovers and reutilizing heat and power systems are notably an efficient ways to mitigate emissions. Additionally, incorporation of carbon capture utilization and storage (CCUS) technology enhances the capacity for achieving net negative emissions.

7 Conclusion and Future Aspects

Valorization of lignocellulosic biomass (LCB) through sustainable approaches provides an opportunity to decrease reliance on fossil resources while mitigating environmental challenges associated with greenhouse gas emissions. LC biorefineries are increasingly recognized as a sustainable substitute for fossil fuels, utilizing nonfood biomass including agricultural wastes, forest refuse, and nonfood-grade crops, to generate energy, chemicals, polymers, and materials. This chapter demonstrates the various pretreatment methods and conversion technologies devised to combat the structural recalcitrance of lignocellulosic biomass and facilitate the efficient production of biofuels, biochemicals, biomaterials, and other value-added products. Significant advancements have been achieved in deconstructing the lignocellulosic matrix by improving enzymatic accessibility using physical, chemical, and advanced biotechnological means. Conversion processes including pyrolysis, gasification, hydrothermal liquefaction, enzymatic hydrolysis, and microbial fermentation have resulted in the production of a diverse array of economically attractive value-added products. Furthermore, the use of tailored microbial systems, consolidated bioprocessing, and simultaneous saccharification and fermentation increased the efficiency of the process up to severalfold yielding high-quality products. Nonetheless, there are a few obstacles that hinder large-scale deployment, which includes high operational cost, enzyme constraints, inhibitor generation, and lignin valorization.

Future developments should focus on increasing pretreatment efficiency with minimal expenditure, augmenting enzyme stability, and designing microbial strains proficient in utilizing both C_5 and C_6 sugars. Integration of data-driven process control, renewable energy sources, and carbon capture technologies can significantly improve the environmental and economic sustainability of lignocellulosic biorefineries. Additionally, constructive legislation policies, infrastructural investment, and regional biomass utilization strategies would play an integral part for full-scale operation of LC biorefineries. In summary, lignocellulosic biorefineries constitute a fundamental component of the burgeoning bioeconomy. Through ongoing research, innovation, and policy endorsements, they possess the capacity to convert agricultural residues and forestry wastes into sustainable fuels and materials, facilitating the worldwide shift toward a circular, carbon-negative future.

References

1. De Bhowmick G, Sarmah AK, Sen R (2018) Lignocellulosic biorefinery as a model for sustainable development of biofuels and value added products. Bioresour Technol 247: 1144–1154. https://doi.org/10.1016/j.biortech.2017.09.163
2. Newell RG, Raimi D (2020) Global energy outlook comparison methods: 2020 update. Report 20-06. Resources for the Future. https://media.rff.org/documents/Global_Energy_Outlook_Comparison_Methods_2020.pdf
3. Ebadian M, van Dyk S, McMillan JD, Saddler J (2020) Biofuels policies that have encouraged their production and use: an international perspective. Energy Policy 147:111906. https://doi.org/10.1016/j.enpol.2020.111906
4. Usmani Z, Sharma M, Awasthi AK, Lukk T, Tuohy MG, Gong L, Nguyen-Tri P, Goddard AD, Bill RM, Nayak SC, Gupta VK (2021) Lignocellulosic biorefineries: the current state of challenges and strategies for efficient commercialization. Renew Sustain Energy Rev 148: 111258. https://doi.org/10.1016/j.rser.2021.111258
5. Hassan NS, Jalil AA, Hitam CN, Vo DV, Nabgan WJ (2020) Biofuels and renewable chemicals production by catalytic pyrolysis of cellulose: a review. Environ Chem Lett 18(5): 1625–1648. https://doi.org/10.1007/s10311-020-01040-7

6. Kumar B, Bhardwaj N, Agrawal K, Chaturvedi V, Verma P (2020) Current perspective on pretreatment technologies using lignocellulosic biomass: an emerging biorefinery concept. Fuel Process Technol 199:106244. https://doi.org/10.1016/j.fuproc.2019.106244
7. O'Connor K, Gaffey J, Gavin E, Stout J, Holden NM (2023) Circular bioeconomy outlook study 2030–2050 in support of climate action, sustainable food and biobased systems. Evidence Synthesis Report 2. https://www.epa.ie/publications/research/epa-research-2030-reports/Evidence-Synthesis-Report-2.pdf
8. Vu HP, Nguyen LN, Vu MT, Johir MA, McLaughlan R, Nghiem LD (2020) A comprehensive review on the framework to valorise lignocellulosic biomass as biorefinery feedstocks. Sci Total Environ 743:140630. https://doi.org/10.1016/j.scitotenv.2020.140630
9. Mankar AR, Pandey A, Modak A, Pant KK (2021) Pretreatment of lignocellulosic biomass: a review on recent advances. Bioresour Technol 334:125235. https://doi.org/10.1016/j.biortech.2021.125235
10. Ubando AT, Felix CB, Chen WH (2020) Biorefineries in circular bioeconomy: a comprehensive review. Bioresour Technol 299:122585. https://doi.org/10.1016/j.biortech.2019.122585
11. Dhanya BS, Mishra A, Chandel AK, Verma ML (2020) Development of sustainable approaches for converting the organic waste to bioenergy. Sci Total Environ 723:138109. https://doi.org/10.1016/j.scitotenv.2020.138109
12. Hendriks AT, Zeeman G (2009) Pretreatments to enhance the digestibility of lignocellulosic biomass. Bioresour Technol 100(1):10–18. https://doi.org/10.1016/j.biortech.2008.05.027
13. Agbor VB, Cicek N, Sparling R, Berlin A, Levin DB (2011) Biomass pretreatment: fundamentals toward application. Biotechnol Adv 29(6):675–685. https://doi.org/10.1016/j.biotechadv.2011.05.005
14. Pimienta JA, Papa G, Rodriguez A, Barcelos CA, Liang L, Stavila V, Sanchez A, Gladden JM, Simmons BA (2019) Pilot-scale hydrothermal pretreatment and optimized saccharification enables bisabolene production from multiple feedstocks. Green Chem 21(11):3152–3164. https://doi.org/10.1039/c9gc00323a
15. Song C, Zhang C, Zhang S, Lin H, Kim Y, Ramakrishnan M, Du Y, Zhang Y, Zheng H, Barceló D (2020) Thermochemical liquefaction of agricultural and forestry wastes into biofuels and chemicals from circular economy perspectives. Sci Total Environ 749:141972. https://doi.org/10.1016/j.scitotenv.2020.141972
16. de Jong E, Jungmeier G (2015) Biorefinery concepts in comparison to petrochemical refineries. In: Industrial biorefineries & white biotechnology. Elsevier, pp 3–33. https://doi.org/10.1016/B978-0-444-63453-5.00001-X
17. Takkellapati S, Li T, Gonzalez MA (2018) An overview of biorefinery derived platform chemicals from a cellulose and hemicellulose biorefinery. Clean Technol Environ Policy 20(7):1615. https://doi.org/10.1007/s10098-018-1568-5
18. Patel R, Rajaraman TS, Rana PH, Ambegaonkar NJ, Patel S (2025) A review on techno-economic analysis of lignocellulosic biorefinery producing biofuels and high-value products. Results Chem 13:102052. https://doi.org/10.1016/j.rechem.2025.102052
19. Akizuki S, Suzuki H, Fujiwara M, Toda T (2023) Impacts of steam explosion pretreatment on semi-continuous anaerobic digestion of lignin-rich submerged macrophyte. J Clean Prod 385:135377. https://doi.org/10.1016/j.jclepro.2022.135377
20. Sha Y, Zhang C, Xu Z, Zhai R, Jin M (2024) Quantitative assessment of ash effects on densifying lignocellulosic biomass with chemicals followed by autoclave (DLCA) pretreatment. Ind Crops Prod 216:118767. https://doi.org/10.1016/j.indcrop.2024.118767
21. Fia AZ, Amorim J (2023) Microwave pretreatment of biomass for conversion of lignocellulosic materials into renewable biofuels. J Energy Inst 106:101146. https://doi.org/10.1016/j.joei.2022.11.006
22. Zhang W, Liu J, Wang Y, Sun J, Huang P, Chang K (2021) Effect of ultrasound on ionic liquid-hydrochloric acid pretreatment with rice straw. Biomass Convers Biorefin 11(5):1749–1757. https://doi.org/10.1007/s13399-019-00595-y

23. Gao B, Yang Q, Xu L, Chen H, Wu S, Cheng H, Zhou H, Wang Y, Shen L, Chen Z (2025) Coupling lignocellulosic ethanol refinery with acid mine drainage treatment: a one-stone-two-birds strategy for waste management. Chem Eng J 503:158600. https://doi.org/10.1016/j.cej.2024.158600
24. Yaverino-Gutierrez MA, Ramos L, Ascencio JJ, Chandel AK (2024) Enhanced production of clean fermentable sugars by acid pretreatment and enzymatic saccharification of sugarcane bagasse. Processes 12(5):978. https://doi.org/10.3390/pr12050978
25. Chakraborty P, Kumar R, Chakrabortty S, Saha S, Chattaraj S, Roy S, Banerjee A, Tripathy SK, Ghosh AK, Jeon BH (2024) Technological advancements in the pretreatment of lignocellulosic biomass for effective valorization: a review of challenges and prospects. J Ind Eng Chem 137:29–60. https://doi.org/10.1016/j.jiec.2024.03.025
26. Liu B, Liu L, Deng B, Huang C, Zhu J, Liang L, He X, Wei Y, Qin C, Liang C, Liu S (2022) Application and prospect of organic acid pretreatment in lignocellulosic biomass separation: a review. Int J Biol Macromol 222:1400–1413. https://doi.org/10.1016/j.ijbiomac.2022.09.270
27. Hermsdorff GB, Escobar EL, da Silva TA, Zandoná Filho A, Corazza ML, Ramos LP (2023) Ethanol organosolv pretreatment of sugarcane bagasse assisted by organic acids and supercritical carbon dioxide. Carbohydr Polym 300:120263. https://doi.org/10.1016/j.carbpol.2022.120263
28. Sun J, Zhang L, Loh KC (2025) Revisiting alkali pretreatment to transform lignocellulose fermentation with integration of bioprocessible lignin. Waste Manag 192:82–90. https://doi.org/10.1016/j.wasman.2024.11.035
29. Tharunkumar J, Arosha VK, Bajhaiya AK, Rakesh S (2025) Optimizing alkaline pretreatment for delignification of paddy straw and sugarcane bagasse to enhance bioethanol production. Biomass Convers Biorefin 15(11):16409–16419. https://doi.org/10.1007/s13399-024-05458-9
30. Chen J, Tan X, Miao C, Zhang Y, Yuan Z, Zhuang X (2021) A one-step deconstruction-separation organosolv fractionation of lignocellulosic biomass using acetone/phenoxyethanol/water ternary solvent system. Bioresour Technol 342:125963. https://doi.org/10.1016/j.biortech.2021.125963
31. Zhang D, Liu J, Xu H, Liu H, He YC (2024) Improving saccharification efficiency of corn stover through ferric chloride-deep eutectic solvent pretreatment. Bioresour Technol 399:130579. https://doi.org/10.1016/j.biortech.2024.130579
32. Wu W, Zhu P, Luo L, Lin H, Tao Y, Ruan L, Wang L, Qing Q (2024) p-Toluenesulfonic acid enhanced neutral deep eutectic solvent pretreatment of soybean straw for efficient lignin removal and enzymatic hydrolysis. Bioresour Technol 395:130338. https://doi.org/10.1016/j.biortech.2024.130338
33. Sunar SL, Oruganti RK, Bhattacharyya D, Shee D, Panda TK (2024) Pretreatment of sugarcane bagasse using ionic liquid for enhanced enzymatic saccharification and lignin recovery: process optimization by response surface methodology. Cellulose 31(4):2151–2173. https://doi.org/10.1007/s10570-024-05768-1
34. Zhu Q, Dong H, Yan D, Gao D, Xu K, Cheng X, Xin J, Lu X (2024) High concentration bioethanol production from corn stalk via enhanced pretreatment with ionic liquids. Chem Eng Sci 283:119375. https://doi.org/10.1016/j.ces.2023.119375
35. Bertran-Llorens S, Perondi F, de Freitas AL, Chen J, van Erven G, Deuss PJ (2024) Supercritical CO_2 as effective wheat straw pretreatment for subsequent mild fractionation strategies. Chem Eng J 497:154491. https://doi.org/10.1016/j.cej.2024.154491
36. Zhang J, Zhang W, Cai Z, Zhang J, Guan D, Ji D, Gao W (2022) Effect of ammonia fiber expansion combined with NaOH pretreatment on the resource efficiency of herbaceous and woody lignocellulosic biomass. ACS Omega 7(22):18761–18769. https://doi.org/10.1021/acsomega.2c01302
37. Wang X, Wei Y, Sun X, Yuan J, Chen H, Sun Y, Luo L (2025) Enzymatic hydrolysis of fermentable sugars prepared from bamboo by pretreatment with hot water and green liquor. Biomass Convers Biorefin 15(5):6615–6628. https://doi.org/10.1007/s13399-024-05506-4

38. Grover C, Choudhury B (2024) Alkaline wet oxidative pretreatment and acid hydrolysis of wheat straw for squalene and monomethyl branched chain fatty acids rich lipid production in *Lentibacillus salarius* BPIITR. Biocatal Agric Biotechnol 57:103062. https://doi.org/10.1016/j.bcab.2024.103062
39. Pérez-Barragán J, García-Depraect O, Maya-Yescas R, Vallejo-Rodríguez R, Palacios-Hinestroza H, Coca M, Castro-Muñoz R, León-Becerril E (2024) Solid and liquid fractionation of sugarcane and Agave bagasse during ozonolysis and enzymatic hydrolysis: impact on biohydrogen and biogas production. Ind Crops Prod 210:118175. https://doi.org/10.1016/j.indcrop.2024.118175
40. Zheng C, Cone JW, van Peer A, Baars JJ, Hendriks WH (2025) Conversion of lignocellulosic biomass into valuable feed for ruminants using white rot fungi. J Anim Physiol Anim Nutr 109(3):800–811. https://doi.org/10.1111/jpn.14099
41. Li Y, Huang C, Mao Y, Zeng W, Zhao X, Zhu Y, Li X (2025) New insights for biomass utilization by brown-rot and white-rot fungi: the differing role of hemicellulose degradation in the incipient decay process. ACS Sustain Chem Eng 13(13):5157–5167. https://doi.org/10.1021/acssuschemeng.5c00882
42. Mehta P, Chelike DK (2024) Utilizing fungal biodegradation for valorisation of lignocellulosic waste biomass and its diverse applications. Appl Res 3(4):e202300119. https://doi.org/10.1002/appl.202300119
43. Ennouri H, Manai I, Ennouri H, Diaz SZ, Güelfo LA, Solera R, Hamdi M, Bouallagui H (2024) Effect of solid-state fermentation by a novel soft-rot fungus *Chaetomium globosum* IMA1 on the anaerobic co-digestion performances of waste activated sludge and municipal solid waste. J Environ Manag 371:123189. https://doi.org/10.1016/j.jenvman.2024.123189
44. Rabi Prasad B, Polaki S, Padhi RK (2024) Isolation of delignifying bacteria and optimization of microbial pretreatment of biomass for bioenergy. Biotechnol Lett 46(2):183–199. https://doi.org/10.1007/s10529-023-03463-y
45. Parveen F, Hussain A, Saxena A, Husain FM, Ashfaque M (2025) Optimized bioethanol production from diverse lignocellulosic biomass sources through integrative approach of plant cell wall degrading enzyme pretreatment. Biomass Bioenergy 197:107837. https://doi.org/10.1016/j.biombioe.2025.107837
46. Seesatat A, Rattanasuk S, Bunnakit K, Maneechot P, Sriprapakhan P, Artkla R (2021) Biological degradation of rice straw with thermophilic lignocellulolytic bacterial isolates and biogas production from total broth by rumen microorganisms. J Environ Chem Eng 9(1):104499. https://doi.org/10.1016/j.jece.2020.104499
47. Raina N, Boonmee R, Kirdponpattara S, Narasingha M, Sriariyanun M, Phitsuwan P, Chuetor S (2024) Process performance evaluation of different chemical pretreatments of lignocellulosic biomass for bioethanol production. Ind Crops Prod 211:118207. https://doi.org/10.1016/j.indcrop.2024.118207
48. Kaur P, Bohidar HB, Pfeffer FM, Williams R, Agrawal R (2023) A comparative assessment of biomass pretreatment methods for the sustainable industrial upscaling of rice straw into cellulose. Cellulose 30(7):4247–4261. https://doi.org/10.1007/s10570-023-05161-4
49. Chang C, Wu H, Zhao S, Zhao X, Ma Q, Li P, Xu G (2023) Chemical conversion of lignocellulosic biomass into platform chemicals for fuels and polymers. In: Advances in bioenergy, vol 8. Elsevier, pp 1–91. https://doi.org/10.1016/bs.aibe.2023.02.003
50. Haq IU, Qaisar K, Nawaz A, Akram F, Mukhtar H, Xu Y, Mumtaz MW, Rashid U, Ghani WA, Choong TS (2021) Advances in valorization of lignocellulosic biomass towards energy generation. Catalysts 11(3):309. https://doi.org/10.3390/catal11030309
51. Ingvarsson S, Odenberger M, Johnsson F (2024) Lignin extraction in chemical pulp mills: the role of flexible operation. Sustain Chem Environ 7:100137. https://doi.org/10.1016/j.scenv.2024.100137
52. Abolore RS, Jaiswal S, Jaiswal AK (2024) Green and sustainable pretreatment methods for cellulose extraction from lignocellulosic biomass and its applications: a review. Carbohydr Polym Technol Appl 7:100396. https://doi.org/10.1016/j.carpta.2023.100396

53. Manikandan S, Vickram S, Sirohi R, Subbaiya R, Krishnan RY, Karmegam N, Sumathijones C, Rajagopal R, Chang SW, Ravindran B, Awasthi MK (2023) Critical review of biochemical pathways to transformation of waste and biomass into bioenergy. Bioresour Technol 372:128679. https://doi.org/10.1016/j.biortech.2023.128679
54. Nanda S, Mohammad J, Reddy SN, Kozinski JA, Dalai AK (2014) Pathways of lignocellulosic biomass conversion to renewable fuels. Biomass Convers Biorefin 4(2):157–191. https://doi.org/10.1007/s13399-013-0097-z
55. Osman AI, Farghali M, Ihara I, Elgarahy AM, Ayyad A, Mehta N, Ng KH, Abd El-Monaem EM, Eltaweil AS, Hosny M, Hamed SM (2023) Materials, fuels, upgrading, economy, and life cycle assessment of the pyrolysis of algal and lignocellulosic biomass: a review. Environ Chem Lett 21(3):1419–1476. https://doi.org/10.1007/s10311-023-01573-7
56. Chopra J, Rangarajan V, Sen R (2022) Recent developments in oleaginous yeast feedstock based biorefinery for production and life cycle assessment of biofuels and value-added products. Sustain Energy Technol Assess 53:102621. https://doi.org/10.1016/j.seta.2022.102621
57. Kan T, Strezov V, Evans TJ (2016) Lignocellulosic biomass pyrolysis: a review of product properties and effects of pyrolysis parameters. Renew Sustain Energy Rev 57:1126–1140. https://doi.org/10.1016/j.rser.2015.12.185
58. Putro JN, Soetaredjo FE, Lin SY, Ju YH, Ismadji S (2016) Pretreatment and conversion of lignocellulose biomass into valuable chemicals. RSC Adv 6(52):46834–46852. https://doi.org/10.1039/C6RA09851G
59. Segers B, Nimmegeers P, Spiller M, Tofani G, Jasiukaitytė-Grojzdek E, Dace E, Kikas T, Marchetti JM, Rajić M, Yildiz G, Billen P (2024) Lignocellulosic biomass valorisation: a review of feedstocks, processes and potential value chains and their implications for the decision-making process. RSC Sustain 2(12):3730–3749. https://doi.org/10.1039/D4SU00342J
60. Jerzak W, Acha E, Li B (2024) Comprehensive review of biomass pyrolysis: conventional and advanced technologies, reactor designs, product compositions and yields, and techno-economic analysis. Energies 17(20):5082. https://doi.org/10.3390/en17205082
61. Kim D (2018) Physico-chemical conversion of lignocellulose: inhibitor effects and detoxification strategies: a mini review. Molecules 23(2):309. https://doi.org/10.3390/molecules23020309
62. Díaz B, Sommer-Márquez A, Ordoñez PE, Bastardo-González E, Ricaurte M, Navas-Cárdenas C (2024) Synthesis methods, properties, and modifications of biochar-based materials for wastewater treatment: a review. Resources 13(1):8. https://doi.org/10.3390/resources13010008
63. Xie W, Zhang Y, Zhang Y, Liu C, Wang Y, Xie Y, Ji G, Li A (2024) Fast pyrolysis of biomass with diverse properties to produce liquid hydrogen storage molecules. Carbon Capture Sci Technol 12:100230. https://doi.org/10.1016/j.ccst.2024.100230
64. Elhassan M, Abdullah R, Kooh MR, Chau YF (2023) Hydrothermal liquefaction: a technological review on reactor design and operating parameters. Bioresour Technol Reports 21:101314. https://doi.org/10.1016/j.biteb.2022.101314
65. Mohammadi M, Najafpour GD, Younesi H, Lahijani P, Uzir MH, Mohamed AR (2011) Bioconversion of synthesis gas to second generation biofuels: a review. Renew Sustain Energy Rev 15(9):4255–4273. https://doi.org/10.1016/j.rser.2011.07.124
66. Gutiérrez-Antonio C, Romero-Izquierdo AG, Gómez-Castro FI, Hernández S (2021) Production processes from lignocellulosic feedstock. In: Production processes of renewable aviation fuel, pp 129–169
67. Liu L, Corma A (2018) Metal catalysts for heterogeneous catalysis: from single atoms to nanoclusters and nanoparticles. Chem Rev 118(10):4981–5079. https://doi.org/10.1021/acs.chemrev.7b00776

68. Guo Y, Wang SZ, Xu DH, Gong YM, Ma HH, Tang XY (2010) Review of catalytic supercritical water gasification for hydrogen production from biomass. Renew Sustain Energy Rev 14(1):334–343. https://doi.org/10.1016/j.rser.2009.08.012
69. Shapiro-Bengtsen S, Hamelin L, Bregnbæk L, Zou L, Münster M (2022) Should residual biomass be used for fuels, power and heat, or materials? Assessing costs and environmental impacts for China in 2035. Energy Environ Sci 15(5):1950–1966. https://doi.org/10.1039/d1ee03816h
70. Mujtaba M, Fraceto LF, Fazeli M, Mukherjee S, Savassa SM, de Medeiros GA, Santo Pereira AD, Mancini SD, Lipponen J, Vilaplana F (2023) Lignocellulosic biomass from agricultural waste to the circular economy: a review with focus on biofuels, biocomposites and bioplastics. J Clean Prod 402:136815. https://doi.org/10.1016/j.jclepro.2023.136815
71. Druzhinina IS, Kubicek CP (2017) Genetic engineering of Trichoderma reesei cellulases and their production. Microb Biotechnol 10(6):1485–1499. https://doi.org/10.1111/1751-7915.12726
72. Zerva A, Pentari C, Ferousi C, Nikolaivits E, Karnaouri A, Topakas E (2021) Recent advances on key enzymatic activities for the utilisation of lignocellulosic biomass. Bioresour Technol 342:126058. https://doi.org/10.1016/j.biortech.2021.126058
73. Shallom D, Shoham Y (2003) Microbial hemicellulases. Curr Opin Microbiol 6(3):219–228. https://doi.org/10.1016/S1369-5274(03)00056-0
74. Xia J, Yang Y, Liu CG, Yang S, Bai FW (2019) Engineering *Zymomonas mobilis* for robust cellulosic ethanol production. Trends Biotechnol 37(9):960–972. https://doi.org/10.1016/j.tibtech.2019.02.002
75. Geijer C, Faria-Oliveira F, Moreno AD, Stenberg S, Mazurkewich S, Olsson L (2020) Genomic and transcriptomic analysis of Candida intermedia reveals the genetic determinants for its xylose-converting capacity. Biotechnol Biofuels 13(1):48. https://doi.org/10.1186/s13068-020-1663-9
76. Zhou H, Cheng JS, Wang BL, Fink GR, Stephanopoulos G (2012) Xylose isomerase overexpression along with engineering of the pentose phosphate pathway and evolutionary engineering enable rapid xylose utilization and ethanol production by *Saccharomyces cerevisiae*. Metab Eng 14(6):611–622. https://doi.org/10.1016/j.ymben.2012.07.011
77. Hong W, Zhang J, Feng Y, Mohr G, Lambowitz AM, Cui GZ, Liu YJ, Cui Q (2014) The contribution of cellulosomal scaffoldins to cellulose hydrolysis by *Clostridium thermocellum* analyzed by using thermotargetrons. Biotechnol Biofuels 7(1):80. https://doi.org/10.1186/1754-6834-7-80
78. Chung D, Cha M, Guss AM, Westpheling J (2014) Direct conversion of plant biomass to ethanol by engineered *Caldicellulosiruptor bescii*. Proc Natl Acad Sci 111(24):8931–8936. https://doi.org/10.1073/pnas.1402210111
79. Brethauer S, Studer MH (2014) Consolidated bioprocessing of lignocellulose by a microbial consortium. Energy Environ Sci 7(4):1446–1453. https://doi.org/10.1039/c3ee41753k
80. Mussatto SI, Yamakawa CK, van der Maas L, Dragone G (2021) New trends in bioprocesses for lignocellulosic biomass and CO_2 utilization. Renew Sustain Energy Rev 152:111620. https://doi.org/10.1016/j.rser.2021.111620
81. Akhlaghi N, Najafpour-Darzi G (2020) A comprehensive review on biological hydrogen production. Int J Hydrogen Energy 45(43):22492–22512. https://doi.org/10.1016/j.ijhydene.2020.06.182
82. Amaya-Santos G, Chari S, Sebastiani A, Grimaldi F, Lettieri P, Materazzi M (2021) Biohydrogen: a life cycle assessment and comparison with alternative low-carbon production routes in UK. J Clean Prod 319:128886. https://doi.org/10.1016/j.jclepro.2021.128886
83. Sen AS, Basak N (2025) Strategies towards fermentative biohydrogen generation using cellulosic pulp and paper waste: challenges, current progress, and future perspectives. Biomass Convers Biorefin 15(8):11443–11474. https://doi.org/10.1007/s13399-024-06149-1

84. Pugazhendhi A, Mathimani T, Varjani S, Rene ER, Kumar G, Kim SH, Ponnusamy VK, Yoon JJ (2019) Biobutanol as a promising liquid fuel for the future-recent updates and perspectives. Fuel 253:637–646. https://doi.org/10.1016/j.fuel.2019.04.139
85. Zhou Z, Jing Y, Wei S, Zhang Q, Peng S, An X, Li H (2023) Enhancement of butanol production in Clostridium acetobutylicum SE25 through oxidation-reduction potential regulation and analysis of its metabolic mechanisms. Fuel 331:125708. https://doi.org/10.1016/j.fuel.2022.125708
86. Poy H, da Costa LAM, Lladosa E, Gabaldón C, Loras S, Silvestre AJ (2023) Enhanced biomass processing towards acetone-butanol-ethanol fermentation using a ternary deep eutectic solvent. Renew Energy 219:119488. https://doi.org/10.1016/j.renene.2023.119488
87. Ding K, Liu D, Chen X, Zhang H, Shi S, Guo X, Zhou L, Han L, Xiao W (2024) Scalable lignocellulosic biorefineries: technoeconomic review for efficient fermentable sugars production. Renew Sustain Energy Rev 202:114692. https://doi.org/10.1016/j.rser.2024.114692
88. Dutta N, Usman M, Ashraf MA, Luo G, Gamal El-Din M, Zhang S (2023) Methods to convert lignocellulosic waste into biohydrogen, biogas, bioethanol, biodiesel and value-added chemicals: a review. Environ Chem Lett 21(2):803–820. https://doi.org/10.1007/s10311-022-01511-z
89. Jain S, Kumar S (2024) A comprehensive review of bioethanol production from diverse feedstocks: current advancements and economic perspectives. Energy 296:131130. https://doi.org/10.1016/j.energy.2024.131130
90. Eswaran S, Subramaniam S, Geleynse S, Brandt K, Wolcott M, Zhang X (2021) Techno-economic analysis of catalytic hydrothermolysis pathway for jet fuel production. Renew Sustain Energy Rev 151:111516. https://doi.org/10.1016/j.rser.2021.111516
91. Tsegaye B, Balomajumder C, Roy P (2019) Microbial delignification and hydrolysis of lignocellulosic biomass to enhance biofuel production: an overview and future prospect. Bull Natl Res Centre 43(1):1–6. https://doi.org/10.1186/s42269-019-0094-x
92. Andharia K, Shilpkar P, Kundu S (2023) Production and optimization of exopolysaccharides (eps) using low-cost bagasse as substrate by *Lysinibacillus macroides*. Asian J Biol Life Sci 12:54–59. https://doi.org/10.5530/ajbls.2023.12.8
93. Raj T, Chandrasekhar K, Kumar AN, Kim SH (2022) Lignocellulosic biomass as renewable feedstock for biodegradable and recyclable plastics production: a sustainable approach. Renew Sustain Energy Rev 158:112130. https://doi.org/10.1016/j.rser.2022.112130
94. Debnath A, Bhowmick S, Sen R (2024) Comparison of microbial and sono-assisted alkaline pre-treatment of sugarcane bagasse, followed by one-pot synthesis of microbial extracellular polymeric substance through simultaneous delignification, saccharification, and fermentation. bioRxiv. https://doi.org/10.1101/2024.06.22.600045
95. Maki M, Leung KT, Qin W (2009) The prospects of cellulase-producing bacteria for the bioconversion of lignocellulosic biomass. Int J Biol Sci 5(5):500. https://doi.org/10.7150/ijbs.5.500
96. Arli SD, Trivedi UB, Patel KC (2011) Curdlan-like exopolysaccharide production by *Cellulomonas flavigena* UNP3 during growth on hydrocarbon substrates. World J Microbiol Biotechnol 27(6):1415–1422. https://doi.org/10.1007/s11274-010-0593-2
97. Young ES, Kenyon WJ (2023) Interaction of the dye Aniline Blue with the Curdlan-type exopolysaccharide from *Cellulomonas flavigena* KU (ATCC 53703). Ind Biotechnol 19(4):204–207. https://doi.org/10.1089/ind.2023.0015
98. Balasubramanian B, Ilavenil S, Al-Dhabi NA, Agastian P, Choi KC (2019) Isolation and characterization of *Aspergillus* sp. for the production of extracellular polysaccharides by response surface methodology. Saudi J Biol Sci 26(3):449–454. https://doi.org/10.1016/j.sjbs.2018.10.015
99. Chowdhury SR, Basak RK, Sen R, Adhikari B (2011) Production of extracellular polysaccharide by *Bacillus megaterium* RB-05 using jute as substrate. Bioresour Technol 102(11):6629–6632. https://doi.org/10.1016/j.biortech.2011.03.099
100. Singh MK, Kumar M, Thakur IS (2017) Proteomic characterization and schizophyllan production by *Schizophyllum commune* ISTL04 cultured on *Leucaena leucocephala* wood under

submerged fermentation. Bioresour Technol 236:29–36. https://doi.org/10.1016/j.biortech.2017.03.170

101. Abdeshahian P, Ascencio JJ, Philippini RR, Antunes FA, Dos Santos JC, da Silva SS (2020) Utilization of sugarcane straw for production of β-glucan biopolymer by *Lasiodiplodia theobromae* CCT 3966 in batch fermentation process. Bioresour Technol 314:123716. https://doi.org/10.1016/j.biortech.2020.123716
102. Singh RS, Kaur N (2019) Understanding response surface optimization of medium composition for pullulan production from de-oiled rice bran by *Aureobasidium pullulans*. Food Sci Biotechnol 28(5):1507. https://doi.org/10.1007/s10068-019-00585-w
103. Koller M, Mukherjee A (2022) Polyhydroxyalkanoates (PHAs) – production, properties, and biodegradation. In: Biodegradable polymers in the circular plastics economy. Wiley, pp 145–204. https://doi.org/10.1002/9783527827589.ch6
104. Reddy CN, Mishra B, Mandal SK, Agrawal DC, Kruthiventi C (2022) An insight into pullulan and its potential applications. In: Polysaccharides of microbial origin: biomedical applications. Springer International Publishing, Cham, pp 247–277. https://doi.org/10.1007/978-3-030-42215-8_15
105. Yuan M, Fu G, Sun Y, Zhang D (2021) Biosynthesis and applications of curdlan. Carbohydr Polym 273:118597. https://doi.org/10.1016/j.carbpol.2021.118597
106. Das S (2022) Genetic regulation, biosynthesis and applications of extracellular polysaccharides of the biofilm matrix of bacteria. Carbohydr Polym 291:119536. https://doi.org/10.1016/j.carbpol.2022.119536
107. DeAngelis PL, Zimmer J (2023) Hyaluronan synthases; mechanisms, myths, and mysteries of three types of unique bifunctional glycosyltransferases. Glycobiology 33(12):1117–1127. https://doi.org/10.1093/glycob/cwad075
108. Shrivastav P, Pramanik S, Vaidya G, Abdelgawad MA, Ghoneim MM, Singh A, Abualsoud BM, Amaral LS, Abourehab MA (2022) Bacterial cellulose as a potential biopolymer in biomedical applications: a state-of-the-art review. J Mater Chem B 10(17):3199–3241. https://doi.org/10.1039/D1TB02709C
109. Venkatramanan V, Shah S, Prasad R, Shah M (2021) Life cycle assessment of lignocellulosic waste biorefinery. In: Bio-valorization of waste: trends and perspectives. Springer, Singapore, pp 327–347. https://doi.org/10.1007/978-981-15-9696-4_15
110. Van Hung N, Migo MV, Quilloy R, Chivenge P, Gummert M (2019) Life cycle assessment applied in rice production and residue management. In: Sustainable rice straw management. Springer International Publishing, Cham, pp 161–174. https://doi.org/10.1007/978-3-030-32373-8_10
111. Jeswani HK, Chilvers A, Azapagic A (2020) Environmental sustainability of biofuels: a review. Proc R Soc A Math Phys Eng Sci 476(2243). https://doi.org/10.1098/rspa.2020.0351

Adv Biochem Eng Biotechnol (2026) 196: 79–112
https://doi.org/10.1007/10_2026_317

Published online: 26 May 2026

Challenging Pretreatment Technologies of Lignocellulosic Biomass

Xinchuan Yuan, Guannan Shen, Qianqian Yang, Yin Cheng, Yiwei Chang, Yao Chen, Xingye Hu, Rui Zhai, and Mingjie Jin

Contents

Abstract Lignocellulosic biomass, derived from agricultural and forestry waste, is a promising renewable resource for biofuels and biochemicals. However, its recalcitrant structure, which is comprised of intertwined cellulose, hemicellulose, and

X. Yuan, G. Shen, Q. Yang, Y. Cheng, Y. Chang, Y. Chen, X. Hu, R. Zhai, and M. Jin (✉)
School of Environmental and Biological Engineering, Nanjing University of Science and Technology, Nanjing, People's Republic of China
e-mail: jinmingjie@njust.edu.cn

lignin, hinders efficient conversion, necessitating pretreatment as a crucial step to disrupt this matrix and enhance enzymatic hydrolysis. Pretreatment methods are categorized based on their target components: hemicellulose depolymerization strategies such as acid pretreatment (e.g., inorganic, organic, and solid acids) and physicochemical methods like steam explosion improve sugar release but face challenges like inhibitor formation and high energy costs; lignin removal strategies, including alkaline, oxidative, and biological pretreatments, effectively degrade lignin yet struggle with reagent recovery and low working efficiency; and selective fractionation strategies, such as organosolv, ionic liquids, and deep eutectic solvents, aim for high-value utilization but encounter issues with solvent sustainability and costs. Emerging technologies like dry acid pretreatment with biodetoxification (DryPB), ammonia-based methods, and densifying lignocellulosic biomass with chemicals (DLC) offer improved efficiency, reduced waste, and high product yields, highlighting the need for future optimization in cost-effectiveness, selectivity, and integration for scalable biorefineries.

Graphical Abstract

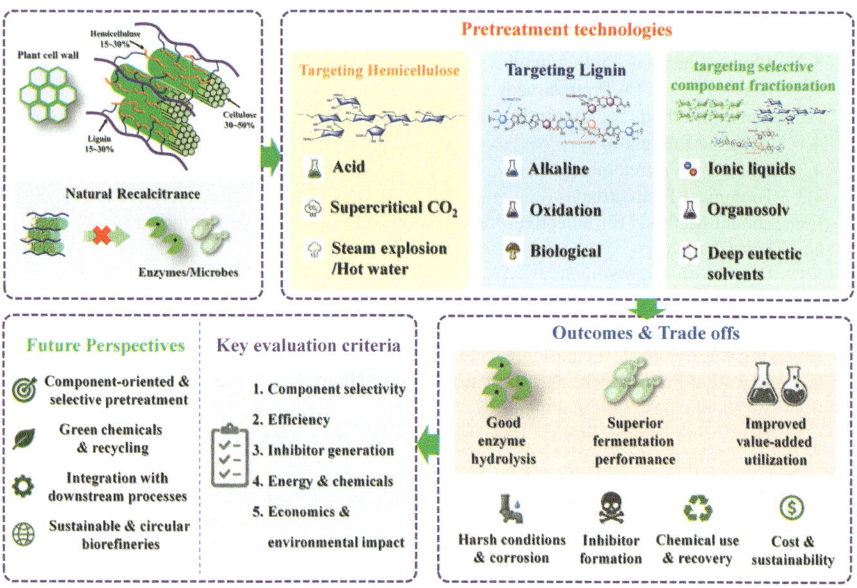

Keywords Biorefinery, Component targeting, Lignocellulosic biomass, Pretreatment strategy, Sustainable conversion

1 Introduction

Lignocellulosic biomass, derived from agricultural residues, forestry waste, and dedicated energy crops, is an abundant and renewable resource with vast potential to replace fossil-based fuels and chemicals. It plays a critical role in the transition toward a bio-based circular economy, offering significant environmental benefits such as reduced greenhouse gas emissions, lower dependence on petroleum-based products, and enhanced energy security. Unlike food-based feedstocks, lignocellulosic biomass is nonfood competitive and widely available, making it a sustainable alternative for producing biofuels, biochemicals, and biomaterials. Given these advantages, lignocellulosic biorefinery has gained increasing attention as a viable pathway for sustainable industrial applications.

However, the efficient conversion of lignocellulosic biomass remains a major challenge due to its intrinsically recalcitrant structure. Composed of cellulose, hemicellulose, and lignin, these components are chemically interlinked, forming a rigid three-dimensional polymeric network that provides mechanical strength and resistance to microbial degradation. While these structural properties are essential for plant integrity, they greatly hinder enzymatic hydrolysis and microbial fermentation, which are key steps in biomass valorization. The complex interactions between lignin and polysaccharides create a physical and chemical barrier that limits enzyme accessibility and sugar release, thereby reducing the overall efficiency of biomass conversion.

To overcome these challenges, pretreatment is a crucial step that modifies the structural and chemical characteristics of lignocellulosic biomass, enhancing its susceptibility to enzymatic hydrolysis and microbial conversion. Pretreatment primarily functions by disrupting the lignin-carbohydrate complex (LCC), reducing cellulose crystallinity, and solubilizing hemicellulose, thereby improving enzymatic accessibility and facilitating the efficient release of fermentable sugars. Various pretreatment strategies have been developed, including physical, chemical, physicochemical, and biological methods, each with distinct mechanisms, advantages, and challenges. However, despite significant advancements, pretreatment remains one of the most cost-intensive and technically demanding steps in lignocellulosic biorefining. Issues such as high reagent costs, inhibitor formation, solvent recovery difficulties, and energy-intensive operations limit the industrial scalability of many existing pretreatment technologies. Furthermore, biomass heterogeneity introduces variability in pretreatment efficiency, requiring customized approaches for different feedstocks.

Pretreatment technologies are often classified based on their process attributes, focusing on their technological properties rather than their selective impact on specific lignocellulosic components. However, an effective strategy must consider how each pretreatment method differentially targets cellulose, hemicellulose, or lignin to optimize component recovery and high-value utilization. This chapter reframes the discussion by categorizing pretreatment methods based on their targeted effects on lignocellulosic components, rather than a technology-driven perspective. By examining the unique attributes and limitations of component-oriented pretreatment strategies, this approach aims to facilitate the development of more efficient, cost-effective, and sustainable biomass conversion processes, ultimately promoting high-value lignocellulosic utilization in biorefineries.

2 Structural Challenges in Lignocellulosic Biomass

As an abundant and renewable versatile natural resource derived from plants, lignocellulosic biomass is regarded as an attractive alternative to fossil fuels and petroleum-based materials, which can reduce the greenhouse gas emissions and consolidate national energy security [1]. Based on these merits, lignocellulosic biorefinery has attracted extensive attention due to its great potential in sustainable industrial use. Lignocellulose is mainly composed of lignin, cellulose, and hemicellulose. These three components are intricately interlinked by chemical linkages, creating a three-dimensional polymeric structure that offers essential rigidity to the interior cells and tissues of plants [2]. Among them, the content of cellulose ranges from 30% to 50%, and the contents of hemicellulose and lignin are about 15–30% and 15–30%, respectively [3]. The contents of the three primary components in various plant varieties differ to some degree (Table 1).

Cellulose consists of elongated chains of glucose molecules linked by β-1,4-glycosidic linkages [4]. These elongated molecules constitute microfibrils, which subsequently assemble into bigger fibers. Some of these fibers are arranged neatly, while others are placed loosely and chaotically. The regularly arranged regions are crystalline, whereas the randomly organized regions are amorphous [5]. Hemicellulose, the second most prevalent carbohydrate in lignocellulose, consists of various monosaccharides (including xylose, arabinose, mannose, glucose, galactose, etc.) and glucuronic acid groups linked by β-D-furanose glycosidic bonds, glucoside bonds, and ester bonds, resulting in branched complex polysaccharides [6]. In contrast to cellulose and hemicellulose, which form linear chains, lignin constitutes a highly branching three-dimensional network. This intricate structure mostly comprises phenylpropanoid units (guaiacyl (G), syringyl (S), and *p*-hydroxyphenyl (H)) linked by several chemical linkages, including ether and carbon-carbon bonds. Lignin primarily functions to confer mechanical support and stiffness to plant cell walls [7, 8] (Fig. 1).

The highly resilient three-dimensional network macromolecules of lignin wrap cellulose and hemicellulose inside. This natural complex cross-linked structure

Table 1 Approximate composition of lignocellulose [3]

Species	Lignin	Cellulose	Hemicellulose
Softwood	29.1 ± 1.7%	42.1 ± 7.1%	25.1 ± 5.2%
Hardwood	19.5 ± 4.1%	50.8 ± 6.9%	29.7 ± 4.3%
Corn stover	15.2 ± 1.6%	34.3 ± 2.5%	20.7 ± 2.0%
Bagasse	15.7 ± 1.7%	28.9 ± 2.9%	16.7 ± 3.9%

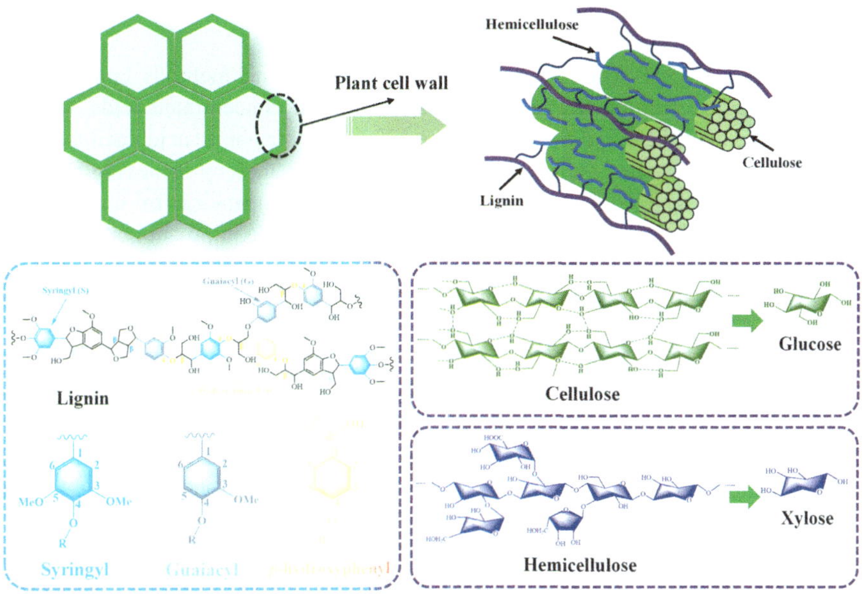

Fig. 1 The main structural compositions of lignocellulose

makes it difficult for the polysaccharides inside the lignocellulose to encounter bioactive substances such as enzymes or microorganisms [9]. Therefore, the lignocellulose needs to be pretreated to destroy its structure, promoting the effective separation and high-value utilization of lignocellulose components.

3 Pretreatment for Hemicellulose Depolymerization and Their Challenges

Hemicellulose depolymerization is a crucial step in biomass conversion. It involves breaking down hemicellulose by physical, chemical, or biological methods to improve cellulose accessibility and enzymatic hydrolysis efficiency. Key technologies include acid, physicochemical (e.g., steam explosion and liquid hot water), and supercritical CO_2 pretreatment.

3.1 Acid Pretreatment

Acid pretreatment, one of the earliest and most extensively studied chemical pretreatment methods, can be categorized based on chemical composition and structural characteristics into four types: inorganic acid, organic acid, solid acid, and sulfite pretreatment to overcome recalcitrance of lignocelluloses (SPORL).

3.1.1 Inorganic Acid Pretreatment

Inorganic acid hydrolysis can be categorized into two distinct methods: concentrated acid method and dilute acid method. The utilization of concentrated acids, such as 72% sulfuric acid, 42% hydrochloric acid, or 77–83% phosphoric acid, facilitates the dissolution of crystalline cellulose at low temperatures, thereby achieving uniform hydrolysis [10, 11]. Nevertheless, the practical implementation of this method remains constrained due to significant concerns regarding equipment corrosion and inefficient acid recovery processes.

In contrast, the more widely adopted dilute acid method in industry employs an acid concentration below 10% and a temperature ranging from 100 to 240°C [11–13]. This process mainly targets the acid hydrolysis of hemicellulose and amorphous cellulose through H^+ ions, effectively modifying the pore structure and significantly improving enzymatic hydrolysis efficiency (Fig. 2) [13]. Recently, Tong et al. [14] found that adding 2-naphthol-7-sulfonate and sodium bisulfite during acid pretreatment increased enzymatic hydrolysis yield from 47.9% to 90.6%, with fermentation achieving about 90% of ethanol yield.

Despite these advantages, inorganic acid pretreatment still faces several critical challenges. The use of strong acids often leads to the formation of inhibitory

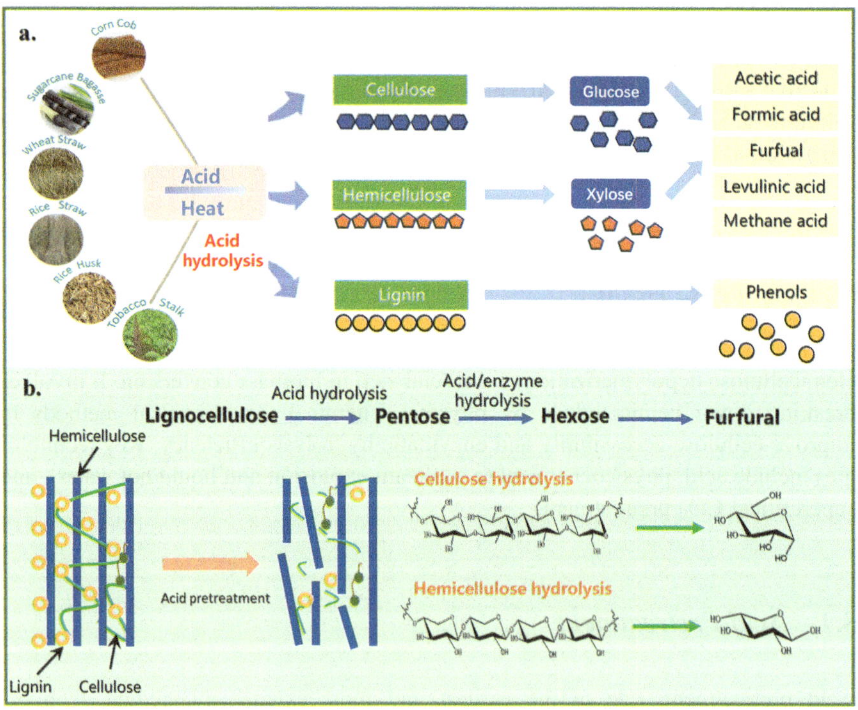

Fig. 2 The acid hydrolysis of lignocellulose. (**a**) Scheme for the degradation process and (**b**) structural change during pretreatment

by-products such as furfural and 5-hydroxymethylfurfural (HMF), which can negatively affect downstream fermentation processes. In addition, the neutralization and recovery of acids posttreatment remain economically and environmentally problematic. The need for corrosion-resistant equipment further increases capital investment, while the treatment and disposal of acidic waste streams pose environmental concerns. Therefore, optimizing reaction conditions, improving acid recovery techniques, and reducing inhibitor formation are essential to enhance the feasibility and sustainability of inorganic acid pretreatment in industrial applications.

3.1.2 Organic Acid Pretreatment

The pretreatment of organic acids has been identified as a novel and efficient method of clean separation, exhibiting merits such as high acid separation efficiency, reduced pollution levels, and mild reaction conditions. Common organic acids include formic [15], acetic [16], oxalic [17], levulinic [18], lactic [19], and citric acids [20]. Due to their relatively low acid dissociation constants, organic acids enable the selective hydrolysis and separation of hemicellulose, often leading to the production of high-value-added xylo-oligosaccharides (XOS). However, the high acid separation and purification cost hinders its large-scale industrial use. However, one of the major challenges impeding the large-scale application of organic acid pretreatment lies in the high costs associated with acid recovery and purification. The separation of organic acids from hydrolysates typically requires energy-intensive processes such as distillation, solvent extraction, or membrane separation, which increase operational complexity and reduce economic viability. To address this issue, Yang et al. [21] developed an innovative strategy using a lactic acid-rich fermentation broth directly as the pretreatment medium for corncob, thereby eliminating the need for costly acid purification. This method achieved an impressive XOS yield of 66.0%, demonstrating the potential of integrating organic acid pretreatment with bioprocessing streams to enhance overall process sustainability and cost efficiency. Despite these advancements, further efforts are needed to lower production costs, optimize process integration, and improve acid recyclability.

3.1.3 Solid Acid Pretreatment

Solid acids, including sulfonated activated carbon and hetero-polyacid, effectively depolymerize hemicellulose and partially remove lignin through their surface active sites [22]. These acids catalyze the hydrolysis of biomass polysaccharides into fermentable sugars using surface active sites [23]. Despite the advantages of solid acid catalysts, including high efficiency and environmental protection, their high preparation cost and low density of active sites may require more catalysts to match the performance of liquid acids [22, 24]. Zhang et al. [25] initially proposed the production of xylo-oligosaccharides (47.7%) by acid hydrolysis of xylan using a strongly acidic cation-exchange resin (NKC-9) as a solid acid catalyst. Subsequent assessment of the recyclability of the solid acid catalyst involved repetition of the process, thus overcoming the problem of difficult separation and recovery of liquid acid.

Despite these promising results, further research is needed to enhance the stability, reusability, and cost-effectiveness of solid acid catalysts. In particular, optimizing catalyst structure to increase the accessibility and density of acid sites and developing scalable, low-cost synthesis methods are essential for advancing the industrial application of solid acid pretreatment in biorefinery processes.

3.1.4 Sulfite Pretreatment to Overcome the Recalcitrance of Lignocelluloses Pretreatment

Sulfite pretreatment to overcome the recalcitrance of lignocelluloses (SPORL) is a specialized acid-based pretreatment method that employs sulfite salts – such as sodium bisulfite or sodium sulfite – to selectively remove lignin and partially hydrolyze hemicellulose. This approach effectively disrupts the structural recalcitrance of lignocellulosic biomass, thereby enhancing cellulose accessibility and significantly improving enzymatic hydrolysis efficiency. Zhang et al. [26, 27] showed that the two-stage process (hot water + SPORL) with sodium bisulfite improved lignin dissolution and hemicellulose depolymerization, increasing saccharification efficiency, while minimizing the generation of inhibitors. However, SPORL faces several practical challenges. The use of multiple chemical reagents, including sulfite salts, increases both the economic and environmental costs of the process. Additionally, the operational complexity associated with chemical handling, process control, and waste treatment makes industrial-scale implementation more difficult. To address these limitations, Su et al. [28] revealed that combined pretreatments (e.g., SPORL + $FeCl_3$) reduced chemical dosage and energy consumption while improving the biodegradability of cellulose and hemicellulose, resulting in a daily methane yield of 131.3 mL/g TS, which provided insights for reducing the cost of biorefinery. Despite its advantages, further optimization of SPORL is needed, particularly in terms of reagent recovery, process simplification, and environmental impact mitigation.

3.2 Physicochemical Pretreatment

The physicochemical pretreatment disrupts biomass structure through physical and chemical processes, improving cellulose accessibility and facilitating enzymatic hydrolysis or fermentation. These methods are categorized into steam explosion (SE) and liquid hot water (LHW) pretreatment based on their mechanisms, conditions, and effects.

3.2.1 Steam Explosion Pretreatment

During the steam explosion (SE) process, biomass is subjected to elevated pressures (0.69–4.83 MPa) and high temperatures (160–260°C) in saturated steam [29]. The high pressure facilitates the penetration of vapor molecules into the biomass structure (Fig. 3). The subsequent release of pressure causes the rapid expansion of entrapped water molecules, resulting in an explosive effect. This process effectively disrupts the biomass structure, promotes hemicellulose depolymerization, and partially removes lignin, thereby significantly improving cellulose accessibility [30]. Hoang et al. [31] successfully applied SE technology to produce xylo-oligosaccharides from bagasse, achieving a yield of up to 35.0% from the initial xylan content. Notably, approximately 50% of the product consists of xylobiose and xylotriose, which are particularly valuable in prebiotic applications. Furthermore, enzymatic hydrolysis of SE-pretreated biomass, particularly acid-catalyzed SE-pretreated biomass, has been reported to yield fermentable sugar with few inhibitory products, which could be employed in various bioproduct fermentations, including bioethanol and lactic acid [32].

Despite its effectiveness, the SE process presents several challenges. The generation of degradation products such as furfural, HMF, and acetic acid, particularly under severe pretreatment conditions, may still inhibit microbial fermentation. In addition, the high energy input required for steam generation and pressure maintenance, along with potential equipment wear due to explosive decompression, can increase operational costs and complicate scale-up. Therefore, further optimization of process parameters, integration with acid/alkali catalysts, and effective detoxification strategies are crucial for improving the overall efficiency and industrial viability of SE pretreatment.

3.2.2 Liquid Hot Water Pretreatment

Liquid hot water (LHW) pretreatment uses high-temperature (160–240°C) and high-pressure (5 MPa) water to dissolve hemicellulose, partially degrade lignin, and

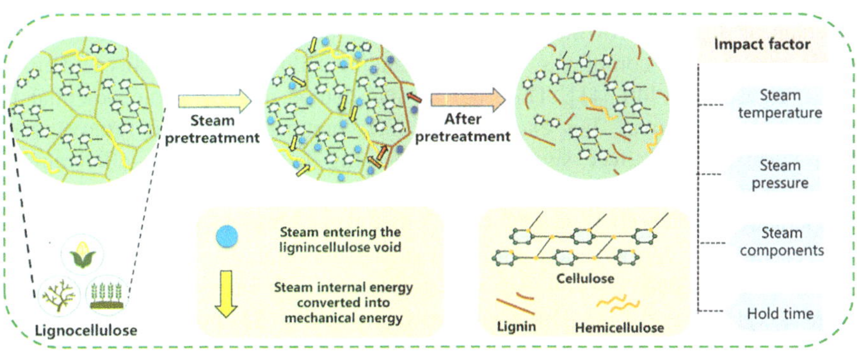

Fig. 3 Process and impact factors of steam pretreatment

enhance cellulose accessibility for enzymatic hydrolysis [33, 34]. This process causes the autoionization of water at elevated temperatures, which generates hydronium ions capable of catalyzing hydrolysis reactions without the need for added chemicals. For instance, LHW-pretreated corncob at 160°C for 10 min achieved 73.1% hydrolysis efficiency, with 58.8% hemicellulose sugar recovery and 60% lignin removal [35, 36]. As a chemical-free method, LHW significantly reduces environmental impact and minimizes equipment corrosion risks, offering a more sustainable and less corrosive alternative to acid-based methods [37]. However, LHW pretreatment is not without challenges. The process is energy intensive due to the need for maintaining high temperatures and pressures, which increases operational costs. Additionally, under certain conditions, degradation of released sugars can lead to the formation of inhibitory compounds such as furfural and 5-hydroxymethylfurfural (5-HMF), which negatively impact downstream microbial fermentation processes [38]. To overcome these limitations, recent studies have explored process optimization strategies and hybrid approaches. For example, combining LHW with microwave-assisted heating has shown potential to enhance sugar yields, shorten reaction times, and reduce the formation of inhibitors [37, 39]. Such integrated methods may improve the efficiency and cost-effectiveness of LHW pretreatment for large-scale biorefinery applications.

Overall, SE and LHW pretreatments effectively disrupt biomass structure, enhance cellulose accessibility, and improve efficiency of saccharification. However, both methods face challenges such as high equipment costs, strict operating conditions, and inhibitor formation. Therefore, it is necessary to optimize pretreatment conditions and integrate complementary technologies to improve pretreatment efficiency and economic feasibility.

3.3 *Supercritical CO_2 Pretreatment*

Supercritical CO_2 (SC-CO_2) pretreatment utilizes the unique physicochemical properties of carbon dioxide in its supercritical state, such as high diffusivity, low viscosity, and gas-like penetration ability, to infiltrate biomass structures. Upon rapid depressurization, the formation and expansion of microscopic bubbles induce mechanical disruption of the biomass cell walls and promote the degradation of lignin-carbohydrate complexes, thereby enhancing cellulose accessibility and enzymatic digestibility [40, 41]. This method has demonstrated notable potential in hemicellulose removal. For example, SC-CO_2 pretreatment could remove over 60% xylan in corn stover, significantly improving the mass transfer efficiency of broth during enzymatic hydrolysis [40]. Additionally, operating at relatively low temperatures (35–70°C) allows SC-CO_2 to minimize xylose degradation and reduce the generation of fermentation inhibitors such as furfural. This temperature range also helps to facilitate the conversion of xylose into high-value platform chemicals such as xylonic acid [42]. Despite its advantages, SC-CO_2 pretreatment presents several technical and economic challenges. The process requires precise control over reaction parameters (temperature, pressure, and depressurization rate) to avoid biomass over-degradation or insufficient disruption.

4 Pretreatment Methods for Lignin Removal and Their Challenges

Lignin has been reported as a major obstacle to enzymatic hydrolysis in lignocellulosic biorefinery [43]. To enhance the enzymatic hydrolysis efficiency of biomass, years of effort have been devoted to removing lignin while preserving sugar components during the pretreatment process. As a result of continuous efforts, a variety of lignin-targeted pretreatment strategies, such as alkali, oxidative, and biological methods, have been developed to improve saccharification performance through efficient delignification of biomass [44].

4.1 Alkaline Pretreatment

Alkaline pretreatment is a mature and typical pretreatment technology that uses alkaline compounds (such as sodium hydroxide, calcium hydroxide, sodium carbonate, etc.) to pretreat lignocellulosic biomass. During alkaline pretreatment, the hydroxyl groups in the alkaline compounds can break the aryl ether bonds and ester bonds associated with lignin, thereby removing lignin and enhancing the accessibility of cellulose to enzymes (Fig. 4) [45]. In alkaline pretreatment, sodium hydroxide pretreatment is the most widely used and effective method for biomass pretreatment, exhibiting strong delignification capability [46]. Zhang et al. pretreated corn stover with 2.05% (w/w) sodium hydroxide, achieving a lignin removal rate of up to 78.8% [47]. Calcium hydroxide is also frequently employed in biomass pretreatment due to its lower cost and satisfactory delignification efficiency [48]. Xu et al. pretreated

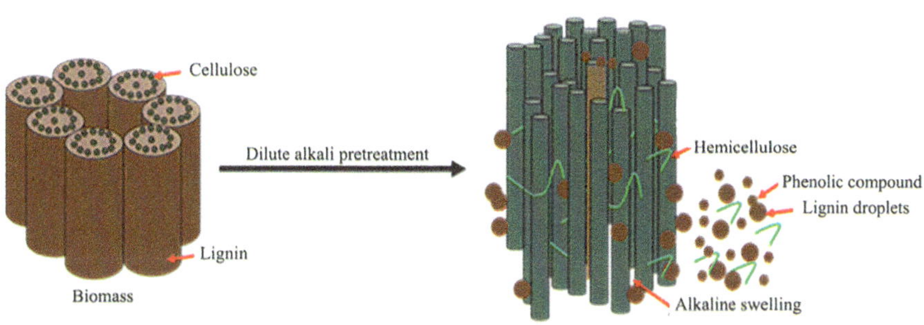

Fig. 4 Mechanism of alkaline pretreatment of biomass [50]

biomass with calcium hydroxide and active carbon, which effectively removed 66.9% lignin and 70.1% furfural under optimized conditions [49].

To further disrupt the biomass structure and improve cellulose accessibility, the deacetylation effect of alkaline compounds has been combined with mechanical refining for biomass pretreatment, called deacetylation and mechanical refining (DMR). Compared with the traditional alkaline pretreatment, DMR requires fewer chemical reagents and operates under milder conditions [51]. Kuhn et al. pretreated corn stover with sodium hydroxide in combination with the DMR process, achieving a glucose yield as high as 95.9% [52].

Despite its effectiveness, alkaline pretreatment faces several significant challenges. The recovery and reuse of alkaline reagents remain technically demanding and economically burdensome. In addition, the treatment and disposal of alkaline wastewater contribute to high operational costs and pose environmental concerns. These factors limit the large-scale industrial deployment of alkaline pretreatment technologies.

4.2 *Oxidative Pretreatment*

Oxidative pretreatment is one of the primary methods for the removal of lignin from biomass; it disrupts the chemical structure of lignin using oxidants such as alkaline hydrogen peroxide (AHP), acidic hydrogen peroxide, and wet oxygen. When pretreating biomass with AHP, hydroxyl radicals are generated from H_2O_2 under alkaline conditions and break the chemical bonds in the aromatic rings and side chains of lignin, leading to its degradation and leaching [53] (Fig. 5). For example, Xia et al. pretreated biomass with calcium hydroxide and hydrogen peroxide at 80° C, achieving lignin removal of approximately 80% [54].

Unlike AHP pretreatment, acidic hydrogen peroxide mainly oxidizes lignin by in situ synthesis of peroxyacids, thereby destroying the structure of lignin and achieving lignin removal [55]. Chang et al. pretreated biomass with acidic hydrogen peroxide in a two-step formic acid–H_2O_2 process, and the lignin removal rate reached 97.97% after pretreatment [56].

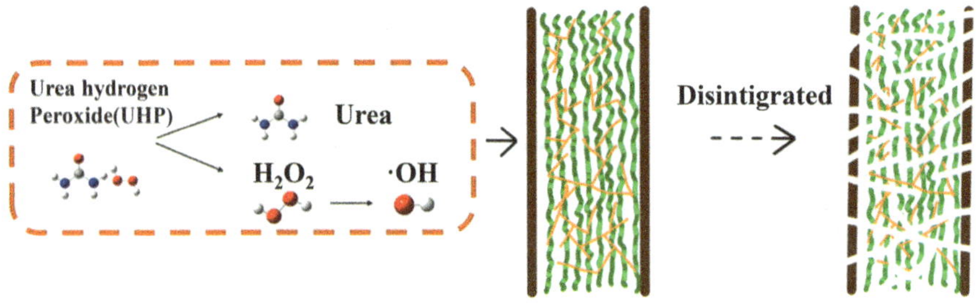

Fig. 5 Alkaline hydrogen peroxide pretreatment of biomass [58]

In addition, ozone is also an oxidant that reacts directly with lignin, oxidizing its double bonds and aromatic rings and thus breaking it down into smaller molecules [55]. Ortega et al. found that the ozonolysis process was 58.5% effective in delignifying pretreated biomass [57]. Wet oxygen pretreatment, on the other hand, uses oxygen to oxidize lignin at high temperature and pressure to degrade its structure and separate it from the feedstock [55]. These methods effectively disrupt the complex structure of lignin through different oxidation mechanisms, facilitating lignin removal and enhancing biomass utilization. Compared to other pretreatment methods, oxidative pretreatment offers strong oxidative capacity, rapid reaction rates, and broad applicability.

Nonetheless, several challenges hinder its industrial implementation. Oxidative pretreatment often involves harsh reaction conditions (e.g., high temperature, pressure, or strongly acidic/alkaline environments), which may lead to higher energy consumption and equipment corrosion. Additionally, the use of strong oxidants can result in the generation of toxic by-products or secondary pollutants, raising environmental and safety concerns. Therefore, future research should focus on developing milder, more selective oxidation strategies and improving oxidant recovery to enhance process sustainability and economic feasibility.

4.3 Biological Pretreatment

Biological pretreatment primarily utilizes microorganisms (such as fungi and bacteria) to degrade lignin and disrupt the structure of biomass; they can secrete a series of extracellular enzymes to degrade lignin, such as peroxidases and laccases [59]. Among these microorganisms, white-rot fungi have been the most extensively studied [60]. Peroxidases can oxidize aromatic ring structures with the help of hydrogen peroxide, thereby disrupting the lignin structure and improving the efficiency of enzymatic saccharification [61]. Laccases, on the other hand, can catalyze the single-electron oxidation of phenolic compounds, generating phenoxy radicals, which play a crucial role in the cleavage of lignin polymers [62]. Chen et al. [63] achieved a 54.7% lignin removal rate by fermenting corn stover with *Pleurotus ostreatus*. Zang et al. [64] pretreated wheat straw with *Inonotus obliquus* after 15 days, resulting in a lignin removal rate of up to 45%. Additionally, certain bacteria can also degrade lignin by secreting specific peroxidases. These enzymes can directly act on specific functional groups in lignin molecules, such as the hydroxylation of aromatic rings and ring-opening reactions [65].

Despite its environmental friendliness and selectivity, biological pretreatment faces several limitations that hinder its industrial scalability. The primary challenges include prolonged incubation periods, relatively low biomass degradation rates, and poor tolerance to harsh environmental conditions (e.g., high temperatures, pH extremes, or presence of inhibitors). Moreover, the performance of lignin-degrading enzymes is often substrate specific, requiring careful optimization of environmental conditions to achieve desirable efficiency [66]. Future research should focus on enhancing microbial strain performance through genetic engineering, optimizing enzyme production, and integrating biological pretreatment with other methods (e.g., mild chemical or physical pretreatments) to shorten processing time and improve efficiency.

5 Pretreatment Methods for Selective Fractionation of Lignocellulose Components and Their Challenges

5.1 Organosolv Pretreatment

Organosolv pretreatment is regarded as an efficient and environmentally friendly technology for the fractionation of lignocellulosic biomass [67]. It effectively depolymerizes lignin into low-molecular-weight fragments that dissolve in the solvent, thereby achieving lignin fractionation. Besides, organosolv pretreatment also drives the degradation of hemicellulose into oligosaccharides and monosaccharides, thereby significantly enhancing cellulose exposure. Currently, various organic solvents are employed, including alcohols (e.g., methanol, ethanol, ethylene glycol), ketones (e.g., acetone, methyl ethyl ketone), ethers (tetrahydrofuran, diethyl ether), and so on (Fig. 6).

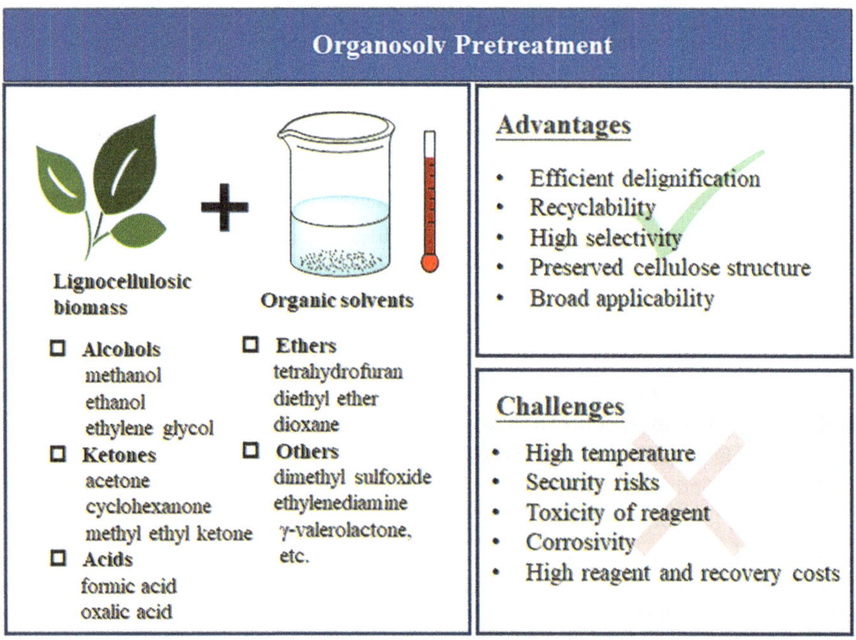

Fig. 6 Different types of organic solvents commonly used in organosolv pretreatment

5.1.1 Alcohol-Based Organosolv

Alcohol-based solvents show advantages in eco-friendliness, low toxicity, high selectivity for lignin fraction, and protecting lignin structure. These solvents, such as methanol and ethanol, enable efficient lignin fractionation while minimizing structural condensation, which is favorable for lignin valorization. However, low-boiling alcohols such as methanol and ethanol are easy to volatilize during high-temperature pretreatment, resulting in solvent loss and increased operational difficulty. In addition, alcohol exhibits poor selectivity for cellulose and hemicellulose.

To address these challenges, various strategies have been developed to improve alcohol-based pretreatment performance, including the use of catalytic additives and hybrid technologies. For instance, microwave-assisted pretreatment of poplar wood using methanol and *p*-toluenesulfonic acid achieved 87.9% lignin yield with 91.8% of ether bonds preserved, creating conditions for the selective depolymerization of lignin and enhancing the enzymatic digestibility of cellulose with a sugar conversion of 98.5% [68]. Acidic catalysts in alcohol pretreatment markedly enhance hemicellulose removal, facilitating subsequent cellulose saccharification. Conversely, alkaline catalysts suppress polysaccharide degradation, enabling high retention of carbohydrate components [69]. Mechanistically, acidic conditions primarily induce cleavage of α-aryl ether bonds in lignin, whereas β-aryl ether bonds are predominantly broken under alkaline conditions [69].

In summary, alcohol-based pretreatment holds strong potential for integrated biorefinery applications due to its lignin-focused selectivity and mild solvent profile. However, solvent recovery, volatility management, and process optimization (such as catalyst selection and energy input) remain key factors for advancing its industrial feasibility.

5.1.2 Ketone-Based Organosolv

Ketone organic solvents have significant advantages in terms of efficient extraction of lignin while preserving cellulose. Acetone, the most widely used ketone, exhibits higher lignin solubility than alcohols [70]. To enhance the selectivity of acetone pretreatment toward the three main biomass components (cellulose, hemicellulose, and lignin), multicomponent organic solvent pretreatment technologies were developed [71]. A ternary solvent system comprising acetone-butanol-ethanol (ABE) was employed to pretreat *Salix schwerinii* at 200°C, achieving a delignification ratio of 58.2% and a hemicellulose fractionation efficiency of 98.5% [72]. In addition to acetone, methyl isobutyl ketone (MIBK), a nonpolar solvent, has also been utilized for lignocellulose pretreatment [73, 74]. MIBK-based pretreatment allows for a phase separation strategy in which cellulose remains in the solid fraction, while lignin is retained in the organic phase. Upon subsequent treatment with water, hemicellulose can be efficiently extracted from the organic mixture, enabling selective and sequential recovery of biomass components. Despite their efficiency, ketone-based solvents pose certain limitations. Many ketones, including acetone and MIBK, are flammable, volatile, and toxic, necessitating stringent operational controls and corrosion-resistant equipment. These safety and material constraints increase capital and operational costs, thereby limiting large-scale industrial deployment.

5.1.3 Emerging Bio-Based Organosolv

Emerging bio-based solvents represented by tetrahydrofuran (THF) and γ-valerolactone (GVL) were also used to facilitate the fractionation of lignocellulose, meeting the requirements of sustainable development [75, 76]. A pretreatment technology termed cosolvent-enhanced lignocellulosic fractionation (CELF) was developed, utilizing the homogeneous amphiphilic solvent formed by THF and water. By cleaving the main lignin intramolecular bonds (β-*O*-4, β–β, and β-5), CELF achieved efficient lignin fractionation and obtained cellulose-rich solids with high digestibility. Nguyen et al. [77] applied CELF to pretreat corn stover followed by ethanol fermentation, achieving an ethanol yield over 80% even at a low enzyme loading (5 mg protein/g glucan). However, the low stability of THF increased the requirements for reaction equipment, and its toxicity also limited its application. Similar to THF, GVL also has the ability to disrupt aryl ether bonds in lignin. But the lignin fractionation efficiency of GVL pretreatment is usually slightly lower than that of THF [78]. In addition, studies indicated that GVL pretreatment also had an effect on the structure of cellulose. Cellulose was induced to undergo esterification during GVL pretreatment under acidic conditions, which may result in the changes of hydrophilicity and crystalline structure [79, 80]. Despite these advances, both THF- and GVL-based pretreatments are hindered by high solvent and recovery costs, which remain significant barriers to their commercial scalability. The practical application requires further optimization in solvent stability, safety, and cost-effectiveness, particularly regarding solvent recycling and compatibility with industrial-scale equipment.

5.2 *Ionic Liquid Pretreatment*

Ionic liquids (ILs) are considered to be green alternatives to volatile organic solvents and have the advantages of high thermal and chemical stability, recyclability, low corrosiveness, and good biodegradability [81]. ILs are composed of organic cations and anions. The types of cations include imidazolium ($[(C_3N_2)X_n]^+$), pyridine ($[(C_5N)X_n]^+$), ammonium ($[NX_4]^+$), and choline, while the types of anions can be either organic or inorganic [82]. Tunable functionality makes ILs attractive for lignocellulosic pretreatment, especially ILs based on imidazolium and pyridine, which show great effects in dissolving cellulose. The aromatic rings in cations reduce electrostatic intensity, weakening interactions with hydrogen-bond-accepting anions and enabling anions to form stronger H-bonds with cellulose hydroxyl groups, which is the determinant factor affecting cellulose dissolution [83]. Studies have indicated that ILs with anions of higher H-bond basicity and dipolarity exhibited superior cellulose dissolution capacity [84]. During IL pretreatment, cellulose crystallinity decreases due to crystal swelling or transformation from cellulose I to II, significantly enhancing its enzymatic digestibility [85]. Lignin fractionation by ILs primarily depends on cation interactions. Through π–π stacking and van der Waals forces, the hydrogen bonds and ether linkages in the lignin are disrupted [86, 87]. It was demonstrated that cholinium-based ILs showed stronger lignin dissolution capabilities and higher biocompatibility with enzymes and microorganisms compared to imidazolium-based ILs.

Despite these advantages, several critical challenges hinder the large-scale application of ILs in biomass pretreatment. These include the high cost of IL synthesis, difficulties in solvent recycling and purification, and, in some cases, limited biocompatibility with enzymes or fermentation microorganisms. Future research should focus on developing low-cost, biodegradable ILs with enhanced biomass fractionation efficiency, simplified recovery strategies, and compatibility with downstream bioconversion processes, in order to improve the techno-economic viability of IL-based biorefineries.

5.3 Deep Eutectic Solvent Pretreatment

Deep eutectic solvents (DESs), composed of two or more chemical components, primarily function through the interaction between a hydrogen bond donor (HBD) and a hydrogen bond acceptor (HBA). Based on the principle of interaction between HBD and HBA, DES may provide an acid-base catalytic mechanism to break the unstable ether bonds between phenylpropane units [88], thereby leading to the depolymerization of lignin and/or separation from biomass (Fig. 7). Among the two components, HBD has a more significant effect on the function of DES. Common HBD types include acid-, base-, and polyol-based compounds, with acid-based and polyol-based HBDs being extensively studied in lignocellulosic biorefining. The most used HBA is represented by choline chloride (ChCl).

5.3.1 DES with Acid-Based HBD

DES with acid-based HBD components (e.g., formic acid (FA), acetic acid (AA), oxalic acid (OA), and lactic acid (LA)) can effectively remove lignin and hemicellulose. Okuofu et al. [89] used ChCl/LA to pretreat Bambara groundnut haulm, achieving 54.5% hemicellulose removal, delignification of 60.7%, and a total sugar yield of 94.8% (at 10% substrate loading). It is currently believed that DESs containing HBDs of higher acidity (greater polarity) exhibit superior performance

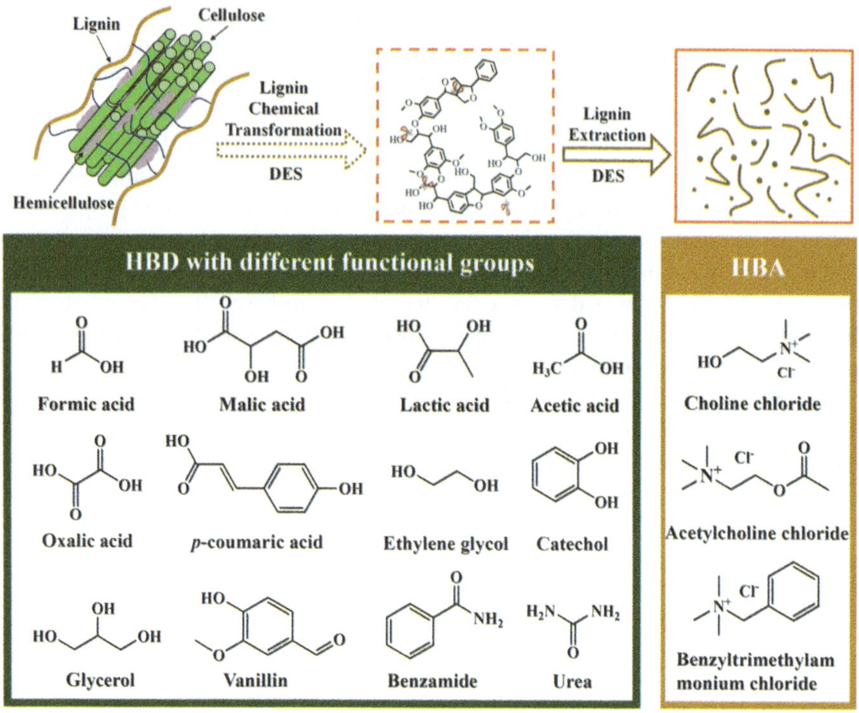

Fig. 7 Different DES systems and the effect of DES pretreatment on lignin [91]

on delignification. Enzymatic digestibility, especially for glucan, strongly correlates with delignification efficiency and/or structural properties of residual lignin after pretreatment [90]. However, DESs with acid-based HBD exhibit poor biocompatibility and are prone to undesirable side reactions during pretreatment, such as polysaccharide degradation and lignin condensation.

5.3.2 DES with Polyol-Based HBD

DESs with polyol-based HBD components (such as glycerol (Gly), urea (Urea), ethylene glycol (EG)) also show the potential to enhance enzymatic digestibility of biomass. The effect of these DES pretreatments does not focus on lignin extraction, but mainly on the effective fibrillation of crystalline cellulose into submicron or nanometer fibers. DESs with polyol-based HBD components have excellent stability, biocompatibility, and low corrosion [92]. Besides, their high hydrophobicity facilitates phase separation in biomass fractionation processes [93]. However, the relatively low acidity of polyol-based DESs limits their ability to break down lignin-carbohydrate complexes, resulting in poor lignin extraction efficiency. Consequently, achieving effective biomass deconstruction often requires prolonged reaction times or harsh operating conditions, such as elevated temperatures or high reagent concentrations. For example, Ai et al. [94] employed ChCl/Gly combined with screw extrusion technology to pretreat sorghum bagasse at 180°C with high solid loadings (30–50%), obtaining over 85% total sugar yield. However, lignin removal was limited, likely due to the insufficient acidity of the DES system.

5.3.3 Strategies for Enhanced DES Pretreatment

Recent studies have demonstrated that acidification strategies can significantly enhance the delignification performance of polyol-based deep eutectic solvent (DES) systems. ChCl/Gly with different heteropoly acids acidified was applied to pretreat *Miscanthus* × *giganteus*; the delignification significantly increased from 6% to 56.5–89.5% compared to the ChCl/Gly pretreatment without acid catalysis [95]. The possible mechanism for enhancing delignification by acidified DES is that acid protons have better availability and reactivity in the DES environment, which facilitates acid-catalyzed cleavage of bonds, especially β-ether bonds, within lignin carbohydrate complexes [96, 97].

Another advantage of acidified DES with polyol-based HBD is that it avoids possible side reactions, such as esterification between carboxylic acid-based DES and cellulose. At the same time, the highly hydrophobic polyol groups in DES can capture benzyl carbonium ions generated during lignin degradation or generate cyclic acetals with active sites generated by cleavage of β-*O*-4 bonds, effectively avoiding lignin recondensation [92].

Nevertheless, despite their promising performance, acidified DES systems still face challenges that limit industrial scalability. The high cost of acidic additives, combined with the complexity and energy demands of solvent recovery, remains a significant barrier to commercialization. Future research should prioritize the development of low-cost, recyclable acidic catalysts, as well as improved recovery systems (e.g., membrane or phase separation-based methods), to enhance the economic and environmental sustainability of DES-based pretreatment technologies.

6 Emerging Pretreatment Technologies and Their Challenges

6.1 Dry Acid Pretreatment and Biodetoxification (DryPB)

Traditional dilute acid pretreatment of lignocellulosic biomass often leads to the formation of various fermentation inhibitors, such as furan aldehydes (e.g., furfural, HMF) and phenolic compounds, due to the partial degradation of sugars and lignin components during high-temperature hydrolysis [98]. This severely impedes subsequent enzymatic hydrolysis and fermentation, thereby greatly limiting the

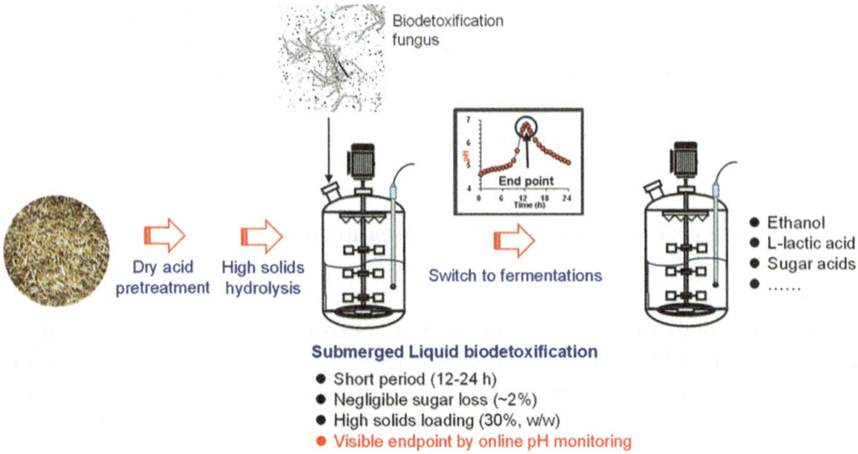

Fig. 8 Schematic diagram of the DryPB process. (Reprinted (adapted) with permission from [103]. Copyright © year 2025, American Chemical Society)

applicability of this method [99]. To mitigate these effects, conventional detoxification strategies such as water washing or over-liming are typically employed. However, these approaches often result in significant sugar losses and generation of large volumes of toxic wastewater, posing both economic and environmental concerns [100].

DryPB represents an innovative technological advancement. Under high solid loading conditions, this method utilizes dilute sulfuric acid coupled with saturated steam to treat lignocellulosic feedstocks [101]. The elevated solid-liquid ratio enhances inhibitor enrichment in the solid matrix, thereby simplifying downstream treatment processes while significantly reducing wastewater generation during pretreatment [102]. Subsequent biological detoxification employing *A. resinae* ZN1 demonstrates distinct advantages, as this fungal strain preferentially metabolizes inhibitors without degrading cellulose [102, 103] (Fig. 8). This dual treatment effectively preserves xylose and glucose released during pretreatment while preventing xylose loss [102]. The integrated approach combining dry acid pretreatment with *A. resinae* ZN1-mediated biodetoxification effectively resolves two critical limitations associated with traditional pretreatment methods: sugar loss during pretreatment and much wastewater generation [104].

DryPB has emerged as a promising approach for bioproduct synthesis, demonstrating remarkable advancements in recent years. The development of wheat straw to ethanol conversion achieved 101.3 g/L ethanol titers [100], optimization of sugar utilization pathways yielded 128.1 g/L D-lactic acid [105], and 31.3 g/L lysine production was enabled from wheat straw [106]. These results demonstrated the superior capability of DryPB in transforming lignocellulosic biomass into high-value bioproducts, but three critical limitations remain: acid-induced equipment corrosion under high-temperature conditions [102], extra detoxification costs

[107], and high energy consumption caused by the severe pretreatment conditions [101].

6.2 *Ammonia-Based Pretreatment*

Ammonia-based pretreatment, as one of the principal technologies in lignocellulosic biorefinery, has attracted extensive research attention. The ammonia fiber explosion (AFEX) method operates by infiltrating biomass with liquid ammonia under high pressure (1.7–2.1 MPa), followed by rapid decompression to induce "explosive" expansion of cellulose fibers, thereby disrupting crystalline structures and enhancing enzymatic hydrolysis efficiency [108]. This method is characterized by negligible wastewater generation, minimal inhibitor production, and reagent recyclability (about 97% ammonia recovery). Current applications demonstrate its efficacy: ethanol production from drought-tolerant AFEX-pretreated switchgrass achieved 40.39 g/L [109], xylitol yield from pretreated sargassum biomass reached 0.41 g/g sargassum [110], and 0.361 g/g sorbitol was produced from pretreated corn stover [110].

The ammonia extraction (EA) pretreatment employs anhydrous liquid ammonia under high liquid-solid ratios. This method enables the selective separation of lignin while preserving carbohydrate fractions [111]. About 40–50% lignin was separated from EA-pretreated corn stover while maintaining over 90% of glucan and 70% of xylan [111]. To further improve process efficiency and reduce ammonia usage, the cyclic ammonia compression biomass refinement approach (COBRA) aims to reduce ammonia consumption [112]. This method pretreated compressed biomass at 100°C, facilitating efficient polysaccharide conversion with low enzyme loading while removing the lignin to enhance yeast metabolic efficiency and ethanol yield [112, 113]. COBRA has many advantages including low ammonia demand, broad biomass compatibility (applicable to bagasse, herbaceous plants, and hardwood), and cost-effectiveness, highlighting its potential for industrial-scale applications [111, 112].

Overall, ammonia-based methods exhibit notable merits, including recyclability of reagents, environmental sustainability, and generation of few toxic degradation products, making them promising candidates for biorefinery industry. Nevertheless, challenges such as high equipment costs and operational risks associated with high-pressure conditions hinder their large-scale use.

6.3 *Densifying Lignocellulosic Biomass with Chemicals (DLC)*

Traditional pretreatment technologies are commonly restricted by biomass inherent limitations such as low bulk density [114] (Fig. 9), logistics challenges, and

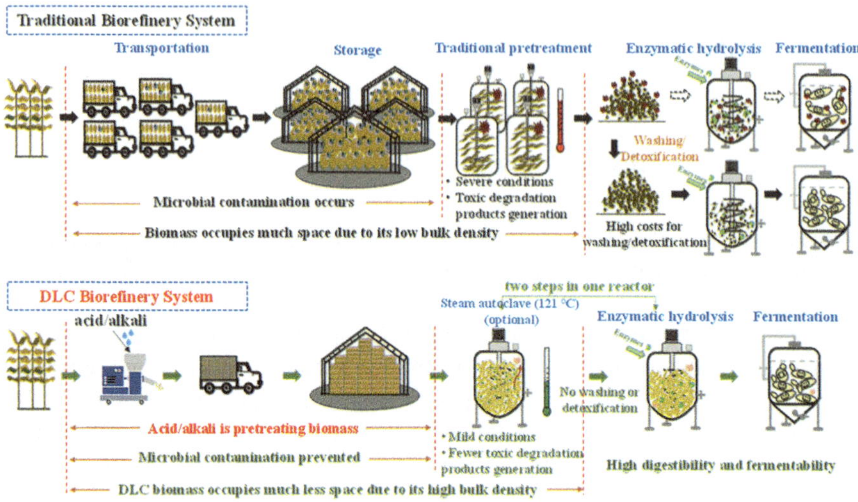

Fig. 9 Comparison of traditional biorefinery system and DLC biorefinery system. (Reprinted (adapted) with permission from [114]. Copyright © year 2022, Elsevier)

generation of many toxic degradation products [115, 116]. A novel pretreatment method, densifying lignocellulosic biomass with chemicals (DLC), was developed in recent years [115, 117, 118]. DLC-pretreated biomass has higher bulk density and uniform shapes, greatly improving the handling, transportation, and storage of biomass (Fig. 9). Besides, due to the chemical reagents contained in DLC-pretreated biomass, the latter can resist microbial contamination, thereby improving biomass durability [116]. Furthermore, the mechanochemical synergism during DLC pretreatment disrupts cell wall integrity, partially hydrolyzes hemicellulose, and exposes internal polysaccharides in biomass [119]. This dual action significantly increases biomass porosity and enzymatic accessible surface area. DLC-pretreated biomass followed by 121°C autoclave treatment exhibited superior digestibility and fermentability without water washing or detoxification, which were unachievable for traditional methods [119] (Fig. 9). Notably, most inter-unit linkages and non-condensed subunits of lignin are largely retained in DLC pretreated biomass, without the formation of pseudo-lignin or many toxic inhibitory phenolic compounds [119].

Recent studies have shown remarkable achievements of DLC pretreatment applied for bioproduct production. For bioethanol production, DLCA(sa) ("sa" refers to sulfuric acid) pretreated corncob achieved a maximum ethanol concentration of 94.76 g/L [120]. Microbial lipid production from DLCA(ch) pretreated rice straw through innovative fermentation strategies reached a lipid concentrations of 38.4 g/L [121]. Significant progress has also been made in organic acid biosynthesis, particularly in lactic acid production, where DLCA(sa)-pretreated corn stover yielded 210.1 g/L lactic acid [122]. Citric acid and succinic acid production from DLCA(sa) pretreated corn stover reached 83.6 g/L [123] and 105.42 g/L [124], respectively. These results further highlighted the technical advantages of DLC pretreatment in converting lignocellulosic biomass into high-value-added bio-based fuels and chemicals, providing substantial support for green and sustainable development of biorefinery industry.

7 Conclusion and Outlook

Biofuels and bio-based chemicals produced from lignocellulosic waste to replace fossil energy and petroleum-based chemicals are regarded as a significant advancement toward realizing a green, low-carbon, and circular economy. As an essential step in biomass valorization, pretreatment determines the efficiency of lignocellulose deconstruction and significantly influences downstream conversion processes. This review has summarized various pretreatment technologies based on their preferential targeting of specific lignocellulosic components, such as hemicellulose or lignin, and discussed their underlying mechanisms, key features, advantages, and limitations (Table 2).

In particular, several emerging pretreatment technologies have demonstrated strong potential to overcome the limitations of traditional methods. Dry acid pretreatment and biodetoxification (DryPB) combines high-solids dilute acid hydrolysis with fungal-based detoxification, significantly reducing wastewater generation and preserving sugar yield, while supporting the production of high-value products such as ethanol, D-lactic acid, and lysine. Ammonia-based pretreatment technologies exhibit excellent recyclability, low inhibitor formation, and broad feedstock compatibility, and have achieved considerable success in producing ethanol, xylitol, and sorbitol from various lignocellulosic sources. Densified Lignocellulosic biomass with Chemicals (DLC) pretreatment improves biomass logistics, biomass durability, and bioconversion efficiency. DLC-pretreated biomass gives high yields of ethanol, organic acids, and lipids without requiring detoxification or water washing, meanwhile minimizing the formation of pseudo lignin and inhibitory compounds.

Future research should aim to continue developing low-cost and highly selective pretreatment strategies tailored to different lignocellulosic feedstocks, while minimizing carbohydrate and lignin degradation. Co-production of high-value chemicals and materials from polysaccharides and lignin residues will be equally important to improve the overall economic benefits of lignocellulosic biorefinery.

Table 2 Comparison of major pretreatment methods for lignocellulosic biomass

Pretreatment method	Main hydrolysate components	Common inhibitors generated	Cost considerations	Waste treatment difficulty	Key advantages	Major limitations
Acid pretreatment	Xylose, oligosaccharides	Furfural, HMF, acetic acid	Moderate (acid consumption, corrosion-resistant equipment)	High (acidic wastewater requires neutralization)	Effective hemicellulose hydrolysis	Equipment corrosion, inhibitor formation, wastewater generation
Steam explosion (SE)	Xylose, oligosaccharides	Furfural, HMF, acetic acid	High (energy-intensive, high-pressure equipment)	Moderate (chemical-free but may require detoxification)	Efficient structure disruption, no chemicals required	High energy input, degradation products formation
Liquid hot water (LHW)	Soluble hemicellulose derivatives	Furfural, HMF, acetic acid	Moderate (high temperature/pressure requirements)	Low (minimal chemical use)	Environmentally friendly, low corrosion risk	Energy-intensive, limited lignin removal
Supercritical CO_2 pretreatment	Xylose, oligosaccharides	Furfural, acetic acid	Moderate (high temperature/high equipment cost)	Low (no require neutralization after CO_2 release)	Low operate temperature, less sugar degradation	High pressure, precise operation, high equipment requirements
Alkaline pretreatment	Lignin fragments, soluble sugar	Phenols, acetic acid	Low-Moderate (alkali cost, but milder conditions)	High (alkaline wastewater, chemical recovery needed)	Effective delignification, improves enzyme access	Long treatment time, wastewater generation
Oxidative pretreatment	Small molecule lignin compounds	Lignin-derived phenols	High (high energy consumption, corrosion-resistant equipment)	High (acidic/alkaline wastewater, Oxidizing agent removal)	Effective lignin removal, fast reaction rate, wide range of applications	High energy consumption, high risk of equipment corrosion
Biological pretreatment	Partially degraded lignin fragments	Low inhibitor generation	Low (low energy, but long duration)	Low (environmentally friendly)	Mild conditions, low energy, eco-friendly	Long processing time, low degradation efficiency
Organosolv	High-purity lignin fragments, soluble suagr	Low inhibitor generation	High (solvent cost, recovery needed)	Moderate (solvent recovery reduces waste)	High-purity lignin recovery, fractionation capability	Solvent volatility, high cost

Ionic liquids (ILs)	Amorphous cellulose, lignin fragments	Low inhibitor generation	High (expensive solvents, complex recovery)	Moderate (recyclable but purification challenging)	High dissolution capacity, tunable properties	High cost, potential toxicity, enzyme incompatibility
Deep eutectic solvents (DES)	High-purity lignin, soluble sugar	Low inhibitor generation	Moderate (low-cost components but energy for recovery)	Low-Moderate (biodegradable, some recyclable)	Biodegradable, design flexibility	Limited scalability, solvent recovery complexity
DryPB	Fermentable sugars, detoxified hydrolysate	Inhibitors biodetoxified	Moderate (acid and biological agent cost)	Low (reduced wastewater, integrated detoxification)	High working efficiency, integrated detoxification, high yield	Equipment corrosion, high energy consumption, extra biodetoxification cost
Ammonia-based pretreatment	Extracted lignin fragments, soluble suagr	Low inhibitor generation	High (high-pressure equipment)	Low (minimal wastewater, no detoxification needed)	Reagent recyclable, environmentally friendly, few toxic degradation products generation	High equipment costs, operational risks under high pressure conditions
DLC Pretreatment	High-density fermentable sugars	Very low inhibitor generation	Low (low chemical cost, milder conditions)	Low (minimal wastewater, no detoxification needed)	High biomass density, improved logistics, high product yield	Insufficient efficiency in polysaccharide deconstruction

References

1. Ali F, Dawood A, Hussain A, Alnasir MH, Khan MA, Butt TM et al (2024) Fueling the future: biomass applications for green and sustainable energy. Discov Sustain 5(1):156. https://doi.org/10.1007/s43621-024-00309-z
2. Mujtaba M, Fernandes Fraceto L, Fazeli M, Mukherjee S, Savassa SM, Araujo de Medeiros G et al (2023) Lignocellulosic biomass from agricultural waste to the circular economy: a review with focus on biofuels, biocomposites and bioplastics. J Clean Prod 402:136815. https://doi.org/10.1016/j.jclepro.2023.136815
3. Sethupathy S, Murillo Morales G, Gao L, Wang H, Yang B, Jiang J et al (2022) Lignin valorization: status, challenges and opportunities. Bioresour Technol 347:126696. https://doi.org/10.1016/j.biortech.2022.126696
4. Nosrati H, Khodaei M, Alizadeh Z, Banitalebi-Dehkordi M (2021) Cationic, anionic and neutral polysaccharides for skin tissue engineering and wound healing applications. Int J Biol Macromol 192:298–322. https://doi.org/10.1016/j.ijbiomac.2021.10.013
5. Wohlert M, Benselfelt T, Wågberg L, Furó I, Berglund LA, Wohlert J (2022) Cellulose and the role of hydrogen bonds: not in charge of everything. Cellulose 29(1):1–23. https://doi.org/10.1007/s10570-021-04325-4
6. Rao J, Lv Z, Chen G, Peng F (2023) Hemicellulose: structure, chemical modification, and application. Prog Polym Sci 140:101675. https://doi.org/10.1016/j.progpolymsci.2023.101675
7. Linger JG, Vardon DR, Guarnieri MT, Karp EM, Hunsinger GB, Franden MA et al (2014) Lignin valorization through integrated biological funneling and chemical catalysis. Proc Natl Acad Sci 111:12013–12018. https://doi.org/10.1073/pnas.1410657111
8. Erfani Jazi M, Narayanan G, Aghabozorgi F, Farajidizaji B, Aghaei A, Kamyabi MA et al (2019) Structure, chemistry and physicochemistry of lignin for material functionalization. SN Appl Sci 1(9):1094. https://doi.org/10.1007/s42452-019-1126-8
9. Worku LA, Bachheti RK, Bachheti A, Milessi TS, Chandel AK (2024) Understanding the biochemical changes at molecular level during biomass pretreatment: a comprehensive analysis. Cellulose 31(12):7281–7312. https://doi.org/10.1007/s10570-024-06081-7
10. Knauf MJ, Moniruzzaman M (2004) Lignocellulosic biomass processing: a perspective. Int Sugar J 106:147–150. https://doi.org/10.1016/j.indcrop.2003.09.005
11. Zhou Z, Liu D, Zhao X (2021) Conversion of lignocellulose to biofuels and chemicals via sugar platform: an updated review on chemistry and mechanisms of acid hydrolysis of lignocellulose. Renew Sustain Energy Rev 146:111169. https://doi.org/10.1016/j.rser.2021.111169
12. Chandel AK, Da Silva SS, Singh OV (2013) Detoxification of lignocellulose hydrolysates: biochemical and metabolic engineering toward white biotechnology. Bioenergy Res 6:388–401
13. du Pasquier J, Paës G, Perré P (2023) Principal factors affecting the yield of dilute acid pretreatment of lignocellulosic biomass: a critical review. Bioresour Technol 369:128439. https://doi.org/10.1016/j.biortech.2023.129167
14. Tong W, Fang H, Song K, Xie X, Wang J, Jin Y et al (2023) Modified acid pretreatment to alter physicochemical properties of biomass for full cellulose/hemicellulose utilization. Carbohydr Polym 299:120182
15. Zhou Y, Jiang D, Wang Z, Yi L, Sun J, Liu D et al (2024) Bandgap engineering of carbon nitride by formic acid assisted thermal treatment for photocatalytic degradation of tetracycline hydrochloride. Chem Eng J 485:149830
16. Pagano M, Hernando H, Cueto J, Cruz PL, Dufour J, Moreno I et al (2023) Insights on the acetic acid pretreatment of wheat straw: changes induced in the biomass properties and benefits for the bio-oil production by pyrolysis. Chem Eng J 454:140206

17. Liu L, Liu B, Li X, Wang Z, Mu L, Qin C et al (2023) Mannitol assisted oxalic acid pretreatment of poplar for the deconstruction and separation of hemicellulose. Ind Crop Prod 200:116811
18. Jeong S-Y, Koo B, Lee J-W (2022) Structural changes in biomass (yellow poplar and empty fruit bunch) during hydrothermal and oxalic acid pretreatments and their effects on enzymatic hydrolysis efficiency. Ind Crop Prod 178:114569
19. Yang Q, Sun F, Zhang RJ (2023) An integrated biorefinery approach to obtain xylo-oligosaccharides from corncob using lactic acid-rich fermentation broth. J Clean Prod 397: 136513. https://doi.org/10.1016/j.biortech.2021.125943
20. Valladares-Diestra KK, de Souza Vandenberghe LP, Torres LAZ, Zandoná Filho A, Woiciechowski AL, Soccol CR (2022) Citric acid assisted hydrothermal pretreatment for the extraction of pectin and xylooligosaccharides production from cocoa pod husks. Bioresour Technol 343:126074
21. Yang Q, Sun F, Meng X, Pu Y, Ragauskas AJ, Zhang J (2023) An integrated biorefinery approach to obtain xylo-oligosaccharides from corncob using lactic acid-rich fermentation broth. J Clean Prod 397:136513
22. Zeng M, Pan X (2022) Insights into solid acid catalysts for efficient cellulose hydrolysis to glucose: progress, challenges, and future opportunities. Catal Rev 64(3):445–490
23. Tyagi U, Sarma AK (2022) Perspectives of biomass based lignin to value added chemicals in biorefineries: challenges, extraction strategies and applications. Biofuels Bioprod Biorefin 16(6):1869–1892
24. Lv Y, Zhang Y, Xu Y (2024) Understanding and technological approach of acid hydrolysis processing for lignocellulose biorefinery: panorama and perspectives. Biomass Bioenergy 183:107133. https://doi.org/10.1016/j.biombioe.2024.107133
25. Zhang Q, Zhou X, Xu Y (2022) Xylooligosaccharides production from Xylan hydrolysis using recyclable strong acidic cationic exchange resin as solid acid catalyst. Appl Biochem Biotechnol 194(8):3609–3620
26. Zhang C (2014) Study on the pretreatment effect of sport method and dilute acid method on woody biomass
27. Pu C (2012) Study on the pretreatment and enzymatic hydrolysis of corn straw by sport and hot water
28. Su X, Wang X, Wang Y, Zhao X, Lu J, Qin G et al (2022) Effect of bisulfite pretreatment on methanogenic potential of yellow corn straw. China Biogas 40(03):50–57. https://doi.org/10.20022/j.cnki.1000-1166.2022030050
29. Yu Y, Wu J, Ren X, Lau A, Rezaei H, Takada M et al (2022) Steam explosion of lignocellulosic biomass for multiple advanced bioenergy processes: a review. Renew Sustain Energy Rev 154:111871. https://doi.org/10.1016/j.rser.2021.111871
30. Hoang AT, Nguyen XP, Duong XQ, Ağbulut Ü, Len C, Nguyen PQP et al (2023) Steam explosion as sustainable biomass pretreatment technique for biofuel production: characteristics and challenges. Bioresour Technol 385:129398
31. Brenelli LB, Bhatia R, Djajadi DT, Thygesen LG, Rabelo SC, Leak DJ et al (2022) Xylo-oligosaccharides, fermentable sugars, and bioenergy production from sugarcane straw using steam explosion pretreatment at pilot-scale. Bioresour Technol 357:127093
32. Cavalaglio G, Gelosia M, Giannoni T, Temporim RBL, Nicolini A, Cotana F et al (2021) Acid-catalyzed steam explosion for high enzymatic saccharification and low inhibitor release from lignocellulosic cardoon stalks. Biochem Eng J 174:108121
33. Treichel H, Fongaro G, Scapini T, Frumi Camargo A, Spitza Stefanski F, Venturin B (2020) Utilising biomass in biotechnology. Springer
34. Li J, Si B, Xu Q, Duan N, Liu Z (2024) Hot water pretreatment. In: Handbook of biorefinery research and technology: biomass logistics to saccharification, pp 315–340
35. Imman S, Laosiripojana N, Champreda V (2018) Effects of liquid hot water pretreatment on enzymatic hydrolysis and physicochemical changes of corncobs. Appl Biochem Biotechnol 184:432–443

36. Scapini T, Dos Santos MS, Bonatto C, Wancura JH, Mulinari J, Camargo AF et al (2021) Hydrothermal pretreatment of lignocellulosic biomass for hemicellulose recovery. Bioresour Technol 342:126033
37. Chen W-H, Nižetić S, Sirohi R, Huang Z, Luque R, Papadopoulos AM et al (2022) Liquid hot water as sustainable biomass pretreatment technique for bioenergy production: a review. Bioresour Technol 344:126207
38. Kammoun M, Margellou A, Toteva VB, Aladjadjiyan A, Sousa AF, Luis SV et al (2023) The key role of pretreatment for the one-step and multi-step conversions of European lignocellulosic materials into furan compounds. RSC Adv 13(31):21395–21420
39. Gallego-García M, Moreno AD, Manzanares P, Negro MJ, Duque A (2023) Recent advances on physical technologies for the pretreatment of food waste and lignocellulosic residues. Bioresour Technol 369:128397
40. Sarkar A, Lassi U (2024) Processing of biomass waste: technological upgradation and advancement. Elsevier
41. Fonseca-Pérez RM, Almena A, Ramírez-Márquez C, Bonilla-Petriciolet A, Martín M (2024) Techno-economic and environmental comparison of processes for the production of grape oil. J Clean Prod 441:141041
42. Zhang M (2018) Analysis of lignocellulosic SC-CO_2 explosion pretreatment process and its effect on component separation
43. Kellock M, Maaheimo H, Marjamaa K, Rahikainen J, Zhang H, Holopainen-Mantila U et al (2019) Effect of hydrothermal pretreatment severity on lignin inhibition in enzymatic hydrolysis. Bioresour Technol 280:303
44. Yoo CG, Meng X, Pu Y, Ragauskas AJ (2020) The critical role of lignin in lignocellulosic biomass conversion and recent pretreatment strategies: a comprehensive review. Bioresour Technol 301:122784
45. Sun Q, Foston M, Meng X, Sawada D, Kumar RR (2014) Effect of lignin content on changes occurring in poplar cellulose ultrastructure during dilute acid pretreatment. Biotechnol Biofuels 150:1–14
46. Zhu J, Wan C, Li Y (2010) Enhanced solid-state anaerobic digestion of corn stover by alkaline pretreatment. Bioresour Technol 101(19):7523–7528. https://doi.org/10.1016/j.biortech.2010.04.060
47. Zhang H, Wu J (2021) Statistical optimization of sodium hydroxide pretreatment and enzymatic hydrolysis of corn stover powder for enhancing sugar production using response surface methodology. Biomass Convers Biorefin. https://doi.org/10.1007/s13399-021-01638-z
48. Chen BY, Chen SW, Wang HT (2012) Use of different alkaline pretreatments and enzyme models to improve low-cost cellulosic biomass conversion. Biomass Bioenergy 39:182–191
49. Xu F, Chen J, Yang G, Ji X, Wang Q, Liu S et al (2019) Combined treatments consisting of calcium hydroxide and activate carbon for purification of xylo-oligosaccharides of pre-hydrolysis liquor. Polymers 11(10):1558
50. Kahar P (2013) Synergistic effects of pretreatment process on enzymatic digestion of rice straw for efficient ethanol fermentation. In: Environmental biotechnology – new approaches and prospective applications. IntechOpen, pp 73–77
51. Chen X, Kuhn E, Jennings EW, Nelson R, Tao L, Zhang M et al (2016) DMR (deacetylation and mechanical refining) processing of corn stover achieves high monomeric sugar concentrations (230 g L^{-1}) during enzymatic hydrolysis and high ethanol concentrations (>10% v/v) during fermentation without hydrolysate purification or concentration. Energy Environ Sci 9(4):1237–1245
52. Kuhn EM, Chen X, Tucker MP (2020) Deacetylation and mechanical refining (DMR) and deacetylation and dilute acid (DDA) pretreatment of corn stover, switchgrass, and a 50/50 corn stover/switchgrass blend. ACS Sustain Chem Eng 8:6734
53. Owolabi AF, Haafiz MM, Hossain MS, Hussin MH, Fazita MN (2017) Influence of alkaline hydrogen peroxide pre-hydrolysis on the isolation of microcrystalline cellulose from oil palm fronds. Int J Biol Macromol 95:1228–1234

54. Xia Y, Liu Q, Hu X, Li X, Huang Y, Li W et al (2022) Structural evolution during corn stalk acidic and alkaline hydrogen peroxide pretreatment. Ind Crop Prod 176:114386. https://doi.org/10.1016/j.indcrop.2021.114386
55. Toghiani J, Malekzadeh S, Jamali N, Afsham N, Fallah N, Mahboubi A et al (2024) Novel advanced oxidation processes (AOPs) as lignocellulosic biomass pretreatment approaches and their sustainability assessment: a review. Curr Pollut Rep 10:207–246
56. Chang M, Wang X, Lin Q, Li R, Zhao L, Ren J et al (2022) Formic acid–hydrogen peroxide treatment of furfural residue for production of nanocellulose, lignin, and nano-scale lignin. Green Chem 24(16):6232–6240
57. Ortega JO, Vargas JAM, Metzker G, Gomes E, da Silva R, Boscolo M (2021) Enhancing the production of the fermentable sugars from sugarcane straw: a new approach to applying alkaline and ozonolysis pretreatments. Renew Energy 164:502–508
58. Wu Y, Liu C, Song X, Liang J, Zhi M, Lu J et al (2023) A novel lignocellulose pretreatment technology by combining KOH, urea peroxide and organosilane to improve glucose yield. Chem Eng J 457:141296
59. Pérez J, Muñoz-Dorado J, de la Rubia T, Martínez J (2002) Biodegradation and biological treatments of cellulose, hemicellulose and lignin: an overview. Int Microbiol. https://doi.org/10.1007/s10123-002-0062-3
60. van Erven G, Hilgers R, Waard P, Gladbeek E-J, van Berkel WJH, Kabel MA (2019) Elucidation of in situ ligninolysis mechanisms of the selective white-rot fungus Ceriporiopsis subvermispora. ACS Sustain Chem Eng. https://doi.org/10.1021/acssuschemeng.9b04235
61. Datta R, Kelkar A, Baraniya D, Molaei A, Moulick A, Meena R et al (2017) Enzymatic degradation of lignin in soil: a review. Sustainability. https://doi.org/10.3390/su9071163
62. Gianfreda L, Xu F, Bollag J-M (1999) Laccases: a useful group of oxidoreductive enzymes. Bioremed J. https://doi.org/10.1080/10889869991219163
63. Chen Y, Fan H, Meng F (2016) Pleurotus ostreatus decreases cornstalk lignin content, potentially improving its suitability for animal feed. J Sci Food Agric. https://doi.org/10.1002/jsfa.7907
64. Zang Q, Chen X, Zhang C, Lin M, Xu X (2021) Improving crude protein and methionine production, selective lignin degradation and digestibility of wheat straw by Inonotus obliquus using response surface methodology. J Sci Food Agric. https://doi.org/10.1002/jsfa.11451
65. Sainsbury PD, Mineyeva Y, Mycroft Z, Bugg TDH (2015) Chemical intervention in bacterial lignin degradation pathways: development of selective inhibitors for intradiol and extradiol catechol dioxygenases. Bioorg Chem. https://doi.org/10.1016/j.bioorg.2015.05.002
66. Xu X, Pang D, Chen J, Luo Y, Zheng M, Yin Y et al (2018) Straw return accompany with low nitrogen moderately promoted deep root. Field Crop Res. https://doi.org/10.1016/j.fcr.2018.02.009
67. Wei Kit Chin D, Lim S, Pang YL, Lam MK (2020) Fundamental review of organosolv pretreatment and its challenges in emerging consolidated bioprocessing. Biofuels Bioprod Biorefin 14(4):808–829. https://doi.org/10.1002/bbb.2096
68. Fan D, Yang J, Xie X, Cao M, Xu L, Qiu X et al (2023) Microwave-assisted fractionation of poplar sawdust into high-yield noncondensed lignin and carbohydrates in methanol/p-toluenesulfonic acid. Chem Eng J 454. https://doi.org/10.1016/j.cej.2022.140237
69. Zhou Z, Lei F, Li P, Jiang J (2018) Lignocellulosic biomass to biofuels and biochemicals: a comprehensive review with a focus on ethanol organosolv pretreatment technology. Biotechnol Bioeng 115(11):2683–2702. https://doi.org/10.1002/bit.26788
70. Zhou Z, Liu D, Zhao X (2021) Deconstruction of lignocellulose recalcitrance by organosolv fractionating pretreatment for enzymatic hydrolysis. In: Liu Z-H, Ragauskas A (eds) Emerging technologies for biorefineries, biofuels, and value-added commodities. Springer International Publishing, Cham, pp 23–56
71. Li W, Tan X, Miao C, Zhang Z, Wang Y, Ragauskas AJ et al (2023) Mild organosolv pretreatment of sugarcane bagasse with acetone/phenoxyethanol/water for enhanced sugar production. Green Chem 25(3):1169–1178. https://doi.org/10.1039/D2GC04404H

72. Yang M, Guo X, Liu G, Nan Y, Zhang J, Noyazzesh H et al (2022) Effect of solvent mixture pretreatment on sugar release from short-rotation coppice Salix schwerinii for biobutanol production. Bioresour Technol 344:126262. https://doi.org/10.1016/j.biortech.2021.126262
73. Sun L-L, Yue Z, Sun S-C, Li Y, Cao X-F, Sun S-N (2023) Microwave-assisted choline chloride/1,2-propanediol/methyl isobutyl ketone biphasic system for one-pot fractionation and valorization of Eucalyptus biomass. Bioresour Technol 369:128392. https://doi.org/10.1016/j.biortech.2022.128392
74. Katahira R, Mittal A, McKinney K, Ciesielski PN, Donohoe BS, Black SK et al (2014) Evaluation of clean fractionation pretreatment for the production of renewable fuels and chemicals from corn stover. ACS Sustain Chem Eng 2(6):1364–1376. https://doi.org/10.1021/sc5001258
75. Zhu Y, Yang J, Mei F, Li X, Zhao C (2022) Bio-based 1,4-butanediol and tetrahydrofuran synthesis: perspective. Green Chem 24(17):6450–6466. https://doi.org/10.1039/D2GC02271K
76. Du X-L, Bi Q-Y, Liu Y-M, Cao Y, Fan K-N (2011) Conversion of biomass-derived levulinate and formate esters into γ-valerolactone over supported gold catalysts. ChemSusChem 4(12):1838–1843. https://doi.org/10.1002/cssc.201100483
77. Nguyen TY, Cai CM, Osman O, Kumar R, Wyman CE (2016) CELF pretreatment of corn stover boosts ethanol titers and yields from high solids SSF with low enzyme loadings. Green Chem 18(6):1581–1589. https://doi.org/10.1039/C5GC01977J
78. Meng X, Bhagia S, Wang Y, Zhou Y, Pu Y, Dunlap JR et al (2020) Effects of the advanced organosolv pretreatment strategies on structural properties of woody biomass. Ind Crop Prod 146:112144. https://doi.org/10.1016/j.indcrop.2020.112144
79. Luterbacher JS, Rand JM, Alonso DM, Han J, Youngquist JT, Maravelias CT et al (2014) Nonenzymatic sugar production from biomass using biomass-derived γ-valerolactone. Science 343(6168):277–280. https://doi.org/10.1126/science.1246748
80. Raj T, Chandrasekhar K, Banu R, Yoon J-J, Kumar G, Kim S-H (2021) Synthesis of γ-valerolactone (GVL) and their applications for lignocellulosic deconstruction for sustainable green biorefineries. Fuel 303:121333. https://doi.org/10.1016/j.fuel.2021.121333
81. Alayoubi R, Mehmood N, Husson E, Kouzayha A, Tabcheh M, Chaveriat L et al (2020) Low temperature ionic liquid pretreatment of lignocellulosic biomass to enhance bioethanol yield. Renew Energy 145:1808–1816. https://doi.org/10.1016/j.renene.2019.07.091
82. Hossain MM, Rawal A, Aldous L (2019) Aprotic vs protic ionic liquids for lignocellulosic biomass pretreatment: anion effects, enzymatic hydrolysis, solid-state NMR, distillation, and recycle. ACS Sustain Chem Eng 7(14):11928–11936. https://doi.org/10.1021/acssuschemeng.8b05987
83. Wang H, Gurau G, Rogers RD (2012) Ionic liquid processing of cellulose. Chem Soc Rev 41(4):1519–1537. https://doi.org/10.1039/C2CS15311D
84. Li Y, Wang J, Liu X, Zhang S (2018) Towards a molecular understanding of cellulose dissolution in ionic liquids: anion/cation effect, synergistic mechanism and physicochemical aspects. Chem Sci 9(17):4027–4043. https://doi.org/10.1039/c7sc05392d
85. Zhang J, Zou D, Zhai S, Yan Y, Yang H, He C et al (2022) Enhancing the interaction between cellulose and dilute aqueous ionic liquid solutions and its implication to ionic liquid recycling and reuse. Carbohydr Polym 277:118848. https://doi.org/10.1016/j.carbpol.2021.118848
86. Zhu Y, Yan J, Liu C, Zhang D (2017) Modeling interactions between a β-O-4 type lignin model compound and 1-allyl-3-methylimidazolium chloride ionic liquid. Biopolymers 107(8):e23022. https://doi.org/10.1002/bip.23022
87. Mohan M, Choudhary H, George A, Simmons BA, Sale K, Gladden JM (2021) Towards understanding of delignification of grassy and woody biomass in cholinium-based ionic liquids. Green Chem 23(16):6020–6035. https://doi.org/10.1039/D1GC01622A
88. Alvarez-Vasco C, Ma R, Quintero M, Guo M, Geleynse S, Ramasamy KK et al (2016) Unique low-molecular-weight lignin with high purity extracted from wood by deep eutectic solvents

(DES): a source of lignin for valorization. Green Chem 18(19):5133–5141. https://doi.org/10.1039/C6GC01007E
89. Okuofu SI, Gerrano AS, Singh S, Pillai S (2022) Deep eutectic solvent pretreatment of Bambara groundnut haulm for enhanced saccharification and bioethanol production. Biomass Convers Biorefin 12(8):3525–3533. https://doi.org/10.1007/s13399-020-01053-w
90. Zhang Z, O'Hara IM, Doherty WOS (2013) Pretreatment of sugarcane bagasse by acidified aqueous polyol solutions. Cellulose 20(6):3179–3190. https://doi.org/10.1007/s10570-013-0068-3
91. Hong S, Shen X-J, Xue Z, Sun Z, Yuan T-Q (2020) Structure–function relationships of deep eutectic solvents for lignin extraction and chemical transformation. Green Chem 22(21): 7219–7232. https://doi.org/10.1039/d0gc02439b
92. Liu Y, Deak N, Wang Z, Yu H, Hameleers L, Jurak E et al (2021) Tunable and functional deep eutectic solvents for lignocellulose valorization. Nat Commun 12(1):5424. https://doi.org/10.1038/s41467-021-25117-1
93. Wang W, Lee DJ (2021) Lignocellulosic biomass pretreatment by deep eutectic solvents on lignin extraction and saccharification enhancement: a review. Bioresour Technol 339:125587. https://doi.org/10.1016/j.biortech.2021.125587
94. Ai B, Li W, Woomer J, Li M, Pu Y, Sheng Z et al (2020) Natural deep eutectic solvent mediated extrusion for continuous high-solid pretreatment of lignocellulosic biomass. Green Chem 22(19):6372–6383. https://doi.org/10.1039/D0GC01560A
95. Guo Z, Zhang Q, You T, Ji Z, Zhang X, Qin Y et al (2019) Heteropoly acids enhanced neutral deep eutectic solvent pretreatment for enzymatic hydrolysis and ethanol fermentation of Miscanthus x giganteus under mild conditions. Bioresour Technol 293:122036. https://doi.org/10.1016/j.biortech.2019.122036
96. Bai Y, Zhang X-F, Wang Z, Zheng T, Yao J (2022) Deep eutectic solvent with bifunctional Brønsted-Lewis acids for highly efficient lignocellulose fractionation. Bioresour Technol 347: 126723. https://doi.org/10.1016/j.biortech.2022.126723
97. Škulcová AaR A, Jablonský M, Šíma J (2018) The pH behavior of seventeen deep eutectic solvents. Bioresources 13(3):5042–5051
98. Tian S-Q, Zhao R-Y, Chen Z-C (2018) Review of the pretreatment and bioconversion of lignocellulosic biomass from wheat straw materials. Renew Sustain Energy Rev 91:483–489. https://doi.org/10.1016/j.rser.2018.03.113
99. Zhou M, Tian X (2022) Development of different pretreatments and related technologies for efficient biomass conversion of lignocellulose. Int J Biol Macromol 202:256–268. https://doi.org/10.1016/j.ijbiomac.2022.01.036
100. Liu G, Zhang Q, Li H, Qureshi AS, Zhang J, Bao X et al (2017) Dry biorefining maximizes the potentials of simultaneous saccharification and co-fermentation for cellulosic ethanol production. Biotechnol Bioeng 115(1):60–69. https://doi.org/10.1002/bit.26444
101. He Y, Zhang J, Bao J (2014) Dry dilute acid pretreatment by co-currently feeding of corn stover feedstock and dilute acid solution without impregnation. Bioresour Technol 158:360–364. https://doi.org/10.1016/j.biortech.2014.02.074
102. Ahmed F, Yan Z, Bao J (2019) Dry biodetoxification of acid pretreated wheat straw for cellulosic ethanol fermentation. Bioresour Bioprocess 6(1). https://doi.org/10.1186/s40643-019-0260-x
103. Han T, Zhang B, Yang H, Liu X, Bao J (2023) Changes in pH values allow for a visible detection of the end point in submerged liquid biodetoxification during biorefinery processing. ACS Sustain Chem Eng 11(46):16608–16617. https://doi.org/10.1021/acssuschemeng.3c04986
104. Gu H, Han X, Zhang J, Bao J (2024) Upgrading dry acid pretreatment by post-hydrolysis for carbon efficient conversion of lignocellulose. Bioresour Technol 394:130261. https://doi.org/10.1016/j.biortech.2023.130261
105. He N, Chen M, Qiu Z, Fang C, Liden G, Liu X et al (2023) Simultaneous and rate-coordinated conversion of lignocellulose derived glucose, xylose, arabinose, mannose, and galactose into

D-lactic acid production facilitates D-lactide synthesis. Bioresour Technol 377:128950. https://doi.org/10.1016/j.biortech.2023.128950
106. Jin C, Bao J (2021) Lysine production by dry biorefining of wheat straw and cofermentation of Corynebacterium glutamicum. J Agric Food Chem 69(6):1900–1906. https://doi.org/10.1021/acs.jafc.0c07902
107. Wang Y, Yang H, Zhang B, Liu X, Bao J (2022) Continuous enzymatic saccharification and its rheology profiling under high solids loading of lignocellulose biomass. Biochem Eng J:186. https://doi.org/10.1016/j.bej.2022.108543
108. Zhao C, Shao Q, Chundawat SPS (2020) Recent advances on ammonia-based pretreatments of lignocellulosic biomass. Bioresour Technol 298:122446. https://doi.org/10.1016/j.biortech.2019.122446
109. Barten LM, Crandall JG, Xie D, Serate J, Handowski E, Jen A et al (2025) pH adjustment increases biofuel production from inhibitory switchgrass hydrolysates. bioRxiv. https://doi.org/10.1101/2025.01.10.632484
110. Galan G, Martin M, Grossmann IE (2021) Integrated renewable production of sorbitol and xylitol from switchgrass. Ind Eng Chem Res 60(15):5558–5573. https://doi.org/10.1021/acs.iecr.1c00397
111. da Costa SL, Jin M, Chundawat SPS, Bokade V, Tang X, Azarpira A et al (2016) Next-generation ammonia pretreatment enhances cellulosic biofuel production. Energy Environ Sci 9(4):1215–1223. https://doi.org/10.1039/C5EE03051J
112. Morais ARC, Zhang J, Dong H, Otto WG, Mokomele T, Hodge D et al (2022) Development of an ammonia pretreatment that creates synergies between biorefineries and advanced biomass logistics models. Green Chem 24(11):4443–4462. https://doi.org/10.1039/D2GC00488G
113. Renders T, Van den Bossche G, Vangeel T, Van Aelst K, Sels B (2019) Reductive catalytic fractionation: state of the art of the lignin-first biorefinery. Curr Opin Biotechnol 56:193–201. https://doi.org/10.1016/j.copbio.2018.12.005
114. Yuan X, Chen X, Shen G, Chen S, Yu J, Zhai R et al (2022) Densifying lignocellulosic biomass with sulfuric acid provides a durable feedstock with high digestibility and high fermentability for cellulosic ethanol production. Renew Energy 182:377–389. https://doi.org/10.1016/j.renene.2021.10.015
115. Yu J, Xu Z, Chen S, Yu Y, Zhang C, Chen X et al (2022) DLCA (ch) pretreatment brings economic benefits to both biomass logistics and biomass conversion for low cost cellulosic ethanol production. Fuel 311. https://doi.org/10.1016/j.fuel.2021.122603
116. Chen X, Yuan X, Chen S, Yu J, Zhai R, Xu Z et al (2021) Densifying lignocellulosic biomass with alkaline chemicals (DLC) pretreatment unlocks highly fermentable sugars for bioethanol production from corn stover. Green Chem 23(13):4828–4839. https://doi.org/10.1039/D1GC01362A
117. Yuan X, Shen G, Chen S, Chen X, Zhang C, Liu S et al (2022) Modified simultaneous saccharification and co-fermentation of DLC pretreated corn stover for high-titer cellulosic ethanol production without water washing or detoxifying pretreated biomass. Energy 247. https://doi.org/10.1016/j.energy.2022.123488
118. Devi A, Bajar S, Kour H, Kothari R, Pant D, Singh A (2022) Lignocellulosic biomass valorization for bioethanol production: a circular bioeconomy approach. Bioenergy Res 15(4):1820–1841. https://doi.org/10.1007/s12155-022-10401-9
119. Yuan X, Shen G, Chen S, Shen W, Chen X, Liu S et al (2023) Elucidating the mechanism of densifying lignocellulosic biomass with acidic chemicals (DLC) for lignocellulosic biorefinery. Green Chem 25(17):6759–6773. https://doi.org/10.1039/D3GC01703F
120. Chen S, Xu Z, Ding B, Zhang Y, Liu S, Cai C et al (2023) Big data mining, rational modification, and ancestral sequence reconstruction inferred multiple xylose isomerases for biorefinery. Sci Adv 9(5):eadd8835. https://doi.org/10.1126/sciadv.add8835
121. Yu Y, Liu S, Zhang Y, Lu M, Sha Y, Zhai R et al (2022) A novel fermentation strategy for efficient xylose utilization and microbial lipid production in lignocellulosic hydrolysate. Bioresour Technol 361:127624. https://doi.org/10.1016/j.biortech.2022.127624

122. Zhang Y, Xu Z, Lu M, Ma X, Chen S, Wang Y et al (2023) High titer (>200 g/L) lactic acid production from undetoxified pretreated corn stover. Bioresour Technol 388:129729. https://doi.org/10.1016/j.biortech.2023.129729
123. Lu M, Sha Y, Zhang Y, Ge M, Xu Z, Jin M (2025) Efficient production of citric acid from lignocellulose hydrolysate by metabolically engineered Yarrowia lipolytica. Green Chem 27(4):1113–1124. https://doi.org/10.1039/D4GC05521G
124. Ge M, Sha Y, Lu M, Zhang Y, Xu Z, Chen S et al (2025) Construction and optimization of efficient glucose–xylose co-fermenting yeast Yarrowia lipolytica for green and sustainable succinic acid production from lignocellulosic biomass. Green Chem 27(2):450–463. https://doi.org/10.1039/D4GC04189E

Adv Biochem Eng Biotechnol (2026) 196: 113–138
https://doi.org/10.1007/10_2026_318

Published online: 26 May 2026

Robust Production of Cellulases to Develop a Low-Cost Sugar Platform

Kehang Li and Yaohua Zhong

Contents

Abstract Bioconversion of lignocellulosic biomass into fermentable sugars using cellulase is indispensable for the production of biofuels and chemicals in the biorefining industry. However, the high cost and low efficiency of cellulase pose major obstacles to commercializing this process. In recent years, some significant efforts have been made to facilitate the application of cellulases in industry. Firstly, engineered *Trichoderma reesei* strains have been developed to enhance extracellular cellulase activities and efficiency. Secondly, the utilization of low-cost inducers and rational control of inducer feeding rates have facilitated economic production of cellulase. Finally, the development of sugar platforms with low cellulase dosage and

K. Li and Y. Zhong (✉)
State Key Laboratory of Microbial Technology, Microbial Technology Institute, Shandong University, Qingdao, China
e-mail: zhongyaohua@sdu.edu.cn

high sugar concentration has enabled efficient lignocellulose conversion. These systematic efforts have collectively propelled lignocellulose biorefining toward a more economical and efficient industrial future.

Graphical Abstract

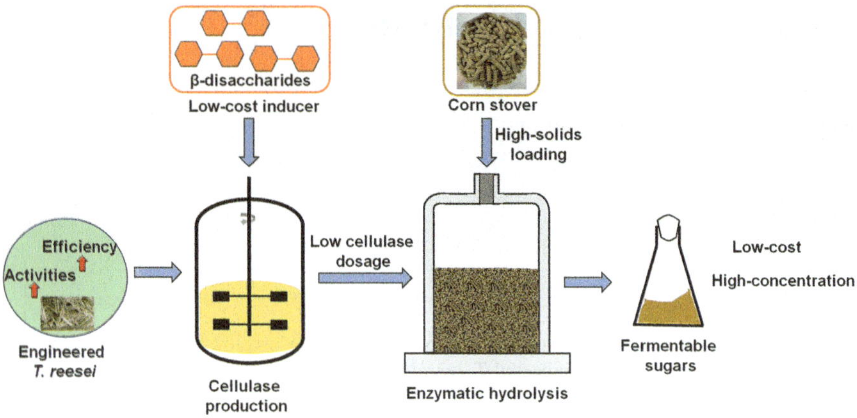

Keywords Cellulase production, High-solid saccharification, Soluble inducers, Transcriptional regulatory, *Trichoderma reesei*

1 Introduction

Lignocellulosic biomass is the bio-renewable and the most abundant resource on Earth, which can be used as an ideal substrate for the biorefining industry to produce bio-based fuels and chemicals [1]. Cellulase is a multienzyme complex that works synergistically to catalyze the hydrolysis of cellulose glycosidic bonds to produce glucose. It plays a key role in fermentable sugar production for lignocellulosic biorefining. However, the high cost of cellulase poses a major obstacle to commercializing this process. The lack of engineered *T. reesei* strains exhibiting high-titer and high-performance cellulase production is the primary reason for the high cost of cellulase [2]. Additionally, the expensive fermentation inducers and the suboptimal cultivation conditions further elevate cellulase production costs [3]. The high cellulase consumption during saccharification of lignocellulosic biomass also leads to enhanced overall cellulase cost. Therefore, reducing cellulase costs requires concerted efforts targeting these aspects.

The development of superior engineered strains of *T. reesei* is essential for robust cellulase production, which can be achieved through several genetic engineering strategies. Researchers have previously isolated a series of hypersecreting *T. reesei* mutants, such as RUT-C30, QM9414, and PC-3-7, through UV or chemical mutagenesis strategies [4–6]. With the in-depth understanding of the transcriptional regulatory mechanism of cellulase genes in recent years, rationally modifying its transcriptional regulatory network has become the primary strategy for enhancing cellulase production in *T. reesei.* Moreover, engineering of protein secretion pathways through modification of endoplasmic reticulum- and vesicle-related genes has been shown to effectively increase cellulase secretion in *T. reesei* [7, 8]. The naturally unbalanced cellulase system of *T. reesei* prevents effective hydrolysis of lignocellulosic biomass, which further leads to high cellulase consumption during saccharification. Therefore, strategies to enhance cellulase efficiency in engineered *T. reesei* strains have been proposed, which primarily include optimizing enzyme component ratios, discovering novel enzymes, and heterologously expressing enzyme components.

T. reesei is the predominant industrial strain for cellulase production, which synthesizes cellulase under induction conditions. However, the high cost of inducers, such as sophorose and lactose, prevents the cost-effective cellulase production. In recent years, some studies have synthesized a low-cost soluble inducer (the mixture of glucose and β-disaccharides) through the transglycosidation reaction, which promoted the efficient cellulase production in *T. reesei* [9, 10]. Furthermore, genetic engineering of *T. reesei* strains also allows them to produce cellulase under non-inducing conditions using the cheaper common carbon sources such as glucose or sugarcane molasses [11, 12]. Developing cost-effective sugar platforms is crucial for bioconversion of lignocellulose. This first requires the establishment of a high-solid saccharification system for lignocellulose to obtain high-concentration fermentable sugars. Moreover, it is also important to reduce cellulase dosage in the saccharification reaction by using additives and recycling enzymes.

2 Cellulase

Cellulase is a multienzyme complex that works synergistically to catalyze the hydrolysis of cellulose glycosidic bonds to produce glucose. It plays a key role in fermentable sugar production for lignocellulosic biorefining. However, the high production costs and low catalytic efficiency of cellulase prevent its application in the biorefining industry.

2.1 Cellulase Classification

Cellulase is a multienzyme complex comprising cellobiohydrolases (CBHs), endoglucanases (EGs), and β-glucosidases (BGLs), which function synergistically to hydrolyze cellulose into glucose [13]. Endocellulase randomly attacks the internal

O-glycosidic bonds in the amorphous regions of the cellulose chain, thereby releasing cello-oligosaccharides and exposing additional reducing and nonreducing ends [14]. Cellobiohydrolases CBHI and CBHII progressively hydrolyze cellulose chains from the reducing and nonreducing ends, respectively, releasing the cellobiose [15]. The β-glucosidase finally catalyzes the hydrolysis of cellobiose into glucose molecules [16]. Newly identified lytic polysaccharide monooxygenases (LPMOs) function as auxiliary enzymes that significantly enhance cellulose degradation [17]. LPMOs are copper-dependent enzymes that catalyze the cleavage of cellulose chains through oxidation at the C1 or C4 position of the glycosidic bonds [18]. In addition, hemicellulases (primarily incorporating exo-xylanases, endo-β-1,4-xylanases, β-xylosidases), cellulose-induced protein (CIPs), and swollenin (SWO1) also contribute to the efficient hydrolysis of lignocellulosic biomass [2, 19].

2.2 *Challenges in Cellulase Application: From Production Costs to Catalytic Efficiency*

The application of cellulase in the lignocellulosic biorefining industry still faces severe challenges, primarily high production costs and low catalytic efficiency. It is well known that the major cost of lignocellulose bioconversion is attributed to cellulase. The cost of cellulase to produce per gallon of bioethanol is reported to range from $0.1 to $0.68 [20]. The high cellulase cost in lignocellulosic biorefining stems from several factors. Although *T. reesei* has been used as the leading cellulase production host in industry for decades, high-yield strains remain scarce, which is the primary reason for high cellulase costs [2]. Additionally, the suboptimal cultivation conditions and the expensive fermentation inducers further elevate cellulase production costs [3, 21]. The higher cellulase dosage required to release sugars from lignocellulose also contributes to the high cellulase cost in the biorefining industry. On a protein weight basis, it takes anywhere from 40 to 100 times more enzyme to break down cellulose than starch [22].

The insufficient degradation efficiency of *T. reesei*'s cellulase system also hinders its application in lignocellulose biorefining. That is mainly caused by the unbalanced enzyme components of the native cellulase system of *T. reesei*. For example, cellobiohydrolases CBH1 and CBH2 constitute 70–90% of *T. reesei*'s secretome, representing its most abundant cellulase components. However, β-glucosidases account for less than 2% of the total proteins in the secretome of *T. reesei* [19]. This makes β-glucosidases the rate-limiting enzyme in the final step of the cellulose hydrolysis process, thereby restricting the degradation efficiency of the entire cellulase system [23]. In addition, compared with the optimal enzyme component ratio determined in some artificial cellulase cocktails, the suboptimal endoglucanase content in the native cellulase system of *T. reesei* may restrict catalytic efficiency [24, 25].

3 Strategies for Developing Engineered *T. reesei* Strains for Robust Cellulase Production

Developing efficient engineered *T. reesei* strains is crucial to achieve robust cellulase production. Some general strategies have been proposed to develop excellent engineered *T. reesei* strains with improved cellulase production and efficiency (Fig. 1). Strategies to improve cellulase production in engineered *T. reesei* strains

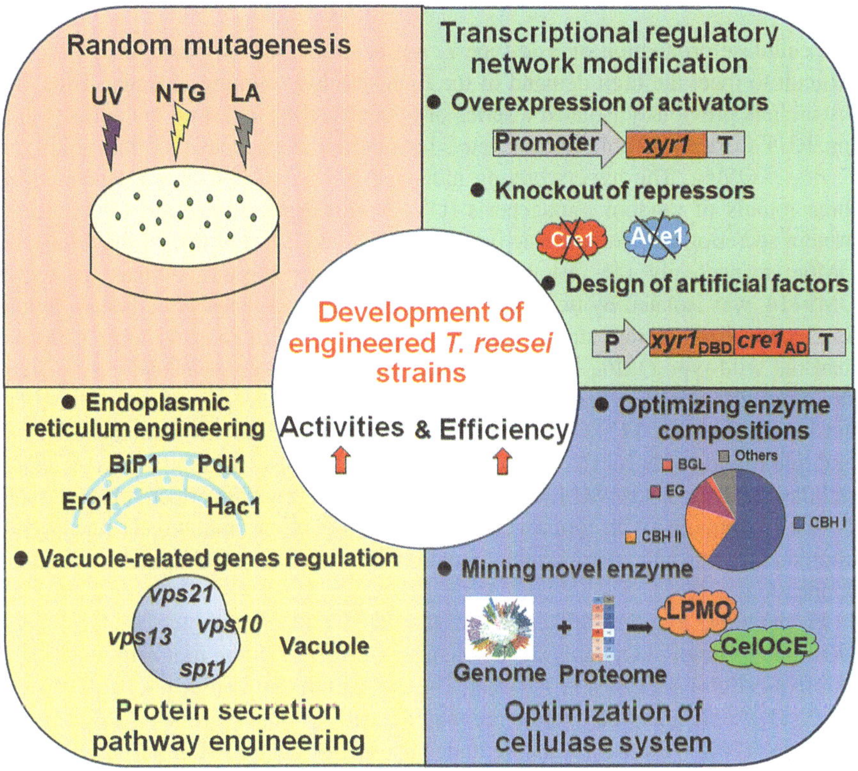

Fig. 1 Genetic engineering strategies to enhance extracellular cellulase activities and efficiency in *T. reesei*

mainly include mutation breeding, transcriptional regulatory network modification, and protein secretion pathway alterations. Meanwhile, strategies to enhance cellulase efficiency in engineered *T. reesei* strains mainly include optimizing enzyme compositions, exploring novel enzyme constituents, and heterologously expressing enzyme components.

3.1 Strategies to Improve Cellulase Activities in T. reesei

The cellulase production of wild-type *T. reesei* strain QM6a is quite low, which is difficult to meet the high demand of the lignocellulose refining industry. Thus, the researchers previously isolated a series of hypersecreting *T. reesei* mutants, including RUT-C30, QM9414, PC-3-7, etc., by random mutagenesis of the wild-type *T. reesei* QM6a. The hypersecreting mutant *T. reesei* RUT-C30 was obtained by three rounds of random mutagenesis (UV, *N*-nitrosoguanidine, and UV), and its protein secretion and cellulase activity were 2.7- and 2.8-fold that of the wild-type QM6a, respectively [4]. Another well-known hypersecreting mutant, *T. reesei* QM9414, was isolated by two rounds of irradiation mutagenesis in a linear particle accelerator. This mutant strain QM9414 produced two- to fourfold more cellulase than the wild-type QM6a [5]. Moreover, *T. reesei* QM9414 was used as a parent strain for further mutagenesis to obtain several hyper-cellulolytic mutant strains. The outstanding mutant PC-3-7, generated after one round of UV mutagenesis and multiple rounds of NTG (N-nitrosoguanidine) mutagenesis, exhibits twice as much cellulase activity as QM9414 [6]. Although mutation breeding effectively generates *T. reesei* mutants with improved cellulase production, mutational phenotypic unpredictability and detrimental mutation accumulation constrain its further application.

Fortunately, a series of key cellulase-related transcription factors have been identified through elucidating the mechanisms involved in the transcriptional regulation of cellulolytic genes in *T. reesei*. The most important transcriptional activators are Xyr1, Ace2, and Ace3, while the major transcriptional repressors include Cre1 and Ace1 [26]. With the advancing understanding of cellulase transcriptional regulation, rational modification of transcriptional regulatory networks has become the primary strategy for enhancing cellulase production in *T. reesei*. This mainly involves overexpression of transcription activators, knockout of transcription repressors, and design of artificial transcription factors. Overexpression of the major transcription activator Xyr1 has become a common strategy to increase cellulase production in *T. reesei* in recent years. For example, Wang et al. reported that the transcription activator Xyr1 was constitutively overexpressed under the control of the strong *T. reesei* pdc promoter, resulting in a significant increase in cellulase activity by 55% compared to the parent strain RUT-C30 [27]. Similarly, Zheng et al. used a copper-repressible promoter P*tcu1* to overexpress the transcription activator Xyr1 and achieved an 82.7% increase in cellulase activity compared to RUT-C30 under Avicel conditions [28]. Moreover, Shen et al. differentially overexpressed the *xyr1* gene under the control of three strong promoters (P*egl2*, P*cbh1*, and P*cdna1*), which increased the cellulase activity by 3.2, 0.9, and 0.8 times compared to the parental strain, respectively [29]. In addition, some studies also used the Xyr1 mutants to improve cellulase production in *T. reesei*. For instance, Derntl et al. reported that the A824V transition in Xyr1 causes elevated basal expression of the cellulase gene [30]. Lv et al. recently demonstrated that constitutive overexpression of a novel Xyr1^{R434L} mutant in *T. reesei*, driven by the pyruvate decarboxylase promoter (P*pdc1*), resulted in a 0.7-fold increase in cellulase activities under glucose conditions [31]. Moreover, Zhao et al. used the Xyr1^{A873Y} mutant predicted by AlphaFold2 to increase the cellulase production of *T. reesei* by 3.3-fold under glucose conditions [32]. The Xyr1^{A824V} mutant was also overexpressed in *T. reesei* using the strong promoter P*cdna1* by Zhao et al., which not only enabled *T. reesei* to produce an FPase activity of 1.9 U/mL under glucose conditions, but also increased its cellulase activity by 0.9 times under microcrystalline cellulose conditions [12].

Compared to single overexpression of Xyr1, co-overexpression of Xyr1 with the transcriptional activators Ace2 or Ace3 also significantly enhanced cellulase production in *T. reesei*. For example, Zhao et al. reported that strain h61, co-overexpressing Xyr1^{A824V} and Ace2, exhibited significantly enhanced cellulase production under both cellulose and glucose conditions compared with strain PX3 expressing only Xyr1^{A824V} [12]. Similarly, Arai et al. demonstrated that co-expression of transcription factors Xyr1^{V821F} and Ace3 resulted in enhanced cellulase productivity, which is 1.5-fold higher than with the single expression of the mutated XYR1^{V821F} [33]. Additionally, that cellulase productivity was 5.5-fold higher than with the single expression of Ace3 [33]. Combining Xyr1 overexpression with downregulation of the transcriptional repressor Ace1 is also an effective strategy to increase cellulase production in *T. reesei*. For instance, Wang et al. additionally downregulated Ace1 based on the strain constitutively overexpressing Xyr1, resulting in a further increase in cellulase production from 55% to 114% compared to the parent *T. reesei* RUT-C30 strain [27]. Likewise, Zhang et al. reported that overexpression of Xyr1^{A824V} at the transcriptional repressor Ace1 locus significantly increased cellulase production by 1-fold and 18-fold under Avicel and glucose conditions, respectively [34]. In addition, modifying the carbon catabolite repressor Cre1 also represents a classic strategy to improve the cellulase production of *T. reesei*. For example, Nakari-Setälä et al. reported that *T. reesei* strains with either deletion or truncation of *cre1* exhibited 1.8-fold and 2.4-fold higher total protein production, respectively, than the parent strain QM6a [35]. Similarly, Rassinger et al. constitutively overexpressed a truncated transcription repressor Cre1 (Cre1–96) in *T. reesei* RUT-C30, resulting in an approximately twofold increase in cellulase activity [36].

Moreover, artificial transcription factor design for enhancing cellulase production of *T. reesei* has been increasingly studied in recent years. For instance, Zhang et al. designed an artificial transcription activator by fusing the Cre1-binding domain to the Xyr1 effector and binding domains, resulting in a 12.75-fold increase in the cellulase production under glucose conditions compared to the parent strain *T. reesei*

RUT-C30 [37]. Similarly, Zhang et al. reported the enhanced transcriptional activators XYR1VP, ACE2VP, and ACE1VP constructed by linking the C-terminus of XYR1, ACE2, or ACE1 with an activation domain of VP16. The expression of XYR1VP resulted in 51% and 80% increase in xylanase I and xylanase II activities, respectively. The expression of ACE2VP and ACE1VP resulted in 52% and 80% increase in cellulase activity, respectively, under Avicel conditions [38]. Zhang et al. also demonstrated that expressing a minimal transcription factor DBDace2-VP16 (VP16 activation domain fused to Ace2 DNA-binding domain) in *T. reesei* enhanced cellulase activity 1.3-fold under Avicel conditions, while expressing DBDcre1-VP16 (VP16 activation domain fused to Cre1 DNA-binding domain) resulted in a 26.5-fold increase in cellulase activity under glucose conditions [38]. Moreover, Derntl et al. introduced two fusion transcription factors consisting of the DNA-binding domain of Xyr1 and the transactivation domain of either Ypr1 or Ypr2 into the *T. reesei*. The fusion of Xyr1 and Ypr1 formed a highly transactivating transcription factor, leading to high-level production of xylanases on glycerol [39]. Meng et al. reported an artificial transcription factor that replaced the activation domain of Gal4 of the artificial zinc-finger proteins (AZFPs) with the endogenous activation domain from Xyr1, thereby increasing cellulase production by 24.6% compared to the parent strain *T. reesei* QS1 [40]. Lv et al. recently reported an artificial transcription factor XT designed by fusing the DNA-binding domain of Xyr1 to the transactivation domain of Tmac1. Overexpression of this transcription factor XT resulted in 1.7- and 8.2-fold increases in pNPCase and xylanase activity, respectively, compared to the parent strain QM9414, while both activities were even 0.9- and 33.8-fold higher than the recombinant strain overexpressing Xyr1 [41].

In addition to the modification of transcriptional regulatory networks, engineering the protein secretion pathway is also an effective strategy to increase cellulase production in *T. reesei*. Several efforts have targeted engineering the endoplasmic reticulum (ER) secretion pathway to increase cellulase production in *T. reesei*. For example, Shen et al. overexpressed the key components of protein folding (Pdi1, Ero1, and BiP1) in *T. reesei*, thereby increasing cellulase activity by 42.3%, 21.8%, and 49.4%, respectively. Moreover, knockout of the endoplasmic reticulum mannosidase I encoding gene mns1 also resulted in a 15.9% increase in cellulase production in *T. reesei* [7]. Likewise, Gao et al. reported that overexpressing Bip1 (an ER-resident molecular chaperone) and Hac1 (a global transcription regulator controlling the unfolded protein response) resulted in increased cellulase production in *T. reesei* strains [42]. Sun et al. achieved downregulation of the relative transcription level of genes (*sed1*, *ymr1*, *pep4*, *doa10*, and *yps1*) involved in protein secretion through RNAi-mediated gene silencing, thereby increasing the endoglucanase activity from 26.0 to 46.2 IU/mL in *T. reesei* [43]. Furthermore, operating vacuole-related genes has been proven to be feasible in improving cellulase secretion in *T. reesei*. For example, Yan et al. found that knocking out vacuolar protein sorting factors such as vps10, vps13, and vps21 could alleviate vacuolar transport, thereby increasing cellulase activity by 1.28-fold, 2.45-fold, and 2.11-fold compared with the parental strain, respectively [8]. Yao et al. also recently reported that overexpression of a novel vacuolar serine protease gene *spt1* promoted a 1.72-fold increase in cellulase activity in *T. reesei* [44]. In addition, Wu et al. overexpressed SNC1, a key protein in vesicle fusion from Spitzenkörper to the plasma membrane in the last step of protein secretion, thereby increasing the production of glucose oxidase by 2.2-fold compared to the parental strain [45].

Comprehensive multi-strategy optimization is becoming an inevitable trend to improve the cellulase production in *T. reesei* engineered strains. Fonseca et al. have introduced six genetic modifications in RUT-C30, which included the constitutive expression of the mutated XYR1^{V821F}, the expression of the β-glucosidase CEL3A from *Talaromyces emersonii* and the invertase SUC1 from *Aspergillus niger*, and the deletion of the cellulase repressor ACE1 and the extracellular proteases SLP1 and PEP1. The engineered strain reached an extracellular protein titer of 80.6 g/L in fed-batch fermentation [11]. As another example, Zhao et al. introduced four genetic modifications in *T. reesei* QM9414 by combinatorial engineering of transcription factors and introduction of heterologous β-glucosidases, which included the constitutive expression of Xyr1^{A824V} and Ace2 and the heterologous expression of the BglA from *A. niger* and the BGL1 from *P. oxalicum*. The engineered strain reached a filter-paper enzyme activity of 60.4 U/mL in fed-batch fermentation [12].

3.2 *Strategies to Enhance Cellulase Efficiency in* T. reesei

The natural cellulase system of *T. reesei* is unbalanced, which severely hinders its efficiency in degrading lignocellulosic biomass. Therefore, early research on improving the efficiency of cellulase system mainly focused on changing the proportion of enzyme components through gene overexpression. For example, β-glucosidases account for less than 2% of the total proteins in the secretome of *T. reesei*, which makes β-glucosidases the rate-limiting enzyme in the cellulose hydrolysis process [19]. Zheng et al. reported that overexpression of the endogenous β-glucosidase gene *bgl1* in *T. reesei* resulted in a 27% increase in the cellulose conversion of its cellulase system [28]. Moreover, Gao et al. heterologously overexpressed the β-glucosidase gene *bglA* from *A. niger*, thereby increasing the β-glucosidase activity by 51.3-fold (up to 103.9 IU/mL) in *T. reesei*. This engineered cellulase system significantly increased the cellulose conversion from 41.7% to 96.6% (2.1 fold) [46]. Similarly, Zhao et al. heterologously overexpressed the BglA from *A. niger* and the BGL1 from *P. oxalicum*, thereby achieving 64.8 U/mL of β-glucosidase activity (106.8-fold higher than parent strain) in *T. reesei*. The cellulase system produced by this engineered strain has increased the hydrolysis efficiency on Avicel by 1.5 times [12]. In addition, the suboptimal content of endoglucanase (the core cellulase component) in *T. reesei* may also restrict the catalytic efficiency of the native cellulase system. Meng et al. overexpressed the endoglucanase gene *egl1* in *T. reesei* RUT-C30, thereby achieving a 132.7% increase in endoglucanase activity. This engineered cellulase system was used to hydrolyze the alkali-pretreated corn stover (APCS) and Jerusalem artichoke stalk (APJAS), increasing glucose yields by 21.5% and 34.6%, respectively

[47]. Co-expression of multiple-enzyme components is a more effective strategy to enhance cellulase efficiency in *T. reesei*. For example, Qian et al. firstly overexpressed the endoglucanase EG2 in *T. reesei* QP4 to obtain the engineered strain QPE36, whose cellulase system can increase the cellulose conversion from 33.4% to 73.4%. Furthermore, the β-glucosidase BGL1 was overexpressed in the engineered strain QPE36 to obtain the *T. reesei* QEB4 strain, whose cellulase system further increased the cellulose conversion to 94.2% [48]. Similarly, Fang et al. co-expressed the β-glucosidase gene *Aabgl1* and the cellobiohydrolase II gene *Trcbh2* to enhance the beta–exo synergism and exo–exo synergism in *T. reesei* cellulase system, which achieved an increase in cellulose conversion from 67.2% to 87.9% [49].

In addition to optimizing the ratio of core cellulase components, overexpression of hemicellulases and auxiliary enzymes is also beneficial to improve the efficiency of cellulase system in *T. reesei*. For example, Li et al. demonstrated that xylanase and cellulase have a significant synergistic effect in hydrolyzing sugarcane bagasse (SCB), replacing 20% of the cellulase by xylanase that enhanced the glucose yield by 6.6%, 8.8%, and 9.5% from SESB-, NaOH-, and H_2O_2-pretreated SCB, respectively [50]. Moreover, the degree of synergism between cellulase and xylanase had a positive relationship with xylan content in the lignocellulosic substrates. Likewise, Jia et al. found that the cellulose-induced protein 1 (Cip1) of *T. reesei* can enhance the enzymatic hydrolysis of pretreated lignocellulose, especially for the substrates with high lignin content. Similar to LPCS and CCR, the addition of Cip1 increased the hydrolysis efficiency of cellulase by 10.75% and 9.68%, respectively [51]. Eibinger et al. also reported that the expansin-like protein (Swollenin SWO1) of *T. reesei* can work synergistically with its cellulase when degrading untreated grass substrate, increasing the reducing sugar release by 1.5-fold [52].

The newly discovered lytic polysaccharide monooxygenase (LPMO) is also a key auxiliary enzyme to improve the efficiency of the cellulase system in *T. reesei*. For example, Harris et al. first heterologously expressed the GH61A protein from *T. aurantiacus* (later defined as LPMO) in *T. reesei*, thereby reducing the protein loading required to reach 91% cellulose conversion by 1.9-fold [53]. Similarly, Zhang et al. heterologously overexpressed TcAA9A (a newly identified AA9 LPMO from *Talaromyces cellulolyticus*) and TaAA9A (a well-studied AA9 LPMO from *Thermoascus aurantiacus*) in *T. reesei*, naming them TrTcAA9A and TrTaAA9A. The cellulase system of TrTcAA9A is more effective than the parent and TrTaAA9A in degrading Avicel and delignified corncob residue [54]. An external electron donor is required for LPMO to oxidatively cleave cellulose. However, when reducing agents such as ascorbic acid (Asc), gallic acid, or pyrogallol are used as the electron donor, LPMO will be inactivated by auto-oxidation. Thus, Adnan et al. heterologously expressed cellobiose dehydrogenase of *Phanerochaete chrysosporium* (*Pc*CDH) in *T. reesei*, which can be used as an effective electron donor to promote the synergistic effect of LPMO and cellulase to increase the degradation efficiency of cellulose [55]. Discovering new auxiliary enzymes is essential for improving the efficiency of the cellulase system in *T. reesei*. Santos et al. recently discovered a metalloenzyme CelOCE by mining the metagenomic "dark matter" of a microbial community specialized in lignocellulose

Table 1 Review of genetic engineering strategies for improving cellulase production and efficiency in *T. reesei*

Parent strain	Genetic engineering strategies	Improvement in cellulase activities/efficiency	References
T. reesei QM6a	Three rounds of random mutagenesis (UV, *N*-nitrosoguanidine, and UV)	2.8-fold higher cellulase activity	[4]
T. reesei QM6a	Two rounds of irradiation mutagenesis	Two- to four fold higher cellulase activity	[5]
T. reesei QM9414	One round of UV mutagenesis and multiple rounds of *N*-nitrosoguanidine mutagenesis	Twice as much cellulase activity	[6]
T. reesei RUT-C30	Overexpressing the transcription activator Xyr1 under the control of the strong *T. reesei* pdc promoter	55% increase in cellulase activity	[27]
T. reesei RUT-C30	Overexpressing the transcription activator Xyr1 under the control of a copper-repressible promoter P*tcu1*	82.7% increase in cellulase activity	[28]
T. reesei QEB4	Overexpressing the *xyr1* gene under the control of three strong promoters (P*egl2*, P*cbh1*, and P*cdna1*)	3.2-, 0.9-, and 0.8-fold higher cellulase activity	[29]
T. reesei QM9414 Δ*xyr1*	Overexpression of a novel $Xyr1^{R434L}$ mutant driven by the pyruvate decarboxylase promoter (P*pdc1*)	0.7-fold increase in cellulase activity under glucose conditions	[31]
T. reesei QMP	Overexpressing the $Xyr1^{A873Y}$ mutant under the control of the native promoter P*xyr1*	3.3-fold increase in cellulase activity under glucose conditions	[32]
T. reesei QMP	Overexpressing the $Xyr1^{A824V}$ mutant using the strong promoter P*cdna1*	0.9-fold increase in cellulase activity under microcrystalline cellulose conditions	[12]
T. reesei PC-3-7	Co-expression of transcription factors $Xyr1^{V821F}$ and Ace3	1.5-fold higher cellulase activity than the single expression of the mutated $XYR1^{V821F}$	[33]
T. reesei RUT-C30	Downregulated Ace1 based on the strain constitutively overexpressing Xyr1	Further increase in cellulase production from 55% to 114%	[27]
T. reesei QM53	Overexpression of $Xyr1^{A824V}$ at the transcriptional repressor Ace1 locus	1-fold and 18-fold increase in cellulase activity under Avicel and glucose conditions, respectively	[34]
T. reesei QM6a	Deletion or truncation of transcriptional repressors *cre1*	1.8-fold and 2.4-fold higher total protein production	[35]
T. reesei RUT-C30	Overexpressing a truncated transcription repressor Cre1 (Cre1–96)	Twofold increase in cellulase activity	[36]

(continued)

Table 1 (continued)

Parent strain	Genetic engineering strategies	Improvement in cellulase activities/efficiency	References
T. reesei RUT-C30	Overexpressing an artificial transcription activator by fusing the Cre1-binding domain to the Xyr1 effector and binding domains	12.75-fold increase in the cellulase production under glucose conditions	[37]
T. reesei RUT-C30	Overexpressing a minimal transcription factor DBDace2-VP16 (VP16 activation domain fused to Ace2 DNA-binding domain)	1.3-fold increase in cellulase activity under Avicel conditions	[38]
T. reesei QS1	Replacing the activation domain of AZFPs with the endogenous activation domain from Xyr1	24.6% increase in cellulase production	[40]
T. reesei QM9414	Overexpression of transcription factor XT (fusing the DNA-binding domain of Xyr1 to the transactivation domain of Tmac1)	1.7- and 8.2-fold increases in pNPCase and xylanase activity	[41]
T. reesei QP4	Overexpressing the key components of protein folding (Pdi1, Ero1, and BiP1) and knockout of the endoplasmic reticulum mannosidase I Mns1	Increasing cellulase activity by 42.3%, 21.8%, 49.4%, and 15.9%, respectively	[7]
T. reesei QM9414	Downregulation of the relative transcription level of genes (*sed1*, *ymr1*, *pep4*, *doa10*, and *yps1*) involved in protein secretion	Increasing the endoglucanase activity from 26.0 to 46.2 IU/mL	[43]
T. reesei RUT-C30	Knocking out vacuolar protein sorting factors vps10, vps13, and vps21	Increasing cellulase activity by 1.28-fold, 2.45-fold, and 2.11-fold, respectively	[8]
T. reesei QM53	Overexpression of a novel vacuolar serine protease gene *spt1*	1.72-fold increase in cellulase activity	[44]
T. reesei QM9414	Overexpression of the endogenous β-glucosidase gene *bgl1*	27% increase in cellulose conversion	[28]
T. reesei SDC11	Heterologous overexpression of the β-glucosidase gene *bglA* from *A. niger*	Increasing the cellulose conversion from 44.7% to 96.6% (2.1-fold)	[46]
T. reesei QMP	Heterologous overexpression of the BglA from *A. niger* and the BGL1 from *P. oxalicum*	Increasing the hydrolysis efficiency on Avicel by 1.5 times	[12]
T. reesei RUT-C30	Overexpressing the endoglucanase gene *egl1*	Improved hydrolysis efficiency by 21.5% and 34.6% for corn stover and Jerusalem artichoke stalk	[47]
T. reesei QP4	Co-overexpressing the endoglucanase EG2 and the β-glucosidase BGL1	Increasing the cellulose conversion from 33.4% to 94.2%	[48]
T. reesei RUT-C30	Co-expressing the β-glucosidase gene *Aabgl1* and the cellobiohydrolase II gene *Trcbh2*	Increasing the cellulose conversion from 67.2% to 87.9%	[49]

(continued)

Table 1 (continued)

Parent strain	Genetic engineering strategies	Improvement in cellulase activities/efficiency	References
/	Replacing 20% of the cellulase by xylanase	Increasing the glucose yield by 6.6%, 8.8%, and 9.5% from SESB-, NaOH-, and H_2O_2-pretreated SCB, respectively	[50]
/	Adding cellulose-induced protein 1 (Cip1) to the cellulase system of *T. reesei*	Increasing the hydrolysis efficiency for LPCS and CCR by 10.75% and 9.68%, respectively	[51]
/	Adding expansin-like protein (Swollenin SWO1) to the cellulase system of *T. reesei*	Increasing the hydrolysis efficiency for untreated grass by 1.5-fold	[52]
T. reesei SMA135	Heterologous expression of the GH61A protein from *T. aurantiacus*	Reducing the protein loading required to reach 91% cellulose conversion by 1.9-fold	[53]
T. reesei Br_TrR03	Expressing a novel metalloenzyme CelOCE and a thermostable AA9 LPMO from the fungus *Lentinus similis*, respectively	Increasing glucose release from pretreated sugarcane bagasse by 21% and 10%, respectively	[56]

degradation, which oxidatively cleaves cellulose. The secretome of an engineered strain of *T. reesei* expressing this metalloenzyme CelOCE increased glucose release from pretreated sugarcane bagasse by 21%, which surpassed the 10% increase of an engineered strain expressing a thermostable AA9 LPMO from the fungus *Lentinus similis* [56]. Finally, the effects of different genetic engineering strategies on improving the cellulase production and efficiency in *T. reesei* are summarized in Table 1.

4 Optimization of Fermentation Conditions for Economic Production of Cellulase

T. reesei is the predominant industrial strain for cellulase production, which synthesizes cellulase under induction conditions. Submerged fermentation (SmF) is the leading technology for commercial cellulase production because it allows for easier control of several factors such as feeding rate, pH, temperature, and dissolved oxygen (DO) in the fermentation process [57]. Thus, several studies on the high-level production of cellulase from *T. reesei* using the low-cost soluble inducers through submerged fermentation have been widely reported in recent years. In addition, the production of cellulase under non-inducing conditions by using genetically engineered *T. reesei* strains has also attracted attention. Meanwhile, rational control of the inducer feeding rate in fed-batch fermentation is crucial to increase the cellulase production of *T. reesei*.

4.1 Low-Cost Inducer Utilization for Cost-Effective Cellulase Production

The utilization of different types of inducers in fed-batch fermentation directly determines the cellulase production of *T. reesei*. It was reported that sophorose, lactose, cellobiose, and some β-oligosaccharides can be used as the soluble inducers for cellulase production in SmF [57]. Sophorose is the most effective inducer for cellulase production in *T. reesei* reported to date, with an induction capacity 2,500 times that of cellobiose and 200 times that of lactose [34]. However, the high cost of pure sophorose prevents its application in the industrial production of cellulase. Therefore, many studies have synthesized a low-cost inducer (the mixture of sophorose and glucose) through transglycosidation reactions, which effectively promote the high-level production of cellulase. For example, Li et al. reported that a low-cost inducer (the mixture of glucose and β-disaccharides) was produced through the transglycosylation reaction catalyzed by the commercial β-glucosidase of SUNSON. When it was used as an inducer in fed-batch fermentation of *T. reesei* RUT-C30, the cellulase activity of 90.3 IU/mL was obtained, which was more than tenfold of that achieved with lactose as inducer [9]. Similarly, Xia et al. also reported that a glucose-sophorose mixture (GSM) containing 7.08% sophorose was synthesized via a transglycosylation reaction catalyzed by β-glucosidase from *A. niger*. When it was used as both carbon source and inducer in fed-batch fermentation, the cellulase activity produced by *T. reesei* reached 95.23 IU/mL, which was 2.1-fold of that achieved by lactose [10]. Moreover, Zhang et al. also prepared a mixture of glucose and sophorose (MGS) by hydrolyzing stevioside with HCl. The cellulase activity induced by MGS was 1.64- and 5.26-fold higher than that of lactose and cellobiose, respectively [58]. Additionally, Huang et al. recently produced a mixture of sugars (ACSM) from glucose through an acid-catalyzed transglycosylation reaction, which contained unconverted glucose, β-disaccharides (sophorose, cellobiose, and gentiobiose), and oligosaccharides with a degree of polymerization ranging from 3 to 7. When the ACSM was used as an inducer for fermentation of *T. reesei* RUT-C30, the cellulase production was comparable to that induced by microcrystalline cellulose, while it was 62.1% higher than that induced by lactose [59].

Furthermore, some studies have also reported that the cellulase production under non-inducing conditions was successfully achieved by using the engineered *T. reesei* strains, which replaced the expensive inducer with a low-cost and common carbon source. For instance, Ellilä et al. introduced the invertase gene of *A. niger* into an engineered strain of *T. reesei*, allowing it to produce cellulase using sugarcane molasses as a carbon source [21]. Furthermore, Fonseca et al. engineered the *T. reesei* strain RUT-C30 using the same strategy, which achieved an extracellular protein titer of 80.6 g/L when the sugarcane molasses were used as the carbon source in fed-batch fermentation [11]. Additionally, Zhao et al. obtained an engineered strain b5 that can produce cellulase under glucose conditions by overexpressing the $Xyr1^{A824V}$ mutant. Via feeding lignocellulosic hydrolysates as a carbon source for fed-batch fermentation, the cellulase activity of the engineered strain reached 60.4 U/mL [12].

4.2 Rational Control of Inducer Feeding Rate to Increase Cellulase Production

Controlling the inducer feeding rate in fed-batch fermentation of *T. reesei* plays an important role in the production of cellulase. Both underfeeding and overfeeding can have a negative effect on microbial growth and enzyme production in fed-batch fermentation [60]. For example, Xiang et al. reported that a too slow feeding rate (0.3 g/L/h) of steam-exploded biomass in fed-batch fermentation of *T. reesei* resulted in lower cellulase activity [61]. That may be attributed to the lack of nutrition of *T. reesei* caused by the underfeeding of the inducer. In addition, the decrease in *T. reesei* cellulase activity due to overfeeding is more commonly observed in fed-batch fermentation. For instance, Rohr et al. reported that when lactose was used as the carbon source in fed-batch culture of *T. reesei* RUT-C30, the highest feed rate of 0.75 g/L/h, which was too high for carbon-limited conditions, resulted in low cellulase activity [62]. Likewise, Zhang et al. also reported that the high-speed continuous feeding of glucose in fed-batch fermentation of *T. reesei* b5 resulted in a gradual accumulation of glucose, which negatively affected the production of cellulase due to the glucose repression effect [63]. The reduction in cellulase activity caused by the high feeding rate of the inducer may be mainly due to the carbon catabolite repression (CCR). Thus, strategies for rationally controlling the inducer feeding rate in fed-batch fermentation of *T. reesei* have been reported in some studies. For example, Li et al. reported that the glucose in the bioreactor was controlled at a low concentration between 0.05 and 0.30 g/L during the fed-batch fermentation of *T. reesei*. This was achieved by adjusting the inducer feeding rate according to the glucose concentration in samples measured every 3 h [9]. Moreover, Zhang et al. recently developed a continuous-pulse feeding model, which maintains glucose concentration below 2 g/L in the fermentation broth by controlling the opening and closing times of the peristaltic pump within each minute. This feeding strategy resulted in a 73.0% increase in FPase activity (from 11.5 to 19.9 U/mL) compared to the continuous feeding strategy at a speed of 1.0 $g_{glucose}$/L/h [63].

5 Developing Cost-Effective Sugar Platforms

Developing cost-effective sugar platforms is essential for economic production of fermentable sugars in the lignocellulosic biorefining industry. This relies on establishing a high-solid saccharification system for lignocellulosic biomass to obtain high-concentration fermentable sugars. Many studies have reported that the fed-batch strategy has been used to establish the efficient high-solid saccharification systems for lignocellulosic biomass. Meanwhile, reducing the cellulase dosage required in the saccharification process is also crucial for lowering overall cellulase costs. Many types of additives have been shown to be effective in promoting enzymatic hydrolysis and thus reducing the cellulase dosage.

5.1 High-Solid Saccharification to Obtain High Concentration of Fermentable Sugars

The high-solid saccharification of lignocellulosic biomass can achieve high concentrations of fermentable sugars to meet requirements for large-scale production of bio-based fuels and chemicals [64]. However, high-solid saccharification also faces several challenges, such as the reduced amount of free water, increased viscosity, severe mass transfer limitation, and decreased sugar yield [65]. Fed-batch strategies have been widely reported to alleviate these challenges, which involves the continuous addition of substrate to maintain a reasonable solid loading throughout the saccharification process. The optimization of the initial solid loading, feed interval, and feed content of the substrate has been shown to significantly improve the glucose yield in high-solid saccharification. For example, Liu et al. determined that the optimal initial solid loading for high-solid saccharification of sugarcane bagasse was 15%. A higher initial solid loading of 18% caused a solid effect and resulted in a glucose conversion rate less than 80% [66]. The optimal initial solid loading determined by Nwamba et al. was 8%, which was screened based on factors such as liquefaction time, glucose yield, and glucose titer [67]. Higher solid loading increased the liquefaction time and reduced the glucose yield, whereas lower solid loading resulted in low glucose titer. In addition, a 10% initial solid loading has also been used in some studies for high-solid saccharification reaction [68, 69]. In addition to the optimal initial solid loading, determining the optimal feeding interval and feeding content of substrate is also crucial to improving the glucose yield in high-solid saccharification. For example, Cai et al. used an optimal feeding mode (10%–3%–3%–3%–3%–3%, every 5 h) to establish a high-solid saccharification system with 25% solid loading for corncobs, which resulted in an 84.4% glucose yield. This is significantly higher than the model 2 (10%–5%–5%–5%, every 8 h) and model 3 (9%–8%–8%, every 12 h), which achieved glucose yields of 66.1% and 64.9%, respectively [68]. This suggested that a "low amount and high frequency" feeding way was much better than a long feeding interval and high feeding amount in high-

Table 2 Review of high-solid saccharification of lignocellulosic biomass for producing high-concentration fermentable sugars

Biomass type	Cellulase type and dosage	Solid loading (%)	Fermentable sugar yield (g/L)	References
Corn stover	Celluclast 1.5 L, 15 mg/g glucan Novozyme 188, 5 mg/glucan	20	126.9	[71]
Corn stover	Cellic CTec2, 10 FPU/g DM	30	130.5	[72]
Corn stover	Cellic CTec2, 10 mg/g glucan	35	173.6	[73]
Corn stover	Cellic CTec3-HS, 10 mg/g glucan	40	255.0	[74]
Sugarcane bagasse	Cellic CTec2, 3 FPU/g DM	20	160.0	[70]
Sugarcane bagasse	Cellic CTec3, 4 FPU/g DM	22	171.0	[69]
Sugarcane bagasse	Cellic CTec, 26 FPU/g DM	27	145.0	[75]
Sugarcane bagasse	Cellic CTec2, 8.5 FPU/g DM	36	231.7	[66]
Corncob residues	Crude cellulase, 8 FPU/g DM	27	205.0	[77]
Poplar	Cellic CTec2, 40 FPU/g DM	40	250.8	[78]

solid saccharification. Conversely, Nwamba et al. reported that the feeding model with a more spaced feeding interval was better than the mode with a short feeding interval. This may be because the spaced short feeding frequency will give enough time for initial substrate hydrolysis [67]. Moreover, Xu et al. developed a high-solid saccharification system with 22% substrate content for sugarcane bagasse, which was performed with an initial substrate content of 10%; subsequently, substrate contents of 5%, 4%, and 3% were added to the saccharification system with a feeding interval of 4 h. This study confirmed that a shorter total feeding time (16 h) favors higher glucose yield, which may be due to the inactivation of cellulase with the prolonged feeding time [69]. Likewise, Mukasekuru et al. demonstrated that feeding the substrate at an earlier stage (18 h) was favorable, while delaying substrate feeding to 30 h resulted in lower hydrolysis yields [70].

Developing high-solid saccharification systems with elevated solid loading for lignocellulosic biomass is critical for producing high concentration of fermentable sugars. High-solid saccharification studies on corn stover with different solid loadings have been widely reported (Table 2). For example, Zabed et al. reported that the high-solid saccharification of corn stover with a 20% solid loading achieved the fermentable sugars of 126.9 g/L [71]. Lu et al. achieved a 30% solid loading in high-solid saccharification of corn stover, and the fermentable sugars of 130.5 g/L were produced [72]. Chen et al. performed high-solid saccharification on corn stover with a 35% solid loading, achieving a fermentable sugar concentration of 173.6 g/L [73]. Li et al. recently reported that the 40% solid loading for high-solid saccharification of corn stover was reached, which produced a fermentable sugar concentration of 255 g/L [74]. In addition, sugarcane bagasse is also a commonly used saccharification substrate to produce high-concentration fermentable sugars. For instance, Mukasekuru et al. developed a high-solid (20%) saccharification system for sugarcane bagasse, achieving a fermentable sugar titer of 160 g/L [70]. Xu et al. reported a high-solid saccharification system with a solid loading of 22% for sugarcane bagasse, achieving a glucose titer of 127 g/L and xylose titer of 44 g/L [69]. Gomes et al. reported a saccharification reaction with higher solid loading of 27%, which obtained a glucose titer of 145 g/L [75]. Furthermore, Liu et al. obtained more than 230 g/L of total fermentable sugars when saccharifying sugarcane bagasse at 36% solid loading [66]. However, too high solid loading may also have a negative impact on the saccharification reaction. For example, Gao et al. reported that the saccharification reaction of corn stover at 40% solid loading achieved only 67% glucan conversion [76]. Therefore, utilizing lignocellulosic biomass with higher glucan and xylan content may also be an effective strategy to enhance fermentable sugar concentration. For example, Huang et al. used the corncob residues (75.1% glucan, 16.7% xylan) for high-solid saccharification, which achieved the fermentable sugar of 205 g/L at only 27% solid loading [77]. Similarly, Ying et al. achieved a maximum glucose concentration of 250.8 g/L at 40% solid loading when the glucan-rich (85.8%) poplar was used as the saccharification substrate [78].

5.2 *Strategies for Reducing Cellulase Dosage During Saccharification*

The nonproductive adsorption of cellulase by lignin is the main reason for the high enzyme dosage in saccharification, which can be alleviated by the use of additives. In recent years, studies on reducing the enzyme dosage required for saccharification of lignocellulosic biomass by using some low-cost additives have been widely reported.

Among them, the nonionic surfactant Tween 80, a typical cheap additive, has been reported to significantly improve saccharification efficiency and thus reduce the cellulase dosage. For example, Zhang et al. reported that adding 150 mg/g DM of Tween 80 to the sugarcane bagasse saccharification reaction reduced the cellulase dosage [79]. Likewise, Ying et al. also reported that the addition of 1 g/L Tween 80 in the saccharification reaction of sugarcane bagasse saved 50% of the cellulase dosage [78]. Another low-cost nonionic surfactant, PEG, has also been reported to significantly reduce the cellulase dosage used in saccharification. For instance, Lou et al. reported that the addition of nonionic surfactant PEG4600 could reduce the inactivation of cellulase caused by shear force and air-liquid interface, especially at low cellulase concentrations. When 25 mg protein/L of cellulase was incubated at 200 rpm for 72 h, adding PEG4600 can increase the enzyme activity from 36.0% to 89.5% [80]. Recently, Hu et al. prepared a low-cost and high-efficiency additive, sulfomethylated tannic acids (STAs). Adding 0.04 g/g-substrate STA can increase the glucose yield from 60.6% to 97.9% at a low cellulase dosage of 5 FPU/g-glucan [81]. Moreover, combining additives may reduce enzyme dosage more effectively than using a single additive. For example, Mukasekuru et al. reported that some additive mixtures (40 mg/g substrate Tween 80 +10 mg/g substrate tea saponin +20 mg/g substrate BSA) can increase glucose yield by 30% in saccharification of sugarcane bagasse, thereby achieving an industrially relevant sugar yield with a low cellulase dosage (3 FPU/g substrate) [70]. Similarly, Xu et al. reported that an optimized additive mixture (whey protein 25 mg/g DM, sophorolipid 25 mL/g DM, Tween 80 40 mL/g DM, and calcium lignosulfonate 10 mg/g DM) could increase the glucose yield of sugarcane bagasse by 17%, ultimately achieving a glucose yield of 80% at a cellulase dosage of only 4 FPU/g dry biomass [69].

Recycling of cellulase is an effective strategy to reduce enzyme dosage in lignocellulose saccharification. Among them, recycling unhydrolyzed solids containing adsorbed cellulase after enzymatic hydrolysis has been used as a promising strategy to reduce cellulase dosage. For example, Yuan et al. reported that cellulase recycling was performed by five successive rounds of recycling of unhydrolyzed solids after 24-h enzymatic hydrolysis, which successfully achieved 40% and 50% reduction in cellulase dosage required for saccharification of dilute alkali-CS and acid-CS, respectively [82]. Likewise, Xin et al. demonstrated that an optimized solid residue recycling process (addition of 5 FPU/g DM cellulases and 2.5 mg/mL BSA) maximized bound enzyme utilization and suppressed the negative effects caused by solid residues, thereby achieving a 50% reduction in cellulase dosage without affecting glucose yield [83]. In addition, cellulase immobilization effectively enhanced enzymatic activity and stability, thus facilitating multiple recycling rounds. For example, Hosseini et al. immobilized cellulase on a magnetic nanocomposite via Fe_3O_4 entrapment into a cross-linked ionic liquid-epoxy polymer, retaining 60% enzymatic activity after six times of recycling [84]. Similarly, Qi et al. immobilized cellulase on a magnetic core-shell metal-organic framework material, UIO-66-NH_2. The immobilized cellulase demonstrated a high enzyme activity recovery of 78.4% and retained most of its hydrolysis ability after five consecutive cycles [85]. In addition, since removing solid residues from crude cellulase produced on-site results in enzyme loss, using crude fermentation broth instead of culture supernatant for lignocellulose saccharification has been reported to be an effective strategy to reduce cellulase dosage. For example, Zhao et al. recently reported that the saccharification of corn stover using the crude fermentation broth without removal of solids did not result in a decrease in sugar concentration, thereby saving the cellulase dosage [12].

6 Future Perspective

The genetic engineering strategies summarized in this chapter, including mutation breeding, transcriptional regulatory network modification, protein secretion pathway engineering, and cellulase system optimization, have significantly improved the activities and efficiency of cellulase in *T. reesei*. However, there are a few studies on combining multiple strategies to construct excellent engineered strains of *T. reesei*. Therefore, it is possible to generate super cellulase-producing strains by rationally combining the above genetic engineering strategies. Although the classic cellulase system of *T. reesei* has been studied for decades, novel enzyme components continue to be identified in recent years. Consequently, mining novel enzymes remains a viable strategy for enhancing cellulase efficiency. Moreover, protein engineering retains considerable potential for enhancing the catalytic efficiency and thermostability of cellulase. Reducing cellulase production costs is essential for commercializing lignocellulosic biorefining. This may rely on the development of novel low-cost inducers that can significantly improve the induction capacity of cellulase production in *T. reesei*. Additionally, further genetic engineering to enhance cellulase production of *T. reesei* under non-inducing conditions using common carbon sources also represents a promising strategy. Numerous studies demonstrate that strategies such as utilizing high-glucan substrates, optimizing fed-batch modes, supplementing additives, and recycling cellulase contribute to the development of cost-effective sugar platforms.

Nevertheless, selecting the optimal combination of above strategies from a cost perspective is necessary for the industrial production of fermentable sugars in lignocellulosic biorefining.

References

1. Patel A, Shah AR (2021) Integrated lignocellulosic biorefinery: gateway for production of second generation ethanol and value added products. J Bioresour Bioprod 6:108–128. https://doi.org/10.1016/j.jobab.2021.02.001
2. Bischof RH, Ramoni J, Seiboth B (2016) Cellulases and beyond: the first 70 years of the enzyme producer *Trichoderma reesei*. Microb Cell Factories 15:106. https://doi.org/10.1186/s12934-016-0507-6
3. Keshavarz B, Khalesi M (2016) *Trichoderma reesei*, a superior cellulase source for industrial applications. Biofuels 7:713–721. https://doi.org/10.1080/17597269.2026.1192444
4. Peterson R, Nevalainen H (2012) *Trichoderma reesei* RUT-C30 – thirty years of strain improvement. Microbiology 158:58–68. https://doi.org/10.1099/mic.0.054031-0
5. Vitikainen M, Arvas M, Pakula T et al (2010) Array comparative genomic hybridization analysis of *Trichoderma reesei* strains with enhanced cellulase production properties. BMC Genomics 11:441. https://doi.org/10.1186/1471-2164-11-441
6. Porciuncula JO, Furukawa T, Mori K et al (2013) Single nucleotide polymorphism analysis of a *Trichoderma reesei* hyper-cellulolytic mutant developed in Japan. Biosci Biotechnol Biochem 77:534–543. https://doi.org/10.1271/bbb.120794

7. Shen L, Gao J, Wang Y et al (2021) Engineering the endoplasmic reticulum secretory pathway in *Trichoderma reesei* for improved cellulase production. Enzym Microb Technol 152:109923. https://doi.org/10.1016/j.enzmictec.2021.109923
8. Yan S, Xu Y, Tao XM et al (2023) Alleviating vacuolar transport improves cellulase production in *Trichoderma reesei*. Appl Microbiol Biotechnol 107:2483–2499. https://doi.org/10.1007/s00253-023-12478-4
9. Li Y, Liu CG, Bai FW et al (2016) Overproduction of cellulase by *Trichoderma reesei* RUT C30 through batch-feeding of synthesized low-cost sugar mixture. Bioresour Technol 216:503–510. https://doi.org/10.1016/j.biortech.2016.05.108
10. Xia Y, Yang LR, Xia LM (2018) High-level production of a fungal β-glucosidase with application potentials in the cost-effective production of *Trichoderma reesei* cellulase. Process Biochem 70:55–60. https://doi.org/10.1016/j.procbio.2018.03.031
11. Fonseca LM, Parreiras LS, Murakami MT (2020) Rational engineering of the *Trichoderma reesei* RUT-C30 strain into an industrially relevant platform for cellulase production. Biotechnol Biofuels 13:93. https://doi.org/10.1186/s13068-020-01732-w
12. Zhao QQ, Zhang Z, Liu ZH et al (2024) A closed-loop strategy for on-site production of saccharolytic enzymes for lignocellulose biorefinery using internal lignocellulosic hydrolysates. Chem Eng J 480:148272. https://doi.org/10.1016/j.cej.2023.148272
13. Thapa S, Mishra J, Arora N et al (2020) Microbial cellulolytic enzymes: diversity and biotechnology with reference to lignocellulosic biomass degradation. Rev Environ Sci Biotechnol 19: 621–648. https://doi.org/10.1007/s11157-020-09536-y
14. Contreras F, Pramanik S, Rozhkova AM et al (2020) Engineering robust cellulases for tailored lignocellulosic degradation cocktails. Int J Mol Sci 21:1589. https://doi.org/10.3390/ijms21051589
15. Fang H, Li C, Zhao J et al (2021) Biotechnological advances and trends in engineering *Trichoderma reesei* towards cellulase hyperproducer. Biotechnol Bioprocess Eng 26:517–528. https://doi.org/10.1007/s12257-020-0243-y
16. Srivastava N, Rathour R, Jha S et al (2019) Microbial beta glucosidase enzymes: recent advances in biomass conversation for biofuels application. Biomolecules 9:220. https://doi.org/10.3390/biom9060220
17. Zhang R (2020) Functional characterization of cellulose-degrading AA9 lytic polysaccharide monooxygenases and their potential exploitation. Appl Microbiol Biotechnol 104:3229–3243. https://doi.org/10.1007/s00253-020-10467-5
18. Singhania RR, Dixit P, Patel AK et al (2021) Role and significance of lytic polysaccharide monooxygenases (LPMOs) in lignocellulose deconstruction. Bioresour Technol 335:125261. https://doi.org/10.1016/j.biortech.2021.125261
19. Adsul M, Sandhu SK, Singhania RR et al (2020) Designing a cellulolytic enzyme cocktail for the efficient and economical conversion of lignocellulosic biomass to biofuels. Enzym Microb Technol 133:109442. https://doi.org/10.1016/j.enzmictec.2019.109442
20. Gomes D, Cunha J, Zanuso E et al (2021) Strategies towards reduction of cellulases consumption: debottlenecking the economics of lignocellulosics valorization processes. Polysaccharides 2:287–310. https://doi.org/10.3390/polysaccharides2020020
21. Ellilä S, Fonseca L, Uchima C et al (2017) Development of a low-cost cellulase production process using *Trichoderma reesei* for Brazilian biorefineries. Biotechnol Biofuels 10:30. https://doi.org/10.1186/s13068-017-0717-0
22. Merino ST, Cherry J (2007) Progress and challenges in enzyme development for biomass utilization. In: Olsson L (ed) Biofuels. Advances in biochemical engineering/biotechnology, vol 108. Springer, pp 1–40. https://doi.org/10.1007/10_2007_066
23. Li N, Li QW, Ge F et al (2025) Immobilization of β-glucosidase and β-xylosidase on inorganic nanoparticles for glycosylated substances conversion. Int J Biol Macromol 292:139173. https://doi.org/10.1016/j.ijbiomac.2024.139173

24. Du J, Liang J, Gao X et al (2020) Optimization of an artificial cellulase cocktail for high-solids enzymatic hydrolysis of cellulosic materials with different pretreatment methods. Bioresour Technol 295:122272. https://doi.org/10.1016/j.biortech.2019.122272
25. Billard H, Faraj A, Ferreira NL et al (2012) Optimization of a synthetic mixture composed of major *Trichoderma reesei* enzymes for the hydrolysis of steam-exploded wheat straw. Biotechnol Biofuels 5:9. https://doi.org/10.1186/1754-6834-5-9
26. Mattam AJ, Chaudhari YB, Velankar HR (2022) Factors regulating cellulolytic gene expression in filamentous fungi: an overview. Microb Cell Factories 21:44. https://doi.org/10.1186/s12934-022-01764-x
27. Wang S, Liu G, Wang J et al (2013) Enhancing cellulase production in *Trichoderma reesei* RUT C30 through combined manipulation of activating and repressing genes. J Ind Microbiol Biotechnol 40:633–641. https://doi.org/10.1007/s10295-013-1253-y
28. Zheng F, Yang R, Cao Y et al (2020) Engineering *Trichoderma reesei* for hyperproduction of cellulases on glucose to efficiently saccharify pretreated corncobs. J Agric Food Chem 68: 12671–12682. https://doi.org/10.1021/acs.jafc.0c04663
29. Shen L, Yan A, Wang Y et al (2022) Tailoring the expression of Xyr1 leads to efficient production of lignocellulolytic enzymes in *Trichoderma reesei* for improved saccharification of corncob residues. Biotechnol Biofuels 15:142. https://doi.org/10.1186/s13068-022-02240-9
30. Derntl C, Gudynaite-Savitch L, Calixte S et al (2013) Mutation of the xylanase regulator 1 causes a glucose blind hydrolase expressing phenotype in industrially used *Trichoderma* strains. Biotechnol Biofuels 6:62. https://doi.org/10.1186/1754-6834-6-62
31. Lv D, Zhang W, Meng X et al (2023) Single mutation in transcriptional activator Xyr1 enhances cellulase and xylanase production in *Trichoderma reesei* on glucose. J Agric Food Chem 71: 11993–12003. https://doi.org/10.1021/acs.jafc.3c03466
32. Zhao Q, Yang Z, Xiao Z et al (2023) Structure-guided engineering of transcriptional activator XYR1 for inducer-free production of lignocellulolytic enzymes in *Trichoderma reesei*. Synth Syst Biotechnol 8:732–740. https://doi.org/10.1016/j.synbio.2023.11.005
33. Arai T, Ichinose S, Shibata N et al (2022) Inducer-free cellulase production system based on the constitutive expression of mutated XYR1 and ACE3 in the industrial fungus *Trichoderma reesei*. Sci Rep 12:19445. https://doi.org/10.1038/s41598-022-23815-4
34. Zhang J, Li K, Sun Y et al (2024) An efficient CRISPR/Cas9 genome editing system based on a multiple sgRNA processing platform in *Trichoderma reesei* for strain improvement and enzyme production. Biotechnol Biofuels 17:22. https://doi.org/10.1186/s13068-024-02468-7
35. Nakari-Setälä T, Paloheimo M, Kallio J et al (2009) Genetic modification of carbon catabolite repression in *Trichoderma reesei* for improved protein production. Appl Environ Microbiol 75: 4853–4860. https://doi.org/10.1128/AEM.00282-09
36. Rassinger A, Gacek-Matthews A, Strauss J et al (2018) Truncation of the transcriptional repressor protein Cre1 in *Trichoderma reesei* Rut-C30 turns it into an activator. Fungal Biol Biotechnol 5:15. https://doi.org/10.1186/s40694-018-0059-0
37. Zhang X, Li Y, Zhao X et al (2017) Constitutive cellulase production from glucose using the recombinant *Trichoderma reesei* strain overexpressing an artificial transcription activator. Bioresour Technol 223:317–322. https://doi.org/10.1016/j.biortech.2016.10.083
38. Zhang J, Zhang G, Wang W et al (2018) Enhanced cellulase production in *Trichoderma reesei* RUT C30 via constitution of minimal transcriptional activators. Microb Cell Factories 17:75. https://doi.org/10.1186/s12934-018-0926-7
39. Derntl C, Mach RL, Mach-Aigner AR (2019) Fusion transcription factors for strong, constitutive expression of cellulases and xylanases in *Trichoderma reesei*. Biotechnol Biofuels 12:231. https://doi.org/10.1186/s13068-019-1575-8
40. Meng QS, Zhang F, Wang W et al (2020) Engineering the effector domain of the artificial transcription factor to improve cellulase production by *Trichoderma reesei*. Front Bioeng Biotechnol 8:675. https://doi.org/10.3389/fbioe.2020.00675

41. Lv D, Zhang W, Meng X et al (2023b) A novel fusion transcription factor drives high cellulase and xylanase production on glucose in *Trichoderma reesei*. Bioresour Technol 370:128520. https://doi.org/10.1016/j.biortech.2022.128520
42. Gao F, Hao Z, Sun X et al (2018) A versatile system for fast screening and isolation of *Trichoderma reesei* cellulase hyperproducers based on DsRed and fluorescence-assisted cell sorting. Biotechnol Biofuels 11:261. https://doi.org/10.1186/s13068-018-1264-z
43. Sun X, Liang Y, Wang Y et al (2022) Simultaneous manipulation of multiple genes within a same regulatory stage for iterative evolution of *Trichoderma reesei*. Biotechnol Biofuels 15:26. https://doi.org/10.1186/s13068-022-02122-0
44. Yao C, Sun N, Gao W et al (2023) Overexpression of a novel vacuolar serine protease-encoding gene (spt1) to enhance cellulase production in *Trichoderma reesei*. Fermentation 9:191. https://doi.org/10.3390/fermentation9020191
45. Wu Y, Sun X, Xue X et al (2017) Overexpressing key component genes of the secretion pathway for enhanced secretion of *Aspergillus niger* glucose oxidase in *Trichoderma reesei*. Enzym Microb Technol 106:83–87. https://doi.org/10.1016/j.enzmictec.2017.07.007
46. Gao J, Qian Y, Wang Y et al (2017) Production of the versatile cellulase for cellulose bioconversion and cellulase inducer synthesis by genetic improvement of *Trichoderma reesei*. Biotechnol Biofuels 10:272. https://doi.org/10.1186/s13068-017-0963-1
47. Meng QS, Liu CG, Zhao XQ et al (2018) Engineering *Trichoderma reesei* Rut-C30 with the overexpression of egl1 at the ace1 locus to relieve repression on cellulase production and to adjust the ratio of cellulolytic enzymes for more efficient hydrolysis of lignocellulosic biomass. J Biotechnol 285:56–63. https://doi.org/10.1016/j.jbiotec.2018.09.001
48. Qian Y, Zhong L, Gao J et al (2017) Production of highly efficient cellulase mixtures by genetically exploiting the potentials of *Trichoderma reesei* endogenous cellulases for hydrolysis of corncob residues. Microb Cell Factories 16:207. https://doi.org/10.1186/s12934-017-0825-3
49. Fang H, Zhao R, Li C et al (2019) Simultaneous enhancement of the beta–exo synergism and exo–exo synergism in *Trichoderma reesei* cellulase to increase the cellulose degrading capability. Microb Cell Factories 18:9. https://doi.org/10.1186/s12934-019-1060-x
50. Li J, Zhou P, Liu H et al (2014) Synergism of cellulase, xylanase, and pectinase on hydrolyzing sugarcane bagasse resulting from different pretreatment technologies. Bioresour Technol 155: 258–265. https://doi.org/10.1016/j.biortech.2013.12.113
51. Jia H, Sun W, Li X et al (2021) Cellulose induced protein 1 (Cip1) from *Trichoderma reesei* enhances the enzymatic hydrolysis of pretreated lignocellulose. Microb Cell Factories 20:136. https://doi.org/10.1186/s12934-021-01625-z
52. Eibinger M, Sigl K, Sattelkow J et al (2016) Functional characterization of the native swollenin from *Trichoderma reesei*: study of its possible role as C1 factor of enzymatic lignocellulose conversion. Biotechnol Biofuels 9:178. https://doi.org/10.1186/s13068-016-0590-2
53. Harris PV, Welner D, McFarland KC et al (2010) Stimulation of lignocellulosic biomass hydrolysis by proteins of glycoside hydrolase family 61: structure and function of a large, enigmatic family. Biochemistry 49:3305–3316. https://doi.org/10.1021/bi100009p
54. Zhang R, Liu Y, Zhang Y et al (2019) Identification of a thermostable fungal lytic polysaccharide monooxygenase and evaluation of its effect on lignocellulosic degradation. Appl Microbiol Biotechnol 103:5739–5750. https://doi.org/10.1007/s00253-019-09928-3
55. Adnan M, Ma X, Xie Y et al (2023) Heterologously expressed cellobiose dehydrogenase acts as efficient electron-donor of lytic polysaccharide monooxygenase for cellulose degradation in *Trichoderma reesei*. Int J Mol Sci 24:17202. https://doi.org/10.3390/ijms242417202
56. Santos CA, Morais MAB, Mandelli F et al (2025) A metagenomic 'dark matter' enzyme catalyses oxidative cellulose conversion. Nature 639:1076–1083. https://doi.org/10.1038/s41586-024-08553-z
57. Meng Q, Abraham B, Hu J et al (2025) Cutting-edge advances in strain and process engineering for boosting cellulase production in *Trichoderma reesei*. Bioresour Technol 419:132015. https://doi.org/10.1016/j.biortech.2024.132015

58. Zhang P, Li Q, Chen Y et al (2022) Induction of cellulase production in *Trichoderma reesei* by a glucose–sophorose mixture as an inducer prepared using stevioside. RSC Adv 12:17392. https://doi.org/10.1039/D2RA01192A
59. Huang XY, Deng L, Zhang X et al (2025) Developing a soluble inducer for robust production of cellulases by *Trichoderma reesei* through submerged fermentation. Green Chem 27:2733–2742. https://doi.org/10.1039/D4GC06001F
60. Li YY, Song WY, Han XY et al (2022) Recent progress in key lignocellulosic enzymes: enzyme discovery, molecular modifications, production, and enzymatic biomass saccharification. Bioresour Technol 363:127986. https://doi.org/10.1016/j.biortech.2022.127986
61. Xiang J, Wang X, Sang T (2021) Cellulase production from *Trichoderma reesei* RUT C30 induced by continuous feeding of steam-exploded *Miscanthus lutarioriparius*. Ind Crop Prod 160:113129. https://doi.org/10.1016/j.indcrop.2020.113129
62. Rohr K, Gremm L, Geinitz B et al (2024) Optimizing microbioreactor cultivation strategies for *Trichoderma reesei*: from batch to fed-batch operations. Microb Cell Factories 23:112. https://doi.org/10.1186/s12934-024-02371-8
63. Zhang Z, Zhang J, Li H et al (2025) The combination of continuous-pulse feeding hydrolysates with soybean hulls induction during fed-batch fermentation of *Trichoderma reesei* b5 significantly elevated the cellulase production and its degradation ability on lignocellulosic biomass. Int J Biol Macromol 307:142244. https://doi.org/10.1016/j.ijbiomac.2025.142244
64. Arora R, Singh P, Sarangi PK et al (2023) A critical assessment on scalable technologies using high solids loadings in lignocellulose biorefinery: challenges and solutions. Crit Rev Biotechnol 44:135–218. https://doi.org/10.1080/07388551.2022.2151409
65. da Silva AS, Espinheira RP, Teixeira RSS et al (2020) Constraints and advances in high-solids enzymatic hydrolysis of lignocellulosic biomass: a critical review. Biotechnol Biofuels 13:58. https://doi.org/10.1186/s13068-020-01697-w
66. Liu Y, Xu J, Zhang Y et al (2015) Optimization of high solids fed-batch saccharification of sugarcane bagasse based on system viscosity changes. J Biotechnol 211:5–9. https://doi.org/10.1016/j.jbiotec.2015.06.422
67. Nwamba MC, Song G, Sun F et al (2021) Efficiency enhancement of a new cellulase cocktail at low enzyme loading for high solid digestion of alkali catalyzed atmospheric glycerol organosolvent pre-treated sugarcane bagasse. Bioresour Technol 338:125505. https://doi.org/10.1016/j.biortech.2021.125505
68. Cai X, Hu CH, Wang J et al (2021) Efficient high-solids enzymatic hydrolysis of corncobs by an acidic pretreatment and a fed-batch feeding mode. Bioresour Technol 326:124768. https://doi.org/10.1016/j.biortech.2021.124768
69. Xu C, Zhang J, Zhang Y et al (2019) Enhancement of high-solids enzymatic hydrolysis efficiency of alkali pretreated sugarcane bagasse at low cellulase dosage by fed-batch strategy based on optimized accessory enzymes and additives. Bioresour Technol 292:121993. https://doi.org/10.1016/j.biortech.2019.121993
70. Mukasekuru MR, Hu J, Zhao X et al (2018) Enhanced high-solids fed batch enzymatic hydrolysis of sugar cane bagasse with accessory enzymes and additives at low cellulase loading. ACS Sustain Chem Eng 6:12787–12796. https://doi.org/10.1021/acssuschemeng.8b01972
71. Zabed HM, Akter S, Dar MA et al (2023) Enhanced fermentable sugar production in lignocellulosic biorefinery by exploring a novel corn stover and configuring high-solid pretreatment conditions. Bioresour Technol 386:129498. https://doi.org/10.1016/j.biortech.2023.129498
72. Lu M, Li J, Han L et al (2020) High-solids enzymatic hydrolysis of ball-milled corn stover with reduced slurry viscosity and improved sugar yields. Biotechnol Biofuels 13:77. https://doi.org/10.1186/s13068-020-01717-9
73. Chen X, Yuan X, Chen S et al (2021) Densifying lignocellulosic biomass with alkaline chemicals (DLC) pretreatment unlocks highly fermentable sugars for bioethanol production from corn stover. Green Chem 23:4828–4839. https://doi.org/10.1039/D1GC01362A

74. Li H, Zhai R, Chen X et al (2024) Modeling and design of fed-batch strategies for achieving 255 g/L sugar concentration from high-solid enzymatic hydrolysis of pretreated corn stover. Chem Eng J 486:150268. https://doi.org/10.1016/j.cej.2024.150268
75. Gomes AC, Moysés DN, Santa Anna LMM et al (2018) Fed-batch strategies for saccharification of pilot-scale mild-acid and alkali pretreated sugarcane bagasse: effects of solid loading and surfactant addition. Ind Crop Prod 119:283–289. https://doi.org/10.1016/j.indcrop.2018.04.026
76. Gao W, Li Z, Liu T et al (2021) Production of high-concentration fermentable sugars from lignocellulosic biomass by using high solids fed-batch enzymatic hydrolysis. Biochem Eng J 176:108186. https://doi.org/10.1016/j.bej.2021.108186
77. Huang J, Li X, Zhao J et al (2024) Fed-batch strategy achieves the production of high concentration fermentable sugar solution and cellulosic ethanol from pretreated corn stover and corn cob. Int J Mol Sci 25:12306. https://doi.org/10.3390/ijms252212306
78. Ying WJ, Zhu JJ, Xu Y et al (2021) High solid loading enzymatic hydrolysis of acetic acid-peroxide/acetic acid pretreated poplar and cellulase recycling. Bioresour Technol 340:125624. https://doi.org/10.1016/j.biortech.2021.125624
79. Zhang HD, Fan MS, Li X et al (2018) Enhancing enzymatic hydrolysis of sugarcane bagasse by ferric chloride catalyzed organosolv pretreatment and Tween 80. Bioresour Technol 258:295–301. https://doi.org/10.1016/j.biortech.2018.03.004
80. Lou H, Zeng M, Hu Q et al (2018) Nonionic surfactants enhanced enzymatic hydrolysis of cellulose by reducing cellulase deactivation caused by shear force and air-liquid interface. Bioresour Technol 249:1–8. https://doi.org/10.1016/j.biortech.2017.07.066
81. Hu S, Zhang T, Jiang B et al (2023) Achieving high enzymatic hydrolysis sugar yield of sodium hydroxide-pretreated wheat straw with a low cellulase dosage by adding sulfomethylated tannic acid. Bioresour Technol 384:129276. https://doi.org/10.1016/j.biortech.2023.129276
82. Yuan Y, Zhai R, Li Y et al (2018) Developing fast enzyme recycling strategy through elucidating enzyme adsorption kinetics on alkali and acid pretreated corn stover. Biotechnol Biofuels 11:316. https://doi.org/10.1186/s13068-018-1315-5
83. Xin D, Yang M, Chen X et al (2020) Improving cellulase recycling efficiency by decreasing the inhibitory effect of unhydrolyzed solid on recycled corn stover saccharification. Renew Energy 145:215–221. https://doi.org/10.1016/j.renene.2019.06.029
84. Hosseini SH, Hosseini SA, Zohreh N et al (2018) Covalent immobilization of cellulase using magnetic poly(ionic liquid) support: improvement of the enzyme activity and stability. J Agric Food Chem 66:789–798. https://doi.org/10.1021/acs.jafc.7b03922
85. Qi B, Luo J, Wan Y (2018) Immobilization of cellulase on a core-shell structured metal organic framework composites: better inhibitors tolerance and easier recycling. Bioresour Technol 268: 577–582. https://doi.org/10.1016/j.biortech.2018.07.115

Adv Biochem Eng Biotechnol (2026) 196: 139–182
https://doi.org/10.1007/10_2026_319

Published online: 7 May 2026

Engineered Microbial Strains for Co-fermentation of C5 and C6 Sugars to Produce Cellulosic Ethanol

Ryo Nasuno, Cai-Yun Xie, Yoshiaki Kawahara, Yue-Qin Tang, and Tomohisa Hasunuma

Contents

R. Nasuno
Engineering Biology Research Center, Kobe University, Kobe, Japan

C.-Y. Xie and Y.-Q. Tang
Department of Environmental Science and Engineering, Sichuan University, Chengdu, China

Y. Kawahara
Graduate School of Science, Technology and Innovation, Kobe University, Kobe, Japan

T. Hasunuma (✉)
Engineering Biology Research Center, Kobe University, Kobe, Japan

Graduate School of Science, Technology and Innovation, Kobe University, Kobe, Japan
e-mail: hasunuma@port.kobe-u.ac.jp

Abstract Efficient fermentation of both hexose (C6) and pentose (C5) sugars derived from lignocellulosic biomass is essential for the sustainable production of cellulosic ethanol. However, conventional yeast *Saccharomyces cerevisiae* lacks native pathways for pentose utilization and is susceptible to various environmental stresses caused by hydrolysate-rich fermentation conditions. This chapter outlines the metabolic engineering strategies for introducing and optimizing C5 pathways alongside native C6 metabolism. It highlights advances in recombinant strain development, including synthetic pathway design, cofactor balancing, and transporter engineering. We also review approaches for consolidated bioprocessing that improve overall productivity by combining saccharification and fermentation in a single batch. These approaches include the secretion or cell surface display of cellulases and hemicellulases and optimizing their expression levels. Furthermore, this chapter introduces tolerance mechanisms against stresses derived from fermentation conditions including high temperature, osmolarity, and ethanol, as well as lignocellulose hydrolysate-derived inhibitors such as acetic acid and furfural. By connecting innovations in metabolic design and stress response, the chapter provides a comprehensive view of the current progress and future directions in developing yeast strains to co-ferment C5/C6 sugars under industrially practical conditions.

Graphical Abstract

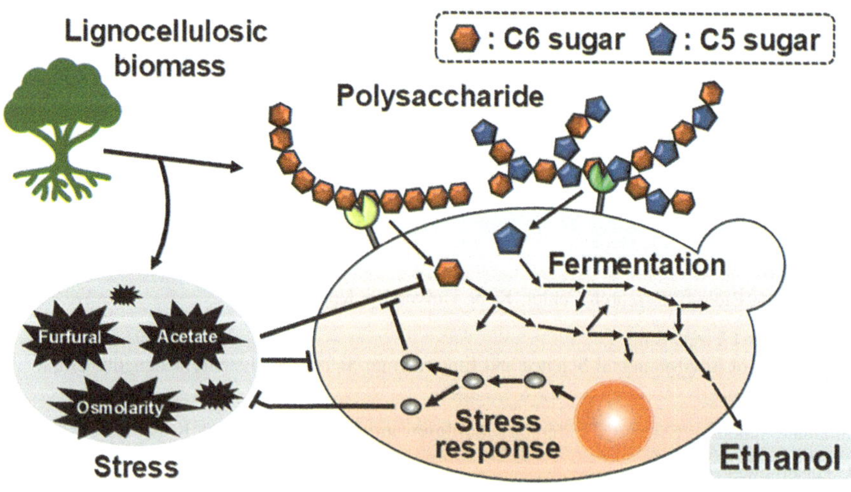

Keywords Cell surface engineering, Lignocellulose, Pentose utilization, Stress response, Yeast

1 Introduction

Climate change, decreasing fossil fuel resources, and increasing energy demands have promoted a global shift toward renewable energy alternatives. Among them, bioethanol has emerged as a leading candidate for sustainable liquid fuel. Not only is it compatible with current transportation infrastructure, but it also has the potential to reduce net carbon emissions, especially when derived from nonfood biomass sources.

First-generation bioethanol, which is produced from sugar- and starch-rich crops such as sugarcane and corn, is commercially established [1]. However, it has received criticism for its competition with food supplies, associated land use changes, and its limited effect on reducing greenhouse gas emissions. Consequently, research and development have increasingly turned to second-generation bioethanol, which is produced from lignocellulosic biomass. Lignocellulosic biomass is a plant-derived material, which is not used for food production, composed primarily of cellulose, hemicellulose, and lignin.

Lignocellulosic biomass is an abundant, diverse, and renewable feedstock, which is represented by agricultural residues such as corn stover and wheat straw, forestry by-products like wood chips and sawdust, and dedicated energy crops including miscanthus and switchgrass. However, its complex structure poses challenges for microbial fermentation [2]. Cellulose is a polymer of glucose (C6), while hemicellulose is heterogeneous, containing both C6 sugars and C5 sugars such as xylose and arabinose [3, 4]. These sugars must be efficiently co-fermented to enable economic production of bioethanol from these ingredients.

Several countries have announced incentives for cellulosic ethanol and offer policies that provide frameworks for scaling advanced bioethanol. However, commercial deployment has faced setbacks, including an inconsistent biomass supply, economic uncertainty, and complex processes [5]. From an industrial perspective, the commercial success of second-generation ethanol has been mixed. A few notable initiatives have highlighted both the technology's potential and the economic challenges. Despite their technological successes, some companies have struggled with financial sustainability due to operational costs and feedstock logistics [6]. Novel technologies and their integrated systems to overcome these challenges are essential for sustainable bioethanol production.

In order to make cellulose and hemicellulose accessible to enzymatic hydrolysis, pretreatment to break down the crystallinity of the cellulose is required. Pretreatment technologies include steam explosion, dilute acid hydrolysis, alkaline treatment, and ammonia fiber expansion [7]. Recently, lignocellulose has also been treated with ionic liquids and deep eutectic solvents. Each method disrupts the lignin barrier and increases degradability; however, there are trade-offs. Steam explosion and acid hydrolysis generate inhibitory compounds such as furfural and 5-hydroxymethylfurfural (5-HMF). Ionic liquid treatment requires costly recovery steps and may pose toxicity challenges, though it is effective. After pretreatment, lignocellulose is degraded into the smallest units of cellulose and hemicellulose, such as glucose and xylose, by exogenously supplied enzymes including β-glucosidase, β-xylanase, and endoglucanase [8]. Through pretreatment and enzymatic saccharification, a mixed sugar hydrolysate is produced. Efficient fermentation requires microorganisms that consume both hexoses like glucose and pentoses such as xylose and arabinose. *Saccharomyces cerevisiae*, the preferred industrial yeast for ethanol production, can naturally metabolize glucose and mannose but is incapable of utilizing xylose or arabinose. Alternative yeast strains like *Scheffersomyces stipitis*, *Spathaspora passalidarum*, *Scheffersomyces shehatae*, and *K. marxianus* can ferment C5 sugars like xylose [9–11]. However, these yeasts have several issues to be addressed, including low ethanol yield, slow growth, and/or difficulty with genetic engineering. Therefore, various approaches have been examined, including the introduction of a xylose utilization pathway into *S. cerevisiae* or enhancing the fermentation ability of xylose assimilation yeasts.

In addition to engineering yeast for mixed sugar utilization and increased fermentation efficiency, the increased stress resistance is another critical area of strain improvement. Fermentation processes generate various stress conditions, not only fermentation inhibitory compounds such as furfural, 5-HMF, and organic acids, but also high temperature, high osmolarity, lower pH, and ethanol itself [7].

These stress conditions lead to cellular dysfunctions, growth defect, and cell death.

Classical approaches enhanced a few gene traits, while contemporary strategies rely on global transcriptomics, proteomics, and genome-wide CRISPR screens to identify stress tolerance networks among multiple genes. Adaptive laboratory evolution also continues to be a complementary method for selecting complex phenotypes under industrially relevant conditions. Recent advances in CRISPR-Cas9 systems, multiplex genome editing, AI-guided design, and systems biology have transformed the landscape of yeast engineering [12]. Automated robotic systems enable rapid prototyping through high-throughput strain construction and screening platforms. Machine learning models, trained on omics datasets, predict beneficial mutations and promising targets for strain regulation. Genome-scale metabolic models enable simulation-driven optimization of metabolic pathways, reducing the need for time-consuming and labor-intensive trial-and-error experiments [13].

Furthermore, the integration of consolidated bioprocessing, where enzymatic hydrolysis and fermentation occur in a single batch, has gained attention. Yeasts expressing cellulases and hemicellulases are suitable microbes for consolidated bioprocessing, especially to utilize lignocellulosic biomass containing cellulose and hemicellulose. Cell surface display, secretion systems, and synthetic cellulosome design improve enzyme localization and activity [14]. However, achieving industrially relevant enzyme expression and maintaining fermentative capacity remain major challenges. Hybrid approaches using microbial consortia or chassis strains combined with functional modular pathways may offer future solutions.

This chapter outlines recent progress in yeast strain development for lignocellulosic ethanol production. Metabolic pathways enabling co-fermentation, construction of recombinant strains tailored for biomass utilization, and yeast stress tolerance mechanisms and their improvement for fermentation are explained and discussed. Future perspectives for advancing this promising biotechnology are also introduced.

2 Metabolic Pathway for Co-fermentation of C5 and C6 Sugars

Xylose and arabinose are the main pentose sugars in lignocellulosic hydrolysates, and their efficient utilization is of great significance for the sustainable production of biofuels and bio-based chemicals. However, *Saccharomyces cerevisiae* naturally lacks metabolic pathways for xylose and arabinose, resulting in low uptake and conversion efficiency. In recent years, researchers have significantly improved the ability of *S. cerevisiae* to utilize xylose and arabinose through strategies such as metabolic engineering, transporter engineering, and evolutionary engineering. This chapter systematically summarizes the research progress, current challenges, and future development directions in C5 sugar metabolic engineering in *S. cerevisiae.*

2.1 Native C5 Sugar-Metabolizing Microorganisms

2.1.1 Native Xylose-Metabolizing Species and Metabolic Pathways

In nature, various microorganisms are capable of utilizing xylose for the production of biofuels and biochemicals. For example, *Rhodosporidium toruloides*, *Lipomyces starkeyi*, *Cryptococcus curvatus*, and *Cutaneotrichosporon curvatus* can convert xylose into lipids [15–18], while *Clostridium tyrobutyricum* is able to transform xylose into butyric acid [19]. Moreover, several yeast species, including *Pachysolen tannophilus*, *Scheffersomyces (Candida) shehatae*, *Scheffersomyces stipitis*, *Kluyveromyces marxianus*, *Ogataea polymorpha*, and *Spathaspora passalidarum*, are notable for their relatively high ethanol yields from xylose [20, 21]. However, most native xylose-fermenting species suffer from slow xylose consumption rates, excessive by-product formation, strong glucose repression effects, and poor stress tolerance. In addition, due to their poorly characterized genetic backgrounds and limited genetic tools, these strains are rarely used directly in ethanol production. Instead, they are more commonly employed as sources of genes, enzymes, and/or sugar transporters for engineering industrial strains aimed at efficient bioethanol production from renewable biomass.

Currently, three distinct xylose metabolic pathways have been identified in microorganisms. In bacteria, xylose is directly isomerized to xylulose by xylose isomerase (XI). In most native xylose-metabolizing yeasts, xylose is converted to xylulose via a redox pathway composed of xylose reductase (XR) and xylitol dehydrogenase (XDH) (Fig. 1). Xylulose is then phosphorylated by xylulokinase (XK) to form xylulose-5-phosphate, which subsequently enters the glycolytic

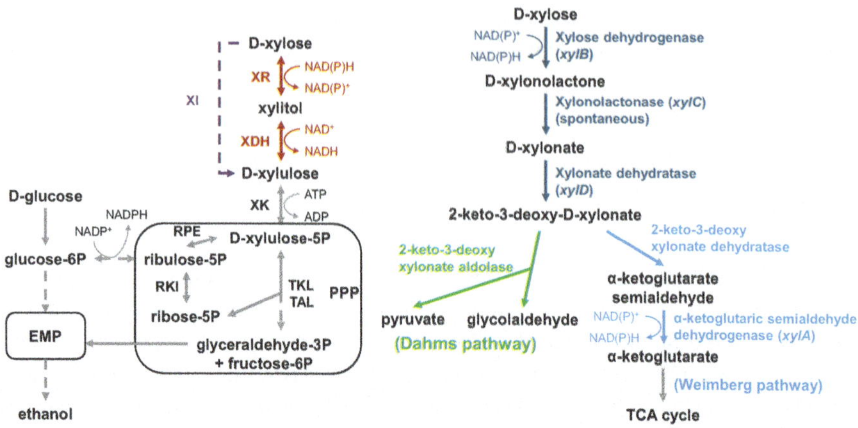

Fig. 1 Xylose metabolic pathways in microorganisms, including the xylose isomerase pathway, the redox pathway, and the xylose oxidative pathway. Abbreviations: *XI* xylose isomerase, *XR* xylose reductase, *XDH* xylitol dehydrogenase, *XK* xylulokinase

pathway through the non-oxidative pentose phosphate pathway (PPP). Additionally, a xylose oxidative pathway (XOP) exists in most archaea, as well as in some bacteria and fungi (Fig. 1). The XOP is initiated by the oxidation of D-xylose to D-xylonolactone, which is subsequently hydrolyzed into D-xylonic acid. D-Xylonic acid is then dehydrated to form 2-keto-3-deoxy-D-xylonic acid (KDX), which may either undergo further dehydration and oxidation to yield α-ketoglutarate (Weimberg pathway) or be cleaved via aldol reaction to produce pyruvate and glycolaldehyde (Dahms pathway) [22]. Compared to the XI and XR-XDH pathways, the XOP offers several advantages: directly converts xylose into key intermediates of the PPP and glycolysis without carbon or ATP loss; operates independently of other sugar metabolic pathways; and provides a platform for the synthesis of various novel molecules [22].

2.1.2 Natural L-Arabinose-Metabolizing Species and Metabolic Pathways

L-Arabinose can be converted into D-xylulose-5-phosphate through two major metabolic pathways and subsequently enters glycolysis via the non-oxidative PPP. In bacteria, L-arabinose is sequentially converted to L-ribulose, L-ribulose-5-phosphate, and finally D-xylulose-5-phosphate through the actions of three key enzymes: L-arabinose isomerase (AI), encoded by *araA*; L-ribulose kinase (RK), encoded by *araB*; and L-ribulose-5-phosphate 4-epimerase (R5PE), encoded by *araD* (Fig. 2) [23]. A major advantage of this pathway is its independence from redox cofactors. However, a key bottleneck is the limited activity of AI, which restricts the overall efficiency of L-arabinose metabolism.

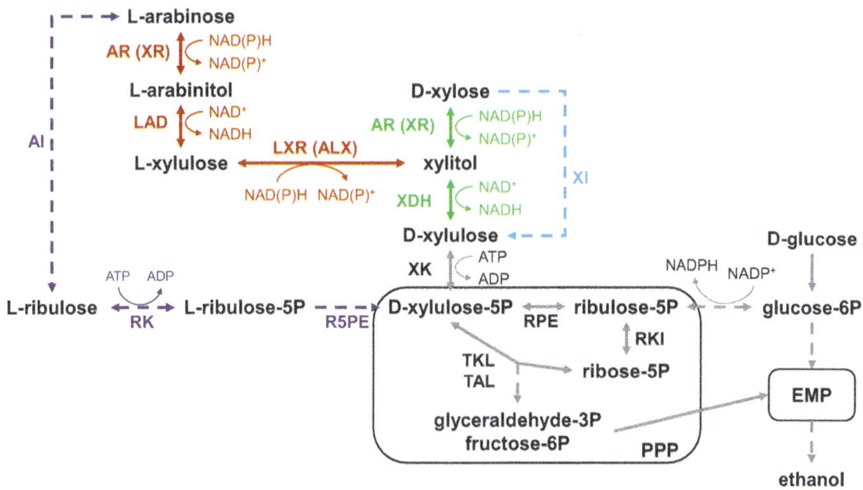

Fig. 2 Metabolic pathways for D-glucose, D-xylose, and L-arabinose in the engineered *S. cerevisiae* strains. Abbreviations: *XI* xylose isomerase, *XDH* xylitol dehydrogenase, *XK* xylulokinase, *AR(XR)* aldose reductase, *LAD* L-arabinitol-4-dehydrogenase, *LXR(ALX)* L-xylulose reductase, *AI* L-arabinose isomerase, *RK* L-ribulokinase, *R5PE* L-ribulose-5-P-4-epimerase

In nature, only a few bacterial species, such as *Sarcina ventriculi*, are capable of producing ethanol from L-arabinose, with a reported yield of 0.31 g/g. Most bacterial species, including *Lactobacillus plantarum*, *Burkholderia sacchari*, *Alicyclobacillus acidocaldarius*, *Bacillus coagulans*, and *Escherichia coli*, utilize L-arabinose primarily as a carbon and energy source.

Some filamentous fungi and yeasts utilize a redox-dependent pathway to convert L-arabinose into D-xylulose-5-phosphate. This pathway involves a series of enzymatic reactions catalyzed by NAD(P)H-specific aldose reductase (AR or XR), which reduces L-arabinose to L-arabinitol; NAD^+-specific L-arabinitol-4-dehydrogenase (LAD), which oxidizes L-arabinitol to L-xylulose; NAD(P)H-specific L-xylulose reductase (LXR or ALX), which reduces L-xylulose to xylitol; NAD^+-specific XDH, which oxidizes xylitol to D-xylulose; and XK, which phosphorylates xylulose to form xylulose-5-phosphate (Fig. 2) [24]. Notably, both AR and LXR typically prefer NADPH as a cofactor, whereas LAD and XDH strictly depend on NAD^+. This cofactor preference leads to redox imbalance, particularly under anaerobic conditions where NAD^+ regeneration is limited, resulting in significant accumulation of L-arabinitol.

In nature, most yeast species and some fungal species are capable of consuming L-arabinose, but rarely produce ethanol. Only a few species, including *Candida* sp. XF217, *Candida* sp. YB-2248, *Candida auringiensis*, *Candida succiphila*, *Candida arabinofermentans*, and *Ambrosiozyma monospora*, are able to ferment L-arabinose to ethanol, albeit at very low concentrations. To date, an NADH-dependent L-xylulose reductase (ALX) has been identified only in the yeast *A. monospora* [24].

2.2 *Construction of C5 Sugar-Fermenting Recombinant* S. cerevisiae *Strains*

2.2.1 Construction of Xylose-Fermenting Strains

Recombinant Strains Harboring the XR-XDH Pathway

Over the years, researchers have expressed the XR-XDH pathway in *S. cerevisiae*, enabling recombinant strains to utilize xylose for ethanol production. Most studies have involved the expression of Ss*XYL1* and Ss*XYL2* from *S. stipitis*, which encode XR and XDH, respectively. Compared to the XI pathway, recombinant strains expressing the XR-XDH pathway exhibit higher xylose consumption rates and ethanol yields. However, XR prefers NADPH, while XDH strictly requires NAD^+, leading to redox cofactor imbalance. This issue is particularly pronounced under anaerobic conditions, where the lack of oxygen hinders the regeneration of NAD^+. The resulting cofactor imbalance leads to the accumulation of xylitol, which decreases both ethanol production and yield.

Several strategies have been employed to improve cofactor balance in engineered *S. cerevisiae* strains expressing the XR-XDH pathway. On the one hand, efforts targeting the XR-XDH pathway have included protein engineering to enhance the NADH preference of *S. stipitis* XR (SsXR) [25] or to confer $NADP^+$ dependency on XDH. Additionally, replacing SsXR with an NADH-dependent XR from *S. passalidarum* has proven effective in reducing xylitol accumulation. Knockout of the endogenous NADPH-preferring aldose reductase gene *GRE3* in *S. cerevisiae* has also shown beneficial effects. Moreover, studies have indicated that a higher XDH/XR enzymatic activity ratio can help minimize xylitol formation.

On the other hand, strategies have focused on modulating intracellular cofactor biosynthesis. For example, deletion of NADPH-generating genes such as *ZWF1*, *GND1*, and *ALD6* has been shown to limit NADPH availability, thereby shifting XR toward greater utilization of NADH and resulting in improved ethanol yields. However, without altering XR's intrinsic affinity for NADH, limiting NADPH supply may reduce overall XR activity, which could subsequently decrease xylose consumption rates. Further studies have overexpressed NADH-dependent glutamate dehydrogenase (*GDH2*) or introduced the *noxE* gene from *Lactococcus lactis*, encoding an NADH oxidase, both of which promote NAD^+ regeneration and effectively reduce xylitol production. Notably, the introduction of *adhE* from *Escherichia coli*, which encodes an NADH-dependent acetylating acetaldehyde dehydrogenase, not only facilitates NAD^+ regeneration but also enhances the co-consumption of xylose and the inhibitory compound acetate [26].

Recombinant Strains Harboring the XI Pathway

Compared to the XR-XDH pathway, the XI pathway does not require cofactors and does not produce xylitol, resulting in higher ethanol yields. However, most bacterial-derived XIs exhibit very low activity when expressed in *S. cerevisiae*, making it difficult to support effective xylose metabolism. Several studies have attempted to improve its expression levels and activity in yeast through codon optimization [27], multi-copy expression [28], and directed evolution [29]. Given that xylitol inhibits XI, some research has focused on knocking out the endogenous *GRE3* gene in yeast to prevent xylitol production [30]. Other studies have explored cell surface display or secretion of XI, allowing the enzyme to function outside the cell, thus avoiding inhibition by xylitol and facilitating extracellular conversion of xylose to xylulose, which is more readily absorbed by the cells [29]. Providing suitable metal ions can also enhance the activity of XI.

Screening for high-activity XIs remains a primary focus in the engineering of xylose-metabolizing *S. cerevisiae* strains. To date, XI-encoding genes from various microbial sources have been successfully expressed in *S. cerevisiae*, including those from anaerobic fungi such as *Piromyces* sp. E2 and *Orpinomyces* sp.; bacteria such as *Prevotella ruminicola*, *Clostridium phytofermentans* [27], *Bacteroides stercoris*, *Bacterium* J10, *Hallella seregens*, *Acetanaerobacterium elongatum*, and *Streptobacillus canis*; and the protist *Reticulitermes speratus* [31]. With the advancement of big data mining technologies, future efforts should focus on broadly identifying high-activity XIs, comprehensively characterizing their functional diversity, and establishing a foundation for machine learning-guided optimization of XI activity.

Recombinant Strains Harboring the XOP

Currently, several studies have introduced partial or complete XOP into *S. cerevisiae* to produce various products. For instance, by introducing a xylose dehydrogenase gene, recombinant *S. cerevisiae* strains can utilize xylose to produce D-xylonic acid [32, 33]. Introducing the first two enzymes of the XOP allows recombinant strains to use 2-keto-3-deoxy-xylonate as a precursor for the production of 1,2,4-butanetriol [34] and 3,4-dihydroxybutyrate [35]. Furthermore, expressing the complete XOP in *S. cerevisiae* enables recombinant strains to convert xylose into glycolic acid and ethylene glycol [36, 37]. However, the accumulation of the key intermediate D-xylonic acid in the XOP often results in low yields of the desired products. Future research should focus on understanding the intrinsic interactions between the XOP and the yeast cellular metabolic network and optimizing critical steps within the pathway to enhance product yields.

2.2.2 Construction of L-Arabinose-Fermenting Strains

To enable *S. cerevisiae* to utilize L-arabinose, researchers have introduced metabolic pathways derived from bacteria or fungi into this yeast. By introducing SsXR and

SsXDH along with LAD and LXR from *Trichoderma reesei* or *Penicillium chrysogenum*, and overexpressing the endogenous XK, recombinant strains were able to grow slowly on L-arabinose and produce trace amounts of ethanol under anaerobic conditions [38, 39]. However, the accumulation of L-arabinitol caused by redox cofactor imbalance remains a major limiting factor. Replacing the NADPH-dependent LXR with the NADH-dependent ALX from *A. monospora* partially alleviates the cofactor imbalance and thereby improves ethanol yield [24, 40, 41].

Alternatively, introducing *araB* and *araD* from *E. coli* and *araA* from *Bacillus subtilis* into *S. cerevisiae* enables the conversion of L-arabinose to xylulose-5-phosphate. This pathway avoids the cofactor imbalance; however, the low activity of arabinose isomerase (encoded by *araA*) limits efficient L-arabinose conversion [42]. To overcome this limitation, researchers have employed several strategies, including identifying novel AraA variants with higher activity, increasing copy number of *araA*, and optimizing codon usage for improved expression [23, 43–46]. Additionally, fine-tuning the expression levels of *araA*, *araB*, and *araD* has also been shown to enhance L-arabinose metabolism [44, 46, 47].

2.2.3 Construction of D-Xylose and L-Arabinose Co-fermenting Strains

In practical applications, the engineered *S. cerevisiae* strains need to ferment both xylose and L-arabinose simultaneously. To achieve co-utilization of the two pentoses, some studies have employed synthetic microbial consortia by combining xylose-fermenting strains with L-arabinose-fermenting strains [43, 48]. More commonly, researchers have introduced both L-arabinose and D-xylose metabolic pathways into a single yeast strain. As shown in Table 1, when the XR-XDH pathway is combined with either the AR or the AI pathway, significant accumulation of by-products such as xylitol and arabinitol occurs. Simultaneous expression of the XR-XDH, AI, and XI pathways in *S. cerevisiae* has been shown to reduce the formation of these by-products [49]. When the XI and AI pathways are combined, deletion of the endogenous *GRE3* has been implemented to prevent xylitol-mediated inhibition of AI activity [44].

2.3 *Optimization of C5 Sugar Metabolic Pathways*

2.3.1 Increasing Flux Through Downstream Metabolic Pathways

Introducing heterologous C5 sugar metabolic pathways into *S. cerevisiae* alone is insufficient for achieving efficient C5 sugar fermentation. The flux through downstream pathways is also a critical limiting factor for C5 sugar utilization. In xylose-metabolizing strains, both the XI and XR-XDH pathways convert xylose to xylulose, which is then phosphorylated by xylulokinase (XK) to form xylulose-5-phosphate before entering the non-oxidative PPP. Consequently, XK and the non-oxidative

Table 1 Engineered arabinose-fermenting and xylose-arabinose co-fermenting *Saccharomyces cerevisiae*

Strains	Genetic modifications	Fermentation condition	q_{xylose} (g/g/h)	$q_{arabinose}$ (g/g/h)	Ethanol (g/L)	$q_{ethanol}$ (g/g/h)	$Y_{ethanol}$ (g/g)	$Y_{xylitol}$ (g/g)	$Y_{arabitol}$ (g/g)	Reference
H2561	**Plasmid**: *P. stipitis XYL1*, *XYL2*; *T. reesei LAD1*, *LXR1*; overexpression of *XKS1*	5% Arabinose, anaerobic	–	–	0.1	0.35×10^{-3}	–	–	–	[39]
IMS0002	**Integration plasmid**: *L. plantarum araB*; **multi-copy plasmid**: *L. plantarum araAD*; *Piromyces xylA*; overexpression of *XKS1*, *TAL1*, *TKL1*, *RPE1*, *RKI1*; evolved on arabinose	2% Arabinose, anaerobic	–	0.27	8.6	0.12	0.43	–	<0.001	[47]
		2% Glucose and 2% arabinose, anaerobic	–	0.19	16.8	0.09	0.42	–	<0.001	[47]
BWY1-4S	**Multi-copy plasmid**: *B. licheniformis araA* (codon-optimized); *E. coli araBD* (codon optimized); overexpression of *GAL2*	3% Arabinose, first in aerobic and then in anaerobic	–	0.11	11.7	0.036	0.39	–	0.11	[23]
BSW3AG	**Chromosomal integration**: *L. plantarum araABD* (codon optimized); **plasmid**: *L. plantarum araA* (codon optimized); overexpression of *TAL1*, *TKL1*, *RPE1*, *RKI1*, *GAL2*; evolved on arabinose	2% Arabinose, anaerobic	–	0.61	8.6	0.27	0.43	–	–	[46]

(continued)

Table 1 (continued)

Strains	Genetic modifications	Fermentation condition	q_{xylose} (g/g/h)	$q_{arabinose}$ (g/g/h)	Ethanol (g/L)	$q_{ethanol}$ (g/g/h)	$Y_{ethanol}$ (g/g)	$Y_{xylitol}$ (g/g)	$Y_{arabitol}$ (g/g)	Reference
TMB3063	**Chromosomal integration**: *P. stipitis XYL1*, *XYL2*; *E. coli araB*^{G361A}; *B. subtilis araA* (multi-copy), *E. coli araD* (multi-copy); overexpression of *XKS1*	2% Glucose, 2% xylose, and 2% arabinose, anaerobic	0.042	0.029	14.7	–	0.16	0.11	0.68	[44]
BWY02.XA	**Multi-copy plasmid**: *B. subtilis araA*; *E. coli araD*, *araB*^{G361A}; **integrative plasmid**: *P. stipitis XYL1*, *XYL2*; overexpression of *XKS1*	2% Glucose, 2% xylose, and 2% arabinose, anaerobic	0.041	0.035	10.4	–	0.26	0.33	1.14	[44]
		2% Glucose and 2% arabinose, anaerobic	–	0.03	9.0	–	0.38	–	0.77	[44]
TMB3664	**Integrative plasmid**: *P. stipitis XYL1*^{K270R}, *XYL2*; **multi-copy plasmid**: *T. reesei LAD1* (codon optimized), *A. monospora ALX1* (codon optimized); overexpression of *TAL1*, *TKL1*, *RKI1*, *RPE1*, *XKS1*; deletion of *GRE3*	2% Glucose, 2% xylose, and 2% arabinose, anaerobic	0.08	0.021	15.3	0.035	0.40	0.09	0.48	[41]
		2% Glucose and 2% arabinose, anaerobic	–	0.011	9.3	0.003	0.42	–	0.39	[41]

BSW4XA3	**Chromosomal integration**: *L. plantarum araABD* (codon optimized); *P. stipitis XYL1*, *XYL2*; **plasmid**: *B. rumen metagenome xylA*, *L. plantarum araA*; overexpression of *XKS1*, *TAL1*, *TKL1*, *RPE1*, *RKI1*; evolved on xylose and arabinose	2% Glucose, 2% xylose, and 2% arabinose, oxygen limited	0.055	0.059	11.95	0.076	0.35	0.09	0.19	[49]
CIBTS0735-14	**Chromosomal integration**: *Piromyces xylA*, 14 copies of *B. licheniformis araA* (codon optimized), *E. coli araBD* (codon optimized); overexpression of *XKS1*, *TAL1*, *TKL1*, *RPE1*, *RKI1*, *GAL2*, *A. thaliana STP2*; deletion of *GRE3*, *ARG1*	8% Glucose, 4% xylose, and 2% arabinose, anaerobic	0.16	0.06	57.69	0.14	0.41	–	–	[45]
		2% Arabinose, anaerobic	–	0.19	8.23	0.08	0.42	–	–	[45]
CIBTS1974	**Chromosomal integration**: *Piromyces xylA*, 1 copy of *B. licheniformis araA* (codon optimized), *E. coli araBD* (codon optimized); overexpression of *XKS1*, *TAL1*, *TKL1*, *RPE1*, *RKI1*, *GAL2*, *A. thaliana STP2*; deletion of *GRE3*, *ARG1*; evolved on arabinose	8% Glucose, 4% xylose, and 2% arabinose, anaerobic	0.17	0.07	57.07	0.14	0.41	–	–	[45]
		2% Arabinose, anaerobic	–	0.16	7.31	0.06	0.40	–	–	[45]

(continued)

Table 1 (continued)

Strains	Genetic modifications	Fermentation condition	q_{xylose} (g/g/h)	$q_{arabinose}$ (g/g/h)	Ethanol (g/L)	$q_{ethanol}$ (g/g/h)	$Y_{ethanol}$ (g/g)	$Y_{xylitol}$ (g/g)	$Y_{arabitol}$ (g/g)	Reference
IMS0010	**Plasmid**: *Piromyces xylA*, *L. plantarum araAD*; **integrative plasmid**: *L. plantarum araB*; overexpression of *TAL1*, *TKL1*, *RPE1*, *RKI1*, *XKS1*; deletion of *GRE3*; evolved on glucose, xylose, and arabinose	3% Glucose, 1.5% xylose, and 1.5% arabinose, anaerobic	0.23	0.23	25.8	0.23	0.43	–	–	[57]
424A (LNH-ST)/ pLXR$_{NAD}$-LAD	**Plasmid**: *A. monospora ALX1*, *T. reesei LAD1*; *P. stipitis XYL1*, *XYL2*; overexpression of *XKS1*	4.5% Arabinose, micro-aerobic	–	0.08	10	0.02	0.22	0.022	0.337	[40]
		2% Glucose, 2% xylose, 2% galactose, 2% arabinose, and 2% mannose, micro-aerobic	0.05	0.05	37	0.09	0.37	0.016	0.277	[40]

q_{xylose} (g/g/h): xylose consumption rate; $q_{arabinose}$ (g/g/h): arabinose consumption rate; $q_{ethanol}$ (g/g dry weight/h): ethanol production rate; $Y_{ethanol}$ (g/g pentose): ethanol yield; $Y_{xylitol}$ (g/g xylose): xylitol yield; $Y_{arabitol}$ (g/g arabinose): arabitol yield

PPP are considered rate-limiting steps in xylose metabolism. Overexpression of genes such as *XKS1*, *TAL1*, *TKL1*, *RPK1*, and *RPE1* has been shown to increase xylose metabolism rates [50]. However, since XK consumes ATP, excessive overexpression can be counterproductive and must be balanced at an optimal expression level. Moreover, the flux through the non-oxidative PPP may be influenced by the strain's genetic background, and overexpression of PPP enzymes is not universally beneficial across all strains.

Similarly, arabinose-metabolizing strains also rely on XK and the non-oxidative PPP to channel intermediates into glycolysis. In the arabinose-fermenting strain IMS0002, deletion of *TKL2* and *YGR043C* (encoding transaldolase) significantly impaired the strain's ability to grow on L-arabinose [47, 51], highlighting the importance of flux through the non-oxidative PPP. When both xylose and arabinose are present, the consumption rate of arabinose is typically lower than that of xylose (Table 1). Since both xylose and arabinose metabolic pathways share the intermediate xylulose-5-phosphate, competition for flux through the non-oxidative PPP may limit the co-metabolism of these sugars.

2.3.2 Optimizing Transport Systems

Xylose Transport

S. cerevisiae lacks specific xylose transporters and relies on hexose transporters such as Hxt1–Hxt5, Hxt7, and Hxt11 and the galactose permease Gal2 for xylose uptake. These transporters exhibit significantly higher affinity for glucose than for xylose, resulting in preferential glucose transport. Moreover, under conditions where only xylose is available, their low affinity for this pentose sugar limits efficient uptake at low concentrations.

To enhance xylose transport, expressing heterologous xylose-specific transporters has proven to be an effective strategy. For example, the expression of *T. reesei* Xyltr1p, *Lipomyces starkeyi* LST1_205437, *Candida intermedia* CiGxf1 and CiGxs1, and *S. stipitis* PsSut1, SsXut1, and SsXut3 in *S. cerevisiae* has been shown to improve xylose absorption and ethanol production [52, 53]. However, these transporters are all subject to glucose repression and generally exhibit low xylose transport rates. Although *Arabidopsis thaliana* AtSWEET7 has been reported to transport both glucose and xylose simultaneously with no inhibition, its transport rates for both sugars are lower than those of LST1_205437 [53].

In contrast, engineering endogenous Hxt transporters in *S. cerevisiae* has yielded more promising results. Through rational design or evolutionary engineering approaches, various mutations have been introduced into native sugar transporters to enhance xylose transport capacity or reduce sensitivity to glucose inhibition. For instance, the N366T and N366M mutations in Hxt11 shift its specificity toward xylose over glucose [54], while Hxt7-N370F and Gal2-N376F exhibit exclusive xylose transport activity. The Hxt36-N367A variant demonstrates enhanced xylose transport rate and affinity, enabling efficient co-consumption of D-glucose and D-xylose. In addition, overexpression of *HXT1*, *HXT2*, or *HXT7* has also been shown to increase the xylose metabolism rate.

L-Arabinose Transport

Due to the lack of efficient L-arabinose transporters in *S. cerevisiae*, L-arabinose uptake represents a significant rate-limiting step. In *S. cerevisiae*, L-arabinose is primarily transported into cells via Gal2, which exhibits higher affinity for glucose than for L-arabinose and is therefore subject to competitive inhibition by glucose [47]. The presence of glucose or low extracellular L-arabinose concentrations can severely restrict arabinose uptake.

To address this challenge, researchers have employed multiple strategies to optimize the L-arabinose transport system: enhancing *GAL2* expression using strong constitutive promoters has been shown to improve L-arabinose uptake efficiency [55]; site-directed mutations (e.g., F85S and N376T) in Gal2 have been introduced to increase its affinity and transport activity toward L-arabinose while simultaneously reducing its glucose affinity; and screening for high-affinity L-arabinose transporters from heterologous sources has led to the identification of promising candidates such as *Arabidopsis thaliana* AtStp2, *S. stipitis* SsAraT, *P. chrysogenum* PcAraT, and *A. monospora* AmLAT1/2. Notably, PcAraT functions as a high-specificity, high-affinity L-arabinose/proton symporter with no detectable transport activity toward xylose or glucose, making it insensitive to glucose repression [56]. Despite these advances, most identified transporters remain susceptible to glucose inhibition.

2.3.3 Global Regulation of C5 Sugar Metabolism

Evolutionary Engineering

Adaptive laboratory evolution (ALE) is a widely and effectively used strategy for improving C5 sugar utilization. By subjecting engineered *S. cerevisiae* strains to continuous cultivation on xylose as the sole carbon source, variants with enhanced growth rates can be enriched. Selected mutant strains exhibit significantly improved xylose fermentation capacity, as well as enhanced co-fermentation performance of glucose and xylose [25, 29]. ALE conducted under various carbon conditions, including L-arabinose as the sole carbon source, xylose plus arabinose, or glucose, xylose, and arabinose in combination, has resulted in evolved strains exhibiting superior arabinose fermentation performance (Table 1). For example, strain BSW3AG, which expresses *Lactobacillus plantarum araABD* and was evolved on arabinose as the sole carbon source, exhibits the highest reported arabinose consumption rate (0.61 g/g/h) and ethanol yield (0.43 g/g arabinose) [46]. Another strain, IMS0010, expressing *Piromyces xylA* and *L. plantarum araABD*, and evolved on a mixed carbon source of glucose, xylose, and arabinose, achieves the highest xylose consumption rate (0.23 g/g/h), arabinose consumption rate (0.23 g/g/h), and ethanol yield (0.43 g/g pentose) [57].

However, adapted strains often display genetic and metabolic instability during long-term fermentation, which may be exacerbated under industrial conditions, thereby compromising process robustness. To obtain genetically stable xylose-fermenting strains, reverse metabolic engineering approaches have been employed. Omics-based approaches have been used to reveal the molecular basis of improved xylose fermentation in evolved strains, facilitating the identification of key genes and regulatory factors as targets for further metabolic engineering.

To date, reverse metabolic engineering has confirmed that deletion of several genes enhances xylose metabolism. These include the phosphatase gene *PHO13*; RAS/PKA pathway-related genes *BCY1* and *IRA2*; mitogen-activated protein kinase (MAPK) signaling-related genes *HOG1*, *KSS1*, and *SMK1*; mitochondrial iron-sulfur cluster biogenesis gene *ISU1*; thiamine biosynthesis-related transcription factor *THI2*; and glucose repression-related regulators such as *MIG1*, *SNF1*, and *HAP4* [58]. In arabinose-fermenting strains, reducing intracellular cAMP levels has been shown to effectively alleviate glucose repression. Although fewer studies have focused on arabinose fermentation compared to xylose, it is likely that some of the genetic modifications beneficial for xylose metabolism may also enhance arabinose utilization.

Regulation of Signal Pathway

The presence of D-glucose leads to the inactivation of catabolic pathways for other carbon sources, a phenomenon known as carbon catabolite repression (CCR). *S. cerevisiae* cells sense sugar through three main signaling networks: the Snf3p/Rgt2p glucose sensor system, the cAMP/PKA pathway, and the Snf1/Mig1p regulatory pathway. Additionally, the high-osmolarity glycerol (HOG) pathway, one of the four MAPK pathways in yeast, has also been implicated in sugar sensing to some extent.

The Rgt2/Snf3-Rgt1-Mth1/Std1 regulatory system senses extracellular glucose and controls the transcription of hexose transporter genes. However, due to limited sensitivity to xylose, Rgt2 and Snf3 fail to effectively activate downstream signaling in the presence of this pentose sugar. Deletion of *RGT1* alleviates repression on hexose transporter gene expression, thereby enhancing xylose consumption under both xylose-only and mixed glucose-xylose conditions [59].

The cAMP/PKA pathway senses glucose through the Gpr1-Gpa2 receptor complex and Ras1/2 proteins, regulating cellular growth, metabolism, and stress responses. Knockout of *PDE1*, *PDE2*, or *IRA2* has been shown to improve xylose utilization [59]. These deletions may lead to constitutive activation of PKA, thereby amplifying xylose signaling to levels similar to those induced by high glucose concentrations [58].

The Snf1 signaling pathway responds to glucose starvation by phosphorylating and regulating key transcription factors such as Mig1 and Rgt1. Studies have shown that *SNF1* is essential for anaerobic fermentation of both glucose and xylose, and deletion of *MIG1* results in reduced xylose metabolism [60, 61]. However, overexpression of a mutant form of Mig1 has also been reported to enhance xylose utilization [62], indicating that the Snf1/Mig1 regulatory axis plays a complex role in xylose metabolism. In addition, attenuation of phosphorylation signaling in the HOG-MAPK pathway reduces Hog1p activity, which in turn alleviates the inhibition of the PPP, promoting xylose metabolism [58].

In addition, ATP-dependent chromatin remodeling complexes (e.g., SWI/SNF, SWR1, and ISW1) play important roles in xylose utilization by regulating the expression of HXT transporter genes and carbon metabolism-related transcription factors. Studies have found that overexpression of SWR1 and ISW1, two global chromatin remodelers, significantly enhances xylose utilization efficiency in the engineered *S. cerevisiae* strain P10-5 [63]. These chromatin remodelers interact with canonical signaling pathways, collectively participating in the fine-tuned regulatory network of xylose metabolism. Notably, different remodelers exhibit distinct mechanisms of transcriptional regulation at the genome-wide level, and their potential impact on xylose metabolism warrants further in-depth investigation.

2.4 New Products Derived from C5 Sugar Fermentation

As mentioned above, expression of the XOP in *S. cerevisiae* provides precursors for the synthesis of a variety of novel molecules. In addition, introduction of XR enables the conversion of xylose to xylitol. To enhance xylitol production, several engineering strategies have been employed, including the use of XR from different sources with varied cofactor preferences [64]; deletion of xylitol-consuming genes such as *XYL2*, *SOR1*, *SOR2*, and *XKS1* [65]; overexpression of *ZWF1* and *ACS1* to increase intracellular NAD(P)H levels [64]; and overexpression of *FPS1* to facilitate xylitol secretion [66].

Moreover, xylose- and arabinose-fermenting strains can utilize intermediates from the PPP and glycolysis as precursors for the synthesis of novel chemicals. For example, pyruvate serves as a precursor for lactate and isobutanol production; acetyl-CoA is used in the biosynthesis of 3-hydroxypropionic acid [67] and isoprenoid [68]; and sedoheptulose 7-phosphate (S7P) acts as a precursor for shinorine synthesis.

Compared to glucose metabolism, xylose metabolism exhibits slower glycolytic flux, but higher cytosolic acetyl-CoA synthesis flux [68], which can be advantageous for the production of these value-added compounds. However, the production of non-ethanol products in *S. cerevisiae* remains challenging due to the need to overcome the Crabtree effect, which favors ethanol formation even under aerobic conditions [69].

2.5 *Prospects and Challenges in Future Research*

Achieving co-fermenting of C5 and C6 sugars has long been a shared goal in the field. Advances in metabolic engineering, evolutionary engineering, and omics technologies have provided a solid foundation for strain development, identified effective genetic targets, and enabled a systems-level understanding of C5 sugar metabolism in yeast. Despite these advancements, several challenges remain, including low pentose metabolic efficiency, glucose repression, and poor stress tolerance. Targeted efforts to rewire signaling and regulatory networks, aimed at mitigating glucose repression and enabling precise carbon flux distribution, offer significant potential for improving C5/C6 sugar co-fermentation in engineered *S. cerevisiae* strains.

3 Development of Recombinant Yeast Strains Capable of Utilizing Lignocellulosic Biomass

Lignocellulose, which is a plant-derived biomass, is one of the most abundant sustainable carbon sources. Lignocellulose consists of three basic structural components: cellulose, hemicellulose, and lignin. Cellulose and hemicellulose, both of which are carbohydrates, account for up to 50% and 20% of lignocellulose, respectively [70]. Therefore, it is highly attractive to utilize these biomasses consisting of glucose and xylose, which are derived from cellulose and hemicellulose, respectively, for microbial production of various compounds through their co-fermentation. However, there are several issues to be overcome to use lignocellulose as carbon sources for bioproduction. In this section, the genetic engineering strategies to develop yeast strains capable of utilizing lignocellulose are introduced, including some case studies.

3.1 *Utilization of Cellulose by Genetically Modified Microorganisms*

3.1.1 Cellulose Utilization by Natural Microbes

The efficient degradation and/or utilization of cellulose has been most actively researched, because it is a major component in lignocellulosic biomass. Cellulose is a homopolymer of glucose linked through β-1,4-glycoside bond. Therefore, the degradation of cellulose requires multiple types of cellulases [71]. Many kinds of fungi and bacteria with cellulolytic activity have been reported. For example,

Trichoderma reesei, a well-known cellulolytic fungus, secretes various cellulolytic enzymes, including endoglucanase (EG), cellobiohydrolase (CBH), and β-glucosidase (BGL) [72, 73]. EG randomly hydrolyzes the internal glycoside bonds within the amorphous regions of cellulose, generating reducing and nonreducing ends. CBH recognizes these ends and hydrolyzes cellulose into cellobiose, a disaccharide consisting of two molecules of glucose. An endo–exo synergetic degradation by EG and CBH is crucial to convert insoluble cellulose chain to soluble cellobiose and cello-oligosaccharides [74, 75]. Finally, cellobiose is hydrolyzed to glucose by BGL. The anaerobic bacteria *Clostridium thermocellum*, *C. cellulovorans*, and *C. cellulolyticum* are reported to produce a functional enzyme complex called cellulosome on the cell surface [76–78]. Cellulosome arranges various types of cellulases closely on a scaffolding backbone, which enhances enzyme synergy, leading to efficient cellulose degradation.

3.1.2 Genetic Engineering of Yeast for Ethanol Fermentation Using Cellulose

For consolidated bioprocessing suitable for cellulosic bioethanol production, the yeast *S. cerevisiae* has gained much attention as a potential host for fermentation due to its high ethanol productivity, high ability to overproduce heterologous proteins, tolerance to harsh environmental conditions, and ease of genetic manipulation [79, 80]. However, *S. cerevisiae* is incapable of degrading cellulosic material by itself; thus, the heterologous expression of cellulase-encoding genes in this yeast is required.

Jeon et al. attempted to produce ethanol from β-glucan, a soluble polysaccharide consisting of glucose with mainly β-1,3-glycoside bond, using a recombinant yeast strain which secretes two cellulolytic enzymes, *C. cellulovorans* EG and *Saccharomycopsis fibuligera* BGL [81]. This yeast strain converted 20 g/L of β-glucan to 9.15 g/L ethanol within 50 h, which corresponds to 80.3% of the theoretical yield.

Due to the complicated pretreatment process of lignocellulosic biomass, the yeast strains for industrial applications are required to possess the high productivity of heterogeneous proteins and excellent robustness such as acid-, osmo-, and thermostolerance [82, 83]. Recently, Minnaar et al. isolated a new strain of the yeast *S. cerevisiae* with tolerance to several fermentation-associated and/or secretion-related stresses from nature and engineered it for cellulose degradation [84]. They introduced EG of *T. reesei*, CBH1 of *Talaromyces emersonii*, CBH2 of *Chrysosporium lucknowense*, and BGL of *Aspergillus aculeatus* with CRISPR/Cas9 tools. The resultant recombinant strain produced 5.5 g/L of ethanol from 20 g/L of Avicel without the addition of exogenous cellulases, which corresponds to approximately 40% of the theoretical yield.

3.1.3 Yeast Cell Surface Engineering for Efficient Utilization of Cellulose

An alternative strategy to confer cellulolytic activity to *S. cerevisiae* is the cell surface engineering to display cellulolytic enzymes on its cell surface (Fig. 3). While a lot of proteins synthesized intracellularly are secreted extracellularly through the endoplasmic reticulum (ER) transport pathway, some are immobilized on the cell surface via the interaction with cell surface components like the cell wall [85]. For example, Flo1, Sed1, Cwp1, Cwp2, Tip1, and Tir1/Srp1 are well-known cell wall-associated proteins that contain the glycosylphosphatidylinositol (GPI)-anchoring domain, whose hydrophobic region at its C-terminus plays a crucial role in their localization to the cell surface [86]. Proteins with GPI-anchoring domains (GPI-anchor proteins) are transported into the ER and then cleaved at the ω site in its GPI-anchoring domain [87]. The cleaved proteins are linked to the ethanolamine moiety of GPI in the ER membrane, via the newly formed amide bond. After transport to the plasma membrane via the Golgi apparatus, the glycan moiety of the GPI anchor is cleaved, resulting in the release of the GPI-anchor proteins from the cell membrane. The released proteins are then covalently linked to β-1,6-glucan, a component of the cell wall, thereby facilitating its stable incorporation into the cell wall. As an application of this GPI-anchor system for cell surface engineering, a number of attempts have been made to construct efficient whole-cell biocatalysts for ethanol production by fusing cellulase genes to GPI-anchor proteins, resulting in cellulolytic enzymes on the yeast cell surface [88]. Whole-cell biocatalysts for cellulosic material degradation have several advantages: co-localization of multiple

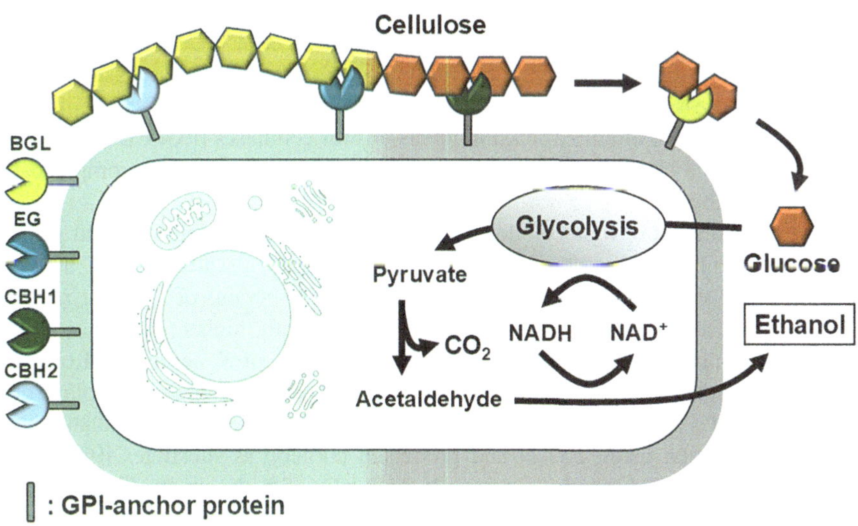

Fig. 3 Direct ethanol production from cellulose by displaying cellulolytic enzymes on the yeast cell surface. Abbreviations: *BGL* β-glucosidase, *EG* endoglucanase, *CBH* cellobiohydrolase, *GPI* glycosylphosphatidylinositol

cellulases on the same cell to enhance synergistic cellulolytic activity, easy separation of the catalyst from the product, and reusability of the recombinant strains, which reduces the cost of microbial production. Yamada et al. developed a diploid *S. cerevisiae* strain displaying three types of cellulases, EG and CBH2 of *T. reesei* and BGL of *A. aculeatus* to produce ethanol directly from rice straw [89]. In this study, rice straw was pretreated with hot water and then processed into a water-insoluble fraction with a particle size of 0.5 mm. As a result, the recombinant strain produced 7.5 g/L of ethanol from 100 g of pretreated rice straw within 72 h, achieving approximately 33% of the theoretical yield.

Despite ongoing efforts to produce ethanol through cell surface engineering, the yield remains insufficient for industrial applications. Therefore, various strategies have been explored to construct more efficient whole-cell biocatalysts [88, 90]. The modification of gene cassette components for immobilizing cellulases on the surface, such as signal peptide (SP) sequences or anchoring proteins, has been considered [91–94]. Inokuma et al. investigated SP sequences to increase the efficiency of extracellular protein secretion and then showed that the SP sequence derived from *S. cerevisiae* Sed1 increased BGL activity by more than 1.3-fold compared to the conventional SP sequences such as *Rhizopus oryzae* glucoamylase and *S. cerevisiae* α-mating pheromone [91]. Phienluphon et al. optimized the anchoring proteins from a number of putative fungal or yeast GPI-anchor proteins [92]. The activity of these putative GPI-anchor proteins to display green fluorescent protein and BGL was comprehensively evaluated. As a result, they successfully identified a novel efficient anchoring protein 6_Kl from *Kluyveromyces lactis*, whose C-terminal sequence was optimal for cell surface engineering. Additionally, Aga1p and Dan4p were also found to be suitable anchoring proteins for the display of relatively large proteins such as BGL [93]. Furthermore, it is crucial for efficient cellulose degradation to optimize the cellulase expression ratio [95]. Yamada et al. developed a cocktail δ-integration system incorporating evolutionary engineering and successfully optimized the expression balance of three cellulases (EG, CBH2, BGL) [95]. Similarly, Liu et al. tuned the ratio of expression levels of four cellulases (EG, CBH1, CBH2, BGL) using rice straw and improved ethanol yield by almost 60% compared to pre-optimization.

The pretreatment of lignocellulose is conducted at high temperatures and generates various inhibitory compounds [96, 97]. Thus, many researchers have tried to introduce cell surface engineering to not only laboratory strains of *S. cerevisiae* but also its industrial strains or other yeast species with higher stress tolerance [98, 99]. Cunha et al. isolated several industrial strains of *S. cerevisiae* from the harsh environment of a bioethanol production plant and attempted to display four types of cellulases – BGL, EG, and CBH1/2 – on the cell surface to evaluate their ethanol production using Avicel and pretreated corncob as substrates [98]. They successfully produced ethanol from both materials, with the most efficient strain yielding approximately 10 g of ethanol from 100 g of glucan derived from pretreated corncob within 24 h. As an example of applying cell surface engineering to other yeast species, Yanase et al. utilized the thermotolerant yeast *Kluyveromyces marxianus* and successfully constructed a strain displaying both BGL and EG on its surface [99]. This engineered strain efficiently hydrolyzed cellulosic material β-glucan at an elevated temperature of 48°C – close to the optimal temperature range for cellulolytic enzymes – achieving an ethanol yield of 0.47 g/g of consumed substrate, which corresponds to 92.2% of the theoretical yield. This result demonstrates the utility of engineering strategies based on cell surface display in thermotolerant yeasts for efficient bioethanol production.

3.2 Direct Ethanol Production from Hemicellulosic Materials

To achieve economic success in ethanol production from lignocellulosic biomass, it is essential to utilize hemicellulose as well as cellulose. Hemicellulose accounts for more than 20% of lignocellulosic biomass, and β-1,4-xylan is its major component, consisting of a linear backbone of β-1,4-linked D-xylosyl residues, partially substituted with acetyl, arabinosyl, and glucuronyl side chains [100]. Thus, efficient degradation of xylan as well as cellulose is required for effective bioethanol production. However, *S. cerevisiae* lacks xylan-degrading capability and is also unable to directly utilize xylose, the major component of xylan. Therefore, the introduction of exogenous xylose metabolic pathways, along with the heterologous expression of xylanolytic enzymes, has been attempted [101–103]. To date, two major xylose metabolic pathways are used for ethanol production [104, 105] (Fig. 4). One is the xylose reductase (XR)/xylitol dehydrogenase (XDH) pathway, which utilizes XR and XDH mainly derived from *Scheffersomyces stipites* to convert xylose into xylulose. Xylose isomerase (XI) from *Piromyces* or *Orpinomyces* is also used for the conversion of xylose to xylulose by its isomerase activity. In both cases, xylulose is further converted into xylulose-5-phosphate by endogenous xylulokinase and subsequently metabolized into ethanol through the pentose phosphate pathway and glycolysis.

Endowing *S. cerevisiae* with xylanolytic capability has been exploited by heterologous expression of xylanolytic enzymes from xylan-degrading organisms such as *Trichoderma* spp. and *Aspergillus* spp. [72]. Efficient xylan degradation requires two types of xylanolytic enzymes, endoxylanase and β-xylosidase; β-1,4-xylan is initially degraded into xylooligosaccharides by endoxylanase, followed by hydrolysis into xylose by β-xylosidase. Therefore, co-expression of these enzymes is required for direct xylan degradation and ethanol production. Mert et al. developed a recombinant *S. cerevisiae* strain co-expressing xylanase of *T. reesei* and xylosidase of *A. niger* [106]. By introducing the XI pathway, the engineered strain converted 20 g/L of xylose into ethanol with approximately 90% of the theoretical yield, growing on xylan as a sole carbon source.

Furthermore, efforts have been made to display xylan-degrading enzymes on the yeast cell surface. Cunha et al. engineered an industrial *S. cerevisiae* strain harboring both the XR/XDH and XI pathways to display *A. oryzae* β-xylosidase, *T. reesei* endoxylanase, and *A. aculeatus* BGL, accelerating the saccharification of glucooligosaccharides present in hemicellulose [107]. This recombinant strain

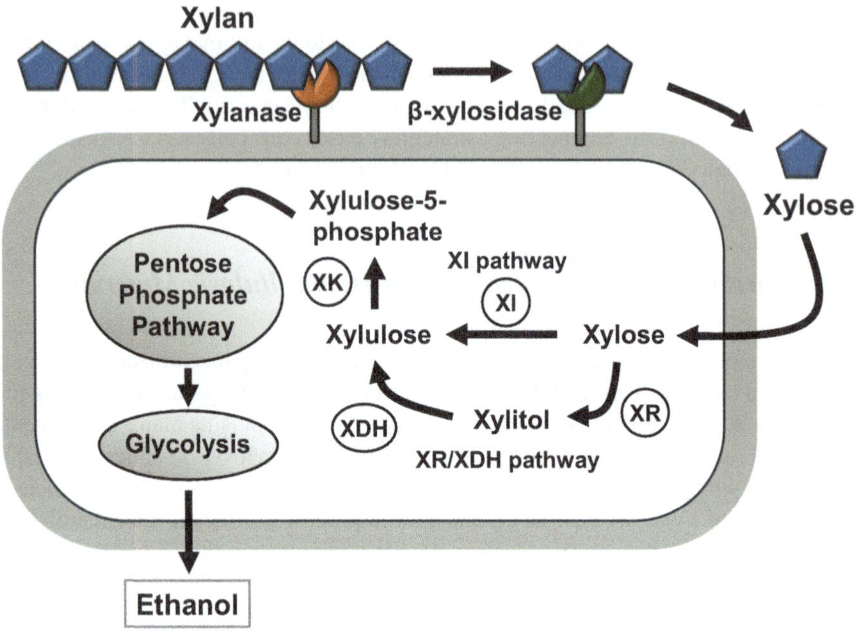

Fig. 4 Direct utilization of xylan by simultaneously introducing xylose metabolic pathways and heterologously expressing xylanolytic enzymes. Abbreviations: *XI* xylose isomerase, *XDH* xylitol dehydrogenase, *XK* xylulokinase, *XR* xylose reductase

successfully produced approximately 11 g/L of ethanol within 72 h from detoxified hemicellulose hydrolysate prepared by hot water pretreatment of corncob, without the external enzyme supplementation. Additionally, alternative strategies have been explored, including displaying a mini-scaffolding derived from *C. thermocellum* fused with hemicellulases on the cell surface [108] and secreting seven different cellulases and hemicellulases to facilitate ethanol production from bagasse [109].

3.3 Summary

Various engineering strategies have been adopted to improve the yeast ability to utilize lignocellulosic biomass. Among these strategies, cell surface engineering is one of the most promising technologies with the potential for further improvement. Maximizing the synergistic effect while minimizing cellular burden will require balancing spatial organization and expression. Furthermore, cell surface display can be extended to binding domains that enhance yeast-substrate interactions. For example, inclusion of cellulose-binding modules could promote efficient substrate utilization under lignocellulose-rich conditions. The integration of cell surface engineering into advanced yeast platforms with efficient fermentation ability and higher robustness could bring consolidated bioprocessing closer to industrial realization.

4 Yeast Stress Tolerance Mechanism and Its Improvement for Ethanol Production from Lignocellulosic Materials

4.1 Utilization of Lignocellulosic Biomass and the Related Stresses

Lignocellulosic biomass (LB), which is a plant-derived biomass, is one of the most abundant sustainable carbon sources. LB mainly consists of cellulose, hemicellulose, and lignin. When LB is used as a carbon source for bioproduction using yeast like *S. cerevisiae*, we have several problems to be solved, including the utilization strategy described in the previous section. Other issues to be overcome are fermentation condition-derived stresses. In addition to general stresses induced under fermentation conditions, such as high temperature, high osmolarity, high concentration of ethanol, and oxidative stress, LB-derived stress compounds including furfural, vanillin, and 5-HMF are generated.

4.2 Fermentation-Related Stresses and the Response Systems in Yeast

4.2.1 High Temperature and Heat Shock Stress

Various factors in nature always fluctuate, including the availability of nutrients, temperature, and concentration of toxic compounds. These factors are sometimes harmful to organisms, which is called environmental stress. High temperatures or heat shock is one of the most common environmental stresses. For example, when yeast cells are involved in ethanol fermentation, high-temperature stress always occurs due to the metabolic heat. Therefore, many organisms, including not only yeast but also humans and mice, possess a response system to high temperature or heat shock stress.

One of the most powerful high-temperature stress-resistance pathways is the heat shock response (HSR) system. The transcriptional activator Hsf1, which recognizes and binds to a pentameric heat shock element (HSE) defined as repeated nGAAn sequences, is a key factor in HSR [110]. Hsf1 is bound to HSE sequences even without stress stimulus; however, when yeast cells are exposed to high temperature, Hsf1 is rapidly activated through its phosphorylation and then transcribes the heat stress-responsive genes [110]. Some of the target genes of Hsf1 encode heat shock proteins (HSPs), including chaperones like Hsp70 and Hsp90 to refold heat-denatured proteins under high-temperature stress conditions [111]. Although Hsf1 isoforms (Hsf2, 3, and 4) have been reported in mammalians, only Hsf1 is conserved in lower eukaryote like yeast. The yeast Hsf1 is also unique in its protein structure. A mammalian Hsf1 consists of a DNA-binding domain (DBD), a leucine zipper motif required for intramolecular interaction and activation, a serine-rich regulatory domain, and a transcriptional activation domain in an order from N-terminus to C-terminus. In contrast, the yeast Hsf1 possesses an additional N-terminal amino acid up to 150 residues, followed by a mammalian-type Hsf1 structure.

In addition to Hsf1, a transcriptional activator Msn2 and its closely related isoform Msn4 (Msn2/4) are involved in the responses to high temperature and heat shock stress [112]. Msn2/4 are important for the expression of various genes in response to several environmental stresses, while they are not required for growth under normal conditions [113]. Msn2/4 recognize the stress-responsive element (STRE), consisting of a pentamer of CCCCT sequences, and induce various genes including a cytosolic catalase-encoding gene *CTT1* [112]. The signaling pathway, which is mediated by cyclic AMP (cAMP) and protein kinase A (PKA), functions as a repressor of Msn2/4 [114]. Under normal growth conditions, PKA activated by cAMP phosphorylates Msn2/4, inhibiting the nuclear localization of Msn2/4 and the transcription of the target genes [115].

When yeast cells are exposed to severe high-temperature stress conditions, the expression of housekeeping genes or heat shock genes is downregulated or upregulated, respectively. In addition, cytoplasmic and nucleoplasmic macromolecular assemblies emerge [116]. These assemblies could be around 10 nm, containing a few macromolecules like messenger ribonucleoprotein particles, and grow to submicron sizes, which consist of more than 100 kinds of macromolecules such as condensates [116]. These assemblies are transient and dissociate after yeast cells are relieved from heat shock conditions [116]. Under heat shock conditions, the yeast genome and intergenic interaction of heat shock genes are also dynamically and spatially reorganized. For example, Hsf1 binds to multiple HSEs in the promoter region of heat shock genes cooperatively. Hsf1 also interacts with chromatin-associated factors like a multiple protein complex Mediator and RNA polymerase II complexes upon heat shock stress [117]. These interactions accelerate the assembly of Hsf1, RNA polymerase II complexes, and Mediator into condensates at heat shock genes, accompanied by rapid synthesis of heat shock genes.

There have been several reports about the sensing mechanism of heat shock stress. One of the most acceptable hypotheses is that the accumulation of heat-misfolded proteins induces the activation of Hsf1. A toxic proline analogue, azetidine-2-carboxylic acid, which is incorporated into newly synthesized proteins and then leads to their misfolding, was shown to induce the transcription of Hsf1 target genes but not Msn2/4 regulons. Sublethal concentration of ethanol also phosphorylates Hsf1 [110] and then induces the expression of Hsf1 regulons [118]. This evidence supports the mechanism in which unfolded proteins function as a sensor of heat shock stress. However, it is still unclear whether the direct trigger is the misfolding of existing proteins, newly synthesized proteins, or both.

4.2.2 High Osmolarity

Yeast is exposed to high-osmolarity stress derived from high concentration of carbon sources and solute during fermentation. One of the important key molecules for the high-osmolarity response is a mitogen-activated protein kinase (MAPK) Hog1. Hog1 is activated through the MAPK phosphorylation cascade [119] (Fig. 5).

Two independent sensor mechanisms have been reported, so far. The primary osmosensor is a membrane protein Sln1, which is homologous to prokaryotic two-component system proteins. Under normal osmotic pressure conditions, Sln1 is autophosphorylated on its conserved histidine residue, which transfers the phosphate to a regulator protein Ypd1, followed by further phosphotransfer to another regulator Ssk1, consistent with the case of prokaryotes [120, 121]. The phosphorylated Ssk1 is an inactive form and thus incapable of activating the downstream MAPK phosphorylation cascade [122]. When yeast is exposed to hyperosmotic stress, the histidine kinase activity of Sln1 is inhibited, resulting in the increased level of unphosphorylated Ssk1. The unphosphorylated Ssk1 is active and thus binds to MAPK kinase kinase (MAPK3) Ssk2 and Ssk22 to induce their autophosphorylation and self-activation, for the activation of MAPK downstream pathway [123].

The other osmosensing mechanism, the SHO1 branch, is mediated by two putative osmosensors, Hkr1 and Msb2, which are transmembrane proteins with a highly glycosylated extracellular domain. These sensor proteins are responsible for the activation of the small rho-like GTPase Cdc42, in cooperation with a membrane-anchoring protein Opy2 and a bifunctional protein Sho1 with scaffolding and osmosensor activities [124]. Under high-osmolarity conditions, the interaction of Hkr1/Msb2 with Sho1 activates the p21-activated kinase (PAK)-like kinases Ste20 and Cla4 through their association with Cdc42. Subsequently, activated Ste20 and Cla4 phosphorylate and activate the MAPK3 Ste11 [125].

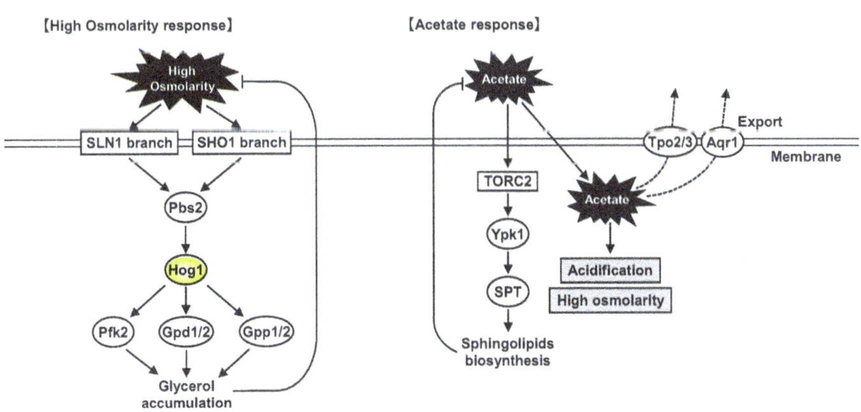

Fig. 5 MAPK cascade-mediated high osmolarity response and acetate stress tolerance, which involves sphingolipid biosynthesis and acetate export, in yeast

Activated MAPK3 Ssk2/22 or Ste11 phosphorylates and activates MAPK kinase (MAPK2) Pbs2, which directly activates Hog1 via its phosphorylation.

Hog1 regulates the expression of more than 300 genes in response to osmotic stress. Among them, some genes are involved in the glycerol synthesis pathway, including *GPD1* and *GPD2* encoding glycerol-3-phosphate dehydrogenases and *GPP1* and *GPP2* encoding glycerol-3-phosphatases [126]. The induction of these genes leads to acceleration of glycerol production from dihydroxyacetone phosphate. Furthermore, the activated HOG1 pathway increases the enzymatic activity of phosphofructokinase 2 (Pfk2), resulting in the accumulation of fructose-2,6-bisphosphate, which activates phosphofructokinase in an allosteric manner, leading to glycerol accumulation [127].

4.2.3 Chemical Stress

Various chemical compounds generated during lignocellulose utilization and fermentation, such as weak acids, furans, phenol derivatives, and ethanol, also induce stress. Acetic acid, which is derived from the acetyl group in the main chain of hemicelluloses, is the most abundant weak acid in lignocellulose hydrolysate. Formic acid and levulinic acid are also present in hemicellulose hydrolysate due to the degradation of furan compounds. Furfural or 5-HMF is generated by dehydration of pentoses or hexoses, respectively, during acid hydrolysis of cellulose and hemicellulose. Furthermore, phenolic compounds like vanillin are produced by depolymerization of lignin. The concentration of these inhibitory compounds in lignocellulosic hydrolysate depends on the source of lignocellulose and pretreatment methods.

Previously, the global responses to acetic acid in yeast were investigated by transcriptomic and metabolomic analyses [128]. Guerreiro et al. also reported the acetic acid tolerance mechanism mediated by lipid biosynthesis [129] (Fig. 5). A protein kinase Ypk1 is phosphorylated and activated by the target of rapamycin complex 2 (TORC2) in response to acetic acid stress. The Ypk1-mediated activation of L-serine:palmitoyl-CoA acyltransferase (SPT), followed by the increased production of complex sphingolipids by SPT, confers acetic acid tolerance to yeast cells. Under acidic conditions, acetic acid exists in its protonated form since it has a pK_a of 4.76 and thus diffuses into the cell across the plasma membrane. Acetic acid is deprotonated under neutral conditions inside the cell and then accumulates in the cell, leading to higher turgor pressure and oxidative stress. Furthermore, proton released from acetic acid molecules causes the intracellular acidification. *S. cerevisiae* possesses specific inducible transporters for acetate anion, such as Tpo2, Tpo3, and Aqr1, as a defense mechanism against acetic acid stress [130] (Fig. 5). The genome-wide comprehensive analysis also indicated that *TRK1* encoding the high-affinity potassium transporter and some genes regulated by the transcriptional activator Haa1 are required for susceptibility and tolerance to formic acid, respectively [131].

Furfural leads to the fragmentation of mitochondria and vacuole, nuclear chromatin disorganization, abnormal morphology of actin cytoskeleton, and oxidative stress [132]. Adaptive laboratory evolution and multi-omics analyses suggested that some genes such as *GAL7*, *PDC6*, and *HXT1*, which encode enzymes involved in carbohydrate metabolism, are required for furfural resistance [133]. Furfural and 5-HMF also induce oxidative stress, which was shown by the accumulation of ROS and upregulation of some antioxidative genes including *YAP1* and *TRX2* [134]. Actually, the growth inhibition caused by furfural or 5-HMF was canceled by the supplementation of reductants like glutathione or dithiothreitol [134]. The genes encoding transcription factors including *PDR1*, *PDR3*, *RPN4*, and *HSF1* have also been suggested to be important for global adaptation to 5-HMF in yeast [135]. Similarly, vanillin also leads to the accumulation of ROS and activation of Yap1 [136]. Two NADPH-dependent medium-chain alcohol dehydrogenases encoded by *ADH6* or *ADH7* have been shown to catalyze the reduction of vanillin to vanillyl alcohol [137, 138]. Nguyen et al. showed that the inducible expression of *ADH7* in response to vanillin stress was important for tolerance against high concentration of vanillin [139].

High concentration of ethanol is an inevitable stress because yeast grows using the glycolysis and fermentation pathway and is used for ethanol production. Ethanol initially interacts with phospholipids in the yeast cell membrane when yeast is exposed to ethanol. Hydrophobic moiety or hydroxyl group of ethanol interacts with a hydrophobic or hydrophilic region of phospholipids, resulting in membrane expansion. Ethanol stress causes a reduction of cell membrane thickness, which can result in increased membrane permeability [140]. The effect of ethanol on the membrane depends on the lipid and sterol content in the membrane. The in vitro analysis of the interaction of ethanol with the cellular membrane using model yeast biomembranes indicated that increased unsaturated lipid and/or ergosterol protects the membrane by inhibiting the formation of the interdigitated phase under exposure to a high concentration of ethanol [141]. High concentration of ethanol also causes protein denaturation. Importantly, invasion of ethanol into the membrane under a higher concentration of ethanol causes the leakage of electrons via the inhibition of cytochrome P450, resulting in the generation of ROS including hydroxyl radical released from hydrogen peroxide through the Fenton reaction mediated by free ferric ion. In addition to protein denaturation, lipid peroxidation and nucleotide oxidation, the latter of which causes mutations and single- or double-strand breaks, are induced by ROS. Protein denaturation by ethanol also causes various phenomena: inhibition of metabolic pathway including glycolysis and sterol biosynthesis, abnormal cell cycle by the upregulation of SWE1 kinase, or disturbance of actin organization [142]. Researchers demonstrated that the ethanol treatment induces the accumulation of aggregated and insoluble proteins at almost the same level as heat shock at 42°C [143, 144]. Therefore, the responses to ethanol stress and heat shock stress have many similarities. In fact, ethanol stress induces the activation of Hsf1 and the induction of Hsf1 target genes.

It was suggested that yeast cells change the concentration of fatty acids including hexadecanoic, octadecanoic, and palmitoleic acids to maintain the membrane integrity under ethanol stress conditions. Sphingolipid biosynthesis is also suggested to be important for ethanol stress tolerance since many genes responsible for sphingolipid biosynthesis were upregulated under ethanol stress conditions. The previous research by Kubota et al. showed that ergosterol is important for yeast ethanol tolerance using the strain lacking *ERG2*, *ERG3*, *ERG5*, *ERG6*, or *ERG26*, which encodes enzymes involved in sterol biosynthesis in yeast.

The yeast cell wall consists of a mannoprotein-rich outer layer and an inner layer, which is made up of β-glucan and chitin connected by glycosylphosphatidylinositol (GPI). It was reported that yeast remodels its cell wall through the regulation of GPI-anchor protein when exposed to ethanol. For example, *SED1* encoding GPI-anchor protein is upregulated under ethanol exposure conditions [145]. Ethanol tolerance is also dependent on cell wall integrity, which is maintained by the upregulation of many signaling proteins encoded by *WSC3*, *SLG1*, and *SLT2*, through the activation of PKC1-MPK1 kinase pathway [145].

More directly, alcohol dehydrogenases encoded by *ADH1*, *ADH2*, *ADH7*, and *SFA1* function for ethanol tolerance. These alcohol dehydrogenases catalyze not only production of ethanol but also utilization of ethanol as an energy source.

Many genes are regulated under ethanol stress conditions by the accumulation of Msn2/4 in the nucleus. Induction of Msn2/4 protein production is mediated by many pathways, including cAMP-dependent protein kinase pathway, Tor kinase pathway, or HOG1 pathway. Under nonstress conditions, glucose increases cAMP level, which activates PKA and inhibits stress response mediated by the transcription factors like Msn2/4 and Yap1. When ethanol stress occurs, the decrease of cAMP level allows the stress-responsive transcription factors like Msn2/4 to accumulate in the nucleus to induce their target genes for stress response, which is similar to HSR. One of the downstream pathways responsible for stress response regulated by Msn2/4 is the biosynthesis of trehalose. Trehalose, which is a disaccharide consisting of two molecules of glucose linked through α(1,1)-glycoside bond, is synthesized from glucose-6-phosphate and UDP-glucose by the enzymes encoded by *TPS1* and *TPS2*. Under high ethanol stress conditions, trehalose biosynthesis is accelerated through the upregulation of *TPS1* and *TPS2* by Msn2/4. Trehalose reduces membrane permeability and protein denaturation. Trehalose functions cooperatively with HSPs, in which the decreased level of protein denaturation by trehalose is accompanied by the reduced aggregation and protein refolding by HSPs.

4.2.4 Oxidative Stress

S. cerevisiae has several response and tolerance systems against oxidative stress: superoxide dismutase degrades superoxide anion to hydrogen peroxide, followed by the detoxification of hydrogen peroxide to water by catalase, which are conserved in a wide variety of organisms from prokaryote to higher eukaryotes.

Furthermore, some reductive detoxification systems have been reported not only in yeast but also in other various organisms. Thioredoxin (Trx)/Trx reductase (Trr) system and glutaredoxin (Grx), both of which function as disulfide oxidoreductases, are antioxidative mechanisms using NADPH as a reducing power. Trx is typically a small reductase with a molecular weight of 12 kDa. Trx transfers electron from its active cysteine residues to disulfide bonds in oxidized protein to reduce them to thiol groups, leading to the intramolecular disulfide formation of active cysteine residues on Trx. The disulfide bond on Trx is reduced by Trr with the aid of NADPH. The most abundant intracellular small compound with a thiol group is glutathione (GSH). GSH reacts with oxidized cysteine residues in proteins, including sulfenic acid and sulfinic acid, to form mixed disulfide bonds between GSH and proteins, called *S*-glutathionylation. *S*-glutathionylated proteins are reduced to proteins with free cysteine residues via the reaction with Grx and GSH. Grx is classified into two types, monothiol and dithiol types. In the case of monothiol-type Grx, GSH molecules on *S*-glutathionylated proteins are transferred to Grx and then unmodified proteins are released. When *S*-glutathionylated proteins react with dithiol-type Grx, free GSH is released and the intramolecular disulfide bond is formed in Grx, which further reacts with an additional free GSH to form *S*-glutathionylated Grx. *S*-glutathionylated Grx is converted to a reduced form of Grx through the reaction with an additional free GSH, forming an oxidized form of glutathione (GSSG), which is reduced to a reduced form of GSH by glutathione reductase and NADPH.

Transcriptional regulation of oxidative stress tolerance genes is also important. A transcriptional factor Yap1 is a major regulator of many antioxidative genes. Under normal conditions, Yap1 is localized in the cytoplasm because it is continuously exported from the nucleus to the cytoplasm. When oxidative stress occurs, Yap1 is oxidized and then the intramolecular disulfide bond is formed. The oxidized Yap is no longer recognized by the nucleus exporter system and thus is localized in the nucleus, activating its regulon, including *SOD1*, *CTT1*, and *TRX2* encoding SOD, catalase, and Trx, respectively [146].

During the life cycle of yeast or fermentation conditions, direct oxidative stress, such as the treatment of hydrogen peroxide, rarely occurs, even though oxidant treatment is a common experimental procedure in a laboratory. However, oxidative stress is induced under other stress conditions including heat shock, high osmolarity, and toxic chemical compounds, as described above (Fig. 6). Thus, these stress response systems of yeast overlap each other and are functionally redundant. Therefore, oxidative stress response mechanisms practically contribute to maintaining the fermentation productivity.

4.3 Improvement of Stress Tolerance for Efficient Ethanol Fermentation Using Lignocellulosic Biomass

Many researchers tried to improve stress tolerance mechanisms to increase ethanol fermentation productivity from lignocellulosic biomass.

The growth inhibition caused by furfural or 5-HMF was canceled by the overexpression of *YAP1* or catalase-encoding genes. Some researchers demonstrated

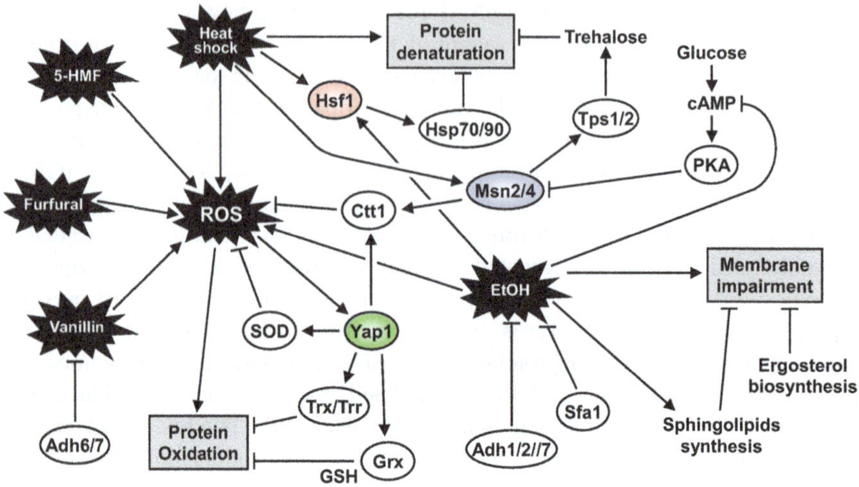

Fig. 6 Multiple cross talk among stress response mechanisms, in which various stress stimuli cause the generation of ROS, indicating that oxidative stress response mechanisms exert pivotal roles in environmental stress tolerance in yeast

that *YAP1* overexpression led to the better growth or higher ethanol yield of laboratory strains in the presence of furfural and 5-HMF [147, 148]. The overexpression of *YAP1* also decreased or increased the lag phase or growth rate of a yeast strain with both industrial strain background and xylose fermentation ability during fermentation of spruce hydrolysate, respectively [149]. Overexpression of *TRX1* encoding Trx also increased the ethanol productivity of a laboratory strain in the culture medium containing acetic acid, furfural, and 5-HMF [150]. It was reported that the overexpression of *RCK1* gene encoding protein kinase led to higher oxidative stress tolerance and improved fermentation ability under the toxic level of acetic acid, probably through the activation of HOG1 pathway via the inhibition of protein phosphatase Ptp2 which is a negative regulator of Hog1 [151]. Qin et al. indicated that the engineered yeast strain with xylose utilization pathway expressing some antioxidative genes and genes responsible for intracellular redox balance under the control of stress-driven promoter showed better growth, ethanol productivity, and xylose utilization ability under high-temperature stress conditions [152]. Overexpression of *MSN2* also led to higher ethanol production, faster growth, and lower intracellular ROS level of the industrial strain than those of its host strain in the presence of furfural [153].

Overexpression of *ALD6* and *ZWF1*, which encode aldehyde dehydrogenase and glucose-6-phosphate dehydrogenase, respectively, increased vanillin resistance by enhancing the reducing power for vanillin detoxification like NADPH [154]. On the other hand, overproduction of the reductases encoded by *ADH6*, *YNL134C*, and *YJR096W* led to the increased reduction activity of vanillin [154]. Heterologous overexpression of *aldE* gene, which encodes acetylating acetaldehyde dehydrogenase, from *Escherichia coli* and the mutant acetyl-CoA synthetase gene from

Salmonella enterica led to the increased acetate consumption and ethanol production [26].

Some amino acids including proline have functions to reduce the toxicity of several stresses. Proline is synthesized from glutamic acid as a starting compound in yeast, in which glutamyl kinase (GK) encoded by *PRO1* functions as a first step and rate-limiting enzyme. GK is susceptible to allosteric feedback inhibition by proline to maintain the appropriate level of proline synthesis. The previous research reported that the yeast strain producing the feedback-insensitive GK variant accumulated proline and exhibited higher ethanol tolerance [155]. Xiong et al. reported that the yeast overproducing arginase encoded by *CAR1* exhibited higher acetic acid tolerance through a spermidine- and proline-independent novel pathway [156].

4.4 Summary

Yeast has various response systems against different stress conditions; however, these mechanisms overlap each other more or less, as described above. In fact, some strategies succeeded in enhancing the fermentation productivity and/or robustness in yeast through multiple mechanisms. This indicates that it is important to understand not only the response mechanism against specific stresses but also the cross talk of multiple stress response mechanisms. Furthermore, it should highly contribute to the construction of an engineered yeast strain harboring desirable characteristics such as higher growth rate, productivity, and robustness and help to clarify the interaction among each stress tolerance system quantitatively.

5 Future Perspectives

In this chapter, metabolic pathways for co-fermentation of C5 and C6 sugars, engineering of yeast strains for lignocellulose utilization, and yeast stress responses under stress conditions derived from fermentation processes, all of which are important for ethanol fermentation by yeast from lignocellulosic biomass, have been discussed in the previous sections. Technologies related to each issue are important and promising for cost-effective and sustainable bioethanol production from lignocellulose. However, the combination and integrated strategies of these technologies are highly desired for further improvement, requiring further tuning of processes among each step.

Sugars from lignocellulose are required to be incorporated efficiently into cells for effective utilization and fermentation. Thus, for example, co-localization or physically close positioning of sugar transporters and cellulases tethered on the cell surface could contribute to effective uptake and utilization of lignocellulose-derived sugars. Not only physical but also temporal synergy is important [157]. In the yeast recombinant strains, which overexpress xylose utilization pathways and tether cellulases on the cell surface, the gene expression timing must be synchronized, for instance, using the same or cooperative expression systems, for effective utilization. However, in response to stress stimuli induced by fermentation processes, stress-responsive genes are upregulated, and some metabolic pathways and housekeeping genes are downregulated. Under such conditions, expression and functional levels of endogenous metabolic pathways are also affected and different from normal conditions. Therefore, the development of yeast strains, in which the expression of xylose utilization pathways and cell surface display of cellulases are responsive to stress conditions, would contribute to simultaneous assimilation and fermentation of lignocellulose-derived sugars under stress conditions, for practical use. Multiple stress tolerance mechanisms in the recombinant yeast strains are also required to cooperate with each other, even if they are engineered and/or exogenously introduced, for improved stress resistance.

Synergetic effects of multiple factors, which are demonstrated to be important for each pathway, including xylose assimilation, cellulose degradation, or stress resistance, are important. However, it is difficult to determine which factors are important and how they need to interact with each other for effective fermentation. Recent approaches include AI-based pathway designing combined with high-throughput analyses including multi-omics, which are operated by automated robotic systems [158]. Machine learning algorithms trained on large-scale fermentation data can predict stress responses, optimize nutrient feeds, forecast production bottlenecks, and propose the optimal culture conditions [159]. Such big data-based machine learning methods would become the mainstream of recombinant microbe designing of not only yeast but also bacteria in the future. In contrast, ALE and/or screening for desirable strains from a randomly mutagenized yeast library should still work as a powerful tool for isolation of strains with expected characteristics, in which automated robotic systems are very useful for high-throughput analyses [160].

Strain development, which is oriented to fermentation under severe stress conditions derived from industrial fermentation processes, is getting more important for practical use, as described above. Furthermore, yeast strains harboring the sensing and response mechanisms toward biomass availability could ferment by adapting to practical conditions more effectively. Process engineering is also making progress for novel fermentation systems, like fed-batch regimes, cell recycle, and membrane-integrated reactors [161]. Ultimately, aligning strain development with process engineering innovations will be key to bridging laboratory findings with industrial performance.

Finally, these enhancements of yeast preferred characteristics for ethanol production from lignocellulose would lead to the redefinition of microbial manufacturing paradigms for biofuels and highly valuable chemicals [162–164]. Within the broader context of biorefineries, engineered yeasts are no longer mere ethanol producers but modular biocatalysts within integrated production systems. Combined with digital infrastructure and social economics, engineered yeast would have a transformative role in the global transition to a bio-based, climate-neutral economy, far beyond their origins in brewing and baking.

References

1. Jain S, Kumar S (2024) A comprehensive review of bioethanol production from diverse feedstocks: current advancements and economic perspectives. Energy 296:131130. https://doi.org/10.1016/j.energy.2024.131130
2. Cunha JT, Soares PO, Baptista SL et al (2020) Engineered Saccharomyces cerevisiae for lignocellulosic valorization: a review and perspectives on bioethanol production. Bioengineered 11:883–903. https://doi.org/10.1080/21655979.2020.1801178
3. Etale A, Onyianta AJ, Turner SR et al (2023) Cellulose: a review of water interactions, applications in composites, and water treatment. Chem Rev 123:2016–2048. https://doi.org/10.1021/acs.chemrev.2c00477
4. Rao J, Lv Z, Chen G, Peng F (2023) Hemicellulose: structure, chemical modification, and application. Prog Polym Sci 140:101675. https://doi.org/10.1016/j.progpolymsci.2023.101675
5. Huang X, Ji L, Yin J et al (2024) Optimal design and robust operational management of regional bioethanol supply chain with various technological choices and uncertainty fusions. Comput Chem Eng 182:108565. https://doi.org/10.1016/j.compchemeng.2023.108565
6. Afedzi AEK, Afrakomah GS, Gyan K et al (2025) Enhancing economic and environmental sustainability in lignocellulosic bioethanol production: key factors, innovative technologies, policy frameworks, and social considerations. Sustainability 17:499. https://doi.org/10.3390/su17020499
7. Mankar AR, Pandey A, Modak A et al (2021) Pretreatment of lignocellulosic biomass: a review on recent advances. Bioresour Technol 334:125235. https://doi.org/10.1016/j.biortech.2021.125235
8. Li Y, Song W, Han X et al (2022) Recent progress in key lignocellulosic enzymes: enzyme discovery, molecular modifications, production, and enzymatic biomass saccharification. Bioresour Technol 363:127986. https://doi.org/10.1016/j.biortech.2022.127986
9. Acevedo A, Conejeros R, Aroca G (2017) Ethanol production improvement driven by genome-scale metabolic modeling and sensitivity analysis in Scheffersomyces stipitis. PLoS One 12:e0180074. https://doi.org/10.1371/journal.pone.0180074
10. Long TM, Su YK, Headman J et al (2012) Cofermentation of glucose, xylose, and cellobiose by the beetle-associated yeast Spathaspora passalidarum. Appl Environ Microbiol 78:5492–5500. https://doi.org/10.1128/AEM.00374-12
11. Preez JCD, Bosch M, Prior BA (1986) Xylose fermentation by Candida shehatae and Pichia stipitis: effects of pH, temperature and substrate concentration. Enzym Microb Technol 8:360–364. https://doi.org/10.1016/0141-0229(86)90136-5
12. Utomo JC, Hodgins CL, Ro DK (2021) Multiplex genome editing in yeast by CRISPR/Cas9 – a potent and agile tool to reconstruct complex metabolic pathways. Front Plant Sci 12:719148. https://doi.org/10.3389/fpls.2021.719148
13. Lawson CE, Martí JM, Radivojevic T et al (2021) Machine learning for metabolic engineering: a review. Metab Eng 63:34–60. https://doi.org/10.1016/j.ymben.2020.10.005
14. Zhang C, Chen H, Zhu Y et al (2022) Saccharomyces cerevisiae cell surface display technology: strategies for improvement and applications. Front Bioeng Biotechnol 10:1056804. https://doi.org/10.3389/fbioe.2022.1056804
15. Osorio-González CS, Saini R, Hegde K et al (2023) Carbon/nitrogen ratio as a tool to enhance the lipid production in Rhodosporidium toruloides-1588 using C5 and C6 wood hydrolysates. J Clean Prod 384:135687. https://doi.org/10.1016/j.jclepro.2022.135687
16. Gong Z, Wang Q, Shen H et al (2012) Co-fermentation of cellobiose and xylose by Lipomyces starkeyi for lipid production. Bioresour Technol 117:20–24. https://doi.org/10.1016/j.biortech.2012.04.063
17. Gong Z, Shen H, Yang X et al (2014) Lipid production from corn stover by the oleaginous yeast Cryptococcus curvatus. Biotechnol Biofuels 7:158. https://doi.org/10.1186/s13068-014-0158-y

18. Di Fidio N, Liuzzi F, Mastrolitti S et al (2019) Single cell oil production from undetoxified Arundo donax L. hydrolysate by Cutaneotrichosporon curvatus. J Microbiol Biotechnol 29: 256–267. https://doi.org/10.4014/jmb.1808.08015
19. Baroi GN, Baumann I, Westermann P et al (2015) Butyric acid fermentation from pretreated and hydrolysed wheat straw by an adapted Clostridium tyrobutyricum strain. Microb Biotechnol 8:874–882. https://doi.org/10.1111/1751-7915.12304
20. Hua Y, Wang J, Zhu Y et al (2019) Release of glucose repression on xylose utilization in *Kluyveromyces marxianus* to enhance glucose-xylose co-utilization and xylitol production from corncob hydrolysate. Microb Cell Factories 18:24. https://doi.org/10.1186/s12934-019-1068-2
21. Semkiv MV, Ruchala J, Tsaruk AY et al (2022) The role of hexose transporter-like sensor hxs1 and transcription activator involved in carbohydrate sensing azf1 in xylose and glucose fermentation in the thermotolerant yeast Ogataea polymorpha. Microb Cell Factories 21:162. https://doi.org/10.1186/s12934-022-01889-z
22. Valdehuesa KNG, Ramos KRM, Nisola GM et al (2018) Everyone loves an underdog: metabolic engineering of the xylose oxidative pathway in recombinant microorganisms. Appl Microbiol Biotechnol 102:7703–7716. https://doi.org/10.1007/s00253-018-9186-z
23. Wiedemann B, Boles E (2008) Codon-optimized bacterial genes improve L-arabinose fermentation in recombinant Saccharomyces cerevisiae. Appl Environ Microbiol 74:2043–2050. https://doi.org/10.1128/AEM.02395-07
24. Verho R, Putkonen M, Londesborough J et al (2004) A novel NADH-linked l-xylulose reductase in the l-arabinose catabolic pathway of yeast. J Biol Chem 279:14746–14751. https://doi.org/10.1074/jbc.M312533200
25. Xie CY, Yang BX, Wu YJ et al (2020) Construction of industrial xylose-fermenting Saccharomyces cerevisiae strains through combined approaches. Process Biochem 96:80–89. https://doi.org/10.1016/j.procbio.2020.05.022
26. Zhang GC, Kong II, Wei N et al (2016) Optimization of an acetate reduction pathway for producing cellulosic ethanol by engineered yeast. Biotechnol Bioeng 113:2587–2596. https://doi.org/10.1002/bit.26021
27. Brat D, Boles E, Wiedemann B (2009) Functional expression of a bacterial xylose isomerase in Saccharomyces cerevisiae. Appl Environ Microbiol 75:2304–2311. https://doi.org/10.1128/AEM.02522-08
28. Zhang YW, Yang JJ, Qian FH et al (2025) Engineering a xylose fermenting yeast for lignocellulosic ethanol production. Nat Chem Biol 21:443–450. https://doi.org/10.1038/s41589-024-01771-6
29. Bae JH, Kim MJ, Sung BH et al (2021) Directed evolution and secretory expression of xylose isomerase for improved utilisation of xylose in Saccharomyces cerevisiae. Biotechnol Biofuels 14:223. https://doi.org/10.1186/s13068-021-02073-y
30. Träff KL, Otero Cordero RR, van Zyl WH et al (2001) Deletion of the GRE3 aldose reductase gene and its influence on xylose metabolism in recombinant strains of Saccharomyces cerevisiae expressing the xylA and XKS1 genes. Appl Environ Microbiol 67:5668–5674. https://doi.org/10.1128/AEM.67.12.5668-5674.2001
31. Katahira S, Muramoto N, Moriya S et al (2017) Screening and evolution of a novel protist xylose isomerase from the termite Reticulitermes speratus for efficient xylose fermentation in Saccharomyces cerevisiae. Biotechnol Biofuels 10:203. https://doi.org/10.1186/s13068-017-0890-1
32. Toivari MH, Ruohonen L, Richard P et al (2010) Saccharomyces cerevisiae engineered to produce D-xylonate. Appl Microbiol Biotechnol 88:751–760. https://doi.org/10.1007/s00253-010-2787-9
33. Toivari M, Nygård Y, Kumpula EP et al (2012) Metabolic engineering of Saccharomyces cerevisiae for bioconversion of D-xylose to D-xylonate. Metab Eng 14:427–436. https://doi.org/10.1016/j.ymben.2012.03.002

34. Bamba T, Yukawa T, Guirimand G et al (2019) Production of 1,2,4-butanetriol from xylose by Saccharomyces cerevisiae through Fe metabolic engineering. Metab Eng 56:17–27. https://doi.org/10.1016/j.ymben.2019.08.012
35. Yukawa T, Bamba T, Matsuda M et al (2023) Enhanced production of 3,4-dihydroxybutyrate from xylose by engineered yeast via xylonate re-assimilation under alkaline condition. Biotechnol Bioeng 120:511–523. https://doi.org/10.1002/bit.28278
36. Salusjärvi L, Toivari M, Vehkomäki ML et al (2017) Production of ethylene glycol or glycolic acid from D-xylose in Saccharomyces cerevisiae. Appl Microbiol Biotechnol 101:8151–8163. https://doi.org/10.1007/s00253-017-8547-3
37. Tanaka K, Yukawa T, Bamba T et al (2025) Engineering Saccharomyces cerevisiae for growth on xylose using an oxidative pathway. Appl Microbiol Biotechnol 109:30. https://doi.org/10.1007/s00253-025-13417-1
38. Richard P, Putkonen M, Väänänen R et al (2002) The missing link in the fungal L-arabinose catabolic pathway, identification of the L-xylulose reductase gene. Biochemistry 41:6432–6437. https://doi.org/10.1021/bi025529i
39. Richard P, Verho R, Putkonen M et al (2003) Production of ethanol from L-arabinose by Saccharomyces cerevisiae containing a fungal L-arabinose pathway. FEMS Yeast Res 3:185–189. https://doi.org/10.1016/S1567-1356(02)00184-8
40. Bera AK, Sedlak M, Khan A et al (2010) Establishment of L-arabinose fermentation in glucose/xylose co-fermenting recombinant Saccharomyces cerevisiae 424A(LNH-ST) by genetic engineering. Appl Microbiol Biotechnol 87:1803–1811. https://doi.org/10.1007/s00253-010-2609-0
41. Bettiga M, Bengtsson O, Hahn-Hägerdal B et al (2009) Arabinose and xylose fermentation by recombinant Saccharomyces cerevisiae expressing a fungal pentose utilization pathway. Microb Cell Factories 8:40. https://doi.org/10.1186/1475-2859-8-40
42. Sedlak M, Ho NWY (2001) Expression of E. coli araBAD operon encoding enzymes for metabolizing L-arabinose in Saccharomyces cerevisiae. Enzym Microb Technol 28:16–24. https://doi.org/10.1016/S0141-0229(00)00282-9
43. Caballero A, Ramos JL (2017) Enhancing ethanol yields through d-xylose and l-arabinose co-fermentation after construction of a novel high efficient l-arabinose-fermenting Saccharomyces cerevisiae strain. Microbiology (Reading) 163:442–452. https://doi.org/10.1099/mic.0.000437
44. Karhumaa K, Wiedemann B, Hahn-Hägerdal B et al (2006) Co-utilization of L-arabinose and D-xylose by laboratory and industrial Saccharomyces cerevisiae strains. Microb Cell Factories 5:18. https://doi.org/10.1186/1475-2859-5-18
45. Wang X, Yang J, Yang S et al (2019) Unraveling the genetic basis of fast l-arabinose consumption on top of recombinant xylose-fermenting Saccharomyces cerevisiae. Biotechnol Bioeng 116:283–293. https://doi.org/10.1002/bit.26827
46. Wang C, Shen Y, Zhang Y et al (2013) Improvement of L-arabinose fermentation by modifying the metabolic pathway and transport in Saccharomyces cerevisiae. Biomed Res Int 2013:461204. https://doi.org/10.1155/2013/461204
47. Wisselink HW, Toirkens MJ, del Rosario Franco Berriel M et al (2007) Engineering of Saccharomyces cerevisiae for efficient anaerobic alcoholic fermentation of l-arabinose. Appl Environ Microbiol 73:4881–4891. https://doi.org/10.1128/AEM.00177-07
48. Verhoeven MD, de Valk SC, Daran JG et al (2018) Fermentation of glucose-xylose-arabinose mixtures by a synthetic consortium of single-sugar-fermenting Saccharomyces cerevisiae strains. FEMS Yeast Res 18:foy075. https://doi.org/10.1093/femsyr/foy075
49. Wang C, Zhao J, Qiu C et al (2017) Coutilization of D-glucose, D-xylose, and L-arabinose in Saccharomyces cerevisiae by coexpressing the metabolic pathways and evolutionary engineering. Biomed Res Int 2017:5318232. https://doi.org/10.1155/2017/5318232
50. Qi X, Zha J, Liu GG et al (2015) Heterologous xylose isomerase pathway and evolutionary engineering improve xylose utilization in Saccharomyces cerevisiae. Front Microbiol 6:1165. https://doi.org/10.3389/fmicb.2015.01165

51. Wisselink HW, Cipollina C, Oud B et al (2010) Metabolome, transcriptome and metabolic flux analysis of arabinose fermentation by engineered Saccharomyces cerevisiae. Metab Eng 12: 537–551. https://doi.org/10.1016/j.ymben.2010.08.003
52. Jiang Y, Shen Y, Gu L et al (2020) Identification and characterization of an efficient d-xylose transporter in Saccharomyces cerevisiae. J Agric Food Chem 68:2702–2710. https://doi.org/10.1021/acs.jafc.9b07113
53. Kuanyshev N, Deewan A, Jagtap SS et al (2021) Identification and analysis of sugar transporters capable of co-transporting glucose and xylose simultaneously. Biotechnol J 16: e2100238. https://doi.org/10.1002/biot.202100238
54. Shin HY, Nijland JG, de Waal PP et al (2015) An engineered cryptic Hxt11 sugar transporter facilitates glucose-xylose co-consumption in Saccharomyces cerevisiae. Biotechnol Biofuels 8:176. https://doi.org/10.1186/s13068-015-0360-6
55. Subtil T, Boles E (2012) Competition between pentoses and glucose during uptake and catabolism in recombinant Saccharomyces cerevisiae. Biotechnol Biofuels 5:14. https://doi.org/10.1186/1754-6834-5-14
56. Bracher JM, Verhoeven MD, Wisselink HW et al (2018) The Penicillium chrysogenum transporter PcAraT enables high-affinity, glucose-insensitive l-arabinose transport in Saccharomyces cerevisiae. Biotechnol Biofuels 11:63. https://doi.org/10.1186/s13068-018-1047-6
57. Wisselink HW, Toirkens MJ, Wu Q et al (2009) Novel evolutionary engineering approach for accelerated utilization of glucose, xylose, and arabinose mixtures by engineered Saccharomyces cerevisiae strains. Appl Environ Microbiol 75:907–914. https://doi.org/10.1128/AEM.02268-08
58. Osiro KO, Borgström C, Brink DP et al (2019) Exploring the xylose paradox in Saccharomyces cerevisiae through in vivo sugar signalomics of targeted deletants. Microb Cell Factories 18:88. https://doi.org/10.1186/s12934-019-1141-x
59. Wu M, Li H, Wei S et al (2020) Simulating extracellular glucose signals enhances xylose metabolism in recombinant Saccharomyces cerevisiae. Microorganisms 8:100. https://doi.org/10.3390/microorganisms8010100
60. Myers KS, Riley NM, MacGilvray ME et al (2019) Rewired cellular signaling coordinates sugar and hypoxic responses for anaerobic xylose fermentation in yeast. PLoS Genet 15: e1008037. https://doi.org/10.1371/journal.pgen.1008037
61. Zheng L, Wei S, Wu M et al (2020) Improving xylose fermentation in Saccharomyces cerevisiae by expressing nuclear-localized hexokinase 2. Microorganisms 8:856. https://doi.org/10.3390/microorganisms8060856
62. Xu JR, Zhao XQ, Liu CG et al (2018) Improving xylose utilization of Saccharomyces cerevisiae by expressing the MIG1 mutant from the self-flocculating yeast SPSC01. Protein Pept Lett 25:202–207. https://doi.org/10.2174/0929866525666180122142609
63. Wang WB, Tang RQ, Yuan B et al (2025) Engineering chromatin regulation of xylose utilization in budding yeast Saccharomyces cerevisiae for efficient bioconversion. ACS Synth Biol 14:794–803. https://doi.org/10.1021/acssynbio.4c00730
64. Baptista SL, Costa CE, Cunha JT et al (2021) Metabolic engineering of Saccharomyces cerevisiae for the production of top value chemicals from biorefinery carbohydrates. Biotechnol Adv 47:107697. https://doi.org/10.1016/j.biotechadv.2021.107697
65. Yang BX, Xie CY, Xia ZY et al (2020) Improving xylitol yield by deletion of endogenous xylitol-assimilating genes: a study of industrial Saccharomyces cerevisiae in fermentation of glucose and xylose. FEMS Yeast Res 20:foaa061. https://doi.org/10.1093/femsyr/foaa061
66. Tani T, Taguchi H, Akamatsu T (2017) Analysis of metabolisms and transports of xylitol using xylose- and xylitol-assimilating Saccharomyces cerevisiae. J Biosci Bioeng 123:613–620. https://doi.org/10.1016/j.jbiosc.2016.12.012
67. Li X, Wang Y, Chen X et al (2025) Modular deregulation of central carbon metabolism for efficient xylose utilization in Saccharomyces cerevisiae. Nat Commun 16:4551. https://doi.org/10.1038/s41467-025-59966-x

68. Kwak S, Kim SR, Xu H et al (2017) Enhanced isoprenoid production from xylose by engineered Saccharomyces cerevisiae. Biotechnol Bioeng 114:2581–2591. https://doi.org/10.1002/bit.26369
69. Dai Z, Huang M, Chen Y et al (2018) Global rewiring of cellular metabolism renders Saccharomyces cerevisiae Crabtree negative. Nat Commun 9:3059. https://doi.org/10.1038/s41467-018-05409-9
70. McKendry P (2002) Energy production from biomass (part 1): overview of biomass. Bioresour Technol 83:37–46. https://doi.org/10.1016/S0960-8524(01)00118-3
71. Barbosa FC, Silvello MA, Goldbeck R (2020) Cellulase and oxidative enzymes: new approaches, challenges and perspectives on cellulose degradation for bioethanol production. Biotechnol Lett 42:875–884. https://doi.org/10.1007/s10529-020-02875-4
72. Druzhinina IS, Kubicek CP (2016) Familiar stranger: ecological genomics of the model saprotroph and industrial enzyme producer Trichoderma reesei breaks the stereotypes. Adv Appl Microbiol 95:69–147. https://doi.org/10.1016/bs.aambs.2016.02.001
73. Rahman Z, Shida Y, Furukawa T et al (2009) Evaluation and characterization of Trichoderma reesei cellulase and xylanase promoters. Appl Microbiol Biotechnol 82:899–908. https://doi.org/10.1007/s00253-008-1841-3
74. Teeri TT (1997) Crystalline cellulose degradation: new insight into the function of cellobiohydrolases. Trends Biotechnol 15:160–167. https://doi.org/10.1016/S0167-7799(97)01032-9
75. Medve J, Karlsson J, Lee D et al (1998) Hydrolysis of microcrystalline cellulose by cellobiohydrolase I and endoglucanase II from Trichoderma reesei: adsorption, sugar production pattern, and synergism of the enzymes. Biotechnol Bioeng 59:621–634. https://doi.org/10.1002/(SICI)1097-0290(19980905)59:5<621::AID-BIT13>3.0.CO;2-C
76. Himmel ME, Xu Q, Luo Y et al (2010) Microbial enzyme systems for biomass conversion: emerging paradigms. Biofuels 1:323–341. https://doi.org/10.4155/bfs.09.25
77. Shoham Y, Lamed R, Bayer EA (1999) The cellulosome concept as an efficient microbial strategy for the degradation of insoluble polysaccharides. Trends Microbiol 7:275–281. https://doi.org/10.1016/S0966-842X(99)01533-4
78. Doi RH, Tamaru Y (2001) The Clostridium cellulovorans cellulosome: an enzyme complex with plant cell wall degrading activity. Chem Rec 1:24–32. https://doi.org/10.1002/1528-0691(2001)1:1<24::AID-TCR5>3.0.CO;2-W
79. Olsson L, Hahn-Hägerdal B (1993) Fermentative performance of bacteria and yeasts in lignocellulose hydrolysates. Process Biochem 28:249–257. https://doi.org/10.1016/0032-9592(93)80041-E
80. Nevoigt E (2008) Progress in metabolic engineering of Saccharomyces cerevisiae. Microbiol Mol Biol Rev 72:379–412. https://doi.org/10.1128/mmbr.00025-07
81. Jeon E, Hyeon JE, Eun LS et al (2009) Cellulosic alcoholic fermentation using recombinant Saccharomyces cerevisiae engineered for the production of Clostridium cellulovorans endoglucanase and Saccharomycopsis fibuligera β-glucosidase. FEMS Microbiol Lett 301:130–136. https://doi.org/10.1111/j.1574-6968.2009.01808.x
82. Gan Y, Lin Y, Guo Y et al (2018) Metabolic and genomic characterization of stress-tolerant industrial Saccharomyces cerevisiae strains from TALENs-assisted multiplex editing. FEMS Yeast Res 18:foy45. https://doi.org/10.1093/femsyr/foy045
83. Deparis Q, Claes A, Foulquié-Moreno MR et al (2017) Engineering tolerance to industrially relevant stress factors in yeast cell factories. FEMS Yeast Res 17:fox036. https://doi.org/10.1093/femsyr/fox036
84. Minnaar L, den Haan R (2023) Engineering natural isolates of Saccharomyces cerevisiae for consolidated bioprocessing of cellulosic feedstocks. Appl Microbiol Biotechnol 107:7013–7028. https://doi.org/10.1007/s00253-023-12729-4
85. Kuroda K, Ueda M (2011) Cell surface engineering of yeast for applications in white biotechnology. Biotechnol Lett 33:1–9. https://doi.org/10.1007/s10529-010-0403-9

86. Fukuda H, Kondo A, Tamalampudi S (2009) Bioenergy: sustainable fuels from biomass by yeast and fungal whole-cell biocatalysts. Biochem Eng J 44:2–12. https://doi.org/10.1016/j.bej.2008.11.016
87. Ueda M, Tanaka A (2000) Cell surface engineering of yeast: construction of arming yeast with biocatalyst. J Biosci Bioeng 90:125–136. https://doi.org/10.1016/S1389-1723(00)80099-7
88. Minnaar LS, Kruger F, Fortuin J et al (2024) Engineering Saccharomyces cerevisiae for application in integrated bioprocessing biorefineries. Curr Opin Biotechnol 85:103030. https://doi.org/10.1016/j.copbio.2023.103030
89. Yamada R, Taniguchi N, Tanaka T et al (2011) Direct ethanol production from cellulosic materials using a diploid strain of Saccharomyces cerevisiae with optimized cellulase expression. Biotechnol Biofuels Bioprod 4:8. https://doi.org/10.1186/1754-6834-4-8
90. Hasunuma T, Ishii J, Kondo A (2015) Rational design and evolutionary fine-tuning of Saccharomyces cerevisiae for biomass breakdown. Curr Opin Chem Biol 29:1–9. https://doi.org/10.1016/j.cbpa.2015.06.004
91. Inokuma K, Bamba T, Ishii J et al (2016) Enhanced cell-surface display and secretory production of cellulolytic enzymes with Saccharomyces cerevisiae Sed1 signal peptide. Biotechnol Bioeng 113:2358–2366. https://doi.org/10.1002/bit.26008
92. Phienluphon A, Mhuantong W, Boonyapakron K et al (2019) Identification and evaluation of novel anchoring proteins for cell surface display on Saccharomyces cerevisiae. Appl Microbiol Biotechnol 103:3085–3097. https://doi.org/10.1007/s00253-019-09667-5
93. Yang X, Tang H, Song M et al (2019) Development of novel surface display platforms for anchoring heterologous proteins in Saccharomyces cerevisiae. Microb Cell Factories 18:85. https://doi.org/10.1186/s12934-019-1133-x
94. Inokuma K, Kurono H, den Haan R et al (2020) Novel strategy for anchorage position control of GPI-attached proteins in the yeast cell wall using different GPI-anchoring domains. Metab Eng 57:110–117. https://doi.org/10.1016/j.ymben.2019.11.004
95. Yamada R, Taniguchi N, Tanaka T et al (2010) Cocktail δ-integration: a novel method to construct cellulolytic enzyme expression ratio-optimized yeast strains. Microb Cell Factories 9:32. https://doi.org/10.1186/1475-2859-9-32
96. Sharma HK, Xu C, Qin W (2019) Biological pretreatment of lignocellulosic biomass for biofuels and bioproducts: an overview. Waste Biomass Valoriz 10:235–251. https://doi.org/10.1007/s12649-017-0059-y
97. Wan Azelee NI, Mahdi HI, Cheng YS et al (2023) Biomass degradation: challenges and strategies in extraction and fractionation of hemicellulose. Fuel 339:126982. https://doi.org/10.1016/j.fuel.2022.126982
98. Cunha JT, Gomes DG, Romaní A et al (2021) Cell surface engineering of Saccharomyces cerevisiae for simultaneous valorization of corn cob and cheese whey via ethanol production. Energy Convers Manag 243:114359. https://doi.org/10.1016/j.enconman.2021.114359
99. Yanase S, Hasunuma T, Yamada R et al (2010) Direct ethanol production from cellulosic materials at high temperature using the thermotolerant yeast Kluyveromyces marxianus displaying cellulolytic enzymes. Appl Microbiol Biotechnol 88:381–388. https://doi.org/10.1007/s00253-010-2784-z
100. Saha BC (2003) Hemicellulose bioconversion. J Ind Microbiol Biotechnol 30:279–291. https://doi.org/10.1007/s10295-003-0049-x
101. Kruger F, den Haan R (2022) Surface-tethered xylosidase activity improved xylan conversion in engineered strains of Saccharomyces cerevisiae. J Chem Technol Biotechnol 97:1099–1111. https://doi.org/10.1002/jctb.7044
102. Sun L, Jin YS (2021) Xylose assimilation for the efficient production of biofuels and chemicals by engineered Saccharomyces cerevisiae. Biotechnol J 16:2000142. https://doi.org/10.1002/biot.202000142
103. Van Vleet JH, Jeffries TW (2009) Yeast metabolic engineering for hemicellulosic ethanol production. Curr Opin Biotechnol 20:300–306. https://doi.org/10.1016/j.copbio.2009.06.001

104. Hou J, Qiu C, Shen Y et al (2017) Engineering of Saccharomyces cerevisiae for the efficient co-utilization of glucose and xylose. FEMS Yeast Res 17:34. https://doi.org/10.1093/femsyr/fox034
105. Jeong D, Oh EJ, Ko JK et al (2020) Metabolic engineering considerations for the heterologous expression of xylose-catabolic pathways in Saccharomyces cerevisiae. PLoS One 15:0236294. https://doi.org/10.1371/journal.pone.0236294
106. Mert MJ, la Grange DC, Rose SH et al (2016) Engineering of Saccharomyces cerevisiae to utilize xylan as a sole carbohydrate source by co-expression of an endoxylanase, xylosidase and a bacterial xylose isomerase. J Ind Microbiol Biotechnol 43:431–440. https://doi.org/10.1007/s10295-015-1727-1
107. Cunha JT, Romaní A, Inokuma K et al (2020) Consolidated bioprocessing of corn cob-derived hemicellulose: engineered industrial Saccharomyces cerevisiae as efficient whole cell biocatalysts. Biotechnol Biofuels 13:138. https://doi.org/10.1186/s13068-020-01780-2
108. Sun J, Wen F, Si T et al (2012) Direct conversion of xylan to ethanol by recombinant Saccharomyces cerevisiae strains displaying an engineered minihemicellulosome. Appl Environ Microbiol 78:3837–3845. https://doi.org/10.1128/AEM.07679-11
109. Claes A, Deparis Q, Foulquié-Moreno MR et al (2020) Simultaneous secretion of seven lignocellulolytic enzymes by an industrial second-generation yeast strain enables efficient ethanol production from multiple polymeric substrates. Metab Eng 59:131–141. https://doi.org/10.1016/j.ymben.2020.02.004
110. Hashikawa N, Yamamoto N, Sakurai H (2007) Different mechanisms are involved in the transcriptional activation by yeast heat shock transcription factor through two different types of heat shock elements. J Biol Chem 282:10333–10340. https://doi.org/10.1074/jbc.M609708200
111. Morano KA, Grant CM, Moye-Rowley WS (2012) The response to heat shock and oxidative stress in Saccharomyces cerevisiae. Genetics 190:1157–1195. https://doi.org/10.1534/genetics.111.128033
112. Martínez-Pastor MT, Marchler G, Schüller C et al (1996) The Saccharomyces cerevisiae zinc finger proteins Msn2p and Msn4p are required for transcriptional induction through the stress-response element (STRE). EMBO J 15:2227–2235. https://doi.org/10.1002/j.1460-2075.1996.tb00576.x
113. Schmitt AP, McEntee K (1996) Msn2p, a zinc-finger DNA-binding protein, is the transcriptional activator of the multistress response in Saccharomyces cerevisiae. Proc Natl Acad Sci USA 93:5777–5782. https://doi.org/10.1073/pnas.93.12.5777
114. Görner W, Durchschlag E, Martinez-Pastor MT et al (1998) Nuclear localization of the C2H2 zinc finger protein Msn2p is regulated by stress and protein kinase A activity. Genes Dev 12:586–597. https://doi.org/10.1101/gad.12.4.586
115. Durchschlag E, Reiter W, Ammerer G et al (2004) Nuclear localization destabilizes the stress-regulated transcription factor Msn2. J Biol Chem 279:55425–55432. https://doi.org/10.1074/jbc.M407264200
116. Desroches Altamirano C, Alberti S (2025) Surviving the heat: the role of macromolecular assemblies in promoting cellular shutdown. Trends Biochem Sci 50:18–32. https://doi.org/10.1016/j.tibs.2024.09.008
117. Chowdhary S, Kainth AS, Paracha S et al (2022) Inducible transcriptional condensates drive 3D genome reorganization in the heat shock response. Mol Cell 82:4386–4399. https://doi.org/10.1016/j.molcel.2022.10.013
118. Rubio LS, Mohajan S, Gross DS (2024) Heat Shock Factor 1 forms nuclear condensates and restructures the yeast genome before activating target genes. eLife 12:RP92464. https://doi.org/10.7554/eLife.92464.4
119. Brewster JL, Gustin MC (2014) Hog1: 20 years of discovery and impact. Sci Signal 7:re7. https://doi.org/10.1126/scisignal.2005458

120. Posas F, Saito H (1997) Osmotic activation of the HOG MAPK pathway via Ste11p MAPKKK: scaffold role of Pbs2p MAPKK. Science 276:1702–1705. https://doi.org/10.1126/science.276.5319.1702
121. Posas F, Wurgler-Murphy SM, Maeda T et al (1996) Yeast HOG1 MAP kinase cascade is regulated by a multistep phosphorelay mechanism in the SLN1-YPD1-SSK1 "two-component" osmosensor. Cell 86:865–875. https://doi.org/10.1016/S0092-8674(00)80162-2
122. Posas F, Saito H (1998) Activation of the yeast SSK2 MAP kinase kinase kinase by the SSK1 two-component response regulator. EMBO J 17:1385–1394. https://doi.org/10.1093/emboj/17.5.1385
123. Auesukaree C (2017) Molecular mechanisms of the yeast adaptive response and tolerance to stresses encountered during ethanol fermentation. J Biosci Bioeng 124:133–142. https://doi.org/10.1016/j.jbiosc.2017.03.009
124. Tatebayashi K, Yamamoto K, Nagoya M et al (2015) Osmosensing and scaffolding functions of the oligomeric four-transmembrane domain osmosensor Sho1. Nat Commun 6:6975. https://doi.org/10.1038/ncomms7975
125. Saito H, Posas F (2012) Response to hyperosmotic stress. Genetics 192:289–318. https://doi.org/10.1534/genetics.112.140863
126. Hohmann S (2002) Osmotic stress signaling and osmoadaptation in yeasts. Microbiol Mol Biol Rev 66:300–372. https://doi.org/10.1128/mmbr.66.2.300-372.2002
127. Dihazi H, Kessler R, Eschrich K (2004) High osmolarity glycerol (HOG) pathway-induced phosphorylation and activation of 6-phosphofructo-2-kinase are essential for glycerol accumulation and yeast-cell proliferation under hyperosmotic stress. J Biol Chem 279:23961–23968. https://doi.org/10.1074/jbc.M312974200
128. Dong Y, Hu J, Fan L et al (2017) RNA-Seq-based transcriptomic and metabolomic analysis reveal stress responses and programmed cell death induced by acetic acid in Saccharomyces cerevisiae. Sci Rep 7:42659. https://doi.org/10.1038/srep42659
129. Guerreiro JF, Muir A, Ramachandran S et al (2016) Sphingolipid biosynthesis upregulation by TOR complex 2-Ypk1 signaling during yeast adaptive response to acetic acid stress. Biochem J 473:4311–4325. https://doi.org/10.1042/BCJ20160565
130. Palma M, Guerreiro JF, Sá-Correia I (2018) Adaptive response and tolerance to acetic acid in Saccharomyces cerevisiae and Zygosaccharomyces bailii: a physiological genomics perspective. Front Microbiol 9:274. https://doi.org/10.3389/fmicb.2018.00274
131. Henriques SF, Mira NP, Sá-Correia I (2017) Genome-wide search for candidate genes for yeast robustness improvement against formic acid reveals novel susceptibility (Trk1 and positive regulators) and resistance (Haa1-regulon) determinants. Biotechnol Biofuels 10:96. https://doi.org/10.1186/s13068-017-0781-5
132. Allen SA, Clark W, McCaffery JM et al (2010) Furfural induces reactive oxygen species accumulation and cellular damage in Saccharomyces cerevisiae. Biotechnol Biofuels 3:2. https://doi.org/10.1186/1754-6834-3-2
133. Ren J, Zhang M, Guo X et al (2024) Furfural tolerance of mutant Saccharomyces cerevisiae selected via ionizing radiation combined with adaptive laboratory evolution. Biotechnol Biofuels Bioprod 17:117. https://doi.org/10.1186/s13068-024-02562-w
134. Kim D, Hahn JS (2013) Roles of the Yap1 transcription factor and antioxidants in Saccharomyces cerevisiae's tolerance to furfural and 5-hydroxymethylfurfural, which function as thiol-reactive electrophiles generating oxidative stress. Appl Environ Microbiol 79:5069–5077. https://doi.org/10.1128/AEM.00643-13
135. Ma M, Liu ZL (2010) Comparative transcriptome profiling analyses during the lag phase uncover YAP1, PDR1, PDR3, RPN4, and HSF1 as key regulatory genes in genomic adaptation to the lignocellulose-derived inhibitor HMF for Saccharomyces cerevisiae. BMC Genomics 11:660. https://doi.org/10.1186/1471-2164-11-660
136. Nguyen TTM, Iwaki A, Ohya Y et al (2014) Vanillin causes the activation of Yap1 and mitochondrial fragmentation in Saccharomyces cerevisiae. J Biosci Bioeng 117:33–38. https://doi.org/10.1016/j.jbiosc.2013.06.008

137. Larroy C, Parés X, Biosca JA (2002) Characterization of a Saccharomyces cerevisiae NADP (H)-dependent alcohol dehydrogenase (ADHVII), a member of the cinnamyl alcohol dehydrogenase family. Eur J Biochem 269:5738–5745. https://doi.org/10.1046/j.1432-1033.2002.03296.x
138. Larroy C, Rosario M, Ndez F et al (2002) Characterization of the Saccharomyces cerevisiae YMR318C (ADH6) gene product as a broad specificity NADPH-dependent alcohol dehydrogenase: relevance in aldehyde reduction. Biochem J 361:163–172. https://doi.org/10.1042/bj3610163
139. Nguyen TTM, Iwaki A, Izawa S (2015) The ADH7 promoter of Saccharomyces cerevisiae is vanillin-inducible and enables mRNA translation under severe vanillin stress. Front Microbiol 6:1390. https://doi.org/10.3389/fmicb.2015.01390
140. Schiavone M, Formosa-Dague C, Elsztein C et al (2016) Evidence for a role for the plasma membrane in the nanomechanical properties of the cell wall as revealed by an atomic force microscopy study of the response of Saccharomyces cerevisiae to ethanol stress. Appl Environ Microbiol 82:4789–4801. https://doi.org/10.1128/AEM.01213-16
141. Vanegas JM, Contreras MF, Faller R et al (2012) Role of unsaturated lipid and ergosterol in ethanol tolerance of model yeast biomembranes. Biophys J 102:507–516. https://doi.org/10.1016/j.bpj.2011.12.038
142. Kubota S, Takeo I, Kume K et al (2004) Effect of ethanol on cell growth of budding yeast: genes that are important for cell growth in the presence of ethanol. Biosci Biotechnol Biochem 68:968–972. https://doi.org/10.1271/bbb.68.968
143. Kato S, Yoshida M, Izawa S (2019) Btn2 is involved in the clearance of denatured proteins caused by severe ethanol stress in Saccharomyces cerevisiae. FEMS Yeast Res 19:foz079. https://doi.org/10.1093/femsyr/foz079
144. Yoshida M, Kato S, Fukuda S et al (2021) Acquired resistance to severe ethanol stress in Saccharomyces cerevisiae protein quality control. Appl Environ Microbiol 87:e02353. https://doi.org/10.1128/AEM.02353-20
145. Udom N, Chansongkrow P, Charoensawan V et al (2019) Coordination of the cell wall integrity and highosmolarity glycerol pathways in response to ethanol stress in Saccharomyces cerevisiae. Appl Environ Microbiol 85:e00551. https://doi.org/10.1128/AEM.00551-19
146. Rowe LA, Degtyareva N, Doetsch PW (2012) Yap1: a DNA damage responder in Saccharomyces cerevisiae. Mech Aging Dev 133:147–156. https://doi.org/10.1016/j.mad.2012.03.009
147. Wu G, Xu Z, Jönsson LJ (2017) Profiling of Saccharomyces cerevisiae transcription factors for engineering the resistance of yeast to lignocellulose-derived inhibitors in biomass conversion. Microb Cell Factories 16:199. https://doi.org/10.1186/s12934-017-0811-9
148. Alriksson B, Horváth IS, Jönsson LJ (2010) Overexpression of Saccharomyces cerevisiae transcription factor and multidrug resistance genes conveys enhanced resistance to lignocellulose-derived fermentation inhibitors. Process Biochem 45:264–271. https://doi.org/10.1016/j.procbio.2009.09.016
149. Wallace-Salinas V, Signori L, Li YY et al (2014) Reassessment of YAP1 and MCR1 contributions to inhibitor tolerance in robust engineered Saccharomyces cerevisiae fermenting undetoxified lignocellulosic hydrolysate. AMB Express 4:56. https://doi.org/10.1186/s13568-014-0056-5
150. Unrean P, Gätgens J, Klein B et al (2018) Elucidating cellular mechanisms of Saccharomyces cerevisiae tolerant to combined lignocellulosic-derived inhibitors using high-throughput phenotyping and multiomics analyses. FEMS Yeast Res 18:foy106. https://doi.org/10.1093/femsyr/foy106
151. Oh EJ, Wei N, Kwak S et al (2019) Overexpression of RCK1 improves acetic acid tolerance in Saccharomyces cerevisiae. J Biotechnol 292:1–4. https://doi.org/10.1016/j.jbiotec.2018.12.013
152. Qin L, Dong S, Yu J et al (2020) Stress-driven dynamic regulation of multiple tolerance genes improves robustness and productive capacity of Saccharomyces cerevisiae in industrial lignocellulose fermentation. Metab Eng 61:160–170. https://doi.org/10.1016/j.ymben.2020.06.003

153. Sasano Y, Watanabe D, Ukibe K et al (2012) Overexpression of the yeast transcription activator Msn2 confers furfural resistance and increases the initial fermentation rate in ethanol production. J Biosci Bioeng 113:451–455. https://doi.org/10.1016/j.jbiosc.2011.11.017
154. Wang X, Liang Z, Hou J et al (2016) Identification and functional evaluation of the reductases and dehydrogenases from Saccharomyces cerevisiae involved in vanillin resistance. BMC Biotechnol 16:31. https://doi.org/10.1186/s12896-016-0264-y
155. Takagi H, Takaoka M, Kawaguchi A et al (2005) Effect of L-proline on sake brewing and ethanol stress in Saccharomyces cerevisiae. Appl Environ Microbiol 71:8656–8662. https://doi.org/10.1128/AEM.71.12.8656-8662.2005
156. Xiong L, Wang YT, Zhou MH et al (2024) Overexpression of arginase gene CAR1 renders yeast Saccharomyces cerevisiae acetic acid tolerance. Synth Syst Biotechnol 9:723–732. https://doi.org/10.1016/j.synbio.2024.05.013
157. Hartline CJ, Schmitz AC, Han Y et al (2021) Dynamic control in metabolic engineering: theories, tools, and applications. Metab Eng 63:126–140. https://doi.org/10.1016/j.ymben.2020.08.015
158. Cheng Y, Bi X, Xu Y et al (2023) Artificial intelligence technologies in bioprocess: opportunities and challenges. Bioresour Technol 369:128451. https://doi.org/10.1016/j.biortech.2022.128451
159. Packiam KAR, Ooi CW, Li F et al (2022) PERISCOPE-Opt: machine learning-based prediction of optimal fermentation conditions and yields of recombinant periplasmic protein expressed in Escherichia coli. Comput Struct Biotechnol J 20:2909–2920. https://doi.org/10.1016/j.csbj.2022.06.006
160. Halle L, Hollmann N, Tenhaef N et al (2023) Robotic workflows for automated long-term adaptive laboratory evolution: improving ethanol utilization by Corynebacterium glutamicum. Microb Cell Factories 22:175. https://doi.org/10.1186/s12934-023-02180-5
161. Dong Y, Zhang Y, Liu D et al (2023) Strain and process engineering toward continuous industrial fermentation. Front Chem Sci Eng 17:1336. https://doi.org/10.1007/s11705-022-2284-6
162. Perrot T, Marc J, Lezin E et al (2024) Emerging trends in production of plant natural products and new-to-nature biopharmaceuticals in yeast. Curr Opin Biotechnol 87:103098. https://doi.org/10.1016/j.copbio.2024.103098
163. Kulagina N, Besseau S, Godon C et al (2021) Yeasts as biopharmaceutical production platforms. Front Fungal Biol 2:733492. https://doi.org/10.3389/ffunb.2021.733492
164. Nandy SK, Srivastava RK (2018) A review on sustainable yeast biotechnological processes and applications. Microbiol Res 207:83–90. https://doi.org/10.1016/j.micres.2017.11.013

Adv Biochem Eng Biotechnol (2026) 196: 183–202
https://doi.org/10.1007/10_2026_321

Published online: 16 July 2026

Producing Advanced Biofuel Butanol Through Acetone–Butanol–Ethanol Fermentation

Xue Chuang, Liu Ziyu, and Cheng Chi

Contents

Abstract Butanol represents a high-energy-density advanced biofuel with superior compatibility to conventional fuel infrastructures and broad industrial significance. As a promising second-generation biofuel, biobutanol is predominantly produced via acetone–butanol–ethanol (ABE) fermentation by solventogenic *Clostridium* species. This chapter focuses on biobutanol production from lignocellulosic agricultural residues, a non-food and sustainable feedstock. It illustrates the composition of lignocellulosic hydrolysates and the inhibitory effects of toxic byproducts on microbial metabolism and butanol biosynthesis. Key strategies including detoxification technologies, metabolic engineering for improved sugar co-utilization and stress tolerance, and consolidated bioprocessing (CBP) are systematically reviewed.

X. Chuang (✉), L. Ziyu, and C. Chi
MOE Key Laboratory of Bio-Intelligent Manufacturing, Frontiers Science Centre for Smart Materials Oriented Chemical Engineering, School of Bioengineering, Dalian University of Technology, Dalian, China

Ningbo Institute of Dalian University of Technology, Ningbo, China
e-mail: xue.1@dlut.edu.cn

Integration of strain engineering and process optimization significantly enhances fermentation efficiency and butanol productivity. The utilization of lignocellulosic biomass enables cost-effective, low-carbon, and sustainable biobutanol production, demonstrating great potential for industrial application and future bioenergy development.

Graphical Abstract

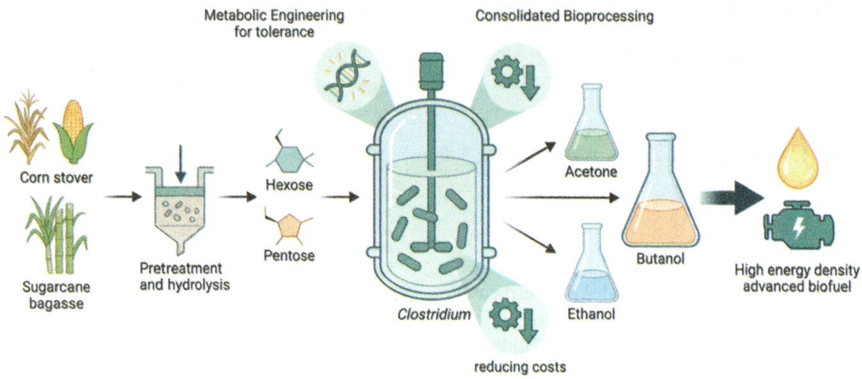

Keywords ABE fermentation, Butanol, Clostridium, Lignocellulosic biomass

1 Introduction: Butanol and ABE Fermentation

Butanol is a C_4 alcohol with the molecular formula C_4H_9OH, existing in four isomeric forms: *n*-butanol, 2-butanol, isobutanol, and *tert*-butanol. These isomers can be produced via thermochemical reactions of fossil fuels in the presence of inorganic catalysts and possess the characteristics of being colorless, flammable, and slightly soluble in water. As a compound with high energy density (29.2 MJ/L), its energy density is close to that of gasoline (32.0 MJ/L), which is much higher than that of ethanol (21.1 MJ/L). Moreover, it has low volatility and strong compatibility with existing fuel infrastructure, enabling it to be used directly as a biofuel or blended with gasoline. Beyond the fuel sector, *n*-butanol also serves as a critical chemical intermediate, widely applied in the synthesis of products such as acrylates, amino resins, and butylamines, with a global market scale of 4.5 million tons/year [1, 2]. In recent years, the demand for bio-based *n*-butanol as a natural ingredient in the food, cosmetics, and pharmaceutical industries has grown rapidly [3]. Compared with petroleum-based butanol produced by chemical synthesis, biobutanol offers advantages of renewability and a low carbon footprint, which aligns with the carbon neutrality strategy. Thus, it has become a research focus in the field of biomass conversion.

Currently, the industrial production of butanol primarily relies on petrochemical routes, such as propylene carbonylation or acetaldehyde condensation [4]. However, the nonrenewable nature of petroleum resources and price fluctuations have driven the development of biosynthetic routes, with lignocellulosic biomass emerging as a pivotal alternative renewable substrate, which is the core research topic of this chapter. The biological production of butanol mainly depends on acetone-butanol-ethanol (ABE) fermentation, which is carried out by anaerobic bacteria such as *Clostridium*. Their natural metabolic pathways can generate a variety of chemicals, including butanol, ethanol, hexanol, acetone, acetic acid, butyric acid, lactic acid, 1,3-propanediol, 2,3-butanediol, and isopropanol. Solvent-producing Clostridia (e.g., *C. acetobutylicum*, *C. beijerinckii*, *C. saccharobutylicum*) produce acetone, *n*-butanol, and ethanol in an approximate weight ratio of 3:6:1 during ABE fermentation [5]. Clostridia exhibit remarkable substrate versatility, being capable of utilizing hexoses like glucose and fructose, pentoses, glycerol, disaccharides like cellobiose and sucrose, polysaccharides (cellulose, hemicellulose), syngas, and amino acids. This substrate versatility makes Clostridia ideal hosts for biofuel production. ABE fermentation has a well-documented industrial development history: originating from the classic Weizmann process in the early twentieth century, it reached its industrial peak during World War II amid global petroleum scarcity and was subsequently largely replaced by cost-competitive petrochemical routes with the rapid rise of the petroleum industry in the postwar era, the approach that addresses the critical limitation of traditional ABE fermentation, which relies on starchy or sucrose-rich substrates such as corn and sugarcane.

As a second-generation biofuel, butanol's compatibility with existing infrastructure and high energy density make it a viable alternative to fossil fuels. The significance of butanol and ABE fermentation lies in their potential to address energy and environmental challenges. Traditional ABE fermentation uses starchy or sucrose-rich substrates such as corn and sugarcane, but these feedstocks compete with food resources, leading to persistently high production costs [6]. To solve this problem, researchers have turned their attention to nonfood biomass such as agricultural residues, including corn fiber, bagasse, and cotton stalks. These feedstocks are of low cost and abundant, which can significantly reduce the substrate cost of butanol production, avoid competition with food supply, and realize utilization of waste resource utilization [7, 8].

This chapter first introduces the characteristics, applications, and advantages of biobutanol, expounds the principle of ABE fermentation and the substrate versatility of solvent-producing Clostridia, and also addresses its inherent technical limitations such as low yield, characterized by butanol titers typically below 12 g/L, and high energy consumption. Then, it analyzes the composition of agricultural residue hydrolysates, their effects on fermentation, and corresponding optimization strategies while exploring the direct utilization methods of agricultural residues. Furthermore, it reviews the progress in improving ABE fermentation efficiency through process engineering, notably consolidated bioprocessing or CBP, which can reduce costs by more than 50%, as well as metabolic engineering approaches like enhancing butanol synthesis and improving strain tolerance. These advances highlight the renewed attention on this technology with the support of synthetic biology, as well as the industrial potential of butanol production from agricultural residues, making it a focus of sustainable bioenergy research.

CBP integrates cellulose degradation, sugar fermentation, and product synthesis into a single step, which can reduce costs by more than 50%. These progresses underscore the industrial potential of ABE fermentation for butanol production from agricultural residues, establishing it as a focal point in sustainable bioenergy research.

2 Utilization of Agricultural Residue Hydrolysate

2.1 Composition of Cellulosic Hydrolysates

Cellulosic agricultural residues, such as corn fiber, cotton stalks, bagasse, and soybean hulls, can be converted into hydrolysates through pretreatment and enzymatic hydrolysis, which serve as potential low-cost substrates for ABE fermentation. The hydrolysates have a complex composition, mainly including fermentable sugars, non-fermentable impurities, and inhibitors generated during pretreatment. The types and proportions of these components directly affect the metabolic activity of Clostridia and butanol yield.

Differences in the lignocellulosic composition of various agricultural residues lead to significant variations in hydrolysate characteristics: cotton stalks and bagasse have a cellulose content of over 40%, with hemicellulose primarily composed of xylan. After hydrolysis, their glucose/xylose ratio ranges from approximately 2 to 6, and the inhibitor content is low, making them suitable for direct use in ABE fermentation. Corn fiber has a high hemicellulose content of around 30%; after hydrolysis, the xylose ratio is about 1.3, and the concentration of levulinic acid reaches 9.44 g/L, significantly higher than that in other feedstocks, requiring detoxification before use. Soybean hulls contain high levels of pectin and protein; in addition to sugars, their hydrolysates also contain amino acids. Although they produce relatively more phenolic compounds, they can partially replace nitrogen sources [7]. These differences in characteristics determine the fermentability of the hydrolysates.

The fermentable sugars in cellulosic hydrolysates are mainly glucose and xylose, along with small amounts of monosaccharides such as arabinose and galactose. Their proportions vary depending on the feedstock type and pretreatment method. Glucose is derived from cellulose hydrolysis and is the preferred carbon source for Clostridia. The glucose content can reach 44.4 g/L in cotton stalk hydrolysates and approximately 32.5 g/L in bagasse hydrolysates. During ABE fermentation, glucose is converted to pyruvate via glycolysis, which is further metabolized to acetyl-CoA and butyryl-CoA, the key carbon skeletons for butanol synthesis [7–9].

As the main component of hemicellulose, xylose accounts for a significant proportion in hydrolysates. The xylose content can reach 40.7 g/L in corn fiber hydrolysates and approximately 17.4 g/L in soybean hull hydrolysates. However, most Clostridia utilize xylose less efficiently than glucose, and glucose-mediated carbon catabolite repression (CCR) leads to low xylose utilization efficiency in mixed sugar fermentation [10].

Other sugars such as arabinose and galactose are present in low concentrations in hydrolysates. Their metabolic pathways are similar to that of xylose, requiring catalysis by isomerases and kinases to enter the pentose phosphate pathway. For example, the small amount of galactose in spent coffee ground hydrolysates can only be efficiently utilized by *C. tyrobutyricum* after the heterologous expression of genes such as *galK* and *galT* [11].

During the pretreatment of cellulose, dilute acid treatment and high-temperature cooking can cause the degradation of sugars in lignocellulose, producing a variety of inhibitory compounds:

Furan compounds include furfural derived from pentose degradation and 5-hydroxymethylfurfural (HMF) derived from hexose degradation, with concentrations typically ranging from 0.1 to 5 g/L in the hydrolysate. These substances can damage the integrity of cell membranes and inhibit the activity of key enzymes such as sugar transporters and aldehyde dehydrogenases [12]. Phenolic compounds are mainly derived from lignin degradation, such as ferulic acid and *p*-coumaric acid, with a concentration of up to 1.41 g/L in corn fiber hydrolysate. Phenols damage DNA and proteins through oxidative stress, significantly reducing the growth rate of Clostridia and the butanol yield [13].

Organic acids such as formic acid, acetic acid, and levulinic acid are mainly derived from the dehydration and oxidation of sugars. Excessive organic acids can reduce the pH of the fermentation system, disrupt the intracellular pH homeostasis, and inhibit energy metabolism. In addition, the hydrolysate also contains small amounts of metal ions such as calcium, magnesium, and ash. Some ions can act as coenzymes for enzymes, but high concentrations can inhibit cell activity [14]. The comprehensive composition of cellulosic hydrolysates and the corresponding source characteristics of each component are visually summarized in Fig. 1, which intuitively reflects the key factors affecting the fermentability of hydrolysates from different agricultural residues.

2.2 Impact of Hydrolysate Components on ABE Fermentation

2.2.1 Impact of Sugar Components

The sugars, inhibitors, and other components in the cellulose hydrolysate directly or indirectly affect the growth, metabolism, and butanol synthesis efficiency of Clostridia through different mechanisms. The fermentable sugars such as glucose and xylose in the hydrolysate are the carbon source basis for butanol synthesis, and their types and proportions directly determine carbon metabolism. As a rapidly utilizable

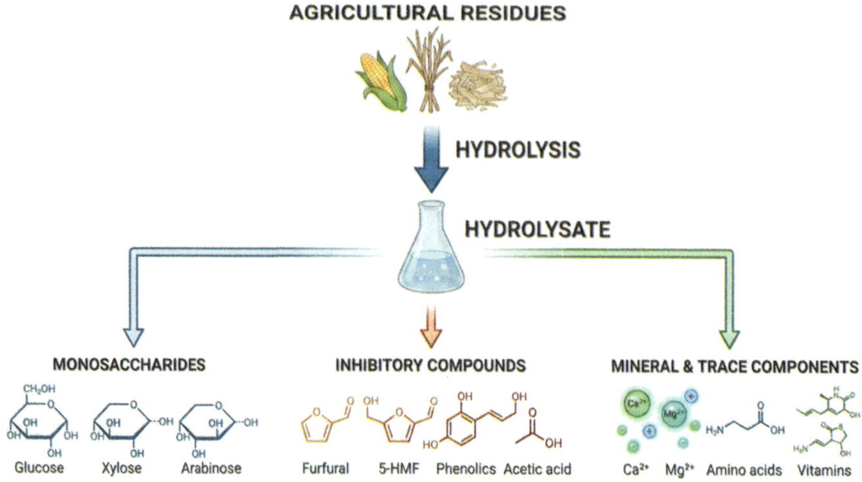

Fig. 1 Schematic of hydrolysis process and key components in agricultural residue hydrolysate

carbon source for Clostridia, glucose rapidly generates pyruvate through glycolysis, which is catalyzed by pyruvate-ferredoxin oxidoreductase to form acetyl-CoA, providing precursors for butyryl-CoA synthesis [9]. For example, *C. tyrobutyricum* can direct more than 70% of the carbon flow to butyryl-CoA during glucose fermentation, with a butanol yield of 0.3 g/g [15]. A high glucose concentration can accelerate sugar metabolism by upregulating the activities of phosphofructokinase *pfkA* and pyruvate kinase *pykA*, thereby increasing the butanol yield [16].

Although the utilization efficiency of xylose is lower than that of glucose, the metabolism of xylose in a mixed sugar system can generate additional NADPH through the pentose phosphate pathway, alleviating the imbalance of reducing power in butanol synthesis. For example, after overexpressing xylose isomerase xylA and xylulokinase xylB in *C. tyrobutyricum*, the butanol yield when co-utilizing glucose and xylose is increased by 15% compared with that using glucose alone [8, 10]. In addition, the slow metabolism of xylose can extend the fermentation cycle and reduce the acute toxicity of butanol to cells [17]. The residual oligosaccharides such as cellobiose and xylobiose in the hydrolysate also make an indirect contribution. Clostridia secrete glycosidases to gradually degrade them, maintaining a continuous supply of carbon sources. *C. cellulovorans* can decompose cellobiose through β-glucosidase in the cellulosome, providing a stable carbon flow for butanol synthesis [18].

2.2.2 Impact of Inhibitors

The inhibitors such as furans, phenols, and organic acids produced during pretreatment, as mentioned earlier, interfere with the fermentation process through multiple pathways and are the main obstacles limiting the direct use of hydrolysates. Furfural and HMF can damage DNA and proteins through oxidative stress and inhibit the activities of sugar transporters, such as the glucose-PTS system, and key enzymes for butanol synthesis. When the furfural concentration exceeds 1 g/L, the growth rate of *C. acetobutylicum* decreases by 50%, and the butanol yield decreases by 40%. In addition, furans can also interfere with the balance of reducing power by consuming NAD(P)H, leading to more carbon flow towards by-products such as acetic acid [12–14].

Phenols and organic acids have a synergistic inhibitory mechanism: phenols such as ferulic acid and p-coumaric acid can disrupt the fluidity of cell membranes, increase the leakage of intracellular butanol, and aggravate product toxicity. When the concentration of phenols in the fiber hydrolysate reaches a certain level, organic acids such as formic acid and acetic acid simultaneously inhibit the energy metabolism of Clostridia by reducing the pH of the fermentation system, resulting in a decrease in ATP production and hindered cell proliferation. When the inhibitor concentration is low, the strain tolerance can be improved by inducing stress response genes such as heat shock protein groESL, but high concentrations can cause irreversible metabolic disorders.

2.2.3 Interactive Regulation in Cells

The coexistence of sugars and inhibitors in the hydrolysate leads to complex interactions, and their comprehensive effects depend on the concentration ratio of each component and the tolerance of the strain. This interaction may either alleviate part of the inhibitory effect through sufficient carbon sources or aggravate metabolic disorders due to the accumulation of inhibitors, ultimately affecting the butanol yield and fermentation stability.

In the hydrolysate with high sugar and low inhibitor content, sufficient carbon sources can promote the rapid proliferation of cells. By increasing the biomass to dilute the inhibitor concentration, part of the toxicity is offset, and at the same time, glucose and xylose are utilized while maintaining a high butanol yield. At this time, the rapid metabolism of glucose provides energy and reducing power (NADH) for the cells, which not only supports butanol synthesis but also converts a small amount of furan inhibitors into nontoxic alcohols through the NADH-dependent reduction pathway, further reducing the inhibitory effect [19].

On the contrary, in a system with low sugar and high inhibitor content, the toxicity of inhibitors is amplified. On the one hand, the slow metabolism of xylose leads to insufficient carbon source supply, limited cell energy production, and inability to effectively activate the detoxification mechanism; on the other hand, high-concentration phenolic compounds damage the integrity of cell membranes, aggravating the functional disorder of xylose transporter xylT and forming a vicious cycle, making carbon source utilization obstruction to insufficient energy with increasing inhibition, which greatly reduces the butanol yield.

For example, low-concentration Mg^{2+}, as a coenzyme of glycolytic enzymes, can enhance the utilization efficiency of xylose by *C. cellulovorans* and at the same time reduce the damage caused by phenolic compounds by stabilizing the cell membrane structure. However, high-concentration Ca^{2+} can combine with organic acids to form precipitates, which can alleviate the pH drop but also inhibit the activity of cellulase, indirectly reducing the release of fermentable sugars.

The metabolic characteristics of the strain also affect its response to component interactions. The solvent-producing *C. acetobutylicum* is more sensitive to inhibitors due to the acid-solvent metabolic transition, and the butanol yield decreases by 50% in the hydrolysate containing 2 g/L HMF, while the engineered acid-producing *Clostridium*, due to the lack of the acid accumulation pathway and a more direct butanol synthesis pathway, can still maintain a 70% yield under the same conditions [20]. This difference stems from the enhanced NADH regeneration capacity of the engineered strain, which can meet the needs of both butanol synthesis and inhibitor reduction.

Metal ions, as minor components in the hydrolysate, play a crucial bidirectional regulatory role in the complex interactive effects between sugars and inhibitors described above. While trace amounts of essential metal ions act as coenzymes for key glycolytic enzymes and stabilize cell membrane structures to mitigate phenolic inhibitor damage, thus enhancing xylose utilization efficiency of *Clostridium* strains, excessive concentrations of metal ions can form insoluble precipitates with organic acids in the hydrolysate. Although this precipitation can partially alleviate the pH drop caused by organic acids, it also inhibits cellulase activity and indirectly reduces the release of fermentable sugars, leading to lower fermentation efficiency. The regulatory effect of metal ions is closely related to their concentration and the specific *Clostridium* strain, which is another key factor to be considered in the subsequent optimization of cellulose hydrolysate for ABE fermentation. The intricate metabolic shifts between acidogenic and solventogenic phases during ABE fermentation, along with the key enzymatic reactions and carbon flow distribution for butanol synthesis, are clearly depicted in Fig. 2, which offers a visual framework for understanding the core biochemical processes underlying clostridial butanol production.

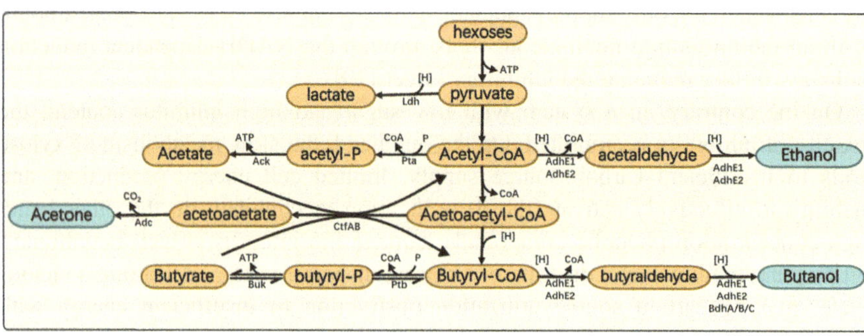

Fig. 2 Metabolic pathways of acidogenesis and solventogenesis in ABE fermentation

2.3 *Optimization Strategies for Cellulose Hydrolysate Fermentation*

2.3.1 Common Detoxification Technologies

The inhibitors in the hydrolysate are mainly derived from the degradation of sugars and lignin during the pretreatment of lignocellulose, which can damage the integrity of *Clostridium* cell membranes, inhibit enzyme activity, and interfere with energy metabolism. Detoxification technologies remove these inhibitors through physical, chemical, or biological methods, which is an important means to improve fermentation performance.

Existing physical detoxification methods mainly use adsorption or separation to remove inhibitors, with simple operation and low cost. Activated carbon adsorption is one of the most commonly used methods, and its porous structure can effectively adsorb furfural, HMF, and phenolic compounds. Studies have shown that after treating corncob acid hydrolysate with activated carbon, the butanol yield of *Clostridium* fermentation increased by 32%, which is because the adsorption rate of activated carbon for phenolic substances can reach more than 85% [21]. In addition, distillation and membrane separation technologies are also used for detoxification: vacuum distillation can remove furfural and HMF through differences in volatility, while ultrafiltration membranes can retain macromolecular phenols while retaining small molecular sugars. However, physical methods have limited effects on removing water-soluble inhibitors and often need to be combined with other methods [13, 22].

Chemical detoxification methods, which convert or neutralize inhibitors through chemical reactions, are suitable for treating high-concentration toxic substances. Excessive lime treatment is the most classic chemical detoxification method: adding $Ca(OH)_2$ to the hydrolysate to adjust the pH to 10–12, and at a temperature of 60–80°C, furfural and HMF can be decomposed into low-toxic furan derivatives, and some phenolic substances can be precipitated at the same time. After treating sugarcane bagasse hydrolysate with this method, the butanol yield of *C. beijerinckii* increased from 3.2 to 8.7 g/L [23]. In addition, adding reducing agents such as sodium sulfite can reduce aldehyde inhibitors to alcohols, reducing their toxicity to Clostridia, while using oxidants such as hydrogen peroxide can oxidize phenolic compounds, reducing their damage to cells [24]. However, chemical methods may lead to the loss of some sugars, so the reaction conditions need to be strictly controlled.

The use of microbial or enzymatic catalytic reactions to convert inhibitors has the advantages of high specificity and environmental friendliness. Many microorganisms can reduce furfural and HMF to corresponding alcohols through their own metabolism, and these products have significantly lower toxicity to Clostridia. For example, after pretreating corn stover hydrolysate with *Saccharomyces cerevisiae*, the butanol yield of *C. acetobutylicum* increased by 40%, and the fermentation lag phase was shortened from 12 to 6 h. Enzymatic detoxification is more targeted:

short-chain dehydrogenase SDR and aldo-keto reductase (AKR) can efficiently catalyze the reduction reactions of furfural and HMF, and engineered bacteria recombinantly expressing these enzymes can convert more than 90% of the aldehyde inhibitors in the hydrolysate within 24 h [19]. The limitation of biological detoxification is that the reaction cycle is relatively long, and the treatment efficiency for high-concentration inhibitors is limited.

2.3.2 Strain Modification

It is difficult to completely solve the complexity of hydrolysates only through detoxification technologies. Currently, the focus of research is to combine strain modification to enhance its ability to utilize mixed sugars and tolerance to inhibitors, thereby achieving efficient fermentation.

At present, the core direction of strain modification is to optimize the sugar metabolism pathway to alleviate carbon catabolite repression. The coexistence of glucose and xylose in lignocellulose hydrolysate triggers the CCR effect of Clostridia, that is, glucose is preferentially utilized while xylose metabolism is inhibited, resulting in low substrate utilization efficiency. Studies have shown that overexpressing key genes for xylose metabolism (xylA: xylose isomerase, xylB: xylulokinase, xylT: xylose transporter) can effectively alleviate this problem. For example, after heterologously expressing xylA, xylB, and xylT of *C. acetobutylicum* in *C. tyrobutyricum*, the co-utilization rate of glucose and xylose increased by 50%, and the butanol yield increased from 8.2 to 12.5 g/L [25]. In addition, knocking out genes related to glucose transport can reduce the preferential uptake of glucose, and introducing a non-phosphorylated xylose transport system can further balance sugar metabolism [26].

Enhancing the strain's tolerance to inhibitors can also reduce the dependence on detoxification steps and lower the process cost. The heat shock protein groESL plays a key role in the cell's response to environmental stress. The butanol yield of *C. tyrobutyricum* overexpressing groESL in the hydrolysate containing 2.5 g/L HMF was 37% higher than that of the wild type, which is because groESL can repair the protein structure damaged by inhibitors. In addition, screening high-tolerance strains through adaptive evolution is also an effective method: in the continuous culture of *C. acetobutylicum* in a fibrous bed bioreactor (FBB), after 100 generations of adaptation, its tolerance concentration to furfural increased from 1 to 3 g/L, and the butanol yield reached 0.35 g/(L h) [27]. Recent studies have also found that regulating the redox balance by introducing ferredoxin-$NAD(P)^+$ oxidoreductase can enhance the cell's resistance to oxidative stress, and the fermentation efficiency of the recombinant strain in the hydrolysate containing phenolic inhibitors is significantly improved.

The effect of a single strategy is often limited. Combining detoxification technology, strain modification, and fermentation process can maximize the fermentation performance of the hydrolysate. For example, using an integrated process of biological detoxification-engineered strain-in situ product separation to produce butanol from corn fiber hydrolysate, the yield reached 15.0 g/L, which was 87% higher than that of the traditional process, and the total energy consumption was reduced by 30%. In addition, fed-batch fermentation can avoid the inhibitory effect of high-concentration sugars and inhibitors. Combined with staged pH regulation, the acidic stage promotes cell growth, and the neutral stage promotes butanol synthesis, so that the butanol yield can reach 0.4 g/(L h) [28].

3 Direct Utilization of Agricultural Residues

3.1 Clostridia with Natural Cellulose-Degrading Ability

Natural cellulose-degrading Clostridia, relying on their unique cellulose decomposition systems, can directly utilize lignocellulosic biomass, providing low-cost substrates for ABE fermentation, and are the core strains of consolidated bioprocessing (CBP). Among them, *C. thermocellum* and *C. cellulovorans* are the most in-depth studied representative strains.

As a thermophilic anaerobic bacterium, *C. thermocellum* has the most significant metabolic feature of its efficient cellulosome system, which is a multienzyme complex that can synergistically degrade crystalline cellulose [14]. The core structure of the cellulosome includes anchor proteins, cellulases (endoglucanases, exoglucanases, β-glucosidases), and dockerin domains, which realize the ordered assembly of enzymes through cohesin-dockerin interactions. This structure enables the high enrichment of enzyme molecules on the substrate surface. Endoglucanases first cut the cellulose chain to produce free ends, exoglucanases further degrade it into cellobiose, and finally β-glucosidases convert it into glucose. The degradation efficiency is 3–5 times higher than that of free enzymes [29].

In terms of the metabolic pathway, glucose generates pyruvate through the EMP pathway, which is then converted into acetyl-CoA by pyruvate:ferredoxin oxidoreductase, while producing reduced ferredoxin. NAD(P)H is regenerated through ferredoxin-NAD(P)$^+$ oxidoreductase, providing reducing power for butanol synthesis. The uniqueness of *C. thermocellum* lies in the coupling of its energy metabolism and cellulose degradation: the cellulosome is combined with the cell membrane, and the degradation products can directly enter the cell, reducing the energy consumption of transportation. At the same time, ATP is generated through the membrane-bound H$^+$-ATPase to meet the energy demand in the high-temperature environment [18]. In addition, the metabolic network of *C. thermocellum* has plasticity, preferentially synthesizing ethanol when the carbon source is sufficient and switching to butanol production under nitrogen-limited conditions. Studies have found that its genome contains a complete butyryl-CoA synthesis pathway (*thl*, *hbd*, *crt*, *bcd* genes), but the butanol yield is low in the natural state, mainly due to the limited activity of aldehyde/alcohol dehydrogenase. This characteristic provides a target for subsequent metabolic engineering.

Although *C. cellulovorans*, as a mesophilic bacterium, also relies on the cellulosome to degrade cellulose, there are differences in its enzyme system composition and metabolic regulation compared with *C. thermocellum*. Its cellulosome contains a variety of cellulases including CelA, CelB, hemicellulase XynA, and pectinase, enabling it to degrade not only cellulose but also hemicellulose and pectin, with a wider substrate spectrum. The cellulosome is bound to the cell surface through anchor proteins, and the degradation product cellobiose enters the cell through the phosphotransferase system and is converted into glucose-6-phosphate by cellobiose phosphorylase, directly entering glycolysis [30].

In central metabolism, the carbon flow distribution of *C. cellulovorans* is more inclined to butyric acid synthesis, and butyrate is the main product in the natural state. Its butyryl-CoA synthesis pathway relies on key enzymes such as thl and 3-hydroxybutyryl-CoA dehydrogenase, and the production of butyrate is completed by butyrate kinase. When heterologous aldehyde/alcohol dehydrogenase is introduced, butyryl-CoA can be further converted into butanol. This process requires the consumption of additional NADH, so the cell regenerates NADH by enhancing FNR activity to balance the reducing power.

Compared with *C. thermocellum*, *C. cellulovorans* has the advantage of stronger tolerance to inhibitors. Studies have shown that it can still maintain more than 60% of cellulose degradation activity in lignocellulose hydrolysate containing furfural and HMF, which is related to the high concentration of glutathione in its cells. In addition, the fermentation temperature of *C. cellulovorans* is closer to the temperature commonly used in industrial fermentation, reducing the temperature control cost and having stronger application potential. The high-temperature adaptability of *C. thermocellum* can reduce the contamination of miscellaneous bacteria, but its cellulosome has high specificity for crystalline cellulose and is difficult to degrade complex lignocellulose. The mesophilic characteristics and broad-spectrum substrate utilization ability of *C. cellulovorans* make it more suitable for treating agricultural wastes, such as corncobs and sugarcane bagasse, but its degradation efficiency is slightly lower [14].

The metabolic mechanism of natural cellulose-degrading Clostridia provides a natural template for the direct use of agricultural wastes to produce butanol. The synergistic effect of its cellulosome system and the adaptability of energy metabolism are the keys to realizing efficient CBP. The metabolic networks of both connect cellulose degradation and butanol synthesis through glycolysis and the butyryl-CoA pathway, but the regulatory mechanisms are different. These differences provide diverse targets for strain modification. For example, introducing the efficient cellulosome genes of *C. thermocellum* into *C. cellulovorans* can construct an engineered bacterium with both high degradation efficiency and broad-spectrum substrate utilization ability.

3.2 Metabolic Engineering Modification of Non-cellulose-Degrading Clostridia

Although non-cellulose-degrading Clostridia cannot naturally degrade cellulose, their mature butanol synthesis pathways and genetic operation systems make them ideal hosts for metabolic engineering modification. By heterologously expressing cellulose degradation-related genes, optimizing metabolic flux distribution, and enhancing substrate transport efficiency, they can be endowed with the ability to directly utilize cellulose, providing a new way for low-cost butanol production.

The core reason why non-cellulose-degrading Clostridia such as *C. acetobutylicum* and *C. tyrobutyricum* lack cellulose decomposition ability is the absence of a functional cellulase system. Therefore, heterologous expression of cellulase or complete cellulosome gene cluster is the primary modification strategy.

In early studies, researchers attempted to express the endoglucanase gene celS of *C. thermocellum* alone, but it could only degrade soluble cellulose derivatives and had no effect on crystalline cellulose. This is because a single enzyme cannot efficiently destroy the crystalline structure of cellulose, and a synergistic enzyme system is required. In subsequent studies, by expressing fusion genes of cellulosome core components such as anchor protein cipA and various cellulases, artificial cellulosomes were constructed, which increased the degradation rate of recombinant strains on microcrystalline cellulose to 35% and the butanol yield to 2.1 g/L [18]. In addition, after introducing the cellulosome genes of *C. cellulovorans* into *C. tyrobutyricum*, its utilization rate of corn stover increased by 28%, but the butanol yield was still low, possibly due to insufficient enzyme assembly efficiency.

To solve the problem of insufficient enzyme activity, researchers improved the transcription efficiency of heterologous genes by optimizing expression elements. For example, using the strong promoter of *C. acetobutylicum* itself to drive the expression of cellulase genes increased the enzyme activity by 2.3 times compared with prokaryotic promoters. At the same time, fusion expression of cellulase and dockerin domain enabled it to be anchored to the cell surface, reducing the free loss of enzymes. The butanol productivity was increased by 40% [9].

At the substrate transport level, *C. acetobutylicum* naturally lacks cellobiose transporters, leading to the accumulation of cellulosome degradation products. By heterologously expressing the cellobiose transport gene cbtA of *C. thermocellum*, the uptake rate of cellobiose by the recombinant strain was significantly increased, and the butanol yield also increased simultaneously. For *C. tyrobutyricum*, overexpressing the xylose transporter xylT can alleviate the metabolic repression of glucose on xylose, increasing the utilization rate of mixed sugars by 50%.

At the carbon flow distribution level, it is necessary to inhibit the synthesis of by-products acetone and acetic acid and enhance the butanol synthesis pathway. Knocking out the acetoacetate decarboxylase gene adc of *C. acetobutylicum* can block acetone production, redirecting the carbon flow to butyryl-CoA. At the same time, overexpressing thiolase thl and 3-hydroxybutyryl-CoA dehydrogenase hbd can enhance butyryl-CoA synthesis. The butanol yield of the recombinant *C. tyrobutyricum* reached 20.5 g/L, which was 37% higher than that of the wild type [21]. In addition, introducing an NADH regeneration system can also significantly increase the yield.

After modification, non-cellulose-degrading Clostridia often face the dual toxic pressure of cellulose hydrolysis inhibitors and butanol, so tolerance modification is required to enhance strain stability. Regarding the impact of inhibitors, overexpressing the heat shock protein gene groESL can repair proteins damaged by inhibitors, and the butanol yield of the recombinant *C. acetobutylicum* strain in the hydrolysate containing furfural was significantly higher than that of the wild type. At the same time, through continuous subculture domestication in a fibrous bed bioreactor, the tolerance concentration of *C. tyrobutyricum* to butanol increased from 12 to 18 g/L, and stable fermentation performance was maintained. At the same time, knocking out spore formation-related genes can reduce strain dormancy, extend the alcohol production period, and shorten the total fermentation time [14].

At present, the metabolic engineering modification of non-cellulose-degrading Clostridia mainly follows the multidimensional optimization of degradation-transport-synthesis-tolerance to gradually realize the direct utilization of cellulose. In the future, synthetic biology tools and systematic metabolic analysis can be combined to further balance the enzyme expression level and metabolic load.

3.3 Consolidated Bioprocessing and Its Optimization

The consolidated bioprocessing (CBP), which directly uses cellulose for ABE fermentation, can couple the cellulose degradation and butanol synthesis processes, eliminating the expensive enzymatic hydrolysis step and significantly reducing the production cost.

Natural cellulose-degrading Clostridia such as *C. cellulovorans* and *C. thermocellum* rely on the cellulosome system to achieve substrate degradation, but their assembly efficiency and substrate specificity seriously restrict the degradation performance. The synergistic effect between the catalytic modules of the cellulosome is crucial for the degradation efficiency. Studies have shown that the cellulosome of *C. thermocellum* has a significantly higher degradation efficiency for crystalline cellulose than for noncrystalline cellulose, but the utilization rate of hemicellulose commonly found in agricultural wastes is less than 30%, resulting in incomplete degradation of mixed substrates. In addition, module mismatch is prone to occur when heterologously expressing the cellulosome, and the catalytic activity of the recombinant cellulosome is generally only 40–50% of that of the natural system.

Glucose and xylose produced by cellulose degradation need to be metabolized into acetyl-CoA and butyryl-CoA through glycolysis and the pentose phosphate pathway, but butanol synthesis consumes a large amount of NADH, leading to the carbon flow tending to generate by-products such as acetic acid and butyric acid to maintain balance. For example, when *C. cellulovorans* directly ferments cellulose, the butanol yield is only 0.15–0.20 g/g, while the proportion of butyric acid can reach more than 40% of the total products. At the same time, cellulose degradation requires the consumption of a large amount of ATP, which competes with the energy demand for butanol synthesis, resulting in poor metabolic coupling between cell growth and product synthesis, further reducing the yield [18].

Inhibitors such as furan compounds and phenolic substances produced during the pretreatment of agricultural wastes not only inhibit the growth of Clostridia but also damage the structural stability of the cellulosome. Studies have found that 0.5 g/L furfural can reduce the cellulose degradation rate of *C. thermocellum* by 50%, and 1.0 g/L HMF can reduce the activity of butyryl-CoA dehydrogenase by 30%. Butanol, as the final product, has a significant destructive effect on the cell membrane. When the concentration exceeds 15 g/L, it will cause cell lysis and fermentation termination; in direct fermentation, the synergistic toxicity of inhibitors and butanol further aggravates the process instability.

Optimizing the cellulosome structure through modular design is the key to improving the degradation efficiency. Overexpressing the cellulosome integrase gene and xylanase gene can increase the degradation rate of corncobs and the butanol yield. For non-cellulose-degrading Clostridia, heterologous expression of the complete cellulosome system can endow them with the ability to directly utilize cellulose. For example, after modifying *C. acetobutylicum* to express *cipA*, *celS*, and *docK* of *C. thermocellum*, it can directly use microcrystalline cellulose to produce butanol up to 4.2 g/L. In addition, knocking out the cellulosome inhibitor gene through CRISPR-Cas9 technology can relieve the transcriptional repression on cellulosome synthesis and further improve the degradation efficiency.

Staged regulation of temperature and pH can coordinate cellulose degradation and butanol synthesis. The optimal degradation temperature of thermophilic Clostridia is 55–60°C, but the butanol synthase has poor stability at this temperature. Using a two-stage temperature control strategy, 55°C in the early stage to promote cellulose degradation and reducing to 37°C in the later stage to maintain the activity of butanol synthase, can increase the butanol concentration by 40%. In terms of pH regulation, neutral conditions are conducive to cellulosome assembly, while slightly acidic conditions can promote butanol synthesis. Online pH switching can reduce butyric acid accumulation, increasing the butanol/acid ratio from 1.2 to 2.8 [18]. In addition, adding electron carriers such as methyl viologen can regulate NADH regeneration, further increasing the butanol yield.

Constructing a co-culture system with complementary functions can overcome the metabolic limitations of a single strain. When *C. thermocellum* and *C. beijerinckii* are co-cultured, the butanol yield of corn stover is significantly higher than that of a single strain. In situ product separation technology can alleviate butanol toxicity. For example, using gas stripping technology or pervaporation to remove butanol in real time can make the butanol concentration exceed 20 g/L. Immobilizing cells in an FBB can increase the cell concentration and enhance inhibitor tolerance. The butanol yield of *C. tyrobutyricum* directly fermenting sugarcane bagasse in FBB reaches 0.3 g/(L h), and it can operate continuously for 30 days without performance degradation [7]. To intuitively compare the butanol production efficiency of different lignocellulosic substrates under various optimization strategies, the key butanol yield data reported in relevant studies are summarized in Table 1. The results show that the integrated strategy combining detoxification, metabolic engineering, and in situ product separation achieves the highest butanol yield, while the direct fermentation of raw biomass by CBP still has room for improvement in yield, which is the key direction for subsequent research.

Table 1 Butanol production yield reported in lignocellulosic substrates' ABE fermentation studies

Microorganism	Cellulosic substrate	Pretreatment/ optimization strategy	Butanol yield	References
C. tyrobutyricum Δack-adhE2	Corn fiber hydrolysate	Undetoxified + fibrous bed bioreactor (FBB)	0.3 g/g substrate	[7]
C. tyrobutyricum	Sugarcane bagasse	FBB immobilization + direct fermentation (CBP)	0.3 g/(L h) productivity	
Engineered *Clostridium tyrobutyricum* (xylA/xylB overexpression)	Lignocellulosic hydrolysate	Glucose-xylose co-utilization engineering	–	[8]
Engineered *C. acetobutylicum* (cipA/celS/docK expression)	Microcrystalline cellulose	Heterologous cellulosome expression (CBP)	4.2 g/L (0.18 g/g substrate)	[18]
Engineered *C. cellulovorans*	Cellulose	Metabolic engineering + CBP	0.20 g/g substrate	
C. acetobutylicum	Corn stover hydrolysate	*Saccharomyces cerevisiae* biological detoxification	40% yield increase vs. control	[19]
Engineered *Clostridium tyrobutyricum* (groESL overexpression)	HMF-containing hydrolysate	Inhibitor tolerance engineering	37% yield increase vs. wild type	
C. acetobutylicum	Corncob acid hydrolysate	Activated carbon physical detoxification	32% yield increase vs. control	
C. beijerinckii	Sugarcane bagasse hydrolysate	Excessive lime detoxification	8.7 g/L (0.29 g/g substrate)	[23]
Engineered *C. tyrobutyricum* (xylA/xylB/xylT heterologous expression)	Lignocellulosic hydrolysate	CCR relief + mixed sugar co-utilization	12.5 g/L (0.35 g/g substrate)	[25]
C. tyrobutyricum	Corn fiber hydrolysate	Biological detoxification + engineered strain + in situ product separation	15.0 g/L (0.42 g/g substrate)	[28]

References

1. Zhao J, Lu C, Chen CC et al (2013) Biological production of butanol and higher alcohols. In: Bioprocessing technologies in biorefinery for sustainable production of fuels, chemicals, and polymers. Wiley, pp 235–261. https://doi.org/10.1002/9781118642047
2. Xue C, Zhao J, Chen LJ et al (2017) Recent advances and state-of-the-art strategies in strain and process engineering for biobutanol production by Clostridium acetobutylicum. Biotechnol Adv 35:310–322. https://doi.org/10.1016/j.biotechadv.2017.01.007
3. Dwidar M, Park JY, Mitchell RJ et al (2012) The future of butyric acid in industry. Sci World J 2012:471417. https://doi.org/10.1100/2012/471417
4. Wang J, Lin M, Xu M et al (2016) Anaerobic fermentation for production of carboxylic acids as bulk chemicals from renewable biomass. Adv Biochem Eng Biotechnol 156:323–361. https://doi.org/10.1007/10_2015_5009
5. Jones DT, Woods DR (1986) Acetone-butanol fermentation revisited. Microbiol Rev 50:484–524. https://doi.org/10.1128/mr.50.4.484-524.1986
6. Kumar M, Goyal Y, Sarkar A et al (2013) Comparative economic assessment of ABE fermentation based on cellulosic and non-cellulosic feedstocks. Appl Energy 93:193–204. https://doi.org/10.1016/j.apenergy.2011.12.079
7. Li J, Du Y, Bao T et al (2019) n-Butanol production from lignocellulosic biomass hydrolysates without detoxification by *Clostridium tyrobutyricum* Δack-adhE2 in a fibrous-bed bioreactor. Bioresour Technol 289:121749. https://doi.org/10.1016/j.biortech.2019.121749
8. Fu H, Yang ST, Wang M et al (2017b) Butyric acid production from lignocellulosic biomass hydrolysates by engineered *Clostridium tyrobutyricum* overexpressing xylose catabolism genes for glucose and xylose co-utilization. Bioresour Technol 234:389–396. https://doi.org/10.1016/j.biortech.2017.11.059
9. Bao T, Feng J, Jiang W et al (2020) Recent advances in n-butanol and butyrate production using engineered *Clostridium tyrobutyricum*. World J Microbiol Biotechnol 36:138. https://doi.org/10.1007/s11274-020-02914-2
10. Fu H, Yu L, Lin M et al (2017a) Metabolic engineering of *Clostridium tyrobutyricum* for enhanced butyric acid production from glucose and xylose. Metab Eng 40:50–58. https://doi.org/10.1016/j.ymben.2016.12.014
11. He F, Qin S, Yang Z et al (2020) Butyric acid production from spent coffee grounds by engineered *Clostridium tyrobutyricum* overexpressing galactose catabolism genes. Bioresour Technol 304:122977. https://doi.org/10.1016/j.biortech.2020.122977
12. Amiri H, Karimi K (2018) Pretreatment and hydrolysis of lignocellulosic wastes for butanol production: challenges and perspectives. Bioresour Technol 270:702–721. https://doi.org/10.1016/j.biortech.2018.08.117
13. Jönsson LJ, Alriksson B, Nilvebrant N-O (2013) Bioconversion of lignocellulose: inhibitors and detoxification. Biotechnol Biofuels 6:16–25. https://doi.org/10.1186/1754-6834-6-16
14. Cheng C, Bao T, Yang ST (2019) Engineering *Clostridium* for improved solvent production: recent progress and perspective. Appl Microbiol Biotechnol 103:5549–5566. https://doi.org/10.1007/s00253-019-09916-7
15. Yu M, Zhang Y, Tang IC et al (2011) Metabolic engineering of *Clostridium tyrobutyricum* for n-butanol production. Metab Eng 13:373–382. https://doi.org/10.1007/s00253-017-8200-1
16. Suo Y, Liao Z, Qu C, Fu H, Wang J (2019) Metabolic engineering of *Clostridium tyrobutyricum* for enhanced butyric acid production from undetoxified corncob acid hydrolysate. Bioresour Technol 271:266–273. https://doi.org/10.1016/j.ymben.2016.12.014
17. Yu L, Zhao J, Xu M, Dong J, Varghese S, Yu M, Tang IC, Yang ST (2015b) Metabolic engineering of *Clostridium tyrobutyricum* for n-butanol production: effects of CoA transferase. Appl Microbiol Biotechnol 99:4917–4930. https://doi.org/10.1007/s00253-015-6566-5
18. Bao T, Hou W, Wu X, Lu L, Zhang X, Yang ST (2021) Engineering *Clostridium cellulovorans* for highly selective n-butanol production from cellulose in consolidated bioprocessing. Biotechnol Bioeng 118:2703–2718. https://doi.org/10.1002/bit.27789

19. Suo YK, Luo S, Zhang YA, Liao ZP, Wang JF (2017) Enhanced butyric acid tolerance and production by Class I heat shock protein-overproducing *Clostridium tyrobutyricum* ATCC 25755. J Ind Microbiol Biotechnol 44:1145–1156. https://doi.org/10.1016/j.jbiotec.2018.02.012
20. Zhang J, Zong W, Hong W et al (2018) Exploiting endogenous CRISPR-Cas system for multiplex genome editing in *Clostridium tyrobutyricum* and engineer the strain for high-level butanol production. Metab Eng 47:49–59. https://doi.org/10.1016/j.ymben.2018.03.007
21. Ezeji TC, Qureshi N, Blaschek HP (2007) Production of acetone butanol ethanol (ABE) from liquefied corn starch, a commercial substrate, using *Clostridium beijerinckii* coupled with product recovery by gas stripping. J Ind Microbiol Biotechnol 34:771–777. https://doi.org/10.1016/j.procbio.2006.07.020
22. Sharma HK, Xu C, Qin W (2019) Biological pretreatment of lignocellulosic biomass for biofuels and bioproducts: an overview. Waste Biomass Valor 10:235–251. https://doi.org/10.1007/s12649-017-0059-y
23. Lu C, Dong J, Yang ST (2013) Butanol production from wood pulping hydrolysate in an integrated fermentation-gas stripping process. Bioresour Technol 143:467–475. https://doi.org/10.1016/j.biortech.2013.06.012
24. Diaz MC et al (2018) Lignocellulosic biomass as renewable feedstock for butanol production: a review. Renew Sustain Energy Rev 87:149–167. https://doi.org/10.1016/j.rser.2022.112130
25. Yu L, Xu M, Tang IC, Yang ST (2015a) Metabolic engineering of *Clostridium tyrobutyricum* for n-butanol production through co-utilization of glucose and xylose. Biotechnol Bioeng 112:2134–2141. https://doi.org/10.1002/bit.25613
26. Xiao H, Gu Y, Ning Y, Yang Y, Mitchell W, Jiang W, Yang S (2011) Confirmation and elimination of xylose metabolism bottlenecks in glucose phosphoenolpyruvate-dependent phosphotransferase system-deficient *Clostridium acetobutylicum* for simultaneous utilization of glucose, xylose, and arabinose. Appl Environ Microbiol 77:7886–7895. https://doi.org/10.1128/AEM.00644-11
27. Jiang L, Wang J, Liang S, Cai J, Xu Z, Cen P, Yang ST, Li S (2011) Enhanced butyric acid tolerance and bioproduction by *Clostridium tyrobutyricum* immobilized in a fibrous bed bioreactor. Biotechnol Bioeng 108:31–40. https://doi.org/10.1002/bit.22927
28. Du Y, Jiang W, Yu M, Tang IC, Yang ST (2015) Metabolic process engineering of *Clostridium tyrobutyricum* Δack-adhE2 for enhanced n-butanol production from glucose: effects of methyl viologen on NADH availability, flux distribution, and fermentation kinetics. Biotechnol Bioeng 112:705–715. https://doi.org/10.1002/bit.25489
29. Tamaru Y, Miyake H, Kuroda K, Nakanishi A, Kawade Y, Yamamoto K, Uemura M, Fujita Y, Doi RH, Ueda M (2010) Genome sequence of the cellulosome-producing mesophilic organism *Clostridium cellulovorans* 743B. J Bacteriol 192:901–902. https://doi.org/10.1128/jb.01450-09
30. Yang X, Xu M, Yang ST (2015) Metabolic and process engineering of *Clostridium cellulovorans* for biofuel production from cellulose. Metab Eng 32:39–48. https://doi.org/10.1016/j.ymben.2015.09.001
31. Lin Z, Cong W, Zhang J (2023) Biobutanol production from acetone–butanol–ethanol fermentation: developments and prospects. Fermentation 9(9):847. https://doi.org/10.3390/fermentation9090847
32. Shanmugam S, Hari A, Pugazhendhi A, Kikas T (2023) Integrated catalytic upgrading of biomass-derived alcohols for advanced biofuel production. Energies 16(13):4998. https://doi.org/10.3390/en16134998
33. Robles A, Sundar SV, Mohana Rangan S, Delgado AG (2023) Butanol as a major product during ethanol and acetate chain elongation. Front Bioeng Biotechnol 11:1181983. https://doi.org/10.3389/fbioe.2023.1181983

34. Poy H, da Costa Lopes AM, Lladosa E, Gabaldón C, Loras S, Silvestre AJD (2023) Enhanced biomass processing towards acetone-butanol-ethanol fermentation using a ternary deep eutectic solvent. Renew Energy 219(Pt 2):119488. https://doi.org/10.1016/j.renene.2023.119488
35. Behera S, Konde K, Patil S (2023) Methods for bio-butanol production and purification. In: Segovia-Hernandez JG, Behera S, Sanchez-Ramirez E (eds) Advances in pollution research: advances and developments in biobutanol production. Woodhead Publishing, pp 279–301. https://doi.org/10.1016/B978-0-323-91178-8.00004-7

Adv Biochem Eng Biotechnol (2026) 196: 203–220
https://doi.org/10.1007/10_2026_322

Published online: 21 July 2026

Metabolic Engineering of Microbial Strains for 1,2,4-Butanetriol Production: Progress and Future Perspectives

Wei-Bin Wang, Meng-Rui Tao, Kai Li, Takahiro Bamba, Cheng Cheng, Tomohisa Hasunuma, Akihiko Kondo, Nuttha Thongchul, Shuai Zhao, Jia-Xun Feng, Xin-Qing Zhao, and Feng-Wu Bai

Contents

W.-B. Wang, M.-R. Tao, K. Li, X.-Q. Zhao (✉), and F.-W. Bai
State Key Laboratory of Microbial Metabolism, Joint International Research Laboratory of Metabolic and Developmental Sciences, School of Life Sciences and Biotechnology, Shanghai Jiao Tong University, Shanghai, China
e-mail: xqzhao@sjtu.edu.cn

T. Bamba, T. Hasunuma, and A. Kondo
Engineering Biology Research Center, Kobe University, Kobe, Japan

C. Cheng
School of Biotechnology and Food Engineering, Hefei Normal University, Hefei, China

N. Thongchul
Institute of Biotechnology and Genetic Engineering, Chulalongkorn University, Bangkok, Thailand

S. Zhao and J.-X. Feng
State Key Laboratory for Conservation and Utilization of Subtropical Agro-bioresources, Guangxi Research Center for Microbial and Enzyme Engineering Technology, College of Life Science and Technology, Guangxi University, Nanning, Guangxi, China

Abstract Sustainable production of biofuels and biochemicals from renewable biomass represents a promising alternative to fossil-based feedstocks, offering significant benefits for low-carbon economies and environmental protection. The biochemical 1,2,4-butanetriol (BT) is an important short-chain chiral triol, which is widely used as pharmaceuticals, industrial foam, anticorrosion additives, and plasticizers. Although biosynthesis of BT has been achieved in various engineered microbes, achieving cost-competitive production remains a significant challenge. This chapter focuses on research progress in metabolic engineering of BT production, which mainly includes (1) design of BT biosynthetic pathways for expression in bacterial and yeast cell factories, (2) current metabolic engineering and process optimization strategies to optimize BT production, and (3) utilizing lignocellulosic biomass as a sustainable feedstock for BT production. In addition, future prospects on enhancing BT production efficiency through advanced metabolic engineering and synthetic biology approaches integrating with artificial intelligence (AI)-based technologies are also presented.

Graphical Abstract

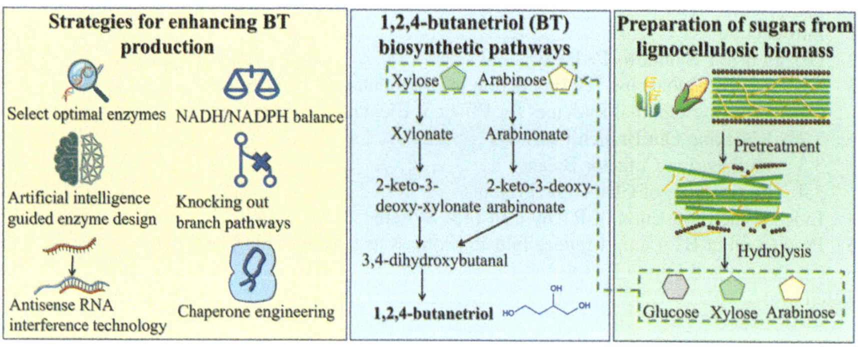

Keywords 1,2,4-Butanetriol (BT), Glucose and xylose, Lignocellulosic biomass, Metabolic engineering, Microbial cell factories

Abbreviations

5-HMF	5-Hydroxymethylfurfural
AI	Artificial intelligence
ALR	Aldehyde reductase
BT	1,2,4-Butanetriol
DESP	Distillation stimulation of bioethanol production
KDC	Ketodecarboxylase
XD	Xylonate dehydratase
XDH	Xylose dehydrogenase

1 Introduction

Natural and engineered microorganisms have been widely used in biomanufacturing of biofuels, biochemicals, nutritional foods, and bioactive natural products, demonstrating enormous potential for sustainable biorefinery [1]. Traditionally, starch and molasses-based substrates are used for biorefining. In recent years, lignocellulosic biomass has garnered significant attention as a cost-effective and abundant renewable resource for microbial biosynthesis of value-added products offering a sustainable alternative to petroleum-based feedstocks [2]. Biorefining of lignocellulosic biomass not only alleviates the reliance on fossil-based feedstocks, but also mitigates environmental burdens associated with traditional chemical manufacturing. The biorefinery process of lignocellulosic biomass generally encompasses several key steps: (1) Pretreatment: This initial crucial procedure aims to disrupt the recalcitrant structure of lignocellulose, enhancing the accessibility of biomass components for subsequent enzymatic or chemical hydrolysis. (2) Enzymatic hydrolysis: The polysaccharides in the pretreated lignocellulosic biomass are converted into fermentable monosaccharides, such as glucose, xylose, and arabinose by enzymes [3]. (3) Fermentation: The hydrolysates are then consumed by fermenting microbes to produce a spectrum of products (Fig. 1).

BT is an important nonnatural chiral polyol with a wide range of industrial and chemical applications. In the pharmaceutical industry, BT is a precursor for the synthesis of various important medical molecules, such as the cholesterol-lowering

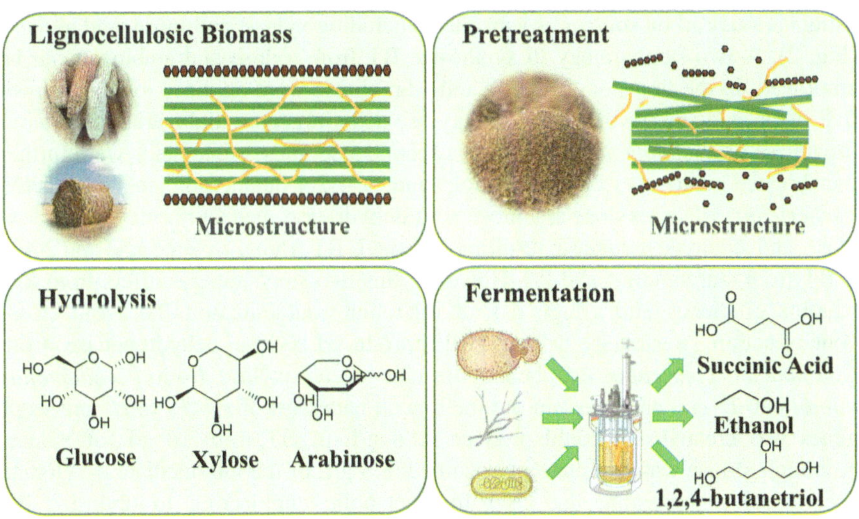

Fig. 1 Schematic diagram of the production of bio-based chemicals from lignocellulosic biomass. The process involves (1) pretreatment to break down the recalcitrant lignocellulosic structure, (2) enzymatic hydrolysis to release fermentable monosaccharides (e.g., glucose, xylose, and arabinose), and (3) microbial fermentation, where engineered or native microorganisms convert sugars into target biochemicals

drugs Lipitor, ezetimibe, and lovastatin [4]. In addition, 1,2,4-butanetriol trinitrate is widely used as a plasticizer due to its low impact sensitivity and high thermal stability, which is a substitute for nitroglycerin [5]. BT is also the precursor of polyurethane foam, which has improved elastic and compression bending properties similar to natural fibers [6]. These diverse applications make BT an important compound with high added value.

At present, large-scale production of BT still mainly relies on traditional chemical synthesis methods, including acid catalysis of 2-butene-1,4-diol and reduction of D, L-malic acid and its esters [7]. These chemical synthesis methods suffer from harsh reaction conditions, poor selectivity, and expensive metal catalysts and are not environmentally friendly. With the increasing demand for green and sustainable development, biosynthetic technology of BT with advantages such as mild reaction conditions, cost-effectiveness, and sustainability has become an important direction of current research.

2 Design of BT Synthesis Pathways

To date, no microorganisms have been found to naturally synthesize BT [8]. Therefore, bioproduction of BT is achieved by heterologous expression of corresponding pathway enzymes in various bacterial and yeast hosts, which will be further summarized in the following section.

Until now, researchers have successfully reconstructed BT biosynthetic pathways using various carbon sources as substrates, including xylose, arabinose, and glucose (Fig. 2). A two-step strategy to synthesize BT from xylose and arabinose can be employed. In the first step, D-xylose and L-arabinose were oxidized to their respective sugar acids by the endogenous D-xylose dehydrogenase and L-arabinose dehydrogenase of *Pseudomonas fragi*. In the second step, these sugar acids were further metabolized into D-1,2,4-butanetriol and L-1,2,4-butanetriol in engineered *Escherichia coli* via xylose/arabinose acid dehydratase, benzoylformate decarboxylase, and aldehyde reductase, resulting in 1.6 g/L BT from D-xylonic acid and 2.4 g/L BT from L-arabinonic acid [9]. However, this two-step strategy suffers from low reaction efficiency and a high risk of microbial contamination. To avoid cross-contamination, a strategy which [10] introduced xylose dehydrogenase from *Caulobacter crescentus* and benzoylformate decarboxylase from *Pseudomonas putida* into *E. coli* and knocked out the branch pathways of *xylAB*, *yjhH*, and *yagE* genes was created and could produce 0.88 g/L of BT from 10 g/L of xylose. Although xylose was completely consumed, 8.8 g/L of the intermediate D-xylonate accumulated, suggesting the need for metabolic engineering to optimize BT production.

Given the high cost of using pure xylose or arabinose as substrates, researchers have explored glucose as an economical substitute for constructing BT biosynthetic pathways. A non-native metabolic pathway for BT synthesis from malate was designed [11], which used glucose as the substrate, achieving BT production through

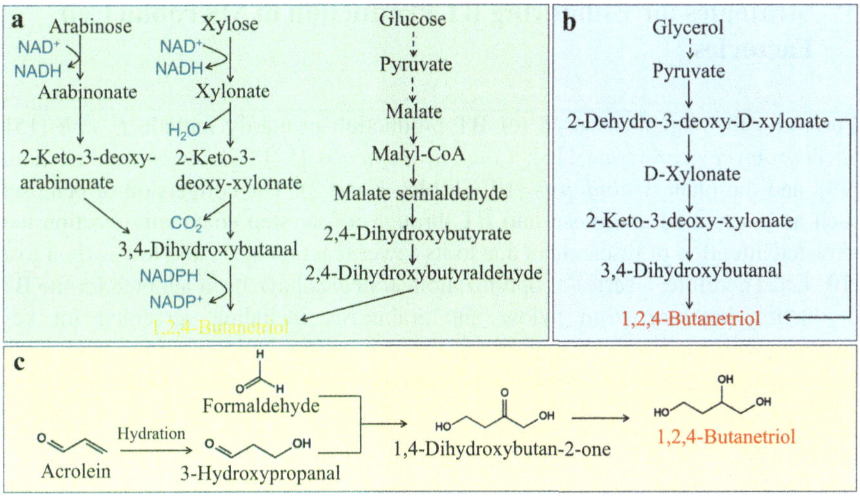

Fig. 2 Design of BT synthesis pathways. (**a**) The BT synthetic pathway constructed and tested in cells using arabinose [12], xylose [9], and glucose [11] as substrates. (**b**) The predicted pathway for BT synthesis using glycerol as the substrate has not yet been tested in microorganisms [13]. (**c**) One-pot chemical enzyme pathway for synthesizing BT from formaldehyde and acrolein [14]

six consecutive enzymatic reactions. Although the yield was only 120 ng/L, this study provided a new approach for BT synthesis using glucose. However, there are some challenges in directly producing BT from glucose on a large scale, including difficulty in balancing cell growth and BT production when using glucose as a carbon source. Additionally, the enzymes designed to catalyze the reactions in the malate-to-BT synthesis pathway exhibit low activity, leading to low metabolic flux [11, 12].

Conversion of carbon dioxide and other C1 compounds into high-value chemicals not only mitigates carbon emissions but also reduces dependency on fossil fuels. C1 compounds such as formaldehyde are used for the synthesis of BT through in vitro enzymatic catalysis. By searching for a new benzaldehyde lyase to catalyze the carbon–carbon coupling reaction between 3-HPA and formaldehyde, and introducing carbonyl reductase to catalyze the reduction reaction of 1,4-dihydroxybutan-2-one, a one-pot chemoenzymatic pathway was established for the synthesis of BT from formaldehyde and acrolein, with a final yield of up to 70% achieved [13]. In addition to the validated feasible BT synthesis routes, potential BT synthesis routes thermodynamically feasible using glycerol as a substrate were predicted through systems biology tools, but so far, this pathway has not been constructed and tested in any microorganisms [14], which warrants further investigation.

3 Strategies for Enhancing BT Production in Microbial Cell Factories

Currently, the organisms used for BT production primarily include *E. coli* [15], *Saccharomyces cerevisiae* [16], *Candida tropicalis* [17], *Gluconobacter oxydans* [18], and the plant *Arabidopsis* [19] (Tables 1 and 2). The conversion of pentoses such as xylose and arabinose into BT through a four-step enzymatic reaction has attracted attention of researchers due to its fewer reaction steps and low carbon loss [10, 12]. Therefore, a series of optimization strategies have been adopted for the BT production pathway from xylose and arabinose, including screening for key enzymes with high activity, optimizing the expression, knocking out branch pathways, enhancing reducing power, and overexpressing chaperone proteins (Fig. 3).

Table 1 Recombinant strains for BT production in prokaryotic organisms

Strain	Carbon source	Fermentation method	Titer (g/L)	Yield (mol/mol)	Reference
E. coli EWBT304	Xylose	Batch fermentation in 300 mL, shake flasks	0.88	0.13	Valdehuesa et al. [10]
E. coli as3KXW004	Xylose	Batch fermentation in 50 mL, shake flasks	10.03	0.47	Jing et al. [30]
E. coli R1	Glucose and xylose	Fed-batch fermentation in 50 mL, shake flasks	14.40	NF	She et al. [38]
E. coli BL21-14	Xylose	Batch fermentation in 500 mL, shake flasks	5.10	0.18	Wang et al. [24]
E. coli BT-10	Glucose, xylose, and arabinose	Fed-batch fermentation in 7.5 L, bioreactor	36.63	0.66	Li et al. [15]
E. coli BL21ΔxylAB/pEmdlCxylBC&pA-adhPyjhG	Glycerol and xylose	Fed-batch fermentation in 5 L-scale bioreactor	3.92	0.28	Cao et al. [33]
E. coli BL21(DE3)/30a-mtkABM. Petroleiphilum-sucD-4hbD/184-abfT-2-adhE2	Glucose	Batch fermentation in 20 mL, shake flasks	120 ng/L	NF	Li et al. [11]
BT-YjhG^{T325F}-SUF	Xylose	Whole-cell transformation reaction	10.36	0.74	Chen et al. [23]
G. oxydans/tGox0313-tppMdlc-tYagf	Xylose	Shake flasks with a 20 mL reaction mixture	5.31	0.376	Mao et al. [18]

NF: not found

Table 2 Recombinant strains for BT production in eukaryotic organisms

Strain	Carbon source	Fermentation method	Titer (g/L)	Yield (mol/mol)	Reference
S. cerevisiae BDδD-2tkdcAΔBOL2-tTYW1	Glucose and xylose	Batch fermentation in 200 mL, baffled Erlenmeyer flask	1.70	0.25	Bamba et al. [21]
S. cerevisiae BDδK6035	Glucose and xylose	Fed-batch fermentation in 1.0 L bioreactor	6.60	0.57	Yukawa et al. [16]
C. tropicalis BT19	Glucose and xylose	Batch fermentation in 50 mL, shake flasks	4.00	NF	Li et al. [17]

NF: not found

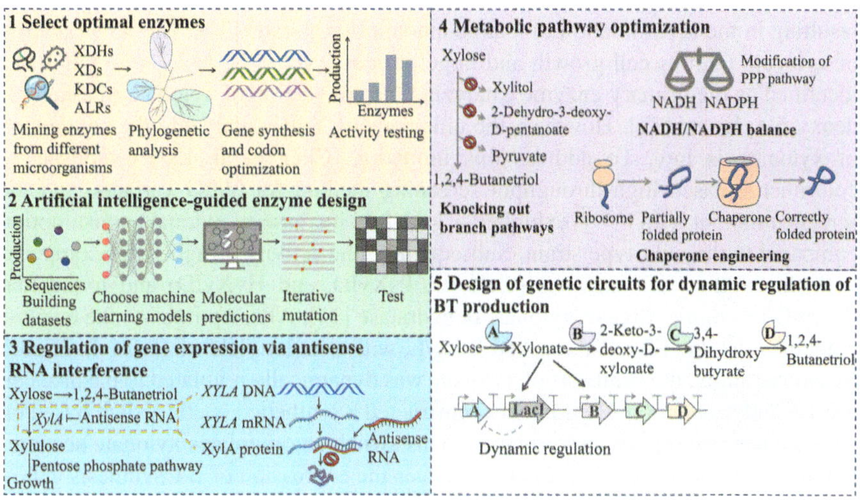

Fig. 3 Strategies for enhancing the biosynthesis of BT. (1) Selecting optimal enzymes. Enzyme candidates are mined from different microorganisms, conduct phylogenetic tree analysis, and then introduce them into hosts for enzyme activity evaluation after gene synthesis and codon optimization [28]. (2) Artificial intelligence (AI) guided enzyme design. Constructing datasets corresponding to the functions of different enzyme sequences, select appropriate machine learning models to train the datasets, and then generate potential mutants that may improve enzyme activity through model prediction, followed by experimental verification. (3) Regulation of gene expression by antisense RNA interference. Antisense RNA can bind to the mRNA sequence of the target gene, thereby preventing protein synthesis and inhibiting the expression of the target gene [30]. (4) Optimization of metabolic pathways includes knocking out branch pathways, balancing NADH and NADPH by regulating the pentose phosphate pathway and NADH kinase, and overexpressing chaperone proteins to enhance enzyme activity and tolerance to BT [16, 17, 32, 34, 35]. (5) Design gene circuits to regulate the accumulation of xylonate. Xylonate can promote the expression of the xylonate-responsive promoter *yjhI* and the expression of downstream genes and the repressor LacI. The expression of downstream genes promotes the conversion of xylonate, while the expression of LacI inhibits the production of xylonate, thereby reducing the accumulation of xylonate [25]

3.1 Selecting Optimal Enzymes for Efficient Biosynthesis of BT

BT biosynthetic pathway using xylose as the substrate involves four enzymes: xylose dehydrogenase (XDH), xylonate dehydratase (XD), ketodecarboxylase (KDC), and aldehyde reductase (ALR). XDH catalyzes the production of xylose lactone from D-xylose and spontaneously hydrolyzes it to xylonate. This enzyme relies on the coenzyme NADH to function. The synthesis rate of intermediate xylonate in the biosynthesis process of BT is much higher than the consumption rate. Recombinant *E. coli* can produce up to 39.2 g/L D-xylonate from 40 g/L D-xylose, with a productivity of up to 1.09 g/L/h [20]. Therefore, Xdh is a non-rate-limiting enzyme upstream of this pathway. On the contrary, the rapid catalytic reaction of the enzyme leads to an imbalance with subsequent enzyme reactions, resulting in the accumulation of a large amount of xylonate [21]. The accumulation of xylonate inhibits cell growth and target product synthesis. In *E. coli*, YjhG was identified as the primary enzyme catalyzing the conversion of xylonate to 2-keto-3-deoxy-D-xylonate [22]. However, the efficiency of YjhG catalyzing the conversion of xylonate is low. To address this limitation, Chen et al. [23] established a colorimetric-based high-throughput screening method for YjhG variants, and the screened mutant YjhG^{T325F} exhibited a 1.82-fold increase in xylonate consumption compared to the wild-type strain. Subsequently, the activity of YjhG was compared with three other XD enzymes (CcXylD, PsXylD, and HvXylD) and found that CcXylD is a more efficient xylonate dehydratase [24]. After optimizing the enzyme combinations, the BT yield reached 5.1 g/L, with a molar yield of 18.0% in *E. coli*. In another study, the synthesis of xylonate was dynamically regulated to balance cell growth and production. Researchers constructed a synthetic genetic switch based on the xylonate-responsive yjhI promoter. This switch autoregulates xylonate accumulation through feedback control and activates the expression of BT synthesis genes, successfully increasing the molar yield of BT in *E. coli* to 31% [25].

During BT biosynthesis, glucose is utilized for microbial growth, whereas the utilization of xylose or lactose is subject to glucose inhibition. To avoid glucose catabolite repression, a method which knocked out the *ptsG* gene encoding glucose transporter and the *pgi* gene encoding glucose-6-phosphate isomerase was applied, and then it increased the expression of *XYLD* and *KDCA* genes induced by lactose. This method reduced the accumulation of xylonate from 12.33 to 4.14 g/L [15]. Therefore, efficient expression of *XYLD* gene is crucial for the conversion of xylonate in the presence of mixed glucose and xylose.

XylD belongs to the IlvD/EDD protein family and requires iron-sulfur clusters for its catalytic activity [26]. The functional expression of XylD in eukaryotes has always been a challenge. And enhancing iron uptake and Fe-S cluster assembly could improve XylD activity [21, 23]. Knocking out the *BOL2* gene and overexpressing a truncated version of the *TYW1* gene achieved a 2.3-fold increase in XylD activity. Ultimately, 10 g/L of xylose was converted into 1.7 g/L of BT, with a molar yield of 24.5% [21]. However, the activity of XylD in *S. cerevisiae* still needs to be further enhanced. Previous studies have shown that XylD is highly active under weakly alkaline conditions (pH = 8.5) [26], and alkaline fermentation could promote the re-consumption of xylonate during the production of 3,4-dihydroxybenzoic acid [27]. The weakly alkaline environment may enhance the activity of heterologous XylD and the re-assimilation of xylonate by increasing the intracellular pH of *S. cerevisiae*. Therefore, re-matching the optimal pH for XylD activity is expected to improve the production of BT. In another study, phylogenetic analysis was used to identify XylD enzymes with higher activity. The XylD enzymes from *Rhizobium miluonense* and *Pseudooceanicola antarcticus* exhibited 29.2- and 30.8-fold higher activity than the traditional XylD from *C. crescentus*, respectively [28].

Decarboxylation is also the rate-limiting step in the BT biosynthetic pathway, primarily due to the low activity of benzylformate decarboxylase (Bfd) towards its nonnatural substrate 2-keto-3-deoxy-D-xylonate (KDX) [29]. The *kivD* gene from *Lactococcus lactis*, encoding 2-ketoisovalerate decarboxylase, is better suited for decarboxylation in BT production. Its expression results in a 72% increase in BT titer compared to the original enzyme [30]. In further comparison, the activities of four different KDC enzymes (MdlC from *P. putida*, KivD from *L. lactis*, KdcA from *L. lactis*, and Aro10 from *S. cerevisiae*) revealed that KdcA exhibited the highest BT production efficiency [24]. Subsequently, computational simulation and structure-guided evolution were employed to screen for KdcA variants to enhance its catalytic activity towards the nonnatural substrate KDX. The catalytic activity of mutants S286L/G402P and V461K was 1.8- and 2.5-fold higher than that of wild-type KdcA, respectively [31].

Aldehyde reductase plays a crucial role in the final step of BT synthesis, reducing 3,4-dihydroxybutyraldehyde to BT. Studies revealed that multiple endogenous enzymes such as YqhD, AdhP, and FucO in *E. coli* are involved in aldehyde reduction during BT production, with YqhD being the main aldehyde reductase [22]. In addition, Adh2 from budding yeast exhibits high activity in BT synthesis, and co-overexpressed with YqhD, BT production is significantly increased in *E. coli* [22]. This indicates that YqhD and Adh2 play a crucial role in BT synthesis. In addition, the knockout of *ADH6* in budding yeast was found to prevent the strain producing BT [27]. Therefore, Adh6 enzyme may play a key role in catalyzing the reduction of 3,4-dihydroxybutyraldehyde to BT in *S. cerevisiae*.

3.2 Knocking Out Branch Pathways

Endogenous reductases and dehydrogenases in microorganisms are capable of catalyzing the intermediates in the BT biosynthetic pathway. For example, 2-keto-3-deoxy-D-xylonate can be cleaved into pyruvate and glycolaldehyde by 2-keto-3-deoxy-D-xylonate aldolase (*yagE/yjhH*) [30] or converted to 2-dehydro-3-deoxy-D-pentonate via glyoxylate reductase (*ycdW/yiaE*) [22]. The intermediate 2-keto-3-deoxy-xylonate can also be catalyzed by transaminase, ketoreductases, and aldolase.

Deletion of the transaminase gene *hisC* increased BT titer by 21.7%, while knockout of ketoreductase genes *dld*, *lldD*, and *ilvC*, as well as aldolase genes *mhpE* and *yfaU*, also showed positive effects [32]. Additionally, 3,4-dihydroxybutanal can be oxidized to 3,4-dihydroxybutyrate by cellular aldehyde dehydrogenases [27]. Inactivation of these competing metabolic pathways is expected to further enhance BT production. However, complete pathway blockage often impairs cell viability. When xylose is used as the sole carbon source, direct knockout of the xylose isomerase gene *xylA* leads to a decrease in biomass. However, interference of *xylA* expression by antisense RNA restores the biomass and BT titer in the mutant strain [30]. In another study, glycerol was supplemented as an additional carbon source to support cell growth, and the engineered strain produced 3.92 g/L of BT from 20 g/L xylose under fed-batch conditions, achieving a molar yield of 27.7% [33].

3.3 Engineering Cofactor Balance

The distinct coenzyme dependencies of aldehyde reductase and xylose dehydrogenase may disrupt the metabolic balance, thereby compromising the BT synthesis rate. Overexpression of NADH kinase, an enzyme that converts NADH to NADPH, effectively alleviates cofactor imbalance. By overexpressing a truncated *POS5* lacking the signal peptide balances the generation and consumption of NADH and NADPH, thereby enhancing BT production [16]. Then, controlling the feeding rates of glucose and xylose during fed-batch fermentation, 16.4 g/L of xylose was ultimately converted into 6.6 g/L of BT, with a molar yield of 57% [16]. The insufficient supply of NADPH impairs the activity of aldehyde reductase. A common strategy to increase NADPH levels is overexpressing key genes in the pentose phosphate pathway. To alleviate NADPH deficiency in BT biosynthesis, a strategy which overexpressed the 6-phosphogluconate dehydrogenase gene *GND1* and knocked out the glucose-6-phosphate isomerase gene *PGI* was operated, achieving a BT yield of 3.2 g/L [17].

3.4 Engineering of Chaperone Proteins

Chaperone proteins play a critical role in enhancing both microbial tolerance to toxic metabolites and functional expression of heterologous enzymes in metabolic engineering. The overexpression of the GroESL chaperone complex in *E. coli* conferred tolerance to various alcohols, including BT, ethanol, and *i*-, *n*-, and 2-butanol, though the tolerance patterns showed complex and unpredictable dynamics [34]. Based on these findings, researchers optimized BT production by expressing different chaperone systems (GroES, GroEL, DnaKJ, and trigger factor) in engineered *E. coli*. Overexpression of the tig gene (trigger factor) has a positive effect on the activity of XDH and benzoyl decarboxylase, increasing BT titers from

0.42 g/L to a maximum of 1.01 g/L. Thus, chaperon proteins have the potential to increase the production of target compounds in microbial engineering [35].

4 Biocatalytic Production of BT by Cell-Free System

Due to the low efficiency of fermentation production of BT using microbial cell factory, there is an urgent need to develop a new biological process for producing BT from xylose. The cell-free catalytic system has the advantages of high substrate conversion rate, simple product separation, and avoiding interference from complex metabolic networks within cells. Therefore, researchers achieved a titer of 125 g/L (S)-BT from 180 g/L D-xylose in a 500 μL laboratory-scale reaction by selecting appropriate enzymes and optimizing reaction conditions, with a yield exceeding 97% [36]. By adopting the "intermediate enhancement" strategy and reducing the dosage of redox cofactors, the yield of (S)-BT reached 170 g/L, which is close to the toxicity level that microorganisms can tolerate. Researchers identified XylB from *C. crescentus*, XylD from *Tranquillimonas alkanivorans*, Kdc from *Staphylococcus aureus*, and AdhP from *Exiguobacterium qingdaonense* as the optimal key enzymes for (S)-BT biosynthesis through enzyme library screening [37]. In a 500 mL scale reaction, efficient conversion of (S)-BT from 200 g/L D-xylose was achieved with a concentration of 129 g/L and a conversion rate of 99%. Therefore, cell-free synthesis of BT is considered a feasible alternative to traditional fermentation.

5 Production of BT Using Lignocellulosic Biomass as a Renewable Feedstock

Lignocellulosic hydrolysates contain abundant glucose and xylose, with xylose being the second most abundant sugar after glucose. Usually, glucose is utilized for the growth of engineered strains, whereas xylose is employed for the synthesis of BT (Table 3). However, inhibitors such as weak acids and furan derivatives produced during the pretreatment process can severely inhibit the production of BT. Recombinant *E. coli* BL21-14 was able to produce 5.1 g/L of BT from pure xylose. However, the strain failed to produce detectable BT levels when applied to corncob hydrolysates under identical fermentation conditions. In contrast, 3.4 g/L of BT was produced through whole-cell bioconversion in corncob hydrolysates [24]. In another study, detoxified corncob hydrolysate was used as a substrate for batch fermentation to produce 43.4 g/L of BT, which is equivalent to 90% of the theoretical yield [15]. *E. coli* exhibits poor resistance to hydrolysis inhibitors commonly present in lignocellulosic hydrolysates. To enhance the tolerance of strains to inhibitors, researchers overexpressed furfural tolerance genes (e.g., *ucpA*, *fucO*) in *E. coli*, reducing reactive oxygen species accumulation and promoting glycolysis,

Table 3 Production of BT using lignocellulosic hydrolysates

Strain	Lignocellulosic biomass	Pretreatment	Titer (g/L)	Yield (mol/mol)	Reference
E. coli BL21-14	Detoxified corncob hydrolysates, whole-cell biocatalysts	Sulfuric acid heat	3.40	NF	Wang et al. [24]
E. coli BT-10	Detoxified corncob hydrolysate	NF	43.40	0.90	Li et al. [15]
E. coli R1	Detoxified corncob cellulose hydrolysate	Sulfuric acid heat	11.20	0.40	She et al. [38]
G. oxydans/tGox0313-tppMdlc-tYagf	Distillation stillage of bioethanol production (DSEP)	Diluted sulfuric acid	4.83	0.34	Mao et al. [18]
S. cerevisiae BDδD-2tkdcAΔBOL2-tTYW1	2.0-fold diluted rice straw hydrolysate	Sulfuric acid heat	1.10	0.21	Bamba et al. [21]
C. tropicalis BT19	1.5-fold diluted corncob hydrolysate	Alkaline heat	3.42	0.16	Li et al. [17]

NF: not found

which increased the BT titer to 14.4 g/L (101% improvement compared to the control) under 0.4 g/L furfural condition [38]. In contrast to *E. coli*, *G. oxydans* exhibits good resistance to furan derivatives such as furfural, HMF, and vanillin. Leveraging this trait, engineered *G. oxydans* achieved a titer of 4.83 g/L BT with a molar yield of 34.2% using distillation stimulation of bioethanol production (DESP) as the substrate [18].

S. cerevisiae is characterized by high robustness and stress tolerance compared to engineered *E. coli* strains [39]. Therefore, employing recombinant yeast strains for the production of BT represents an economically viable and sustainable bioprocess. When undiluted sulfuric acid-pretreated rice straw hydrolysate was used as the substrate for fermentation, the strain produced only 0.62 g/L of BT, with a molar yield of 5.5%. However, fermentation using two-fold diluted rice straw hydrolysate resulted in a higher cell concentration and a molar yield of 20.5% for BT [21]. This indicates that the inhibitors present in hydrolysates severely inhibit the fermentation and growth of strain. *C. tropicalis* demonstrates the capability to utilize diverse lignocellulosic feedstocks, exhibits excellent growth performance, and emerges as a promising host for BT production [17]. When alkaline heat-pretreated corncob hydrolysate was used as the substrate to fermentation, 3.38 g/L of BT was produced, with a molar yield of 13%. Although the hydrolysates contained 0.52 mM furfural and 0.16 mM 5-hydroxymethylfurfural (5-HMF), the final biomass of the recombinant *C. tropicalis* was not adversely affected [17].

6 Challenges and Future Perspectives

In the past years, significant research efforts have been devoted to the microbial biosynthesis of BT, such as key enzyme screening, enzyme engineering, synthetic pathway optimization, and enhancement of reducing power. However, yield of BT using xylose and arabinose as substrates is still much lower than the theoretical yield (Tables 1 and 2), and large-scale industrial production still lacks competitiveness. The key challenges primarily focus on the accumulation of intermediate metabolite, insufficient reducing power, and complex branching pathways.

Firstly, the suboptimal activity and expression levels of key enzymes – xylonate dehydrogenase and 2-ketoacid decarboxylase, represent primary bottlenecks in the pathway. This enzymatic limitation leads to significant accumulation of the intermediate metabolite xylonate, which exerts dual inhibitory effects: (1) impairing cellular growth and metabolic activity and (2) substantially reducing the overall BT conversion efficiency. Improving XylD activity has always been a huge challenge in yeast. Although XylD with increased activity in vitro has been identified, the activity remains poor in vivo [28]. As the optimal pH for CcXylD is 8.5, alkaline fermentation has been demonstrated to promote the reassimilation of xylonate [27], though it has not yet been applied to BT production.

The rapid development of artificial intelligence (AI) can accelerate the process of enzyme engineering. AI-driven approaches enable rapid mining of enzyme candidates from vast gene libraries by establishing sequence-function relationships through machine learning algorithms [40, 41]. For instance, transformer-based models like UniKP have demonstrated exceptional accuracy in predicting enzymatic properties such as catalytic efficiency and thermostability by analyzing primary sequences, significantly accelerating the screening process compared to traditional experimental methods [40]. Through machine learning-driven parallel iterative exploration of protein sequence space, the directed evolution of high-performance biocatalytic enzymes has been further achieved [41]. These results highlight the great potential of AI in accelerating the directed evolution of enzymes and improving strain performance. Therefore, integrating artificial intelligence models to guide rational design of xylonate dehydrogenase and 2-keto decarboxylase may enhance its catalytic activity in yeast, synergistically boosting metabolic efficiency towards BT biosynthesis.

In addition, machine learning was also employed to predict optimal promoter combinations and fine-tune gene transcription levels, achieving precise balancing of metabolic fluxes and significantly enhancing the production of target compounds such as naringenin and ethanol [42, 43]. Thus, balancing the expression of the four enzyme genes in the BT pathway and leveraging machine learning tools to identify optimal promoter combinations hold promise for improving BT production.

In the future, AI could serve as a crucial enabling tool for enzyme engineering and metabolic pathway design. On the other hand, by linking with standardized libraries of genetic elements in synthetic biology, AI can simulate the regulatory effects of different component combinations on gene expression level, thereby achieving optimal matching of multienzyme reactions.

Secondly, in the context of bioprocess engineering, optimizing the fermentation process is essential to improve microbial production of BT. Controlling key parameters such as pH, temperature, aeration, and substrate feeding strategy during batch or fed-batch fermentations can significantly influence BT titers, yields, and by-product formation. Implementation of advanced bioreactor control systems and real-time monitoring technologies can further enhance process consistency and scalability.

Thirdly, insufficient reducing power can affect cell growth and BT synthesis. This metabolic constraint primarily arises from the imbalance between NADH production and NADPH utilization fluxes, creating redox bottlenecks that significantly compromise BT yield. Overexpression of genes related to the pentose phosphate pathway and NADH kinase can alleviate NADPH deficiency [16, 17], and continuous glucose supplementation also provides the cell with the cofactor NADH and NADPH. However, in the co-fermentation of glucose and xylose, due to the lack of xylose-specific transporters in cells, glucose strongly inhibits xylose uptake [44]. Further alleviation of carbon catabolite repression effects to promote xylose consumption and optimization of the feeding concentration and ratio of glucose and xylose could be considered.

Finally, the presence of inhibitors in lignocellulosic hydrolysates, such as acetic acid [45], furfural, and 5-hydroxymethylfurfural, poses a significant challenge to yeast fermentation [46]. The combined effect of different inhibitors can negatively affect cell growth, xylose utilization, and metabolic activity, thereby reducing the efficiency of BT production. Although increasing stress tolerance of the production strain to a single inhibitor slightly improves BT production, the performance of BT-producing strain in hydrolysis solutions with complex inhibitors is still poor [38]. Thus, the development of inhibitor-tolerant strains via adaptive laboratory evolution and metabolic engineering represents a promising approach to improve BT production from lignocellulosic hydrolysates.

In conclusion, to meet the challenges for biosynthesis of BT from mixed sugars, multiple strategies including strain development and process engineering are needed. With the continuous advancement of artificial intelligence and synthetic biology, more efficient microbial sustainable production of BT can be achieved, and further scale-up of the fermentation process using low-cost cellulosic hydrolysate will also be explored in the future for sustainable production.

References

1. Saxena A, Hussain A, Parveen F, Ashfaque M (2023) Current status of metabolic engineering of microorganisms for bioethanol production by effective utilization of pentose sugars of lignocellulosic biomass. Microbiol Res 276:127478. https://doi.org/10.1016/j.micres.2023.127478

2. Zhao XQ, Liu CG, Bai FW (2024) Making the biochemical conversion of lignocellulose more robust. Trends Biotechnol 42:418–430. https://doi.org/10.1016/j.tibtech.2023.09.014
3. Yang J, Yue HR, Pan LY, Feng JX, Zhao S, Suwannarangsee S, Chempreda V, Liu CG, Zhao XQ (2023) Fungal strain improvement for efficient cellulase production and lignocellulosic biorefinery: current status and future prospects. Bioresour Technol 385:129449. https://doi.org/10.1016/j.biortech.2023.129449
4. Mangaloglu L, Van-Iderstine S, Chen B, Taghibiglou C, Cheung R, Adeli K (2000) Effect of atorvastatin (Lipitor™) on VLDL-apoB and VLDL-triglyceride overproduction in vivo in an insulin resistant hamster model. Atherosclerosis 151:45. https://doi.org/10.1016/S0021-9150(00)80202-6
5. Liu MH, Liu CW (2017) Comparative simulation study of chemical synthesis of energetic (R)-1,2,4-butanetriol trinitrate plasticizer. Int J Quantum Chem 117:e25402. https://doi.org/10.1002/qua.25402
6. Lau MK (2007) Syntheses and downstream puriffcation of 1,2,4-butanetriol. Dissertation, Michigan State University. https://www.proquest.com/dissertations-theses/syntheses-downstream-purification-1-2-4/docview/304850015/se-2
7. Hanessian S, Ugolini A, Dubé D, Glamyan A (1984) Facile access to (S)-1,2,4-butanetriol and its derivatives. Can J Chem 62:2146–2147. https://doi.org/10.1139/v84-367
8. Ma X, Sun C, Xian M, Guo J, Zhang R (2024) Progress in research on the biosynthesis of 1,2,4-butanetriol by engineered microbes. World J Microbiol Biotechnol 40(2):68. https://doi.org/10.1007/s11274-024-03885-4
9. Niu W, Molefe MN, Frost JW (2003) Microbial synthesis of the energetic material precursor 1,2,4-butanetriol. J Am Chem Soc 125:12998–12999. https://doi.org/10.1021/ja036391+
10. Valdehuesa KNG, Liu H, Ramos KRM, Park SJ, Nisola GM, Lee W-K, Chung W-J (2014) Direct bioconversion of d-xylose to 1,2,4-butanetriol in an engineered *Escherichia coli*. Process Biochem 49:25–32. https://doi.org/10.1016/j.procbio.2013.10.002
11. Li X, Cai Z, Li Y, Zhang Y (2014) Design and construction of a non-natural malate to 1,2,4-butanetriol pathway creates possibility to produce 1,2,4-butanetriol from glucose. Sci Rep 4:5541. https://doi.org/10.1038/srep05541
12. Wang J, Chen Q, Wang X, Chen K, Ouyang P (2022) The biosynthesis of D-1,2,4-butanetriol from d-arabinose with an engineered *Escherichia coli*. Front Bioeng Biotechnol 10:844517. https://doi.org/10.3389/fbioe.2022.844517
13. Li Y, Yao P, Zhang S, Feng J, Su H, Liu X, Sheng X, Wu Q, Zhu D, Ma Y (2023) Creating a new benzaldehyde lyase for atom-economic synthesis of chiral 1,2,4-butanetriol and 2-aminobutane-1,4-diol from formaldehyde. Chem Catal 3:100467. https://doi.org/10.1016/j.checat.2022.11.006
14. Ghosh D (2017) Synthetic and systems biology approach towards designing metabolic bypass and identifying novel enzymes for cholesterol lowering drug precursor (bto) biosynthesis from crude glycerol. J Appl Pharm Sci 7:10. https://doi.org/10.7324/JAPS.2017.71007
15. Li P, Wang M, Di H, Du Q, Zhang Y, Tan X, Xu P, Gao C, Jiang T, Lü C, Ma C (2024) Efficient production of 1,2,4-butanetriol from corn cob hydrolysate by metabolically engineered *Escherichia coli*. Microb Cell Factories 23:49. https://doi.org/10.1186/s12934-024-02317-0
16. Yukawa T, Bamba T, Guirimand G, Matsuda M, Hasunuma T, Kondo A (2021) Optimization of 1,2,4-butanetriol production from xylose in *Saccharomyces cerevisiae* by metabolic engineering of NADH/NADPH balance. Biotechnol Bioeng 118:175–185. https://doi.org/10.1002/bit.27560
17. Li J, Xia Y, Wei B, Shen W, Yang H, Chen X (2024) Metabolic engineering of *Candida tropicalis* for efficient 1,2,4-butanetriol production. Biochem Biophys Res 710:149876. https://doi.org/10.1016/j.bbrc.2024.149876
18. Mao X, Qian X, Lin J, Wei D (2023) Engineering *Gluconobacter oxydans* for efficient production of 3,4-dihydroxybutunate or 1,2,4-butanetriol from D-xylose. Biochem Eng J 195:108936. https://doi.org/10.1016/j.bej.2023.108936

19. Abdel-Ghany SE, Day I, Heuberger AL, Broeckling CD, Reddy AS (2013) Metabolic engineering of *Arabidopsis* for butanetriol production using bacterial genes. Metab Eng 20:109–120. https://doi.org/10.1016/j.ymben.2013.10.003
20. Liu H, Valdehuesa KN, Nisola GM, Ramos KR, Chung WJ (2012) High yield production of D-xylonic acid from D-xylose using engineered *Escherichia coli*. Bioresour Technol 115:244–248. https://doi.org/10.1016/j.biortech.2011.08.065
21. Bamba T, Yukawa T, Guirimand G, Inokuma K, Sasaki K, Hasunuma T, Kondo A (2019) Production of 1,2,4-butanetriol from xylose by *Saccharomyces cerevisiae* through Fe metabolic engineering. Metab Eng 56:17–27. https://doi.org/10.1016/j.ymben.2019.08.012
22. Zhang N, Wang J, Zhang Y, Gao H (2016) Metabolic pathway optimization for biosynthesis of 1,2,4-butanetriol from xylose by engineered *Escherichia coli*. Enzym Microb Technol 93–94: 51–58. https://doi.org/10.1016/j.enzmictec.2016.07.007
23. Chen J, Liu Z, Mai D, Xu S, Wang X, Chen K (2025) Enzyme and cofactor engineering to increase d-xylonate dehydratase activity for improved d-1,2,4-butanetriol production from d-xylose. Synth Syst Biotechnol 10:1234–1241. https://doi.org/10.1016/j.synbio.2025.07.003
24. Wang X, Xu N, Hu S, Yang J, Gao Q, Xu S, Chen K, Ouyang P (2018) D-1,2,4-Butanetriol production from renewable biomass with optimization of synthetic pathway in engineered *Escherichia coli*. Bioresour Technol 250:406–412. https://doi.org/10.1016/j.biortech.2017.11.062
25. Bañares AB, Valdehuesa KNG, Ramos KRM, Nisola GM, Lee WK, Chung WJ (2019) Discovering a novel D-xylonate-responsive promoter: the PyjhI-driven genetic switch towards better 1,2,4-butanetriol production. Appl Microbiol Biotechnol 103(19):8063–8074. https://doi.org/10.1007/s00253-019-10073-0
26. Andberg M, Aro-Kärkkäinen N, Carlson P, Oja M, Bozonnet S, Toivari M, Hakulinen N, O'Donohue M, Penttilä M, Koivula A (2016) Characterization and mutagenesis of two novel iron-sulphur cluster pentonate dehydratases. Appl Microbiol Biotechnol 100:7549–7563. https://doi.org/10.1007/s00253-016-7530-8
27. Yukawa T, Bamba T, Matsuda M, Yoshida T, Inokuma K, Kim J, Won Lee J, Jin YS, Kondo A, Hasunuma T (2023) Enhanced production of 3,4-dihydroxybutyrate from xylose by engineered yeast via xylonate re-assimilation under alkaline condition. Biotechnol Bioeng 120:511–523. https://doi.org/10.1002/bit.28278
28. Tanaka K, Yukawa T, Bamba T, Wakiya M, Kumokita R, Jin YS, Kondo A, Hasunuma T (2025) Engineering *Saccharomyces cerevisiae* for growth on xylose using an oxidative pathway. Appl Microbiol Biotechnol 109:30. https://doi.org/10.1007/s00253-025-13417-1
29. Yep A, Kenyon GL, McLeish MJ (2008) Saturation mutagenesis of putative catalytic residues of benzoylformate decarboxylase provides a challenge to the accepted mechanism. Proc Natl Acad Sci 105:5733–5738. https://doi.org/10.1073/pnas.0709657105
30. Jing P, Cao X, Lu X, Zong H, Zhuge B (2018) Modification of an engineered *Escherichia coli* by a combined strategy of deleting branch pathway, fine-tuning xylose isomerase expression, and substituting decarboxylase to improve 1,2,4-butanetriol production. J Biosci Bioeng 126: 547–552. https://doi.org/10.1016/j.jbiosc.2018.05.019
31. Lv K, Cao X, Pedroso MM, Wu B, Li J, He B, Schenk G (2024) Structure-guided engineering of branched-chain α-keto acid decarboxylase for improved 1,2,4-butanetriol production by in vitro synthetic enzymatic biosystem. Int J Biol Macromol 255:128303. https://doi.org/10.1016/j.ijbiomac.2023.128303
32. Zhang L, Wang J, Gu S, Liu X, Hou M, Zhang J, Yang G, Zhao D, Dong R, Gao H (2024) Biosynthesis of D-1,2,4-butanetriol promoted by a glucose-xylose dual metabolic channel system in engineered *Escherichia coli*. New Biotechnol 83:26–35. https://doi.org/10.1016/j.nbt.2024.06.003
33. Cao Y, Niu W, Guo J, Xian M, Liu H (2016) Biotechnological production of 1,2,4-butanetriol: an efficient process to synthesize energetic material precursor from renewable biomass. Sci Rep 5(1):18149. https://doi.org/10.1038/srep18149

34. Zingaro KA, Terry Papoutsakis E (2013) GroESL overexpression imparts *Escherichia coli* tolerance to i-, n-, and 2-butanol, 1,2,4-butanetriol and ethanol with complex and unpredictable patterns. Metab Eng 15:196–205. https://doi.org/10.1016/j.ymben.2012.07.009
35. Lu X, He S, Zong H, Song J, Chen W, Zhuge B (2016) Improved 1,2,4-butanetriol production from an engineered *Escherichia coli* by co-expression of different chaperone proteins. World J Microbiol Biotechnol 32(9):149. https://doi.org/10.1007/s11274-016-2085-5
36. Sutiono S, Zachos I, Paschalidis L, Pick A, Burger J, Sieber V (2023) Biocatalytic production of 1,2,4-butanetriol beyond a titer of 100 g/l: boosting by intermediates. ACS Sustain Chem Eng 11:6592–6599. https://doi.org/10.1021/acssuschemeng.2c07418
37. Ji Y, Cui Y, Liu X, Li J, Yao P, Zhao Q, Wu Q, Zhu D (2024) Biocatalytic production of (S)-1,2,4-butanetriol from d-xylose by whole cells of engineered *Escherichia coli*. Mol Catal 562: 114230. https://doi.org/10.1016/j.mcat.2024.114230
38. She D, Wang S, Zong H, Lu X, Zhuge B (2023) Genetic modification of *Escherichia coli* to improve 1,2,4-butanetriol production from cellulose hydrolysate. Syst Microbiol Biomanuf 9: 1–9. https://doi.org/10.1007/s43393-023-00177-0
39. Qin L, Dong S, Yu J, Ning X, Xu K, Zhang SJ, Xu L, Li BZ, Li J, Yuan YJ, Li C (2020) Stress-driven dynamic regulation of multiple tolerance genes improves robustness and productive capacity of *Saccharomyces cerevisiae* in industrial lignocellulose fermentation. Metab Eng 61: 160–170. https://doi.org/10.1016/j.ymben.2020.06.003
40. Yu H, Deng H, He J, Keasling JD, Luo X (2023) UniKP: a unified framework for the prediction of enzyme kinetic parameters. Nat Commun 14(1):8211. https://doi.org/10.1038/s41467-023-44113-1
41. Landwehr GM, Bogart JW, Magalhaes C, Hammarlund EG, Karim AS, Jewett MC (2025) Accelerated enzyme engineering by machine-learning guided cell-free expression. Nat Commun 16:865. https://doi.org/10.1038/s41467-024-55399-0
42. Deng H, Yu H, Deng Y, Qiu Y, Li F, Wang X, He J, Liang W, Lan Y, Qiao L, Zhang Z, Zhang Y, Keasling JD, Luo X (2024) Pathway evolution through a bottlenecking-debottlenecking strategy and machine learning-aided flux balancing. Adv Sci (Weinh) 11(14): e2306935. https://doi.org/10.1002/advs.202306935
43. Khamwachirapithak P, Sae-Tang K, Mhuantong W, Tanapongpipat S, Zhao XQ, Liu CG, Wei DQ, Champreda V, Runguphan W (2023) Optimizing ethanol production in *Saccharomyces cerevisiae* at ambient and elevated temperatures through machine learning-guided combinatorial promoter modifications. ACS Synth Biol 12(10):2897–2908. https://doi.org/10.1021/acssynbio.3c00199
44. Jiang Y, Shen Y, Gu L, Wang Z, Su N, Niu K, Guo W, Hou S, Bao X, Tian C, Fang X (2020) Identification and characterization of an efficient d-xylose transporter in *Saccharomyces cerevisiae*. J Agric Food Chem 68:2702–2710. https://doi.org/10.1021/acs.jafc.9b07113
45. Yuan B, Zhu YF, Li K, Zhao XQ (2025) Chromatin regulation of acetic acid stress tolerance by ino80 in budding yeast *Saccharomyces cerevisiae*. J Agric Food Chem 73(5):2951–2960. https://doi.org/10.1021/acs.jafc.4c09994
46. Sjulander N, Kikas T (2020) Origin, impact and control of lignocellulosic inhibitors in bioethanol production – a review. Energies 13:4751. https://doi.org/10.3390/en13184751

Adv Biochem Eng Biotechnol (2026) 196: 221–248
https://doi.org/10.1007/10_2026_323

Published online: 3 June 2026

Sustainable Production of Succinic Acid from Lignocellulosic Hydrolysate

Rui-Qi Tang, Du-Wen Zeng, Xin-Qing Zhao, and Du Zhu

Contents

R.-Q. Tang (✉)
Key Laboratory of Natural Microbial Medicine Research of Jiangxi Province, College of Life Sciences, Jiangxi Science and Technology Normal University, Nanchang, Jiangxi, China
e-mail: rq_tang@jxstnu.edu.cn

D.-W. Zeng
School of Life Sciences and Biotechnology, Shanghai Jiao Tong University, Shanghai, China
e-mail: zengduwen@sjtu.edu.cn

X.-Q. Zhao
State Key Laboratory of Microbial Metabolism, School of Life Sciences and Biotechnology, Shanghai Jiao Tong University, Shanghai, China
e-mail: xqzhao@sjtu.edu.cn

D. Zhu (✉)
Key Laboratory of Natural Microbial Medicine Research of Jiangxi Province, College of Life Sciences, Jiangxi Science and Technology Normal University, Nanchang, Jiangxi, China

Key Laboratory of Natural Microbial Medicine Research of Jiangxi Province, Jiangxi Science and Technology Normal University, Nanchang, Jiangxi, China
e-mail: zhudu@jxstnu.edu.cn

Abstract Biosynthesis of succinic acid (SA) represents one of the largest sectors in industrial biotechnology for biochemical production. Currently, the markets for fossil-derived and bio-based SA are comparable in scale, with the bio-based segment expanding particularly rapidly. In response to the growing demand for sustainable and renewable production routes, diverse types of lignocellulosic biomass have been employed as feedstocks. This chapter provides an overview of the microorganisms commonly used in SA production, including native producers such as *Actinobacillus succinogenes*, *Anaerobiospirillum succiniciproducens*, *Basfia succiniciproducens*, *Mannheimia succiniciproducens*, as well as engineered producers like *Corynebacterium glutamicum*, *Escherichia coli*, *Saccharomyces cerevisiae*, *Yarrowia lipolytica*, *Kluyveromyces marxianus*, and *Issatchenkia orientalis*. Their production performance, including titer, yield, and productivity, on conventional substrates such as glucose and on mixed sugars derived from lignocellulosic biomass is reviewed. The chapter concludes with a discussion of future prospects, highlighting the importance of integrating designed enzymes, robust strains, and fermentation optimization for sustainable SA production.

Graphical Abstract

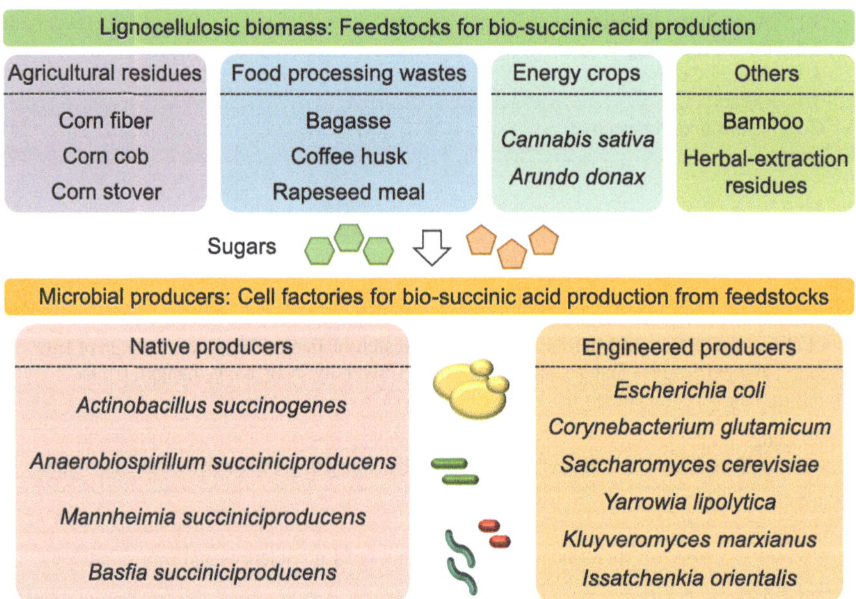

Keywords Lignocellulosic biomass, Succinic acid Native producers Engineered producers

1 Introduction

Succinic acid (SA), also known as amber acid or 1,4-butanedioic acid ($C_4H_6O_4$, CAS registry number 110-15-6), is an important platform chemical in the synthesis of various valuable compounds, including 1,4-butanediol (BDO), adipic acid, tetrahydrofuran, N-methylpyrrolidone, γ-butyrolactone, and 2-pyrrolidone [1, 2]. SA and its derivatives are used across diverse industries to produce a wide array of products, including solvents, surfactants, food additives, biodegradable polymers, and more [3, 4]. This broad applicability is a key driver behind the substantial market demand for SA. Traditionally, SA was produced from fossil-based feedstocks. Today, it has emerged as one of the most important bio-based organic acids, commercially produced through microbial fermentation.

The transition toward bio-based routes for SA production, as opposed to fossil-based methods, is increasingly seen as an inevitable direction for the sector [5, 6]. The bio-based production offers pivotal advantages, particularly in reducing the carbon footprint and the cost. Compared to the fossil-based route, bio-based production of SA achieves a 58% reduction in greenhouse gas emissions per ton [7] and results in a lower environmental cost (1.48 vs. 2.24 USD/kg) [8]. In terms of economics, the production cost for the biomass-based route ranges from 85 to 1,045 €/MT across different hypothetical scenarios, which is significantly lower than that of the petrochemical route (2,554 €/MT) [9]. These compelling advantages are driving the expansion of the bio-SA market.

According to Grand View Research [10], the global bio-SA market size was estimated at USD 126.04 million in 2023 and is predicted to grow at a compound annual growth rate (CAGR) of 12.1% from 2024 to 2030, reaching an estimated value of USD 272.36 million by 2030. As the report demonstrated, the BDO application dominated the market with a revenue share of 34.50% in 2023. Other major applications include polyester polyols, plasticizers, polybutylene succinate/poly(butylene succinate-co-butylene terephthalate) (PBS/PBST), and alkyd resins. Notably, BDO and SA are essential monomers for producing polyester polyols [2]. These polyols are further processed into polymers utilized across a wide spectrum of sectors, from automotive and construction to packaging, agriculture, and pharmaceuticals. By replacing conventional plastics and offering biodegradable solutions, they contribute significantly to the green transformation of these industries. According to Grand View Research [11], the global polyester polyol from bio-SA market size was estimated at USD 183.5 million in 2024 and is projected to reach USD 386.8 million by 2033, growing at a CAGR of 14.9% from 2025 to 2033. To satisfy the rapidly growing demand for the bio-SA and derived polyester polyol markets, the industry needs to develop microbial strains and production processes capable of utilizing low-cost, eco-friendly feedstocks while achieving higher SA yields.

Glucose, crude glycerol, maltose syrup, cassava, food processing waste, agricultural residues, and other renewable biomass resources are the raw materials for the

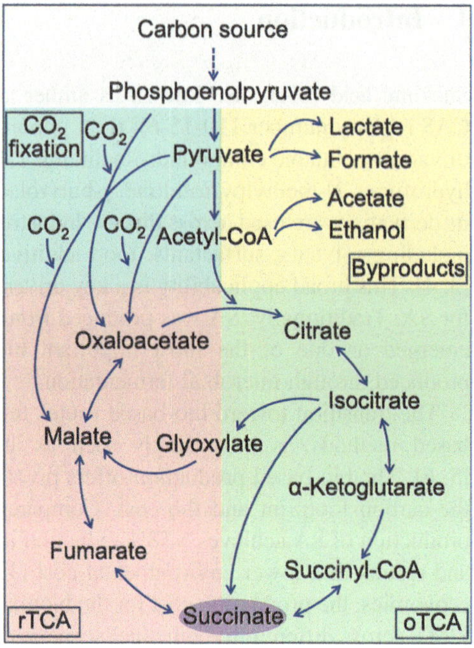

Fig. 1 Three major biosynthesis pathways of succinic acid. *rTCA* reductive branch of tricarboxylic acid cycle, *oTCA* oxidative branch of tricarboxylic acid cycle. (This figure is adapted from literature [12])

biosynthesis of SA [12]. Among these diverse feedstocks, lignocellulosic biomass represents a promising category since it is low-cost and abundantly available. It has been successfully used in fermentation processes with various microorganisms, including both native or natural producers and non-native or engineered producers. The native producers mainly include *Actinobacillus succinogenes* [13], *Anaerobiospirillum succiniciproducens* [14], *Mannheimia succiniciproducens* [15], and *Basfia succiniciproducens* [16]. The engineered producers mainly include *Escherichia coli* [17], *Corynebacterium glutamicum* [18], *Saccharomyces cerevisiae* [19], *Yarrowia lipolytica* [20], *Kluyveromyces marxianus* [21], and *Issatchenkia orientalis* [22].

The biosynthesis pathways of SA are mainly divided into three categories, the reductive branch and the oxidative branch of the tricarboxylic acid (TCA) cycle, and the glyoxylate shunt pathway (Fig. 1). Notably, the utilization of carbon dioxide (CO_2) in the reductive branch of the TCA cycle (rTCA) is an environmentally beneficial process, as it contributes to CO_2 fixation. To improve the production performance of microorganisms, various metabolic engineering and process optimization strategies have been implemented. Metabolic engineering strategies vary based on the host strain used, the SA biosynthesis pathway inherent to the host, and the feedstock selected [3]. Furthermore, process optimization, especially the CO_2 supply and culture media, is frequently performed to realize highly efficient and economical production of SA from lignocellulosic feedstocks [3, 23].

This chapter focuses on the commonly used microbial producers, their highest SA production capacity, and their production performance using lignocellulosic biomass as feedstock. Finally, future prospects are discussed to outline promising research directions.

2 Microbial Producers of Succinic Acid

2.1 Native Producers

2.1.1 *Actinobacillus succinogenes*

A. succinogenes is the most widely studied native producer of SA so far. It is a Gram-negative, facultatively anaerobic, capnophilic bacterium isolated from bovine rumen [24]. *A. succinogenes* has strong tolerance to high SA concentrations and is capable of utilizing a wide spectrum of substrates, including glucose, glycerol, arabinose, xylose, fructose, lactose, sucrose, and more [25]. Given these properties, *A. succinogenes* has been explored for SA production under diverse fermentation conditions, with many studies achieving high titers (Table 1). In batch fermentation, *A. succinogenes* 130Z (ATCC 55618) produced 69 g/L SA from glucose [25], while its variant FZ53 reached a higher titer of 105.8 g/L [26]. By employing immobilized cells in a repeated batch fermentation, a comparable SA concentration of 107 g/L was obtained from glucose [27]. In addition to glucose, other low-cost substrates have been used. Duckweed was converted to 75.46 g/L SA via batch fermentation with a semi-simultaneous saccharification and fermentation process [28]. The fed-batch fermentation of pretreated cane molasses yielded 83.67 g/L SA [29]. Furthermore, a higher titer was achieved by *A. succinogenes* ATCC 55618, which produced 151.44 g/L SA using fresh cassava root as the substrate [30].

2.1.2 *Anaerobiospirillum succiniciproducens*

The first *A. succiniciproducens* strain was isolated from the throat of a beagle dog [40]. It is a Gram-negative, spiral-shaped, and strictly anaerobic bacterium capable of fermenting various substrates, including glucose, lactose, fructose, and sucrose, with SA being the major product [40]. As reported in a patent, a variant strain *A. succiniciproducens* FA-10 produced 55 g/L SA from glucose [31]. Additionally, using the inexpensive feedstock light steep water, an SA titer of 32 g/L was achieved by another strain through controlling the pH and optimizing the CO_2 sparging rate [32]. *A. succiniciproducens* achieved its highest SA titer (83 g/L) through a three-stage continuous process that integrated membrane cell recycling and an electrodialysis system with a bioreactor [33]. A later study reported the development of a feed

Table 1 Production capacity of succinic acid by native strains

Strain	Substrate	Fermentation type	Titer (g/L)	Yield (g/g)	Productivity (g/L/h)	Reference
A. succinogenes 130Z	Glucose	Batch	68.00	NA	NA	[25]
A. succinogenes FZ53	Glucose	Batch	105.80	0.82	1.36	[26]
A. succinogenes 130Z	Glucose	Repeated batch (immobilized cell)	107.00	0.73	1.49	[27]
A. succinogenes GXAS137	Duckweed	Batch	74.46	0.42	1.35	[28]
A. succinogenes 130Z	Cane molasses	Fed batch	83.67	0.93	1.74	[29]
A. succinogenes ATCC 55618 (i.e., 130Z)	Cassava root	Fed batch	151.44	1.51	3.22	[30]
A. succiniciproducens FA-10	Glucose	Batch	55.00	NA	NA	[31]
A. succiniciproducens ATCC 53488	Light steep water	Batch	32.00	0.99	1.19	[32]
A. succiniciproducens ATCC 53488	Glucose	Three-stage process	83.00	0.89	10.40	[33]
A. succiniciproducens DSMZ 6400	Glucose	Fed batch	32.00	0.88	0.49	[14]
M. succiniciproducens PALK	Glucose, glycerol	Fed batch	90.68	0.75	3.49	[34]
M. succiniciproducens PALK (pMS3-pelB-cti)	Glucose, glycerol	Fed batch	97.10	0.83	3.01	[35]
M. succiniciproducens PALKcgmdh	Glucose, glycerol	High-density fed batch	134.25	0.82	21.30	[36]
M. succiniciproducens PALK (pMS3-mgtB)	Glucose, glycerol	High-density fed batch	152.23	0.85	11.71	[37]
B. succiniciproducens	Glucose	Batch	20	0.44	0.53	[38]
B. succiniciproducens JF4016	Glucose	Batch	30.9	0.31	0.26	[16]
E. aerogenes LU2	Whey	Batch	57.7	0.62	0.34	[39]

NA not available

control strategy for SA production [14]. However, this approach achieved a titer of only 32 g/L. In summary, as a native SA producer, *A. succiniciproducens* has been studied for decades since its discovery. However, the overall performance of *A. succiniciproducens* is significantly lower than that of *A. succinogenes*. Consequently, it has attracted diminishing research focus over time.

2.1.3 *Mannheimia succiniciproducens*

M. succiniciproducens is recognized as one of the best native producers of SA. The strain *M. succiniciproducens* MBEL55E is the only representative of this species and was isolated from bovine rumen [41]. This strain is proprietary to a South Korean research group. Since its isolation, it has been intensively studied by this group, resulting in well-defined characteristics and highly efficient SA production. It can utilize diverse substrates, including glucose, fructose, xylose, sucrose, maltose, and lactose [42]. Although it is a natural SA producer, SA production in the wild-type strain was relatively low, and thus, genetic modification and process optimization have been conducted to improve its performance. To reduce the formation of byproducts, genes encoding lactate dehydrogenase (*ldhA*), phosphate acetyltransferase (*pta*), and acetate kinase (*ackA*) were deleted in MBEL55E [34]. The resulting strain PALK produced 90.68 g/L SA using glucose and glycerol as carbon sources. The damage of cell membrane caused by SA is one of the factors inhibiting the achievement of high titer of organic acid. Therefore, cis–trans isomerase from *Pseudomonas aeruginosa* was expressed in strain PALK to enhance the rigidity of cell membrane [35]. The resulting strain produced 97.1 g/L SA. Furthermore, *M. succiniciproducens* produces SA through the rTCA cycle, in which malate dehydrogenase (MDH) serves as a crucial enzyme. Accordingly, the effect of various MDHs on SA production in strain PALK was investigated by replacing the endogenous *mdh* gene [36]. The best performance was achieved upon introduction of MDH from *C. glutamicum*, resulting in a titer of 134.25 g/L in high-cell-density fed-batch fermentation. Additionally, given the importance of magnesium ion (Mg^{2+}) in SA production, various Mg^{2+} transporters were introduced into strain PALK to enhance its performance [37]. The strain harboring a plasmid carrying the Mg^{2+} transporter gene *mgtB* from *Salmonella enterica* exhibited the best performance, achieving a titer of 152.23 g/L.

2.1.4 *Basfia succiniciproducens*

B. succiniciproducens is a Gram-negative, facultatively anaerobic, coccoid bacterium isolated from bovine rumen [43]. *B. succiniciproducens* belongs to the Pasteurellaceae family, which it shares with *M. succiniciproducens*. They also possess comparable genomic characteristics, such as genome size, open reading frame count, and G+C content [42, 43]. Similar to other native producers, it has the ability to utilize a variety of substrates, such as glucose, galactose, mannose, sucrose, and xylose. The wild-type *B. succiniciproducens* produced around 20 g/L SA with acetic acid and formic acid as the major byproducts [38]. Deletion of genes encoding pyruvate-formate lyase (*pflD*) and *ldhA* significantly reduced the formation of byproducts and improved the yield of SA [38]. Another study reported that *B. succiniciproducens* JF4016 produced 30.9 g/L SA with lactic acid as the major byproduct [16]. Compared to the a forementioned native producers, *B. succiniciproducens* exhibits significantly lower efficiency in SA production. Further research is essential to enhance its competitiveness in the bio-based SA market.

2.1.5 Others

Due to its high CO_2 content, bovine rumen is widely used to isolate SA-producing strains that fix CO_2 and further produce SA through the rTCA cycle. Many native SA-producing strains have been isolated from the bovine rumen, for instance, *M. succiniciproducens* MBEL55E. Recently, an SA-producing strain *Enterobacter aerogenes* LU2, newly isolated from bovine rumen, was found to be capable of producing SA from lactose as the major product [39]. Using whey permeate as substrate, the strain produced 57.7 g/L SA.

2.2 *Engineered Producers*

Although some native producers can achieve high SA titers, they suffer from the major drawback of underdeveloped genetic tools and limited understanding of physiological backgrounds, hindering their metabolic engineering. The high-producing engineered variants of *M. succiniciproducens* stand as compelling evidence for the necessity and benefit of metabolic engineering (Table 2). Therefore, to construct efficient and robust microbial cell factories for SA production, it is advantageous to employ well-characterized chassis such as *E. coli*, *C. glutamicum*, *S. cerevisiae*, *Y. lipolytica*, and *K. marxianus*. These platforms offer established genetic modification systems and predictable physiology, which are critical for achieving economically viable industrial-scale fermentation.

2.2.1 *Escherichia coli*

Among reported SA-producing *E. coli* strains, AFP111 stands out as a representative strain. It is a mutant of strain NZN111, in which a spontaneous *ptsG* mutation restored the ability of the parent strain to ferment glucose into SA [44, 45]. Gene *ptsG* encodes a component of the phosphoenolpyruvate:carbohydrate phosphotransferase system, which plays a critical role in carbon catabolite repression. The isolation of AFP111 has greatly facilitated SA production using *E. coli* as a

Table 2 Production capacity of succinic acid by engineered strains with metabolic engineering

Strain	Genotype	Substrate	Fermentation type	Titer (g/L)	Yield (g/g)	Prod. (g/L/h)	Reference
E. coli							
AFP111	*ΔpflAB ΔldhA ptsG*^{mut}	Glucose	Two-stage fed batch	101.20	1.07	1.89	[47]
MLB45-05	*ΔpflB ΔldhA*	Glucose	High-density batch	111.00	0.74	1.29	[48]
SD121	*ΔpflB ΔldhA ptsG*^{mut} *cppc*	Glucose	High-density batch	116.20	1.13	1.55	[49]
FZ661T	*ΔldhA ΔadhE ΔiclR ΔackA-pta ΔptsG ΔlacI* 2× *pycA-fdh1 galR::galP*	Glucose, xylose	Fed batch	107.00	NA	0.60	[51]
C. glutamicum							
R ΔldhA-pCRA717	*ΔldhA pyc*	Glucose	High-density fed batch	146.00	0.92	3.17	[53]
SA5	*ΔldhA Δpqo Δcat ΔackA pyc ppc aceA aceB gltA*	Glucose	Dual-phase fed batch	109.27	0.87	1.11	[54]
K3	*ΔldhA Δpta ΔackA Δcat ppc pyc cgl0275*	Glucose	Dual-phase fed batch	117.80	0.59	1.04	[55]
S. cerevisiae							
PYC2oe-mpc3Δ sdh1Δ	*Δgut1 CjFPS1 Opgdh ScDAK1 ScMDH3 RoFUM TbFRD AnDCT02 ScPYC2 ubr2*^{mut} *Δmpc3 Δsdh1*	Glycerol	Batch	45.50	NA	0.31	[57]
SUC-632	*PCKa FRDg MDH3 DCT_02 FUMR Δadh1 PYC2 ICL1 MLS1*	Glucose	Fed batch	~73.34	NA	~0.79	[59]
Y. lipolytica							
PGC01003	*Δsdh5*	Crude glycerol	Fed batch	160.20	0.40	0.40	[61]
Hi-SA2	*Δsdh5 ΔAch1 YlPyc TbFrd EcFum YlMdh1 YlMdh1* ALE *SpMae1*	Glucose	Fed batch	111.90	0.79	1.79	[65]

(continued)

Table 2 (continued)

Strain	Genotype	Substrate	Fermentation type	Titer (g/L)	Yield (g/g)	Prod. (g/L/h)	Reference
E501XF	Hi-SA2 ALE *YlHxk1 YlPfk1*	Glucose	Fed batch	112.54	0.67	2.08	[66]
Ncr12	2× *cAAT PfAAL YlGDH EcPPC YlG6PD EcAA* 2× *SpMAE* Δ*SDH5 TbFRD YlPYC*	Glycerol Glucose	Fed batch Fed batch	117.74 98.16	0.74 0.91	1.23 1.02	[67]
K. marxianus							
Z6	Δ*mae1*, Δ*pdc1* Δ*jen1* Δ*adh2* Δ*gpd1* Δ*ald6* 2× *m7FUM1-R* 2× *FRDg PYC2 MDH3-R SthA*	Glucose	Two-stage fed batch	103.90	0.51	1.08	[21]
KmSA12	Δ*gpd1 Tbfrd Ropyc Kmmdh Ttfum Spmae1*	Glucose	Fed batch	50.60	0.31	0.42	[73]
YZH43	Δ*sdh1,2,3,4A,5* Δ*gpd1* Δ*ach1* Δ*adh2A NDE1*	Glucose Cellulose	Batch SSF, batch	32.38 20.51	NA NA	NA 0.57	[74]
I. orientalis							
*g3473*Δ*/PaGDH-DAK/g3837*Δ	*SpMAE1* Δ*pdc* Δ*gpd* Δ*g3473 PaGDH DAK* Δ*g3837*	Glucose, glycerol	Fed batch	109.50	0.65	0.54	[77]
EcGS-FF-gltA	Δ*g3473 PaGDH DAK PYC MDH IoPDH* 2× *FUMR* 2× *TbFRD* 2× *gltA AcnA AceA AceB*	Glucose	Fed batch	104.28	0.85	0.97	[78]
V. natriegens							
SUCC1	Δ*lldh* Δ*dldh* Δ*pfl* Δ*ald* Δ*dns::pycCg*	Glucose	Dual-phase batch	60.40	0.75	8.60	[80]

Prod. productivity, *NA* not available, ~ estimated according to the reference host. Using a dual-phase fermentation process, the strain AFP111/pTrc99A-*pyc*, which harbors the pyruvate carboxylase (*pyc*) gene from *Rhizobium etli*, produced 99.2 g/L SA [46]. The optimization of glucose feeding during the aerobic phase of the two-stage (or dual-phase) fermentation process allowed strain AFP111 to reach a titer of 101.2 g/L [47]. In addition to AFP111, another high-producing strain derived from MG1655 has been constructed [48]. Adaptive laboratory evolution (ALE) of the strain MLB (MG1655 Δ*pflB* Δ*ldhA*) was performed in acetate. The resulting isolate MLB46–05 exhibited significantly increased SA titer compared to its parent strain (111 vs. 55 g/L). Another high-cell-density fermentation was performed using strain SD121, resulting in a titer of 116.2 g/L [49]. Additionally, to enable the strains to utilize low-cost and renewable sources, engineered *E. coli* strains that can produce high SA from mixed sugars have been developed. Strain FZ661T, with the deletion of *ptsG* gene and the modification of the *gal* operon, produced 95.8 g/L SA from a mixture of glucose, galactose, and fructose [50] and 107.0 g/L from a mixture of glucose and xylose [51].

2.2.2 *Corynebacterium glutamicum*

C. glutamicum is a Gram-positive, facultatively anaerobic, fast-growing bacterium, which possesses the ability to utilize a diverse array of sugars [52]. An engineered *C. glutamicum* R produced up to 146 g/L SA through disrupting the *ldhA* gene and overexpressing the *pyc* gene [53]. By reconstructing the glyoxylate pathway and overexpressing the SA exporter, the final SA concentration produced by the engineered strain was 109.27 g/L [54]. Another engineered strain, named K3, with multiple gene disruptions and overexpressions, produced 117.8 g/L SA from glucose [55]. Along with the great progress made in producing high-titer SA from glucose using *C. glutamicum*, the efficient valorization of lignocellulosic biomass is now attracting considerable research attention. These advances are discussed in Sect. 3.

2.2.3 *Saccharomyces cerevisiae*

S. cerevisiae is a facultatively anaerobic yeast that has been widely used as a eukaryotic cell factory in industrial applications. It is more robust than bacterial hosts, especially in term of its low pH tolerance and phage resistance. The multiply modified strain DHA-SA-PYC_{oe} achieved a maximum SA titer of 35 g/L at 96 h, using glycerol as a substrate [56]. However, beyond this time point, the SA concentration decreased as the byproduct malic acid accumulated rapidly. In a subsequent investigation, disruption of the mitochondrial pyruvate carrier complex and succinate dehydrogenase complex was implemented to diminish the intramitochondrial consumption of pyruvate and succinate [57]. The engineered strain produced 45.5 g/L SA from glycerol at 144 h, with no significant decline thereafter. This production level is comparable to that reported in a patent for strain SUC-297, which produced 43 g/L SA from glucose while co-generating substantial byproducts, including

16.4 g/L ethanol and 14.9 g/L glycerol [58]. In a study on the recovery of SA from yeast fermentation broth, the titer of SA was reported to reach approximately 630 mM (74.34 g/L) [59]. This represents the highest value reported in the literature for *S. cerevisiae*. Although *S. cerevisiae* produces a relatively low concentration of SA, it remains a promising host for organic acid production, since its high tolerance to low pH is beneficial to the downstream acid separation and purification. According to a patent, SA production was achieved using *S. cerevisiae* strains without pH control, resulting in a final fermentation pH of 3.0 [60]. Thus, further efforts should be directed toward elucidating the genetic bottlenecks in SA production by *S. cerevisiae*, with the subsequent aim of developing robust strains for application.

2.2.4 *Yarrowia lipolytica*

Y. lipolytica is an aerobic, oleaginous nonconventional yeast that has been used to produce SA, and significant progress has been achieved [20]. It exhibits several advantageous properties for SA production, including tolerance to low pH, high carbon flux toward the TCA cycle, and high intracellular supply of acetyl-CoA. Disrupting the *Ylsdh5* gene encoding a subunit of the succinate dehydrogenase complex in strain Po1f generated the strain PGC01003, which was able to accumulate SA, but exhibited weak growth on glucose [61]. Using crude glycerol as substrate, after optimization of the fed-batch fermentation process, the strain produced 160 g/L SA, with a yield and productivity of 0.40 g/g and 0.40 g/L/h, respectively. However, the accumulation of acetic acid by this *SDH-deficient* strain hindered the improvement of SA production. By further deleting the acetyl-CoA transferase gene and overexpressing the phosphoenolpyruvate carboxykinase gene from *S. cerevisiae* and endogenous succinyl-CoA synthase gene to reduce byproduct formation and increase product yield, the yield and productivity reached 0.53 g/g and 0.80 g/L/h, respectively [62]. To enable efficient SA production from glucose by the engineered strain, further genetic modifications were performed to optimize its biosynthesis and transport [63]. The resulting strain achieved an SA titer of 101.4 g/L, with a yield of 0.37 g/g and a productivity of 0.70 g/L/h. In a separate study, overexpression of a novel SA stress tolerance gene, *YALI0_A07997g*, combined with disruption of the succinate dehydrogenase subunit gene *SDH5*, resulted in an engineered strain that produced 88.5 g/L SA [64]. Another study subjected an SDH-deficient strain PGC91-rT to ALE to restore its ability to grow normally on glucose [65]. The evolved strain showed good growth on YPD medium. Subsequently, to enhance SA production from low-cost, glucose-based substrates, it was subjected to further genetic modifications and medium optimization. Ultimately, the engineered strain Hi-SA2 produced 111.9 g/L SA, with an overall yield and productivity of 0.79 g/g and 1.79 g/L/h, respectively. Notably, this high titer was achieved without pH control during the fermentation. In a subsequent study, ALE was employed to improve the SA tolerance of the Hi-SA2 strain [66]. The evolved strain then underwent further genetic modifications targeting the glycolytic pathway.

The final engineered strain achieved an SA titer of 112.54 g/L at pH 3.5. Recently, a highly efficient strain employing a noncanonical rTCA cycle was constructed through a series of metabolic engineering strategies, producing 117.74 g/L of SA from glycerol [67]. In addition to the conventional bioreactor, a fibrous bed bioreactor, which consists of a main fermenter and a fibrous matrix, has been utilized for SA production from glycerol using strain PGC01003 [68, 69]. Through optimization of fed-batch fermentation conditions, SA titers reached 198.2 and 209.7 g/L when cotton towel [68] and sugarcane bagasse [69] were used as matrices, respectively.

2.2.5 *Kluyveromyces marxianus*

K. marxianus is a nonconventional yeast with excellent thermotolerance, high growth rate, and broad substrate spectrum [70]. It has the natural ability to consume xylose, the second most abundant sugar in cellulosic hydrolysates after glucose. This makes it an attractive candidate for the utilization of lignocellulosic biomass. *K. marxianus* has been developed as a microbial cell factory for bioethanol [71] and lactic acid [72]. Recently, it has also been utilized as a promising host for the production of SA. A study revealed that the oTCA pathway could not support the efficient production of SA in *K. marxianus* [21]. In contrast, the implementation of the rTCA pathway, followed by multidimensional optimization, ultimately led to significantly higher titer and yield. Importantly, artificial intelligence (AI)-based fumarate hydratase mutation design was employed to reduce the generation of malic acid, which is commonly co-produced with SA. Combining enzyme and cellular engineering as well as fermentation optimization, a high SA titer of 103.9 g/L was achieved [21]. In a separate study, the rTCA pathway for SA production was constructed in *K. marxianus* through the heterologous expression of enzymes sourced from diverse species. Following the optimization of key enzyme combinations and the expression of a succinate transporter, the engineered strain produced SA at a titer of 50.6 g/L. Notably, the strain produced 41.2 g/L SA at high temperature (45°C) with only a limited supply of neutralizing agents [73]. A recent study reported the development of an SA-producing *K. marxianus* strain employing a CRISPR/Cas12a genome editing toolbox. The engineered strain achieved an SA titer of 20.51 g/L from cellulose at 46°C [74]. In summary, efficient high-temperature SA fermentation by *K. marxianus* using cellulosic hydrolysates is expected in the near future. However, research to date has relied on pure substrates, neglecting the essential issue of inhibitor tolerance. Future work should focus on engineering strains with this crucial trait.

2.2.6 *Issatchenkia orientalis*

I. orientalis, also known as *Pichia kudriavzevii* or *Candida krusei*, is a nonconventional yeast with exceptional tolerance to extreme pH and high concentrations of organic acids, making it a promising platform host for producing organic acids [22]. Previously, a patented *P. kudriavzevii* strain 13723 that possesses a high glucose consumption rate was constructed and produced 48.2 g/L SA [75]. In a separate study, *I. orientalis* SD108 was identified and demonstrated to produce SA through the rTCA pathway [76]. In a further study, metabolic engineering to redirect flux toward SA production was performed. The engineered *I. orientalis* produced 109.5 g/L SA from a mixture of glucose and glycerol at low pH [77]. In a recent study by this group, another engineered *I. orientalis* strain was constructed, which produced 104.28 g/L SA with glucose as the sole carbon source [78].

2.2.7 Others

Currently, bio-based SA is predominantly produced by bacteria and yeasts. A marine bacterium, *Vibrio natriegens*, which has rapid growth in minimal media, was used to produce SA [79]. Metabolic engineering strategies have been employed to minimize the formation of byproducts and enhance the anaplerotic flux. Additionally, dual-phase fermentation processes have been implemented to sustain both rapid biomass accumulation and efficient product synthesis. As a result, the engineered strain achieved an SA titer of 60.4 g/L [80].

In summary, engineered microbial strains have achieved high SA production exceeding 100 g/L in various hosts, with the exception of *S. cerevisiae* and the less commonly used *V. natriegens*. Among the synthetic pathways employed, the rTCA pathway was the most frequently adopted for SA biosynthesis, which provides the highest SA yield because CO_2 is fixed in this pathway (Fig. 1). The metabolic engineering strategies implemented via the rTCA pathway are illustrated in Fig. 2. Gene *pyc*, whose product converts pyruvate to oxaloacetate, is the most important overexpression target in engineered strains and has been successfully applied in both bacteria and yeasts. Moreover, *mae*, *mdh*, *frd*, and *fum* are the crucial targets for overexpression in yeast strains. Notably, *sdh5* gene is a key genetic deletion target in *Y. lipolytica*, which is essential for a functional succinate dehydrogenase (SDH) complex. SDH inactivation through this strategy ensures the accumulation of SA. In bacterial strains, the *ldh* gene is a common deletion target. It encodes lactate dehydrogenase, which converts pyruvate to lactate, a major competing byproduct during SA biosynthesis. Notably, the TCA cycle comprises multiple reversible reactions. It is necessary to modify key reversible enzymes to kinetically favor the forward reaction toward SA, thereby minimizing flux through backward pathways.

Regarding the substrates, glucose and glycerol are the two most widely employed carbon sources for SA production. Low-cost glycerol is primarily sourced as crude glycerol, a major byproduct of the biodiesel industry. Despite its low cost, the limited availability of crude glycerol constrains its large-scale industrial application. In contrast, glucose can be derived from a variety of low-cost feedstocks, such as lignocellulosic biomass, which is diverse and abundant. Notably, achieving a high utilization rate of the total sugars from lignocellulosic biomass requires the efficient conversion of non-glucose monosaccharides as well. However, limited studies have been conducted on the simultaneous conversion of mixed sugars into SA. One study reported SA production from a mixture of glucose and xylose by an *E. coli* strain

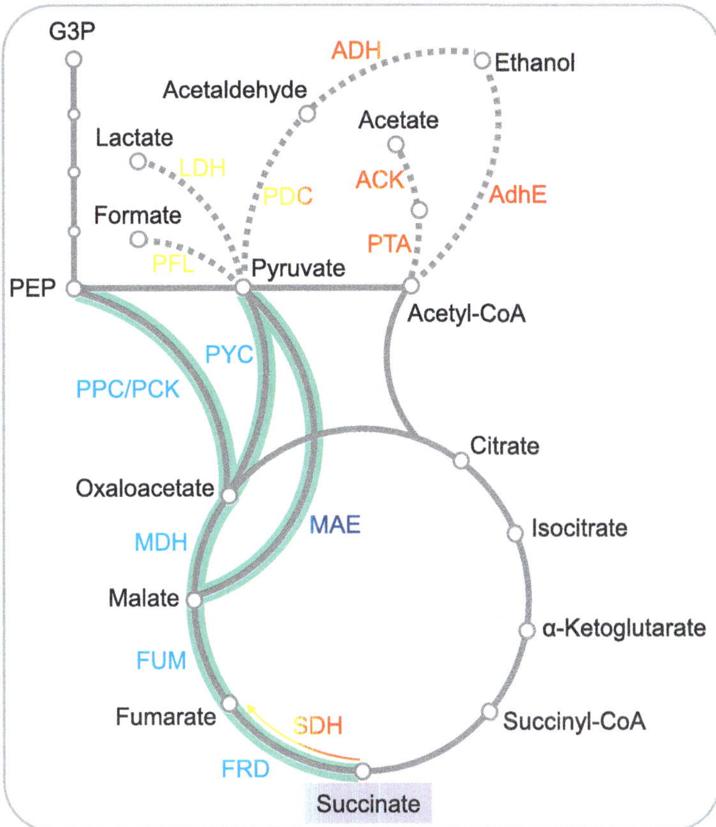

Fig. 2 Reductive TCA pathway for succinic acid biosynthesis. *Light blue* rTCA pathway, *dashed line* byproduct pathway, *red* deletion targets, *sapphire blue* overexpression targets, *PFL* pyruvate formate-lyase, *LDH* lactic dehydrogenase, *PDC* pyruvate decarboxylase, *PTA* phosphotransacetylase, *ACK* acetate kinase, *ADH* alcohol dehydrogenase, *AdhE* acetaldehyde alcohol dehydrogenase, *PYC* pyruvate carboxylase, *PPC* phosphoenolpyruvate carboxylase, *PCK* phosphoenolpyruvate carboxykinase, *MDH* malate dehydrogenase, *FUM* fumarate hydratase, *FRD* fumarate reductase, *MAE* malic enzyme, *SDH* succinate dehydrogenase

whose *ptsG* gene was deleted and the *gal* operon was modified [51]. The strain was further employed to produce SA from woody biomass hydrolysate (Sect. 3.4).

3 Succinic Acid Production from Lignocellulosic Biomass

To enable microbial fermentation, lignocellulosic biomass typically requires pretreatment and hydrolysis to liberate sugars. The separate hydrolysis and fermentation (SHF) strategy and the simultaneous saccharification and fermentation (SSF) process are two commonly employed strategies for the bioconversion of lignocellulose. In the SHF process, feedstocks undergo hydrolysis prior to being added as substrates for fermentation, while in SSF, hydrolysis and fermentation occur simultaneously. The advantages and drawbacks of these two methodologies have been summarized previously [81, 82]. In the microbial production of SA, SHF or SSF processes have been employed for the conversion of diverse lignocellulosic feedstocks, such as agricultural residues, food processing wastes, energy crops, and other biomass types.

3.1 Agricultural Residues

Agricultural activities offer a continuously available source of diverse residues in large quantities, which are promising candidates for use as low-cost feedstocks for producing SA. Owing to their abundant supply and high contents of fermentable sugars after hydrolysis, corn residues, including fiber, cob, and stover, constitute one of the most widely used lignocellulosic feedstocks in SA production. Using corn fiber hydrolysate as substrate, the high-producing *A. succinogenes* strain 130Z produced 70.6 g/L SA [26]. *A. succinogenes* strain NJ113 produced 35.4 g/L SA from the corn fiber hydrolysate treated with $CaCO_3$ neutralization combined with activated carbon adsorption [83]. Using corncob hydrolysate as feedstock, engineered *C. glutamicum* strain NC-2 produced 40.8 g/L SA [84]; engineered *Y. lipolytica* strain Hi-SA2-YlGsh2 produced 45.34 g/L SA [85]. Using corn stover hydrolysate prepared through a sequential process of mild alkaline washing, deacetylation, and dilute acid pretreatment, *A. succinogenes* strain 130Z achieved an SA titer of 42.8 g/L [86], while *B. succiniciproducens* strain JF4016 reached 30.6 g/L [16]. Using corn straw hydrolysate, *A. succinogenes* strain CGMCC1593 produced 45.5 and 53.2 g/L of SA in batch and fed-batch fermentation, respectively [87]. Using corn stalk hydrolysate and a low-cost alkaline neutralizer, the engineered *E. coli* strain KMG111 achieved a high SA titer of 129.21 g/L and a productivity of 2.15 g/L/h [88]. Using corn stover hydrolysate as substrate, the *C. glutamicum* strain K5 with two heterologous xylose-utilizing pathways produced 64.16 g/L SA, with a productivity of 1.07 g/L/h [55]. The corn stover was pretreated with concentrated alkali under steam-assisted conditions, then washed, and hydrolyzed to obtain the hydrolysate used as the substrate. Another xylose-utilizing *C. glutamicum* strain CGS5 produced 98.6 g/L SA from cornstalk hydrolysate, with a high productivity of 4.29 g/L/h [89]. The xylose-fermenting yeast *Y. lipolytica* strain Hi-SA0-G4 produced 83.78 g/L SA from corn stover hydrolysate, with a productivity of 1.21 g/L/h [90]. In contrast to the commonly utilized SHF strategy, limited studies have investigated the SSF process for microbial SA production. One study demonstrated the feasibility of SSF for SA production, using diluted alkaline-pretreated corn stover as the substrate and *A. succinogenes* strain as the microbial producer [91]. The process achieved a maximal titer of 47.4 g/L and a productivity of 0.99 g/L/h. By adopting SSF for the utilization of the hydrolysate liquor and the solid fraction of rice straw, another abundant agricultural residue, *E. coli* strain AS1600a, produced SA up to 85.6 g/L [92].

3.2 Food Processing Wastes

Beyond the agricultural residues, processing waste from the food industry is another important low-cost feedstock for industrial application. Various food processing wastes have been utilized for SA production, among which bagasse is a prominent example. The bagasse hydrolysate from *Agave tequilana* was used to produce SA using an immobilized *A. succinogenes* strain ZT-130, achieving a titer of 33.6 g/L after repeated batch fermentation [93]. Bagasse hydrolysate from sweet sorghum was fermented by *A. succinogenes* 130Z, producing 17.8 g/L SA [94]. Sugarcane bagasse hydrolysate was fermented by the engineered *E. coli* strain BA305, yielding 39.3 g/L of SA in fed-batch fermentation [95] and 83 g/L in repetitive fermentation [96]. The engineered strain KJ122 produced 98.63 g/L of SA via fed-batch SSF using cassava pulp, a byproduct of the cassava processing industry [97].

Several less common residues generated from food processing have also been employed as feedstocks for SA production, although they exhibited low production levels. For instance, an *A. succinogenes* strain produced 18.97 g/L SA from the water extracts of carob pods [98] and 19.3 g/L from coffee husk hydrolysate [99]. Using diluted acid-pretreated rapeseed meal as substrate in SSF, *A. succinogenes* ATCC 55618 produced SA with a titer of 23.4 g/L and a productivity of 0.33 g/L/h in fed-batch fermentation [100]. A modified SSF method was developed for the production of SA from oil palm empty fruit bunch, yielding a final titer of 50.5 g/L [101]. *E. coli* AFP184 produced 18.2 g/L SA from dried distiller grains with solubles [102].

3.3 Energy Crops

Dedicated energy crops, such as miscanthus, switchgrass, and giant reed, represent another major category of abundant lignocellulosic feedstocks for biochemical conversion. *A. succinogenes* strain 130Z produced 21.9 g/L SA from industrial hemp (*Cannabis sativa*) that was pretreated by a thermochemical method [103]. *B. succiniciproducens* strain BPP7 produced 37 g/L SA from the hydrolysate of *Arundo donax* [104].

3.4 Others

Fruit and vegetable waste disposal from the market could cause environmental problems and human health issues. SA was produced by *A. succinogenes* strain 130Z using the hydrolysate of fruit and vegetable waste, reaching a titer of 27.03 g/L

[105]. The organic part of municipal solid waste, mainly collected from households and businesses, was used for SA production, reaching a titer of 29.4 g/L by *A. succinogenes* 130Z [106] and 66.7 g/L by *Y. lipolytica* PSA02004 [107]. Recently, the production of SA by engineered *C. glutamicum* was investigated using hydrolysates derived from bamboo, oak, poplar, and pine [108]. Among these feedstocks, bamboo yielded the highest SA titer of 24.12 g/L. In another study, a woody hydrolysate was utilized by *E. coli* strain FZ661T, obtaining 54.5 g/L SA [51]. The herbal extraction residues from six Chinese herbal medicines were used as substrates for SA production, and the hydrolysate of *Glycyrrhiza uralensis* residue demonstrated the highest titer of 65 g/L [109].

As shown in Table 3, the highest SA titer from lignocellulosic biomass was achieved by *E. coli* strains KMG111 and KJ122 [88, 97], followed by *C. glutamicum* CGS5 [89] and *Y. lipolytica* Hi-SA0-G4 [90]. The productivity achieved by *C. glutamicum* CGS5 (4.29 g/L/h) was the highest among these strains, attributed to the two-stage batch fermentation strategy with an extremely high cell density ($OD_{600} = 150$) [89]. *E. coli* strains KMG111 and KJ122 were capable of simultaneously consuming glucose, xylose, and arabinose, the three most abundant monosaccharides in the hydrolysates from cornstalk and cassava pulp. In *C. glutamicum* CGS5, simultaneous utilization of glucose and xylose was enabled by introducing a heterologous xylose isomerase pathway and modifying native xylose metabolic routes [89]. Similarly, *Y. lipolytica* Hi-SA0-G4 was engineered to co-utilize glucose and xylose for SA production through the introduction of a xylose reductase–xylitol dehydrogenase pathway combined with adaptive evolution [90]. Alongside the ability to ferment sugars, stress tolerance plays an equally critical role in the production of lignocellulosic SA.

Pretreatment of lignocellulosic biomass inevitably co-produces undesirable byproducts that can inhibit fermentation. Therefore, additional downstream processing steps, such as detoxification, are often implemented prior to the fermentation process, adding to the overall cost and environmental footprint. In addition, although the accumulation of SA and the decline in broth pH during fermentation always have a negative impact on cell growth and fermentation performance, they offer several key advantages. It maintains the fully protonated acid form to simplify downstream recovery and lower extraction costs, suppresses microbial contamination during fermentation, and reduces the need for alkaline neutralizers, thereby cutting operational expenses. Therefore, it is of great importance to develop strains with combined tolerance to SA, pretreatment-generated inhibitors, and low pH. This can be achieved through strategies such as overexpressing reported stress tolerance genes. Furthermore, the optimization of industrial production necessitates the application of process engineering, such as scale-up strategies, culture medium formulation, and fermentation process parameters. Figure 3 shows the SA production process using lignocellulosic biomass as feedstock.

Table 3 Production capacity of succinic acid by microbial strains using lignocellulosic biomass as substrate

Biomass	Strain	Method	Titer (g/L)	Productivity (g/L/h)	Reference
Agricultural residues					
Corn fiber	*A. succinogenes* 130Z	SHF, batch	70.60	0.70	[26]
Corn fiber	*A. succinogenes* NJ113	SHF, batch	35.40	0.98	[83]
Corncob	*C. glutamicum* NC-2	SHF, two-stage batch	40.80	0.85	[84]
Corncob	*Y. lipolytica* Hi-SA2-YlGsh2	SHF, fed batch	45.34	1.42	[85]
Corn straw	*A. succinogenes* CGMCC1593	SHF, batch SHF, fed batch	45.50 53.20	0.95 1.21	[87]
Corn stover	*A. succinogenes* 130Z	SHF, batch	42.80	0.59	[86]
Corn stover	*A. succinogenes* CGMCC1593	SSF, batch	47.40	0.99	[91]
Corn stover	*B. succiniciproducens* JF4016	SHF, batch	30.60	0.43	[16]
Cornstalk	*E. coli* KMG111	SHF, two-stage fed-batch	129.21	2.15	[88]
Corn stover	*C. glutamicum* K5	SHF, batch	64.16	1.07	[55]
Cornstalk	*C. glutamicum* CGS5	SHF, two-stage batch	98.60	4.29	[89]
Corn stover	*Y. lipolytica* Hi-SA0-G4	SHF, fed-batch	83.78	1.21	[90]
Rice straw	*E. coli* AS1600a	SSF, Batch	85.60	0.59	[92]
Food processing residues					
Rapeseed meal	*A. succinogenes* ATCC 55618	SSF, fed batch	23.40	0.33	[100]
Carob pod	*A. succinogenes* 130Z	SSF, fed batch	18.97	1.43	[98]
Coffee husk	*A. succinogenes*	SHF, batch	19.30	0.54	[99]
Actinobacillus succinogenes bagasse	*A. succinogenes* ZT-130	SHF, repeated batch (immobilized cell)	33.60	1.32	[93]
Sweet sorghum bagasse	*A. succinogenes* 130Z	SHF, batch	17.80	0.89	[94]
Food processing residues					
Sugarcane bagasse	*E. coli* BA305	SHF, fed batch SHF, repetitive	39.30 83.00	0.33 2.31	[95] [96]
Oil palm empty fruit bunch	*A. succinogenes* ATCC 55618	SSF, batch	50.5	0.38	[101]
Cassava pulp	*E. coli* KJ122	SSF, fed batch	98.63	1.03	[97]

(continued)

Table 3 (continued)

Biomass	Strain	Method	Titer (g/L)	Productivity (g/L/h)	Reference
Dried distiller grains with solubles	*E. coli* AFP184	SHF; batch	18.20	NA	[102]
Energy crops					
Cannabis sativa	*A. succinogenes* 130Z	SHF, batch	21.90	0.46	[103]
Arundo donax	*B. succiniciproducens* BPP7	SHF, fed batch	37.00	0.86	[104]
Others					
Municipal solid waste	*A. succinogenes* 130Z	SHF, batch	29.40	0.89	[106]
Municipal solid waste	*Y. lipolytica* PSA02004	SHF, fed batch	66.70	0.78	[107]
Bamboo	*C. glutamicum* Psod: SucE12-ΔldhA	SHF, batch	24.12	1.01	[108]
Herbal extraction residues	*E. coli* MG-PYC	SHF, fed batch	65.00	0.51	[109]
Woody plant	*E. coli* FZ661T	SHF, dual-phase batch	54.50	1.81	[51]

NA not available, *SHF* separate hydrolysis and fermentation, *SSF* simultaneous saccharification and fermentation

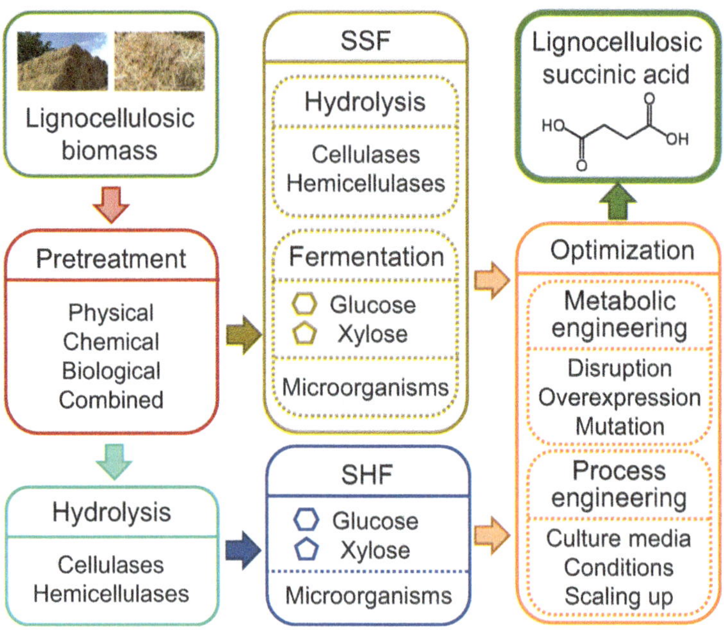

Fig. 3 Schematic of the succinic acid production process using lignocellulosic biomass

4 Conclusions and Perspectives

Microbial SA production has been achieved by native producers and engineered producers. Despite the progress in SA production using lignocellulosic biomass as a renewable and low-carbon feedstock, great challenges still remain to achieve economic and large-scale production. Firstly, microbial strains with higher SA yield and productivity when fermenting lignocellulosic hydrolysate are required. Therefore, efficient xylose utilization, especially in the presence of various inhibitors, is desired. In addition, strains that can thrive under industrial conditions, for example, high temperatures, low pH, and poor nitrogen sources, are required. Addressing these issues requires integrated efforts directed toward several key directions: (1) developing efficient genetic engineering platforms, especially for native microorganisms, to enable rapid and precise metabolic rewiring; (2) elucidating the metabolic networks of strains to identify and resolve metabolic bottlenecks that limit yield and productivity based on multi-omics data and fermentation parameters; (3) harnessing artificial intelligence for *in silico* engineering, especially of pivotal metabolic enzymes, to enable precise functional tailoring and enhanced metabolic flux toward target product; (4) improving the pretreatment and hydrolysis methods for lignocellulosic biomass to minimize inhibitor generation and lower process costs; and (5) performing systematic process engineering to further increase production performance of strains and decrease overall production costs, especially at larger scale. With the development of AI and synthetic biology platforms, as well as advanced fermentation technologies, sustainable SA production can be achieved using microbial cell factories.

References

1. Ahn JH, Jang Y-S, Lee SY (2016) Production of succinic acid by metabolically engineered microorganisms. Curr Opin Biotechnol 42:54–66. https://doi.org/10.1016/j.copbio.2016.02.034
2. Silva DAS, Luque R, Delgado-Arcaño Y, Perpetuo EA, Ruy ADDS, De Souza ALB, Pontes LAM (2025) Prospecting technology and economic potential to produce bio-succinic acid: a review. Biotechnol Rep (Amst) 48:e00920. https://doi.org/10.1016/j.btre.2025.e00920
3. Kumar V, Kumar P, Maity SK, Agrawal D, Narisetty V, Jacob S, Kumar G, Bhatia SK, Kumar D, Vivekanand V (2024) Recent advances in bio-based production of top platform chemical, succinic acid: an alternative to conventional chemistry. Biotechnol Biofuels Bioprod 17:72. https://doi.org/10.1186/s13068-024-02508-2
4. Zhou S, Zhang M, Zhu L, Zhao X, Chen J, Chen W, Chang C (2023) Hydrolysis of lignocellulose to succinic acid: a review of treatment methods and succinic acid applications. Biotechnol Biofuels Bioprod 16:1. https://doi.org/10.1186/s13068-022-02244-5
5. Grand View Research (2021) Succinic acid market (2022–2030). https://www.grandviewresearch.com/industry-analysis/succinic-acid-market. Accessed 26 Dec 2025
6. Maximize Market Research (2024) Succinic acid market – global industry analysis and forecast (2025–2032). https://www.maximizemarketresearch.com/market-report/global-succinic-acid-market/25839/. Accessed 26 Dec 2025

7. Musonda F, Millinger M, Thrän D (2020) Greenhouse gas abatement potentials and economics of selected biochemicals in Germany. Sustainability 12:2230. https://doi.org/10.3390/su12062230
8. Kosamia N, Sanchez A, Rakshit S (2023) Scenario-based life cycle assessment and environmental monetary valuation of biosuccinic acid production from lignocellulosic biomass. Ind Crop Prod 202:117101. https://doi.org/10.1016/j.indcrop.2023.117101
9. Pinazo JM, Domine ME, Parvulescu V, Petru F (2015) Sustainability metrics for succinic acid production: a comparison between biomass-based and petrochemical routes. Catal Today 239: 17–24. https://doi.org/10.1016/j.cattod.2014.05.035
10. Grand View Research (2024) Bio-succinic acid market (2024–2030). https://www.grandviewresearch.com/industry-analysis/bio-succinic-acid-market. Accessed 26 Dec 2025
11. Grand View Research (2025) Polyester polyol from bio-succinic acid market (2025–2033). https://www.grandviewresearch.com/industry-analysis/polyester-polyol-bio-succinic-acid-market-report. Accessed 26 Dec 2025
12. Liu X, Zhao G, Sun S, Fan C, Feng X, Xiong P (2022) Biosynthetic pathway and metabolic engineering of succinic acid. Front Bioeng Biotechnol 10:843887. https://doi.org/10.3389/fbioe.2022.843887
13. Dessie W, Xin F, Zhang W, Jiang Y, Wu H, Ma J, Jiang M (2018) Opportunities, challenges, and future perspectives of succinic acid production by *Actinobacillus succinogenes*. Appl Microbiol Biotechnol 102:9893–9910. https://doi.org/10.1007/s00253-018-9379-5
14. Bretz K, Kabasci S (2011) Feed-control development for succinic acid production with *Anaerobiospirillum succiniciproducens*. Biotechnol Bioeng 109:1187–1192. https://doi.org/10.1002/bit.24387
15. Lee SY, Kim JM, Song H, Lee JW, Kim TY, Jang YS (2008) From genome sequence to integrated bioprocess for succinic acid production by *Mannheimia succiniciproducens*. Appl Microbiol Biotechnol 79:11–22. https://doi.org/10.1007/s00253-008-1424-3
16. Salvachúa D, Smith H, St. John PC, Mohagheghi A, Peterson DJ, Black BA, Dowe N, Beckham GT (2016) Succinic acid production from lignocellulosic hydrolysate by *Basfia succiniciproducens*. Bioresour Technol 214:558–566. https://doi.org/10.1016/j.biortech.2016.05.018
17. Thakker C, Martínez I, San KY, Bennett GN (2011) Succinate production in *Escherichia coli*. Biotechnol J 7:213–224. https://doi.org/10.1002/biot.201100061
18. Tsuge Y, Hasunuma T, Kondo A (2015) Recent advances in the metabolic engineering of *Corynebacterium glutamicum* for the production of lactate and succinate from renewable resources. J Ind Microbiol Biotechnol 42:375–389. https://doi.org/10.1007/s10295-014-1538-9
19. Li C, Ong KL, Cui Z, Sang Z, Li X, Patria RD, Qi Q, Fickers P, Yan J, Lin CSK (2021) Promising advancement in fermentative succinic acid production by yeast hosts. J Hazard Mater 401:123414. https://doi.org/10.1016/j.jhazmat.2020.123414
20. Sun T, Sun ML, Lin L, Gao J, Wang K, Ji XJ (2024) Advancing succinic acid biomanufacturing using the nonconventional yeast *Yarrowia lipolytica*. J Agric Food Chem 73:100–109. https://doi.org/10.1021/acs.jafc.4c09990
21. Zeng DW, Zhu YF, Jiang YH, Ding YH, Wang ZZ, Li K, Liao S, Fan YC, Zhang L, Wang YG, Zhao XQ (2025) Establishment of succinic acid production in recombinant *Kluyveromyces marxianus* enhanced by AI-guided enzyme specificity engineering and redox-controlled fermentation optimization. Chem Eng J 523:168210. https://doi.org/10.1016/j.cej.2025.168210
22. Tan SI, Liu Z, Tran VG, Martin TA, Zhao H (2025) *Issatchenkia orientalis* as a platform organism for cost-effective production of organic acids. Metab Eng 89:12–21. https://doi.org/10.1016/j.ymben.2025.02.003
23. Jiang M, Ma J, Wu M, Liu R, Liang L, Xin F, Zhang W, Jia H, Dong W (2017) Progress of succinic acid production from renewable resources: metabolic and fermentative strategies. Bioresour Technol 245:1710–1717. https://doi.org/10.1016/j.biortech.2017.05.209

24. Guettler MV, Rumler D, Jain MK (1999) *Actinobacillus succinogenes* sp. nov., a novel succinic-acid-producing strain from the bovine rumen. Int J Syst Evol Microbiol 49:207–216
25. Guettler MV, Jain MK, Soni BK (1996) Process for making succinic acid, microorganisms for use in the process and methods of obtaining the microorganisms. US Patent 359370
26. Guettler MV, Jain MK, Rumler D (1996) Method for making succinic acid, bacterial variants for use in the process, and methods for obtaining variants. US Patent 520348
27. Corona-González RI, Miramontes-Murillo R, Arriola-Guevara E, Guatemala-Morales G, Toriz G, Pelayo-Ortiz C (2014) Immobilization of *Actinobacillus succinogenes* by adhesion or entrapment for the production of succinic acid. Bioresour Technol 164:113–118. https://doi.org/10.1016/j.biortech.2014.04.081
28. Shen N, Zhang H, Qin Y, Wang Q, Zhu J, Li Y, Jiang MG, Huang R (2018) Efficient production of succinic acid from duckweed (*Landoltia punctata*) hydrolysate by *Actinobacillus succinogenes* GXAS137. Bioresour Technol 250:35–42. https://doi.org/10.1016/j.biortech.2017.09.208
29. Wang Z, Li H, Feng J, Zhang A, Ying H, He X, Jiang M, Chen K, Ouyang P (2017) Enhanced succinic acid production from polyacrylamide-pretreated cane molasses in microbial electrolysis cells. J Chem Technol Biotechnol 93:855–860. https://doi.org/10.1002/jctb.5440
30. Thuy NTH, Kongkaew A, Flood A, Boontawan A (2017) Fermentation and crystallization of succinic acid from *Actinobacillus succinogenes* ATCC55618 using fresh cassava root as the main substrate. Bioresour Technol 233:342–352. https://doi.org/10.1016/j.biortech.2017.02.114
31. Guettler MV, Jain MK (1996) Method for making succinic acid, *Anaerobiospirillum succiniciproducens* variants for use in process and methods for obtaining variants. US Patent 359152
32. Nghiem NP, Davison BH, Suttle BE, Richardson GR (1997) Production of succinic acid by *Anaerobiospirillum succiniciproducens*. Appl Biochem Biotechnol 63–65:565–576
33. Meynial-Salles I, Dorotyn S, Soucaille P (2007) A new process for the continuous production of succinic acid from glucose at high yield, titer, and productivity. Biotechnol Bioeng 99:129–135. https://doi.org/10.1002/bit.21521
34. Choi S, Song H, Lim SW, Kim TY, Ahn JH, Lee JW, Lee MH, Lee SY (2016) Highly selective production of succinic acid by metabolically engineered *Mannheimia succiniciproducens* and its efficient purification. Biotechnol Bioeng 113:2168–2177. https://doi.org/10.1002/bit.25988
35. Ahn JH, Lee JA, Bang J, Lee SY (2018) Membrane engineering via trans-unsaturated fatty acids production improves succinic acid production in *Mannheimia succiniciproducens*. J Ind Microbiol Biotechnol 45:555–566. https://doi.org/10.1007/s10295-018-2016-6
36. Ahn JH, Seo H, Park W, Seok J, Lee JA, Kim WJ, Kim GB, Kim KJ, Lee SY (2020) Enhanced succinic acid production by *Mannheimia* employing optimal malate dehydrogenase. Nat Commun 11:1970. https://doi.org/10.1038/s41467-020-15839-z
37. Kim JY, Lee JA, Ahn JH, Lee SY (2024) High-level succinic acid production by overexpressing a magnesium transporter in *Mannheimia succiniciproducens*. Proc Natl Acad Sci USA 121:e2407455121. https://doi.org/10.1073/pnas.2407455121
38. Becker J, Reinefeld J, Stellmacher R, Schäfer R, Lange A, Meyer H, Lalk M, Zelder O, Von Abendroth G, Schröder H, Haefner S, Wittmann C (2013) Systems-wide analysis and engineering of metabolic pathway fluxes in bio-succinate producing *Basfia succiniciproducens*. Biotechnol Bioeng 110:3013–3023. https://doi.org/10.1002/bit.24963
39. Szczerba H, Komoń-Janczara E, Dudziak K, Waśko A, Targoński Z (2020) A novel biocatalyst, *Enterobacter aerogenes* LU2, for efficient production of succinic acid using whey permeate as a cost-effective carbon source. Biotechnol Biofuels 13:96. https://doi.org/10.1186/s13068-020-01739-3
40. Davis C, Cleven D, Brown J, Balish E (1976) *Anaerobiospirillum*, a new genus of spiral-shaped bacteria. Int J Syst Bacteriol 26:498–504

41. Lee S, Lee S, Hong H, Chang P (2002) Isolation and characterization of a new succinic acid-producing bacterium, *Mannheimia succiniciproducens* MBEL55E, from bovine rumen. Appl Microbiol Biotechnol 58:663–668. https://doi.org/10.1007/s00253-002-0935-6
42. Hong SH, Kim JS, Lee SY, In YH, Choi SS, Rih JK, Kim CH, Jeong H, Hur CG, Kim JJ (2004) The genome sequence of the capnophilic rumen bacterium *Mannheimia succiniciproducens*. Nat Biotechnol 22:1275–1281. https://doi.org/10.1038/nbt1010
43. Kuhnert P, Scholten E, Haefner S, Mayor D, Frey J (2010) *Basfia succiniciproducens* gen. nov., sp. nov., a new member of the family Pasteurellaceae isolated from bovine rumen. Int J Syst Evol Microbiol 60:44–50. https://doi.org/10.1099/ijs.0.011809-0
44. Donnelly MI, Millard CS, Clark DP, Chen MJ, Rathke JW (1998) A novel fermentation pathway in an *Escherichia coli* mutant producing succinic acid, acetic acid, and ethanol. Appl Biochem Biotechnol 70-72:187–198. https://doi.org/10.1007/BF02920135
45. Chatterjee R, Millard CS, Champion K, Clark DP, Donnelly MI (2001) Mutation of the *ptsG* gene results in increased production of succinate in fermentation of glucose by *Escherichia coli*. Appl Environ Microbiol 67:148–154. https://doi.org/10.1128/aem.67.1.148-154.2001
46. Vemuri GN, Eiteman MA, Altman E (2002) Succinate production in dual-phase *Escherichia coli* fermentations depends on the time of transition from aerobic to anaerobic conditions. J Ind Microbiol Biotechnol 28:325–332. https://doi.org/10.1038/sj.jim.7000250
47. Jiang M, Liu SW, Ma JF, Chen KQ, Yu L, Yue FF, Xu B, Wei P (2010) Effect of growth phase feeding strategies on succinate production by metabolically engineered *Escherichia coli*. Appl Environ Microbiol 76:1298–1300. https://doi.org/10.1128/aem.02190-09
48. Jiang J, Luo Y, Fei P, Zhu Z, Peng J, Lu J, Zhu D, Wu H (2024) Effect of adaptive laboratory evolution of engineered *Escherichia coli* in acetate on the biosynthesis of succinic acid from glucose in two-stage cultivation. Bioresour Bioprocess 11:34. https://doi.org/10.1186/s40643-024-00749-5
49. Wang D, Li Q, Song Z, Zhou W, Su Z, Xing J (2010) High cell density fermentation via a metabolically engineered *Escherichia coli* for the enhanced production of succinic acid. J Chem Technol Biotechnol 86:512–518. https://doi.org/10.1002/jctb.2543
50. Zhu F, San KY, Bennett GN (2020) Improved succinate production from galactose-rich feedstocks by engineered *Escherichia coli* under anaerobic conditions. Biotechnol Bioeng 117:1082–1091. https://doi.org/10.1002/bit.27254
51. Zhu F, Wang C, San K-Y, Bennett GN (2020) Metabolic engineering of *Escherichia coli* to produce succinate from woody hydrolysate under anaerobic conditions. J Ind Microbiol Biotechnol 47:223–232. https://doi.org/10.1007/s10295-020-02259-7
52. Li M-H, Li H, Zhang X, Liang Y-C, Li C, Sun M-L, Li K, Liu C-G, Sinskey AJ (2024) Metabolic engineering of *Corynebacterium glutamicum*: unlocking its potential as a key cell factory platform for organic acid production. Biotechnol Adv 77:108475. https://doi.org/10.1016/j.biotechadv.2024.108475
53. Okino S, Noburyu R, Suda M, Jojima T, Inui M, Yukawa H (2008) An efficient succinic acid production process in a metabolically engineered *Corynebacterium glutamicum* strain. Appl Microbiol Biotechnol 81:459–464. https://doi.org/10.1007/s00253-008-1668-y
54. Zhu N, Xia H, Yang J, Zhao X, Chen T (2013) Improved succinate production in *Corynebacterium glutamicum* by engineering glyoxylate pathway and succinate export system. Biotechnol Lett 36:553–560. https://doi.org/10.1007/s10529-013-1376-2
55. Li K, Li C, Zhao XQ, Liu CG, Bai FW (2023) Engineering *Corynebacterium glutamicum* for efficient production of succinic acid from corn stover pretreated by concentrated-alkali under steam-assistant conditions. Bioresour Technol 378:128991. https://doi.org/10.1016/j.biortech.2023.128991
56. Malubhoy Z, Bahia FM, De Valk SC, De Hulster E, Rendulić T, Ortiz JPR, Xiberras J, Klein M, Mans R, Nevoigt E (2022) Carbon dioxide fixation via production of succinic acid from glycerol in engineered *Saccharomyces cerevisiae*. Microb Cell Factories 21:102. https://doi.org/10.1186/s12934-022-01817-1

57. Rendulić T, Perpelea A, Ortiz JPR, Casal M, Nevoigt E (2024) Mitochondrial membrane transporters as attractive targets for the fermentative production of succinic acid from glycerol in *Saccharomyces cerevisiae*. FEMS Yeast Res 24:foae009. https://doi.org/10.1093/femsyr/foae009
58. Van De Graaf MJ, Valianpoer F, Fiey G, Delattre L, Schulten EAM (2015) Progress for the crystallization of succinic acid. US Patent 14/528321
59. Wahl SA, Bernal Martinez C, Zhao Z, Van Gulik WM, Jansen MLA (2017) Intracellular product recycling in high succinic acid producing yeast at low pH. Microb Cell Factories 16: 90. https://doi.org/10.1186/s12934-017-0702-0
60. Jansen MLA, Heijnen JJ, Verwaal R (2014) Progress for preparing dicarboxylic acids employing fungal cells. US Patent 14/130347
61. Gao C, Yang X, Wang H, Rivero CP, Li C, Cui Z, Qi Q, Lin CSK (2016) Robust succinic acid production from crude glycerol using engineered *Yarrowia lipolytica*. Biotechnol Biofuels 9: 179. https://doi.org/10.1186/s13068-016-0597-8
62. Cui Z, Gao C, Li J, Hou J, Lin CSK, Qi Q (2017) Engineering of unconventional yeast *Yarrowia lipolytica* for efficient succinic acid production from glycerol at low pH. Metab Eng 42:126–133. https://doi.org/10.1016/j.ymben.2017.06.007
63. Jiang Z, Cui Z, Zhu Z, Liu Y, Tang YJ, Hou J, Qi Q (2021) Engineering of *Yarrowia lipolytica* transporters for high-efficient production of biobased succinic acid from glucose. Biotechnol Biofuels 14:145. https://doi.org/10.1186/s13068-021-01996-w
64. Sun ML, Fang D, Zeng D, Wang ZZ, Huang CC, Zhang L, Fan YC, Liao S, Cao Y, Li K, Zhao XQ, Bai FW (2025) Leveraging stress tolerance property to enhance succinic acid production and cell metabolic activity of *Yarrowia lipolytica*. ACS Sustain Chem Eng 13:5210–5219. https://doi.org/10.1021/acssuschemeng.4c10404
65. Cui Z, Zhong Y, Sun Z, Jiang Z, Deng J, Wang Q, Nielsen J, Hou J, Qi Q (2023) Reconfiguration of the reductive TCA cycle enables high-level succinic acid production by *Yarrowia lipolytica*. Nat Commun 14:8480. https://doi.org/10.1038/s41467-023-44245-4
66. Zhong Y, Shang C, Tao H, Hou J, Cui Z, Qi Q (2024) Boosting succinic acid production of *Yarrowia lipolytica* at low pH through enhancing product tolerance and glucose metabolism. Microb Cell Factories 23:291. https://doi.org/10.1186/s12934-024-02565-0
67. Tao H, Hao A, Pan X, Zhong Y, Cui Z, Qi Q (2026) Engineered non-canonical reductive TCA pathway drives high-yield succinic acid biosynthesis in *Yarrowia lipolytica*. Metab Eng 94: 36–44. https://doi.org/10.1016/j.ymben.2025.11.003
68. Li C, Yang X, Gao S, Wang H, Lin CSK (2017) High efficiency succinic acid production from glycerol via in situ fibrous bed bioreactor with an engineered *Yarrowia lipolytica*. Bioresour Technol 225:9–16. https://doi.org/10.1016/j.biortech.2016.11.016
69. Li C, Gao S, Yang X, Lin CSK (2018) Green and sustainable succinic acid production from crude glycerol by engineered *Yarrowia lipolytica* via agricultural residue based *in situ* fibrous bed bioreactor. Bioresour Technol 249:612–619. https://doi.org/10.1016/j.biortech.2017.10.011
70. Wang H, Sun Z, Ni Z, Qi Y, Feng X, Xiao H, Wang L, Hong J, Liang Y, Li C (2025) From genetic constraints to process hurdles: a review of *Kluyveromyces marxianus* for efficient bio-manufacturing and bio-processing in food industry. Curr Res Food Sci 11. https://doi.org/10.1016/j.crfs.2025.101221
71. Zhang K, Jiang Z, Li X, Wang D, Hong J (2024) Enhancing simultaneous saccharification and co-fermentation of corncob by *Kluyveromyces marxianus* through overexpression of putative transcription regulator. Bioresour Technol 399:130627. https://doi.org/10.1016/j.biortech
72. Smets J, Godoy HE, Goossenaerts J, Van Bun E, Deparis Q, Bauwens J, Ortiz-Merino RA, Mancera E, Deluna A, Verstrepen KJ (2025) Metabolic engineering and adaptive laboratory evolution of *Kluyveromyces Marxianus* for lactic acid production. Microb Cell Factories 24: 179. https://doi.org/10.1186/s12934-025-02805-x

73. Gu Z, Ding D, Shan Z, Tang Y, Chen X (2025) Metabolic engineering of the non-conventional yeast *Kluyveromyces marxianus* for enhancing the biosynthesis of succinic acid. Biofuel Res J 12:2503–2516. https://doi.org/10.18331/brj2025.12.3.5
74. Zha H, Li Y, Hu Z, Li J, Xie Y, Zhao M, Ren L, Zhang B (2026) Development of a CRISPR/Cas12a genome editing toolbox in *Kluyveromyces marxianus* and its application in succinic acid biosynthesis. Synth Syst Biotechnol 11:193–204. https://doi.org/10.1016/j.synbio.2025.09.015
75. Brian JR, Arlene MF (2014) Methods for succinate production. US Patent 14/374464
76. Xiao H, Shao Z, Jiang Y, Dole S, Zhao H (2014) Exploiting *Issatchenkia orientalis* SD108 for succinic acid production. Microb Cell Factories 13. https://doi.org/10.1186/s12934-014-0121-4
77. Tran VG, Mishra S, Bhagwat SS, Shafaei S, Shen Y, Allen JL, Crosly BA, Tan SI, Fatma Z, Rabinowitz JD, Guest JS, Singh V, Zhao H (2023) An end-to-end pipeline for succinic acid production at an industrially relevant scale using *Issatchenkia orientalis*. Nat Commun 14:6152. https://doi.org/10.1038/s41467-023-41616-9
78. Tran VG, Tan SI, Xu H, Weilandt DR, Li X, Bhagwat SS, Zhu Z, Guest JS, Rabinowitz JD, Zhao H (2025) Decompartmentalization of the yeast mitochondrial metabolism to improve chemical production in *Issatchenkia orientalis*. Nat Commun 16:7110. https://doi.org/10.1038/s41467-025-62304-w
79. Hoffart E, Grenz S, Lange J, Nitschel R, Müller F, Schwentner A, Feith A, Lenfers-Lücker M, Takors R, Blombach B, Drake HL (2017) High substrate uptake rates empower *Vibrio natriegens* as production host for industrial biotechnology. Appl Environ Microbiol 83:e01614-17. https://doi.org/10.1128/aem.01614-17
80. Thoma F, Schulze C, Gutierrez-Coto C, Hädrich M, Huber J, Gunkel C, Thoma R, Blombach B (2021) Metabolic engineering of *Vibrio natriegens* for anaerobic succinate production. Microb Biotechnol 15:1671–1684. https://doi.org/10.1111/1751-7915.13983
81. Akhtar J, Idris A, Aziz RA (2013) Recent advances in production of succinic acid from lignocellulosic biomass. Appl Microbiol Biotechnol 98:987–1000. https://doi.org/10.1007/s00253-013-5319-6
82. Kawaguchi H, Hasunuma T, Ogino C, Kondo A (2016) Bioprocessing of bio-based chemicals produced from lignocellulosic feedstocks. Curr Opin Biotechnol 42:30–39. https://doi.org/10.1016/j.copbio.2016.02.031
83. Chen K, Jiang M, Wei P, Yao J, Wu H (2008) Succinic acid production from acid hydrolysate of corn fiber by *Actinobacillus succinogenes*. Appl Biochem Biotechnol 160:477–485. https://doi.org/10.1007/s12010-008-8367-0
84. Wang C, Zhang H, Cai H, Zhou Z, Chen Y, Chen Y, Ouyang P (2013) Succinic acid production from corn cob hydrolysates by genetically engineered *Corynebacterium glutamicum*. Appl Biochem Biotechnol 172:340–350. https://doi.org/10.1007/s12010-013-0539-x
85. Zhong Y, Gu J, Shang C, Deng J, Liu Y, Cui Z, Lu X, Qi Q (2024) Sustainable succinic acid production from lignocellulosic hydrolysates by engineered strains of *Yarrowia lipolytica* at low pH. Bioresour Technol 408:131166. https://doi.org/10.1016/j.biortech.2024.131166
86. Salvachúa D, Mohagheghi A, Smith H, Bradfield MFA, Nicol W, Black BA, Biddy MJ, Dowe N, Beckham GT (2016) Succinic acid production on xylose-enriched biorefinery streams by *Actinobacillus succinogenes* in batch fermentation. Biotechnol Biofuels 9:28. https://doi.org/10.1186/s13068-016-0425-1
87. Zheng P, Dong JJ, Sun ZH, Ni Y, Fang L (2009) Fermentative production of succinic acid from straw hydrolysate by *Actinobacillus succinogenes*. Bioresour Technol 100:2425–2429. https://doi.org/10.1016/j.biortech.2008.11.043
88. Wang J, Zhang B, Zhang J, Wang H, Zhao M, Wang N, Dong L, Zhou X, Wang D (2014) Enhanced succinic acid production and magnesium utilization by overexpression of magnesium transporter *mgtA* in *Escherichia coli* mutant. Bioresour Technol 170:125–131. https://doi.org/10.1016/j.biortech.2014.07.081

89. Mao Y, Li G, Chang Z, Tao R, Cui Z, Wang Z, Tang YJ, Chen T, Zhao X (2018) Metabolic engineering of *Corynebacterium glutamicum* for efficient production of succinate from lignocellulosic hydrolysate. Biotechnol Biofuels 11:95. https://doi.org/10.1186/s13068-018-1094-z
90. Zhong Y, Shang C, Gu J, Tao H, Lu X, Hou J, Cui Z, Qi Q (2025) Xylose metabolism perturbation in *Yarrowia lipolytica* for efficient succinic acid bioproduction from lignocellulosic biomass. Adv Sci 12:e07999. https://doi.org/10.1002/advs.202507999
91. Zheng P, Fang L, Xu Y, Dong JJ, Ni Y, Sun ZH (2010) Succinic acid production from corn stover by simultaneous saccharification and fermentation using *Actinobacillus succinogenes*. Bioresour Technol 101:7889–7894. https://doi.org/10.1016/j.biortech.2010.05.016
92. Jampatesh S, Sawisit A, Wong N, Jantama SS, Jantama K (2019) Evaluation of inhibitory effect and feasible utilization of dilute acid-pretreated rice straws on succinate production by metabolically engineered *Escherichia coli* AS1600a. Bioresour Technol 273:93–102. https://doi.org/10.1016/j.biortech.2018.11.002
93. Corona-González RI, Varela-Almanza KM, Arriola-Guevara E, Martínez-Gómez ÁDJ, Pelayo-Ortiz C, Toriz G (2016) Bagasse hydrolyzates from Agave tequilana as substrates for succinic acid production by *Actinobacillus succinogenes* in batch and repeated batch reactor. Bioresour Technol 205:15–23. https://doi.org/10.1016/j.biortech.2015.12.081
94. Lo E, Brabo-Catala L, Dogaris I, Ammar EM, Philippidis GP (2020) Biochemical conversion of sweet sorghum bagasse to succinic acid. J Biosci Bioeng 129:104–109. https://doi.org/10.1016/j.jbiosc.2019.07.003
95. Liu R, Liang L, Li F, Wu M, Chen K, Ma J, Jiang M, Wei P, Ouyang P (2013) Efficient succinic acid production from lignocellulosic biomass by simultaneous utilization of glucose and xylose in engineered *Escherichia coli*. Bioresour Technol 149:84–91. https://doi.org/10.1016/j.biortech.2013.09.052
96. Liang L, Liu R, Li F, Wu M, Chen K, Ma J, Jiang M, Wei P, Ouyang P (2013) Repetitive succinic acid production from lignocellulose hydrolysates by enhancement of ATP supply in metabolically engineered *Escherichia coli*. Bioresour Technol 143:405–412. https://doi.org/10.1016/j.biortech.2013.06.031
97. Sawisit A, Jantama SS, Kanchanatawee S, Jantama K (2014) Efficient utilization of cassava pulp for succinate production by metabolically engineered *Escherichia coli* KJ122. Bioprocess Biosyst Eng 38:175–187. https://doi.org/10.1007/s00449-014-1257-7
98. Carvalho M, Roca C, Reis MAM (2016) Improving succinic acid production by *Actinobacillus succinogenes* from raw industrial carob pods. Bioresour Technol 218:491–497. https://doi.org/10.1016/j.biortech.2016.06.140
99. Dessie W, Zhu J, Xin F, Zhang W, Jiang Y, Wu H, Ma J, Jiang M (2018) Bio-succinic acid production from coffee husk treated with thermochemical and fungal hydrolysis. Bioprocess Biosyst Eng 41:1461–1470. https://doi.org/10.1007/s00449-018-1974-4
100. Chen K, Zhang H, Miao Y, Wei P, Chen J (2011) Simultaneous saccharification and fermentation of acid-pretreated rapeseed meal for succinic acid production using *Actinobacillus succinogenes*. Enzym Microb Technol 48:339–344. https://doi.org/10.1016/j.enzmictec.2010.12.009
101. Khairil Anwar NAK, Hassan N, Idris A, Ramli S, Malek RA (2021) Large scale production of succinic acid by fermentation of sequential inorganic salt pretreated oil palm empty fruit bunch. Biomass Bioenergy 155:106307. https://doi.org/10.1016/j.biombioe.2021.106307
102. Nghiem NP, Montanti J, Kim TH (2016) Pretreatment of dried distiller grains with solubles by soaking in aqueous ammonia and subsequent enzymatic/dilute acid hydrolysis to produce fermentable sugars. Appl Biochem Biotechnol 179:237–250. https://doi.org/10.1007/s12010-016-1990-2
103. Gunnarsson IB, Kuglarz M, Karakashev D, Angelidaki I (2015) Thermochemical pretreatments for enhancing succinic acid production from industrial hemp (*Cannabis sativa* L.). Bioresour Technol 182:58–66. https://doi.org/10.1016/j.biortech.2015.01.126
104. Cimini D, Zaccariello L, D'ambrosio S, Lama L, Ruoppolo G, Pepe O, Faraco V, Schiraldi C (2019) Improved production of succinic acid from *Basfia succiniciproducens* growing on

A. donax and process evaluation through material flow analysis. Biotechnol Biofuels 12:22. https://doi.org/10.1186/s13068-019-1362-6
105. Dessie W, Zhang W, Xin F, Dong W, Zhang M, Ma J, Jiang M (2018) Succinic acid production from fruit and vegetable wastes hydrolyzed by on-site enzyme mixtures through solid state fermentation. Bioresour Technol 247:1177–1180. https://doi.org/10.1016/j.biortech.2017.08.171
106. Stylianou E, Pateraki C, Ladakis D, Cruz-Fernández M, Latorre-Sánchez M, Coll C, Koutinas A (2020) Evaluation of organic fractions of municipal solid waste as renewable feedstock for succinic acid production. Biotechnol Biofuels 13:72. https://doi.org/10.1186/s13068-020-01708-w
107. Stylianou E, Carvajal-Arroyo JM, Ladakis D, Lin CSK, Eßmann V, Dörr S, Marbach J, Rabaey K, Koutinas A, Pateraki C (2023) Development of an electrochemical membrane bioreactor for succinic acid production and in situ separation with engineered *Yarrowia lipolytica* cultivated on municipal biowaste hydrolysate. Chem Eng J 466:142877. https://doi.org/10.1016/j.cej.2023.142877
108. Lee D-S, Cho EJ, Kim S, Nguyen DT, Bae H-J (2025) Succinic acid production from monosaccharides and woody and herbaceous plant hydrolysates using metabolically engineered *Corynebacterium glutamicum*. Food Technol Biotechnol 63:134–148. https://doi.org/10.17113/ftb.63.02.25.8808
109. Wang C, Su X, Sun W, Zhou S, Zheng J, Zhang M, Sun M, Xue J, Liu X, Xing J, Chen S (2018) Efficient production of succinic acid from herbal extraction residue hydrolysate. Bioresour Technol 265:443–449. https://doi.org/10.1016/j.biortech.2018.06.041

Adv Biochem Eng Biotechnol (2026) 196: 249–282
https://doi.org/10.1007/10_2026_324

Published online: 20 June 2026

Composting of Agricultural Residues into Organic Fertilizers for Sustainable Agriculture

Sun Zhaoyong, Iwabuchi Kazunori, Xu Yang, Zhang Wenshuai, and Li Xiaoxing

Contents

S. Zhaoyong (✉), X. Yang, Z. Wenshuai, and L. Xiaoxing
College of Architecture and Environment, Sichuan University, Chengdu, Sichuan, China
e-mail: szy@scu.edu.cn

I. Kazunori
Research Faculty of Agriculture, Hokkaido University, Sapporo, Hokkaido, Japan

Abstract The global generation of vast quantities of agricultural residues poses a significant environmental and logistical challenge. Composting offers an economically feasible and environmentally sustainable solution by efficiently transforming these waste materials into high-value organic fertilizers, thereby enhancing soil fertility and crop productivity. This chapter presents a comprehensive review of recent advances in the composting of agricultural residues, covering the entire process – from fundamental principles and microbial mechanisms to process optimization strategies and agricultural applications of the final compost product. It begins by addressing the challenges associated with the structural complexity and compositional heterogeneity of agricultural residues. Key composting principles, critical physicochemical parameters influencing process efficiency, and successional dynamics of microbial communities are then examined in detail. This chapter further evaluates various strategies to enhance composting performance, including pretreatment methods, co-composting, and integration of emerging additives such as biochar, metal oxides, and microbial inoculants. Special emphasis is placed on the agronomic benefits of mature compost, including its role in improving soil health, ensuring effective inactivation of pathogens and weed seeds, and supporting sustainable, long-term crop production. In summary, this review provides a robust theoretical and technological foundation for the effective valorization of agricultural residues into safe, nutrient-rich organic fertilizers, contributing to the advancement of circular agriculture and more sustainable agroecosystems.

Graphical Abstract

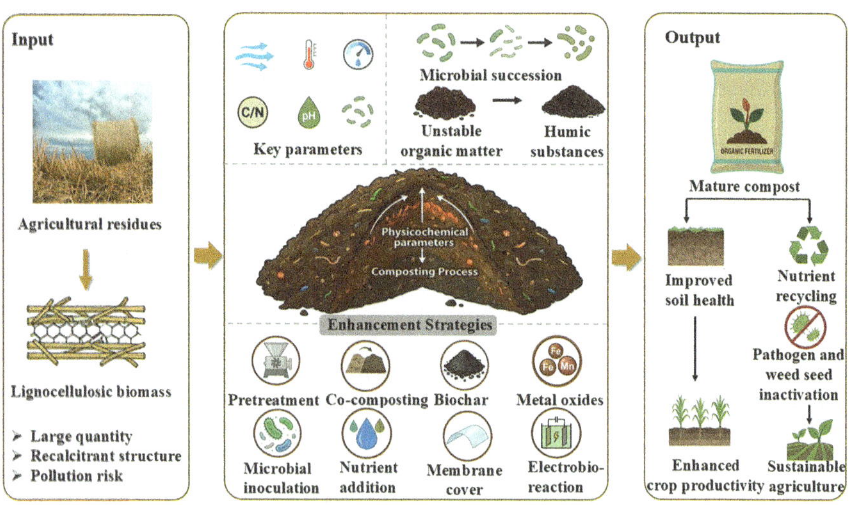

Keywords Agricultural residues, Compost product, Composting, Microbial dynamics, Process optimization, Sustainable agriculture

Nomenclature

W_M	Mass-based moisture content, %
M_S	Solid mass, mg
M_L	Liquid mass, mg
V	Bulk volume, m^3
V_L	Liquid volume, m^3
r_L	Liquid density, mg/m^3
r_S	Solid particle density, mg/m^3
R_{VS}	Volumetric solid ratio, –
R_{VL}	Volumetric liquid ratio, –
R_{VG}	Volumetric gas ratio, –

1 Introduction

Recycling agricultural residues into valuable products, such as fertilizer, is crucial for mitigating environmental pollution, reducing resource waste, and advancing sustainable agriculture. Traditionally, substantial annual residues are often discarded, burned, or directly returned to the field, causing air pollution, pathogen dissemination, greenhouse gas emissions, and resource depletion [1]. Composting technology not only converts agricultural residues into valuable resources but also achieves harmless treatment. During this process, microorganisms degrade harmful substances, transforming residues into stable organic fertilizers. These fertilizers ameliorate soil quality, enhance crop productivity, and help mitigate greenhouse gas emissions. Composting presents significant advantages including low cost, resource recovery, and environmental sustainability. Consequently, as global emphasis on sustainable agriculture and environmental protection intensifies, composting plays an increasingly vital role in agricultural residue management.

2 Agricultural Residue Composting: Mechanism and Affecting Parameters

2.1 Microbial Mechanism Underlying Agricultural Residue Composting

Composting process fundamentally depends on microbial decomposition and transformation of various organic substrates, where distinct microbial taxa play specialized roles at different composting stages, thus enabling waste valorization and efficient nutrient recycling. This section comprehensively reviews the functional roles, microbial community diversity, and successional dynamics involved in the composting process.

Composting is an autothermal, dynamic, and intricate process characterized by distinct microbial community succession correlated with temperature fluctuations. During the mesophilic stage of composting (25–50°C), readily degradable substrates such as sugars, proteins, and amino acids are rapidly metabolized by indigenous microorganisms [2]. Proteobacteria and Bacteroidetes dominate this phase, including genera such as *Pseudomonas*, *Flavobacterium*, and *Chitinophaga*, which efficiently utilize these easily available organic compounds and proliferate rapidly. Furthermore, fungi, notably *Aspergillus* species, are abundant and significantly contribute to substrate decomposition during this stage.

When the compost pile temperature exceeds 50°C due to microbial metabolic heat production, it enters the thermophilic stage. During this stage, the microbial community structure undergoes significant changes, becoming dominated by thermophilic taxa, primarily within Firmicutes and Actinobacteria, such as *Bacillus* and *Thermobifida* genus. *Bacillus* genus, belonging to Firmicutes, secretes enzymes including laccase, cellulase, and peroxidase, facilitating the degradation of complex organic substrates such as cellulose and lignin in agricultural residue [3]. Studies have indicated that Firmicutes can constitute over 50% of the microbial community during the thermophilic stage. Furthermore, as temperature declines slightly, Actinobacteria populations proliferate, further decomposing recalcitrant organic compounds.

Microbial community diversity markedly increases as compost processes into the maturation stage. The gradual temperature decrease promotes the proliferation of actinomycetes, which significantly contribute to the transformation of recalcitrant organic matter and the formation of humic precursors. *Actinomycetales*, a predominant order within actinomycetes during compost maturation, can account for 25–40% of the microbial community [4]. These organisms are highly associated with humus formation and stabilization of compost maturity.

Recent advancements in sequencing technologies, including 16S rRNA gene sequencing, metagenomics, and functional genomics, have significantly enhanced the understanding of microbial community dynamics during agricultural residue composting. Functional gene prediction analyses, such as those performed using PICRUSt2, have revealed distinct stage-specific variations in genes associated with carbohydrate metabolism, nitrogen transformation, and humus synthesis throughout composting. Furthermore, quantitative PCR (qPCR) analyses have corroborated these findings by demonstrating elevated expression levels of key enzymes, including cellulases, ligninolytic enzymes (*LiP* and *MnP*), and laccases, particularly during the thermophilic stage, which closely aligns with the rapid degradation of organic matter observed at this stage.

In summary, the succession of microbial communities and associated functional dynamics during composting underpin the effective transformation of agricultural residue, and a comprehensive understanding of these microbial processes can help optimize composting efficiency and enhance resource utilization.

2.2 Main Parameters Affecting the Composting Process

The composting process is inherently complex and sensitive to multiple interacting factors, such as aeration, temperature, moisture content, C/N ratio, pH, inoculation, microbial composition, and nutrient availability. Effective composting of agricultural residue hinges on the precise regulation of multiple physicochemical parameters that dictate microbial activity and decomposition rates.

2.2.1 Aeration (Oxygen)

Adequate oxygen supply is essential for aerobic microorganisms to efficiently decompose organic matter. To maintain optimal oxygen conditions and avoid anaerobic microenvironments, composting straw must be regularly aerated or turned. Insufficient aeration may lead to localized anaerobic conditions, hindering decomposition rates and generating unpleasant odors (e.g., sulfides). Conversely, excessive aeration may cause a decrease in pile temperature and excessive moisture loss, negatively affecting composting efficiency. Previous studies have demonstrated that intermittent aeration (5 min on/15 min off) at 1,200 L/h during the composting of corn straw and dehydrated sludge significantly accelerated temperature rise, extended the thermophilic stage, improved germination index (GI, up to 95%), and enhanced nitrogen conversion efficiency compared with continuous aeration [5]. Thus, carefully regulating aeration frequency and intensity, such as adjusting turning intervals, is critical to sustain optimal composting conditions for aerobic microbial transformation.

2.2.2 Temperature

Temperature is one of the most critical indicators for assessing the composting efficiency. As shown in Fig. 1, composting is typically classified into three distinct stages: mesophilic, thermophilic, and maturation. During the mesophilic stage (<40°C), easily degradable organic substrates (such as sugars, starches, and proteins) are rapidly metabolized by microorganisms including *Pseudomonas* and *Flavobacterium*, resulting in an initial temperature rise. In the subsequent thermophilic stage, thermophilic bacteria and actinomycetes predominate, efficiently decomposing over 65% of lignocellulose. To satisfy hygienic standards and effectively inactivate pathogens, compost piles must maintain temperatures above 50°C for at least five consecutive days. Recently, hyperthermophilic composting, defined by peak temperatures exceeding 80°C, has attracted considerable attention for its

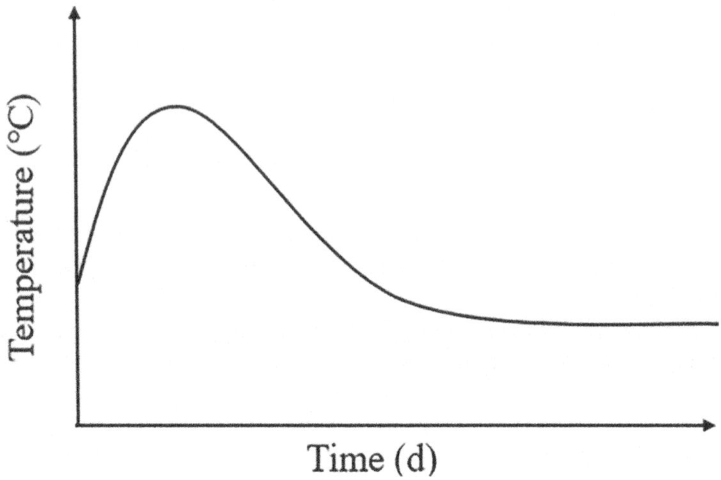

Fig. 1 Changes in composting temperature profiles over time

ability to enhance lignocellulose degradation and accelerate compost maturation [6]. Previous studies have demonstrated that hyperthermophilic composting could achieve compost maturation within 25 days, with high- and ultrahigh-temperature periods comprising over one-third of the total composting cycle, and significantly reduced N_2O emission rates to approximately one-tenth those of traditional composting [7]. Overall, temperature not only serves as a pivotal parameter in evaluating composting efficiency but also directly impacts compost hygiene, safety, and environmental emissions.

2.2.3 Moisture Content

Moisture is crucial for microbial metabolism and nutrient transport, making it one of the most influential parameters in composting process. The optimal moisture content for efficient composting typically ranges between 50% and 65%. Within this moisture range, microorganisms can effectively interact with substrates through a thin water film, establishing an equilibrium between oxygen diffusion and microbial activity. Notably, excessively low moisture content (<40%) can significantly inhibit microbial metabolism, increase heat dissipation due to elevated porosity, and result in biologically unstable compost. Conversely, excessive moisture levels (>70%) saturate the pore spaces, leading to localized anaerobic conditions. Such anoxic conditions not only produce undesirable odors but also increase nutrient loss. Moreover, surplus moisture often generates leachate enriched with nutrients (e.g., nitrogen and phosphorus), heavy metals, ammonia nitrogen, and recalcitrant organic compounds, causing nutrient depletion and potential secondary pollution risks. Therefore, carefully managing moisture levels throughout composting is essential for maintaining robust microbial activity, ensuring compost maturity, and mitigating environmental impacts.

2.2.4 Carbon-to-Nitrogen (C/N) Ratio

The C/N ratio significantly influences microbial activity, composting efficiency, and final product quality. Microorganisms primarily utilize carbon as an energy source and require nitrogen for synthesizing cellular proteins and enzymes. Controlling the initial C/N ratio at a ratio of 25:1–35:1 effectively accelerates organic matter decomposition and enhances microbial metabolic activity. However, typical agricultural residues, such as wheat straw, possess inherently high C/N ratios ranging from 70:1 to 90:1, resulting in nitrogen deficiency and restricted microbial proliferation, thus prolonging the composting process. Additionally, Shen et al. (2024) observed that an initial C/N ratio exceeding 35:1 significantly delayed compost maturation, leaving substantial amounts of undecomposed lignocellulose in the final product [8]. Conversely, initial C/N ratios lower than 15:1 promote rapid microbial ammonification, causing substantial ammonia volatilization, intense odors, and notable nitrogen losses. Hence, maintaining an optimal initial C/N ratio of approximately 25:1–35:1 is critical, which can be achieved through co-composting strategies. For example, mixing wheat straw with poultry manure reduced the final C/N ratio to around 25:1, significantly accelerating organic matter degradation, mitigating odor emissions, and enhancing total nitrogen content by approximately 25%. In summary, careful regulation of the initial C/N ratio is essential for achieving efficient composting, reducing environmental load, and producing high-quality compost product.

2.2.5 pH

Microbial activity during composting is highly sensitive to pH variations, with most microorganisms favoring neutral or slightly alkaline conditions. Throughout composting, pH typically undergoes several stages linked to microbial metabolism. Initially, the pH increases rapidly due to the accumulation of ammonium nitrogen (NH_4^+–N) produced by ammonification. As composting progresses, volatilization of NH_4^+–N and accumulation of acidic nitrification products (H^+) gradually reduce the pH. Eventually, the compost stabilizes at a neutral or slightly alkaline level. Environments that are excessively acidic ($pH < 5$) or strongly alkaline ($pH > 9$) inhibit microbial enzyme activity and significantly reduce organic matter decomposition efficiency. Previous studies indicate that the optimal pH for composting typically ranges between 7 and 8, within which microbial colonization and metabolic activity are maximized [9]. At lower pH values, acid-tolerant bacteria may predominate, yet overall microbial efficiency is compromised; at higher pH levels, ammonia volatilization is enhanced, and fungal activity is inhibited. For straw composting, the initial pH usually presents as neutral or mildly acidic, rising to approximately 8 during the thermophilic phase before gradually returning toward neutral. Generally, explicit pH adjustment is unnecessary during composting due to the process's intrinsic self-regulation mechanisms.

2.2.6 Inoculation

Lignocellulose, a primary component of straw waste, is often identified as the rate-limiting organic substrate in conventional composting processes. Therefore, introducing exogenous lignocellulose-degrading bacteria represents an effective strategy to enhance compost maturity and quality. Previous studies demonstrated that inoculating rice straw compost with thermophilic bacteria increased microbial enzyme activities (particularly cellulase and hemicellulase) by approximately 40–50% compared with uninoculated treatments. Furthermore, bacterial inoculation positively influences nitrogen retention during composting [10]. Previous studies reported that compost inoculated with the nitrogen-fixing bacterium *Bacillus amyloliquefaciens* exhibited a total nitrogen content approximately 25% higher than uninoculated controls [11]. Additionally, mixed inoculants, comprising multiple microbial species targeting diverse substrates, have attracted growing interest due to their synergistic potential. Previous studies have utilized a hybrid microbial inoculant consisting of thermophilic bacteria (*Bacillus subtilis*, *Bacillus cereus*) and the actinomycete *Thermobifida fusca* in corn straw composting [12]. The results indicated significant reductions in composting duration, enhanced enzymatic activity, and accelerated humification, highlighting the synergistic advantages of mixed inoculants for improving compost quality and process efficiency.

2.3 Difficulties in Agricultural Residue Composting

2.3.1 Residue Composition Differences and Their Effects

Agricultural residue composition varies considerably due to geographic and crop-specific differences, significantly influencing composting characteristics. For example, in China, wheat straw and corn straw, predominantly generated in northern China, typically pose high carbon content and low nitrogen content, with C/N ratios ranging from 50 to 80:1, along with elevated lignin levels (15–20%). Compared to other cereal residues, rice straw, primarily produced in southern China, contains notably higher silica content (12–16%) but lower lignin content. Thus, silica rather than lignin represents the main barrier to rice straw decomposition. Additionally, agricultural residues generally exhibit nitrogen deficiencies (<1%), resulting in high C/N ratios and limited protein availability for microbial activity, further hindering decomposition efficiency during composting.

Silica forms a physical barrier surrounding plant fibers, thereby impeding microbial attachment and restricting enzyme accessibility to substrates. As silica is nonbiodegradable, it can complex with enzymes, subsequently reducing enzymatic activity. In contrast, lignin found in wheat and corn straw is recalcitrant to degradation, necessitating specialized hydrolytic enzymes such as peroxidases and ligninases, typically produced in the later stages of composting. Easily degradable organic matter, including sugars and cellulose, is preferentially metabolized by microorganisms early in composting, whereas lignin degradation proceeds significantly slower. Although rice straw contains lower lignin content compared to wheat and corn straw, its decomposition rate remains relatively slower due to silica impeding initial cellulose breakdown. Studies have demonstrated notably slow degradation in rice straw containing approximately 20% silica [13]. In summary, the composting of wheat and corn residues prevalent in northern China is challenged by high carbon content, nutrient limitations, and elevated lignin levels, while rice straw from southern China poses additional complexity due to its high silica content. Therefore, optimizing nutrient balance and employing microbial supplementation strategies are critical for efficient composting.

2.3.2 Climatic Challenges in Different Regions

Diverse climatic conditions significantly influence composting practices, particularly due to marked variations between cold, high-latitude regions and warm, humid areas. In northeast China, such as Heilongjiang Province, prolonged cold winters severely restrict composting efficiency. Low ambient temperatures substantially inhibit microbial metabolic activities, making it difficult for mesophilic and thermophilic microorganisms to proliferate or generate adequate metabolic heat when substrate and ambient air temperatures approach freezing points. Consequently, compost piles that could otherwise warm rapidly under favorable conditions remain cold or experience slow temperature increases. Studies have reported that straw decomposition is notably delayed at temperatures below −20°C, as sustained heat production necessary for decomposition is insufficient [14]. Typically, composting can be completed within 1–2 months in warmer climates; however, in colder environments, the process may stall or remain incomplete throughout winter without interventions such as supplementary heating or frequent pile turning. Although reducing turning frequency or covering compost piles is commonly practiced to retain heat, these methods may inadvertently enhance anaerobic conditions, resulting in increased CH_4 and N_2O emissions.

In cold climates, dealing with these problems requires a delicate balance between aeration and insulation. On the one hand, adequate aeration helps to provide oxygen and prevent anaerobic conditions; on the other hand, excessive aeration can lead to excessive cooling and drying of the pile, which can inhibit the composting process. Cold air entering the pile carries away heat and reduces moisture content through evaporation. It has been shown that excessive aeration intensity can actually cause the compost substrate to lose a lot of heat and become dry, which in turn weakens microbial activity. Therefore, in winter composting practices in Northeast China, "over-aeration" should be avoided and the heat of the compost pile should be maintained through gentle or intermittent aeration strategies.

The warm and humid subtropical regions of southern China (e.g., Guangdong and Guangxi) present distinct conditions for composting agricultural residues compared to the colder climates of northeastern China. Elevated ambient temperature and humidity facilitate composting efficiency, yet also introduce unique challenges. Specifically, higher ambient humidity can rapidly lead to oversaturation within compost piles, potentially creating localized anaerobic zones that inhibit aerobic microbial activity and slow decomposition rates. Nonetheless, once humidity stabilizes at optimal levels, rapid aerobic decomposition of straw typically resumes. In contrast, composting in the colder northeastern climates requires extended time periods and significant insulation efforts due to lower temperatures and limited microbial activity. Overall, subtropical conditions in southern China offer a conducive environment for efficient composting, distinctly differing from the prolonged durations and insulation demands characteristic of higher-latitude cold regions.

Overall, efficient composting of agricultural residues necessitates consideration of both intrinsic compositional properties of the residues and prevailing climatic conditions of the composting regions. Residues characterized by high silica content and elevated C/N ratios require supplemental nutrients or inoculation with specialized microorganisms to overcome their slow decomposition rates; these approaches must also be customized according to regional climate variability. Specifically, in cold northern climates, strategies should prioritize thermal insulation and careful management of aeration intensity to prevent anaerobic conditions, whereas in warm southern regions, maintaining adequate aeration and effective moisture regulation is essential to mitigate rapid decomposition and potential overheating of compost piles. Ultimately, the development of effective composting strategies tailored to specific regions demands an integrated understanding of processes ranging from microbial enzymatic activities to thermal dynamics and gas exchange within compost systems.

3 Strategies for Enhancing Agricultural Residue Composting

Enhancing the composting of agricultural residues requires strategies that improve decomposition efficiency and quality of the final compost. These strategies can be classified into physical, chemical, and biological measures.

3.1 Particle Size Reduction (Pulverization)

Pulverization reduces the particle size of agricultural residues, thereby increasing the surface area available for microbial decomposition. Although pulverization is not necessarily required for soft biomass, such as grass-based crop residues, it has been observed that treated crop residues decompose more rapidly than whole residues. In contradistinction, the pulverization of hard biomass, such as wood, generally requires a substantial amount of energy [15]. Consequently, it is often recommended that by-products from the lumber industry, such as sawdust, bark, and wood shavings, be utilized as alternative materials.

3.2 Free Air Space (FAS) Optimization

Aerobic microbial activity plays a crucial role in composting, necessitating sufficient FAS to facilitate oxygen penetration into the composting mass. FAS is a physical parameter representing the gas phase fraction, which denotes the gas volume fraction relative to the total apparent volume [16].

Practically composting feedstock is often prepared by gently layering to prevent excessive compaction and by employing physical pretreatment techniques to maximize the proportion of the FAS within the compost pile layer. However, measuring FAS in field conditions remains challenging, and its direct control is often impractical. Despite the recognized significance of FAS, reports on its application in field settings are limited, with most studies relying solely on material moisture content as an indicator of composting readiness.

Under certain assumptions, the volumetric gas ratio (R_{VG}), which represents FAS using mass-based moisture content (W_{M}), is predictable.

The W_{M} is expressed as Eq. (1), and the liquid mass (M_{L}) can be represented by Eq. (2):

$$W_{\mathrm{M}} = \frac{M_{\mathrm{L}}}{M_{\mathrm{S}} + M_{\mathrm{L}}} \tag{1}$$

$$M_{\mathrm{L}} = \frac{M_{\mathrm{S}} \cdot W_{\mathrm{M}}}{1 - W_{\mathrm{M}}} \tag{2}$$

Since (M_{L}) can also be expressed as

$$M_{\mathrm{L}} = \rho_{\mathrm{L}} V_{\mathrm{L}},$$

The liquid volume (V_{L}) is given by

$$V_{\mathrm{L}} = \frac{M_{\mathrm{S}} \cdot W_{\mathrm{M}}}{\rho_{\mathrm{L}}(1 - W_{\mathrm{M}})} \tag{3}$$

The liquid phase ratio (R_{VL}) is obtained from V_{L}/V as follows:

$$R_{VL} = \frac{M_S \cdot W_M}{V \cdot \rho_L(1 - W_M)} = \frac{R_{VS} \cdot \rho_S \cdot W_M}{\rho_L(1 - W_M)} \tag{4}$$

The gas phase ratio (R_{VG}) is given by

$$R_{VG} = 1 - (R_{VS} + R_{VL}) \tag{5}$$

Thus, assuming a constant solid phase ratio (R_{VS}), the gas phase ratio (R_{VG}) becomes a function solely of the mass-based moisture content (W_M) as follows:

$$R_{VG} = 1 - \left\{ R_{VS} + \frac{R_{VS} \cdot \rho_S \cdot W_M}{\rho_L(1 - W_M)} \right\} \tag{6}$$

The volumetric solid ratio (R_{VS}) can be considered constant, as experimental studies on compost feedstock mixtures of dairy cattle manure and sawdust have demonstrated that R_{VS} remains approximately 0.1 with minimal variation.

The R_{VG} presented in Fig. 2 represents the calculated results based on Eq. (6). The relationship between volumetric gas ratio (R_{VG}) and material moisture content (W_M)

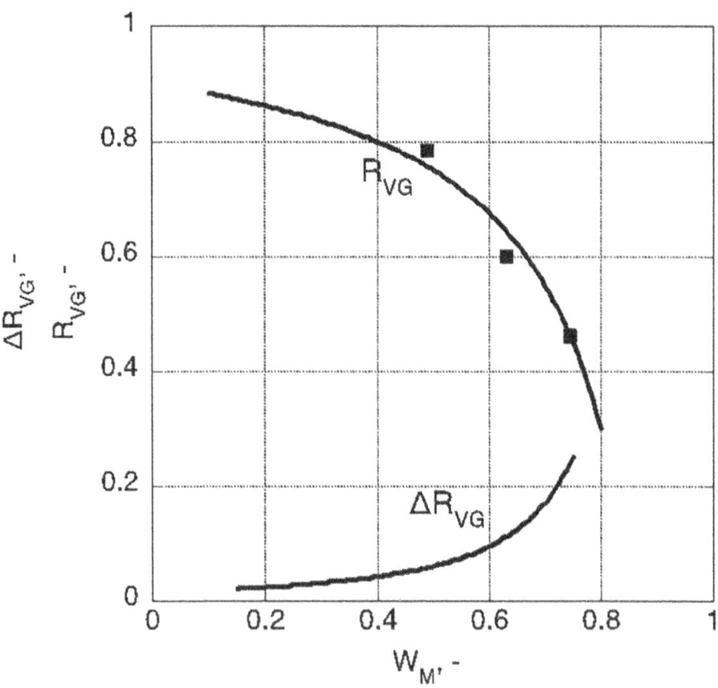

Fig. 2 Relationship between the volumetric gas ratio (R_{VG}), difference in R_{VG} (ΔR_{VG}), and material moisture content (W_M). The graph was derived by solving Eq. (6) with the parameters $R_{VS} = 0.1$, $\rho_L = 1.0$, and $\rho_S = 1.5$

can be clearly illustrated. So far, few studies have derived the volumetric gas ratio (R_{VG}) from material moisture content (W_M), and predicting the R_{VG} of a compost layer in advance has been largely unfeasible. However, Eq. (6) offers an estimation of the relationship between W_M and R_{VG}. When W_M is below approximately 0.6, R_{VG} maintains a relatively large free air space (FAS) of over 0.6. However, as W_M exceeds 0.6, R_{VG} decreases sharply, indicating a rapid decline in available air space. This trend is more clearly understood by examining ΔR_{VG} (difference in R_{VG}), which highlights the pronounced reduction in gas volume at higher moisture levels. These findings suggest that as composting feedstock moisture content increases, FAS decreases drastically, leading to potential oxygen supply limitations and a consequent reduction in oxygen uptake by microorganisms.

The solid black square markers (■) in Fig. 2 represent the measured data. These results demonstrate that the measured R_{VG} closely aligns with the calculated values, validating the model's accuracy.

Developing a physically structured matrix with high porosity is crucial for facilitating oxygen diffusion within the composting layer. Enhanced oxygen diffusion supports the activity of aerobic microorganisms and accelerates the composting process.

3.3 Moisture Content Management

The moisture conditions that facilitate composting reactions have been discussed in previous sections. This section introduces moisture control techniques for composting feedstocks [17].

Mixing low- and high-moisture-content materials: Blending low- and high-moisture materials is a simple yet effective method for moisture regulation [18]. High-moisture materials, such as fruits, vegetables, and livestock waste, typically contain more than 70% moisture. In contrast, brown biomass consists of dry, carbon-rich organic materials such as dried leaves, straw, and rice husks, which help reduce the overall moisture content. Under certain conditions, such as delayed harvesting or in-field drying, the moisture content of these materials can be reduced to below 30%. In-field drying of crops such as rice and wheat allows natural desiccation under sunlight, thereby minimizing postharvest drying costs.

Water spraying: Controlled water spraying is an effective method for increasing moisture content to the desired level. This technique is particularly beneficial during the initial mixing phase and for materials prone to drying out during composting. However, excessive water application can result in anaerobic conditions, which may impede the composting process. Therefore, water spraying should be carefully managed to prevent oversaturation and development of anaerobic environments.

3.4 Adjustment of Nitrogen Components in Composting Materials

Agricultural residues typically exhibit a higher initial C/N ratio. A deficiency in nitrogen severely limits microbial growth and metabolic processes, thereby reducing the rate of organic matter degradation and extending the compost maturation period. Consequently, optimizing the nutrient composition of the composting substrate through nutrient supplementation is essential. Notably, in straw composting, nutrient addition primarily involves the supplementation of nitrogen sources.

Supplementing nitrogen sources can adjust the C/N ratio of raw materials to an optimal range, significantly promoting microbial growth and organic matter decomposition. For instance, when the C/N ratio of the straw and sludge co-composting mixture is adjusted to 25 by adding corn stalks, not only is the temperature higher during the thermophilic phase, but the GI also improves compared to the control group [19]. Additionally, the addition of an appropriate amount of urea to wheat straw composting accelerates the degradation of organic matter. This is because a sufficient nitrogen supply enhances microbial reproduction and increases the activity of degradation enzymes, thereby accelerating the composting process and maturation. Moreover, Wang et al. (2023) evaluated the feasibility of composting rice straw mixed with nitrogen-rich wastes, such as food waste, cow manure, and sewage sludge [20]. The study found that all three nitrogen sources could complete composting the straw within 51 days. However, the food waste treatment group exhibited the fastest straw degradation rate and the highest-quality compost. This improvement was attributed to accelerated bacterial community succession, optimized microbial metabolism, and thus enhanced organic matter degradation and compost maturation. Furthermore, leather protein hydrolysate, a novel nitrogen supplement, performed well in straw composting. Compared to traditional urea treatment, the composting system with leather protein hydrolysate showed a faster heating rate (1.85°C/h), a higher organic matter degradation rate (48.04%), and a significantly higher GI (111.72%), demonstrating its substantial advantages in accelerating compost maturity and improving compost quality [21].

An adequate supply of nitrogen is essential for the successful decomposition and humification of high-carbon materials. Adding an appropriate amount of nitrogen can promote microbial reproduction and degradation of cellulose and lignin, thereby significantly enhancing the efficiency of organic matter degradation and precursor conversion in composting. For example, the addition of chicken manure (rich in organic nitrogen and microorganisms) to straw, combined with pretreatment measures, accelerates the decomposition of crude fiber and organic matter, significantly improves the conversion rate of humus precursors, and increases the final HA content by 19.54% compared to the control group. This improvement is attributed to the pretreatment process, which promotes the decomposition of crude fiber and provides precursors for humus, while the addition of chicken manure facilitates the conversion of organic matter into HA.

The type and form of nitrogen supply play a significant role in the humification process. Organic nitrogen sources, such as livestock and poultry manure, are rich in amino acid precursors and natural microbial communities, which are more conducive to humus synthesis. Although inorganic nitrogen sources like urea can rapidly supply nitrogen, excess amounts may lead to ammonia volatilization and an increase in pH, which may inhibit certain fungal activities. This is supported by research by Dong et al. (2024), which investigated the effect of fungal inoculation on the humification of cornstalks under two nitrogen sources [22]. The study found that the HA content increased by 71.76% compared to the uninoculated test when pig manure was used and by 53.01% when urea was applied, indicating that organic nitrogen sources better promote humus accumulation. Furthermore, the addition of nitrogen sources can accelerate humus formation by activating the enzyme system involved in lignin degradation. Field composting experiments have shown that the application of nitrogen fertilizers alone can increase lignin peroxidase and laccase activity to some extent, with a more significant effect when applied in combination with functional microbial inoculants [23]. Higher levels of laccase and lignin peroxidase activity correlate with faster lignin conversion and aromatic polymerization, which are closely related to the accumulation of HA.

Comprehensive calculations for C/N ratio adjustments are available in the Cornell Composting Guide [24], which serves as a valuable reference. This guide also outlines methodologies for simultaneously adjusting both moisture content and C/N ratio, providing practical insights for optimizing the composting process.

3.5 *Biological Inoculation*

An effective method of biological measures in composting is microbial inoculation. This approach involves the deliberate addition of specific microorganisms to organic waste to enhance the decomposition process and improve the quality of the resulting compost product. Recent studies in this area have identified several key trends, particularly the use of microbial consortia, which combine several beneficial microbial species to establish a functionally effective composting process. For example, the study presented below has shown that the combination of different microbial strains can accelerate the degradation of complex organic materials such as lignocellulosic waste.

López-González et al. (2024) investigated the application of microbial consortia as a strategy to enhance the biodegradability of lignocellulosic waste in composting [25]. The study aimed to improve composting efficiency by pretreating vegetable waste with lignocellulolytic microorganisms to accelerate organic matter decomposition. A microbial screening process identified *Bacillus safensis*, *Bacillus licheniformis*, and *Fusarium oxysporum* as effective inoculants due to their enzymatic capabilities. The study employed a 50:50 mixture of tomato plant and pruning waste, adjusting moisture content to approximately 65%. The effectiveness of pure cultures and microbial consortia was assessed based on various parameters, including moisture content, organic matter, ash composition, C/N ratio, and degradation rates of cellulose, hemicellulose, and lignin. The results demonstrated that the application of bacterial strains at low doses (10^4 CFU/g) significantly increased lignocellulose degradation within 10 days, surpassing fungal treatments and untreated controls.

The findings suggest that microbial consortia serve as an effective bioaugmentation tool for optimizing industrial composting processes, particularly in enhancing the degradation of lignocellulosic material. The study highlights that composting with microbial inoculants not only improves decomposition efficiency but also enhances the agronomic quality of the resulting compost. By integrating microbial consortia with existing composting practices, this approach provides a sustainable solution for managing plant-based waste and producing high-quality organic amendments. Consequently, the research contributes to advancing composting technology by demonstrating the potential of bioaugmentation strategies to improve compost quality and efficiency in large-scale composting operations.

3.6 Co-composting

Although composting can be conducted using a single biomass feedstock, this approach may result in an imbalanced nutrient composition, potentially inhibiting the growth and activity of composting microorganisms. Therefore, by appropriately blending various feedstocks with diverse carbon, nitrogen, moisture content, trace chemical elements, and microbial populations, it is possible to optimize the physical, chemical, and biological properties of the composting process, leading to enhanced efficiency and stability.

Fujiwara (1986) conducted extensive research on practical composting techniques and published detailed data on the composting of both plant-based and animal-derived waste materials [26]. Plant-based compost and manure-based compost exhibit distinct characteristics in terms of nutrient composition (Table 1). The chemical composition of plant-based compost, primarily derived from rice straw, differs significantly from manure-based compost, which is produced from cattle, swine, and poultry manure. Compared to plant-based compost, manure-based compost contains higher levels of essential fertilizer components. Specifically, plant-based compost tends to be lower in calcium oxide (CaO), magnesium oxide (MgO), and phosphoric acid (P_2O_5), whereas manure-based compost is richer in these

Table 1 Nutrient content of plant-based and manure-based compost products (% DM)

Compost products	T-N	C/N	P_2O_5	K_2O	CaO	MgO
Plant-based compost	1.64	18.7	0.77	1.76	1.99	0.55
Manure-based compost (cattle)	2.10	16.5	2.06	2.19	2.31	0.99
Manure-based compost (swine)	2.86	13.2	4.31	2.23	3.96	1.35
Manure-based compost (poultry)	2.89	12.5	5.13	2.68	11.32	1.36

Table 2 Chemical composition of compost feedstock (% DM)

Compost feedstocks	T-C	T-N	C/N (−)
Wheat straw	40–45	0.5–0.7	60–80
Rice straw	40–45	0.7–0.9	50–60
Leaf litter	40–45	0.8–1.5	60–80
Cattle manure	35–40	1.5–2.0	15–20
Swine manure	40–45	4.0–4.5	8–10
Chicken manure	30–35	5.0–5.5	6–8

elements. By combining different types of feedstocks, it is possible to produce compost with a more balanced nutrient profile.

The incorporation of plant-based composting materials, such as wheat straw, with manure-based feedstocks serves not only to enhance mineral content but also to adjust the C/N ratio, a key parameter influencing composting efficiency (Table 2). While the total carbon (T-C) content of plant-based materials, such as wheat straw, is comparable to that of manure-based feedstocks, their total nitrogen (T-N) content varies considerably. Even cattle manure, which contains the lowest T-N content among livestock manures, has more than twice the T-N content of wheat straw. This difference highlights the importance of co-composting different feedstocks to achieve a more favorable nutrient balance and optimize microbial activity during decomposition.

4 Emerging Technology for Improving Agricultural Residue Composting

Agricultural residue composting represents a crucial strategy for sustainable development and resource recycling. However, conventional composting methods are often hindered by challenges such as prolonged decomposition, nutrient losses, and greenhouse gas emissions. In recent years, emerging approaches – including the use of biochar, metal oxides, microbial inoculation, nutrient addition, functional membrane covering, and electrochemical assistance – have garnered significant attention by offering novel solutions to enhance composting efficiency, improve product quality, and mitigate environmental impacts (Table 3).

4.1 Biochar Addition

Biochar is a carbon-rich, porous material produced via the pyrolysis of organic matter (e.g., agricultural and forest residues) under anaerobic conditions. During composting, biochar can significantly enhance the physical, chemical, and biological properties of the system due to its large specific surface area and high porosity.

Table 3 Emerging technologies for improving the composting process of agricultural residue

Composting	Technology	Effect of emerging technologies on composting.	References
Distilled grain waste	Biochar	The addition of biochar increased carbon retention by 13.5% and nitrogen retention by 33.8%	[20]
Rice straw	Biochar	Biochar can effectively inhibit CH_4 and CO_2 emissions during composting	[49]
Rice straw + pig manure	Metal oxide	Iron oxide helps to raise the temperature of the composting to promote decomposition	[34]
Wheat straw	Metal oxide	Iron oxide helps to enhance the humification of composting	[50]
Sorghum straw	Microbial inoculation	Mature compost inoculation significantly promotes organic matter degradation	[36]
Cattle manure + wheat straw	Microbial inoculation	Microbial inoculation promotes cellulose and lignin degradation	[38]
Cattle manure + bagasse	Microbial inoculation	After inoculation, the humus content can reach 86.59 mg/g and HA can reach 25.80 mg/g	[39]
Tomato straw + mushroom residue	Membrane covering	The membrane cover extended the high-temperature period and cut CH_4 and N_2O emissions	[43]
Vegetable residues	Membrane covering	Membrane-covered composting accelerates the degradation rate of 70% of lignocellulose	[44]
Sludge + corn straw	Electrobioreaction	Electric field application accelerates organic matter-degrading microbial community succession	[46]
Food waste	Electrobioreaction	The compost maturation cycle was reduced by 40% and the HA content was increased by 33%	[47]

Consequently, the decomposition of organic matter and the formation of humus are accelerated [27].

The efficiency and quality of composting depend closely on the physicochemical characteristics of the composting pile, including key parameters such as pH, moisture content, temperature, and aeration. Due to its porous structure, large specific surface area, high cation-exchange capacity (CEC), and generally weakly alkaline nature, biochar is considered as an ideal additive to improve the composting environment. During the early stages of composting, the decomposition of organic matter produces organic acids that can lead to acidification, thereby inhibiting microbial activity. The weak alkalinity of biochar neutralizes these acids, buffering the decline in pHs. Tran et al. (2021) reported that the addition of biochar increased the pH of food waste during the initial stages of composting [28]. Moreover, composting systems supplemented with biochar exhibited more stable pH values. For example, 5% biochar addition to cow dung and food waste composting maintained the pH within a range of 6.3–8.7, with less fluctuation compared to the control group [29]. These results indicate that biochar effectively stabilizes the composting pH, thereby creating a more favorable environment for microbial growth. The unique porous structure and surface functional groups of biochar confer exceptional water retention and humidity regulation properties. Both excessively high and low moisture contents can hinder the composting process. The porous nature of biochar adsorbs and stores free water, thereby enhancing the water retention capacity of the compost pile and preventing excessive evaporation and desiccation. Maintaining optimal moisture conditions promotes sustained microbial activity and mitigates fluctuations in degradation rates.

In terms of microbial activity, biochar can effectively enhance the activity of key enzymes involved in carbon metabolism during composting and promote overall microbial function. Several studies have demonstrated that the addition of biochar significantly increased the activities of enzymes such as urease and peroxidase [30]. This enhanced enzymatic activity accelerates the decomposition of lignocellulose and facilitates humus formation, thereby improving the carbon sequestration capacity of composting. Furthermore, Liu et al. (2023) reported that biochar markedly stimulated the metabolism of heterocyclic compounds and amino acids by modulating the metabolic pathways of microbial communities during composting [30]. This process significantly accelerates humic acid (HA) production in the composting, resulting in a 39.4% increase compared to the control group. Additionally, incorporating biochar alters the accumulation dynamics of microbial residues, thereby influencing carbon fixation. Jia et al. (2024) confirmed this effect by demonstrating that the addition of 5% biochar to chicken manure composting increased microbial residue carbon from 24.9 to 77.9 g/kg, markedly enhancing the long-term stabilization of organic carbon [31]. These findings further underscore biochar's potential to promote the accumulation of microbial residue carbon and stabilize organic carbon.

In summary, biochar is an ideal additive for the composting process. It enhances microbial activity by improving the physical and chemical properties of the composting environment (e.g., pH and moisture retention) and significantly increases carbon sequestration by modulating microbial metabolism and promoting humus formation. Moreover, its porous structure and surface functional groups are critical for improving composting efficiency and stabilizing organic carbon over the long term.

4.2 Metal Oxide Addition

Metal oxides, such as iron and manganese oxides, have emerged as promising additives to enhance both the efficiency and quality of composting. These compounds naturally occur as minerals and can interact with adjacent organic components [32]. Their mechanisms in composting are multifaceted. For example, metal oxides can catalyze the polymerization of small organic molecules into stable macromolecules while also forming complexes that inhibit microbial degradation.

Moreover, metal ions can stimulate microbial activity and accelerate composting by serving as essential nutrients or cofactors for microbial growth. Similar to the effect of biochar's porous structure, metal oxides also improve the composting environment, thereby facilitating the degradation process.

Among various metal oxides, iron oxide is a well-studied additive widely applied in composting. Their incorporation significantly enhances humus production during composting, particularly by facilitating the conversion of fulvic acid (FA) to HA. For instance, Noor et al. (2024) demonstrated that increasing the amount of iron oxide added corresponds with a greater conversion of FA to HA [33]. These results indicate that iron oxides accelerate the formation of complex humus components, thereby improving compost maturity and quality. Additionally, iron oxides can elevate the composting temperature, thereby promoting the composting process. Studies have shown that adding biochar-loaded nano-iron oxides to pig manure and straw composting increased the temperature by up to 3.6°C relative to the control, and the final C/N ratio reached 12.9 [34]. Moreover, this treatment enhanced phosphorus stabilization, with the proportion of residual phosphorus in the compost increasing from 19% in the no-additive group to 26–52%. These findings indicate that the incorporation of iron oxides promotes phosphorus fixation and reduces nutrient loss, ultimately enhancing the fertilizer efficiency of the composting.

In addition to iron oxides, the role of manganese oxides in composting has also been confirmed. Previous studies have demonstrated that MnO_2 enhances lignocellulose degradation and promotes humification by modulating the abundance of microorganisms in sludge composting, including genera such as *Nonomuraea*, *Actinomadura*, and *Streptomyces* [35]. Consequently, organic matter degradation increased by up to 19.2%, and HA content rose by up to 20.6%. These results indicate that MnO_2 directs the microbial community toward the humus synthesis pathway, thereby significantly enhancing compost humification. Furthermore, metal oxides demonstrate potential in composting organic waste contaminated with heavy metals. They reduce the bioavailability of heavy metal ions through passivation, thereby mitigating the risk of environmental pollution. Overall, metal oxides represent promising additives that enhance both composting efficiency and quality via multiple mechanisms, including the accelerated degradation and transformation of organic matter and increased humus production.

4.3 Microbial Inoculation

Microbial inoculation, an effective strategy to improve the composting process, optimizes the microbial community structure by introducing functional microorganisms with potent degradation and humification capabilities, thereby accelerating organic matter decomposition and overall composting. Li et al. (2024) demonstrated that the addition of 10% mature compost during the composting of sorghum stalks significantly enriched the key microbial genera involved in organic matter degradation, including *Thermobifida*, *Lysinibacillus*, and *unclassified_o_Bacillales*

[36]. Upon completion of the composting process, the degradation efficiencies of volatile solids and seed GI were enhanced by 30.34% and 34.64%, respectively, compared to the control treatment without microbial inoculant. These findings suggest that the incorporation of exogenous microbial inoculants not only accelerated the composting process but also facilitated the maturation of the compost by reshaping the microbial community structure. Studies have demonstrated that incorporating microbial inoculants in food waste composting can achieve thermophilic phase within 1 week, with the seed GI rapidly exceeding 80%, indicating that the compost has reached nontoxic maturity [37]. In contrast, uninoculated control groups typically require a longer duration to attain comparable levels of maturity. These findings indicate that the addition of exogenous microorganisms facilitates the decomposition of refractory organic matter – such as lignin, hemicellulose, and cellulose – thereby accelerating the transition to the high-temperature decomposition stage. In addition, a study on the co-composting of straw and cow dung revealed that the inoculation with a microbial consortium consisting of three strains of *Bacillus* and *Geobacillus* effectively extended the thermophilic phase of composting. This inoculation also resulted in an increase in the degradation rates of cellulose, hemicellulose, and lignin by 11.1–16.7% [38]. These results underscored that microbial inoculation accelerates organic matter conversion by increasing the abundance of cellulolytic bacteria, highlighting the critical role of functional strains in the composting process.

Microbial inoculation significantly influences the humification process during composting. Following inoculation, the complexity and diversity of the microbial community increase, which in turn markedly enhances the aromaticity and stability of soluble organic matter and HA. In the co-composting of cow dung and bagasse, a two-stage inoculation strategy with lignocellulose-degrading microorganisms – applied during the heating and cooling phases – resulted in humus and HA contents of 86.59 mg/g and 25.80 mg/g, respectively, and a GI of 128.77% [39]. This approach stimulated an increase in the abundance of amino acid synthesis pathways within the composting, thereby effectively improving compost quality. Studies have demonstrated that microbial inoculants significantly increased HA content in compost by as much as 43.85% by enhancing both the Maillard reaction and phenolic protein condensation. Moreover, the maximum temperature of the composting reached 72.5°C [40]. In summary, microbial inoculation markedly enhances humification, increases microbial community diversity, and promotes HA production while improving compost quality and temperature, thereby accelerating compost maturation.

In addition to accelerating the degradation and humification of organic matter, microbial inoculation exhibits significant potential for optimizing nitrogen transformation and reducing NH_3 volatilization during composting. For instance, studies have shown that inoculating extreme thermophilic microorganisms during the aerobic composting of refinery sludge not only enhanced compost maturity but also reduced NH_3 and N_2O emissions by 32.36% and 10.17%, respectively [41]. These findings indicate that microbial inoculants promote nitrogen stabilization, reduce nitrogen loss, and improve overall composting quality. *Bacillus* inoculation enhanced ammonia assimilation and nitrification, thereby reducing NH_3 emissions during composting and increasing the nitrogen content of the final product. Moreover, microbial inoculation often exhibits synergistic effects with other technologies. For example, when combined with biochar, microbial inoculation provided the catalytic drive for degradation, while biochar improved the composting environment, jointly enhancing compost quality and reducing greenhouse gas emissions. Overall, the appropriate application of microbial inoculants can effectively shorten the composting cycle and improve maturity.

4.4 Membrane-Covered Composting

Membrane-covered composting optimizes the composting environment by covering the surface of the compost heap with a breathable membrane, typically a semipermeable or controlled-permeability material. This technique creates a relatively closed microenvironment that retains heat and moisture while allowing limited gas exchange, thereby providing favorable conditions for microbial activity. Although early examples of membrane-covered technology date back to the 1980s, systematic research into its mechanisms has emerged only in the past decade. In recent years, semi-permeable membrane-covered composting has been widely implemented at large-scale composting sites in Europe, the United States, and China, demonstrating its effectiveness in reducing odor and greenhouse gas emissions as well as improving composting efficiency [42].

Membrane-covered composting significantly influences the temperature dynamics and degradation of organic matter. The covering membrane reduces heat and moisture loss from the pile surface, thereby increasing the maximum compost temperature and extending the thermophilic phase, which is crucial for pathogen inactivation and accelerated decomposition of refractory organic matter (e.g., lignin and cellulose). Studies have shown that the peak core temperature of a membrane-covered compost heap was approximately 3.9°C higher than that of an uncovered test, with the thermophilic phase extended by 4 days. Concurrently, CH_4 and N_2O emissions were effectively reduced [43]. Furthermore, the composting process was significantly accelerated; the treated composting reached the same maturity level as the control group 12 days earlier, with a GI of 70% achieved in 18 days. Additionally, membrane coverage markedly enhanced the decomposition of refractory organic matter such as lignin, cellulose, and hemicellulose. Studies have shown that membrane-covered composting not only accelerated the degradation rate of lignocellulose by 70% but also achieved a high humus conversion rate, a GI exceeding 100%, and a higher degree of maturity [44]. Furthermore, some studies have combined membrane coverage with microbial inoculants, resulting in a significant improvement in lignin degradation and humification. These results show that membrane-covered composting accelerated the degradation of organic matter and increased the rate of humus production, which is conducive to obtaining a compost product with higher maturity and stability.

In addition to accelerating composting degradation and maturation, membrane-covered composting can also capture some of the volatile gases (e.g., NH_3, CH_4, N_2O) released during the process, thereby reducing direct emissions of odorous and harmful compounds. Cao et al. (2022) demonstrated that membrane coverage reduced NH_3 emissions by 25.8% and N_2O emissions by 13.1% compared to the control group [45]. This effect is likely due to the hydrogel film formed during membrane-covered composting, which effectively adsorbs NH_3 and reduces nitrogen loss. The micro-positive pressure environment also suppresses denitrification by enhancing oxygen distribution, which in turn reduces N_2O emissions. Furthermore, membrane-covered composting regulates the abundance of genes associated with carbon and nitrogen emissions (e.g., *mcrA*, *nirS*, *nirK*, and *nosZ*), effectively decreasing CH_4 and N_2O emissions. Overall, membrane-covered composting significantly reduces total greenhouse gas emissions compared to open-air composting, representing a critical advancement for achieving cleaner agricultural residue treatment.

In summary, membrane-covered composting accelerates the degradation of organic matter, enhances compost maturity, and improves environmental performance by optimizing thermal and moisture conditions within the compost pile and reducing gas emissions. The key to this technology lies in selecting an appropriate membrane material and maintaining proper ventilation to trap heat and moisture while preventing the formation of anaerobic zones. Membrane-covered composting has proven to be practical in large-scale applications, shortening the composting cycle and effectively controlling odors and emissions, which aligns with the requirements for sustainable agricultural development.

4.5 Electrobioreaction

Electrobioreaction technology is an emerging approach that employs an electric field or current to influence microbial activity and metabolic pathways during composting. This is typically achieved by constructing a bioelectrochemical system or embedding electrodes within the compost. The fundamental principle involves introducing an electron exchange interface into the composting system, which enables microorganisms to transfer electrons to the anode during organic matter decomposition and to receive electrons at the cathode, thereby generating a current or electric potential. The externally applied electric field or current alters the redox environment of the composting system and modulates the metabolic pathways of the microbial community.

Low composting temperatures, prolonged fermentation cycles, and insufficient degradation of organic matter have long been major challenges to the efficiency of composting agricultural residues. However, the application of an electric field significantly enhances microbial activity and accelerates compost maturity. Studies have shown that applying a 5 V electric field to the aerobic composting of sludge and straw not only prolonged the thermophilic phase but also increased the GI by 13.38%, effectively promoting pathogen inactivation and product maturation [46]. Moreover, the electric field treatment increased the abundance of the phyla Proteobacteria and Bacteroidetes, enriched organic matter-degrading bacteria (e.g., *Pseudomonas*), and accelerated the succession of beneficial microbial communities. In addition, the degree of compost maturity under an electric field exhibits spatial differences, with the cathode region maturing faster than the anode region. For example, under the influence of an electric field, a moisture gradient and corresponding variations in microbial activity develop from the anode to the cathode. The cathode area showed higher water content, more vigorous microbial activity, and superior compost maturity indicators [41]. At the end of the composting process, the moisture content and GI in the cathode region were 13.75% and 54.47% higher than those in the anode region, respectively. These findings indicate that an electric field can direct the migration of water and nutrients within the compost, thereby locally enhancing microbial metabolism and maturation. A judicious arrangement of electrodes is expected to promote more uniform and rapid composting throughout the entire pile.

Electric field-assisted composting can significantly accelerate humus production, thereby improving compost quality. Mi et al. (2023) demonstrated remarkable outcomes by applying a novel arrayed electric field-assisted aerobic composting process of food waste: the maximum temperature reached 69.3°C, the maturation cycle was shortened by 40%, and the contents of HA and FA increased by 33% and 37%, respectively [47]. This enhancement is attributed to the electric field ability to accelerate the biodegradation of organic matter and hasten the humification process. Moreover, the impact of the electric field varies at different stages of composting. At the beginning of composting, the application of an electric field can significantly enhance the mineralization of organic matter, which may lead to increased CO_2 emissions. However, during the later stages, the electric field effectively activates extracellular enzymes associated with refractory substances such as cellulose and lignin, thereby accelerating their degradation and conversion into HA [48]. Moreover, the electric field not only facilitates HA formation but also significantly enhances heavy metal passivation. Studies have demonstrated that applying an external electric field during livestock and poultry manure composting promotes the complexation of HA with heavy metals, ultimately increasing the complexation of Cu, Zn, As, and Cd by 29–135.1%.

Electro-bioreactor technology optimizes organic matter degradation and shortens the ripening period through electrical stimulation, exhibiting significant potential to enhance degradation efficiency, reduce emissions, and increase HA production. Although engineering challenges remain, further technological optimization and in-depth mechanistic research are expected to advance the efficient composting and resource utilization of agricultural residues.

5 Agricultural Utility of Agricultural Residue Compost Products

5.1 Effectiveness of Organic Fertilizers in Sustainable Crop Production

Sustainable agricultural practices seek to enhance food security while minimizing environmental impact and reducing dependence on synthetic fertilizers and pesticides. The application of organic fertilizers not only influences food production and soil quality but also contributes to the enhancement of soil organic carbon (SOC) sequestration.

Despite these benefits, organic farming faces challenges, primarily due to its lower yield compared to conventional farming, which relies on synthetic fertilizers. This limitation raises concerns regarding its sustainability and widespread adoption. Additionally, research on organic farming requires long-term studies to assess its effectiveness accurately, yet such studies remain relatively scarce. Two key challenges described below are associated with organic fertilization.

5.1.1 Slow Nutrient Release

Nutrients in organic fertilizers are released gradually through microbial decomposition, potentially leading to deficiencies, particularly during the early growth stages of crops. For instance, nitrogen in organic fertilizers primarily exists in organic forms and must undergo microbial conversion into ammonium (NH_4^+) and nitrate (NO_3^-) before becoming available to plants. As microbial activity is influenced by soil pH, moisture, and temperature, nutrient availability can be inconsistent. In contrast, synthetic fertilizers dissolve readily in water, providing immediate nutrient availability, which is advantageous during the initial growth phase.

5.1.2 Challenges in Nutrient Ratio Adjustment

Synthetic fertilizers allow precise adjustments of nitrogen, phosphorus, and potassium ratios, ensuring optimal nutrient supply for crops. In contrast, organic fertilizers have fixed nutrient compositions, making it difficult to fine-tune nutrient ratios, which may result in imbalances. Therefore, careful nutrient management is required to maintain an appropriate balance in organic farming systems.

Despite these challenges, organic fertilizers play a crucial role in sustainable agriculture, and their adoption is expected to increase. A meta-analysis conducted by Shi et al. (2024) on the global comparison of organic and synthetic fertilizers highlighted several key findings [47]. Organic fertilization significantly enhances aboveground biomass, increasing it by 56% compared to ambient conditions, surpassing the 42% increase observed with synthetic fertilization. Additionally, while synthetic fertilization often reduces plant diversity, organic fertilization promotes biodiversity, particularly in moisture-rich grasslands. Organic fertilization also improves soil health by increasing SOC by 19% relative to ambient conditions and by 15% compared to synthetic fertilization, underscoring its role in carbon sequestration. These findings highlight organic fertilization as a nature-based solution that simultaneously enhances agricultural productivity, supports biodiversity, and mitigates the negative effects associated with synthetic fertilizers.

5.2 Long-Term Study on Organic Rice Cultivation

A long-term study on organic rice farming demonstrated that stable yields can be achieved using only organic fertilizers. A 15-year comparative study between organic and synthetic fertilizer-based rice farming revealed that while organic yields were initially lower, the gap gradually diminished over time. Such long-term studies, often conducted by public institutions, provide valuable insights into the viability of organic farming practices.

Asai et al. (2016) conducted a 15-year organic rice cultivation trial at Shizuoka University, Japan [51]. The study comprised four experimental plots, with the study period divided into three 5-year phases: the early phase (1996–2000), the middle phase (2001–2005), and the late phase (2006–2010).

Rice husk compost plot – cultivated using compost primarily derived from rice husks, with rapeseed oil cake as a supplementary component, without pesticide application

Cattle manure compost plot – cultivated using compost made from cattle manure, without pesticide application

Synthetic fertilizer plot – cultivated using synthetic fertilizers with pesticide application

No fertilizer plot – cultivated without fertilizers but with pesticide application

The results indicated that the synthetic fertilizer plot consistently produced the highest and most stable yield. The organic plots (rice husk and cattle manure compost) required 5–6 years from the initiation of the experiment to achieve stability (Table 4). The cattle manure compost plot reached stability by year 5, while the rice husk compost plot stabilized by year 6. Although organic yields remained slightly lower than those obtained with synthetic fertilizers, statistical analysis revealed no significant differences between them. Furthermore, the organic rice demonstrated superior taste quality, suggesting a higher market value potential.

To establish sustainable agricultural systems, organic farming requires a transition period of at least 5 years to mitigate initial yield fluctuations. Long-term compost application gradually increases total soil nitrogen, highlighting the importance of continuous soil nutrient monitoring and management. By adjusting compost application rates accordingly and maintaining appropriate soil and drainage management, sustainable organic farming practices can be effectively implemented.

Table 4 Sampled rice yield (g/m^2)

Experimental period	Rice husk-based compost plot	Cattle manure-based compost plot	Chemical fertilizer plot	Non-fertilizer plot
1996–2000 (early phase)	397 ± 35^b	427 ± 27^b	538 ± 29^a	393 ± 22^b
2001–2005 (middle phase)	447 ± 34^b	$470 \pm 33^{a,b}$	543 ± 63^a	351 ± 39^c
2006–2010 (late phase)	437 ± 32^a	430 ± 50^a	523 ± 49^a	329 ± 53^b

Average ± standard deviation. Different letters indicate significant differences at the 5% level based on Tukey's method

5.3 Synergistic Use of Organic and Inorganic Fertilizers

Beyond the independent benefits of organic fertilizers, their combined application with inorganic fertilizers has demonstrated synergistic effects. Studies indicate that integrating compost with mineral fertilizers can enhance agricultural productivity. Diacono et al. (2010) reviewed multiple studies and concluded that optimal agronomic performance is achieved when compost is applied at high rates and frequency, functioning as a slow-release nitrogen fertilizer with long-term residual effects [52].

5.4 Utilization of Composting Heat for Weed Seed Inactivation and Prevention of Ecosystem Disturbance

The application of compost containing viable weed seeds to agricultural fields has been demonstrated to promote weed proliferation, thereby inhibiting the growth and development of cultivated crops. This issue poses a significant challenge for organic farmers committed to sustainable agricultural practices, as it conflicts with efforts to maintain weed-free cultivation without relying on synthetic herbicides.

The probability of weed seed presence in plant-based compost materials is significant. Additionally, manure-based compost is also vulnerable to weed seed contamination, as livestock forage, such as hay and silage, may inadvertently contain weed seeds. Certain weed species can survive passage through the digestive systems of ruminant livestock, retaining their germination capacity and persisting in manure. When such manure is applied as organic fertilizer, it introduces viable weed seeds into agricultural fields, increasing the risk of weed infestations.

Inadequate composting, where temperatures do not reach sufficiently high levels, may fail to inactivate weed seeds effectively, allowing viable seeds to be introduced into fields. This can lead to uncontrolled weed growth, reduced crop yields, and increased management costs. Moreover, in countries that import hay, there is an additional risk of ecosystem disruption due to the introduction of invasive species.

The proliferation of highly competitive non-native plants can threaten native flora and destabilize local ecosystems.

To prevent the spread of weed seeds through composting, it is essential to implement appropriate composting practices. Maintaining high composting temperatures and ensuring prolonged heat exposure are critical measures to effectively inactivate weed seeds, thereby reducing their dissemination and minimizing ecological risks.

Larney and Blackshaw (2003) investigated the impact of composting on weed seed viability in beef cattle manure in southern Alberta, Canada, where manure composting is increasingly practiced as an alternative to direct manure application [53]. The study addressed concerns that fresh manure contains viable weed seeds, potentially leading to infestations in agricultural fields. While composting is generally promoted as a "weed-free" alternative due to the high temperatures expected to inactivate seeds, previous studies have indicated that some seeds survive, suggesting that temperature fluctuations and composting conditions influence seed mortality. This study aimed to examine the relationship between composting duration, temperature, and seed viability; identify species-specific lethal temperature thresholds; and assess the influence of moisture and oxygen on seed degradation. The results demonstrated that although composting significantly reduced weed seed viability, it did not completely inactivate all species. Some seeds remained viable despite losing their ability to germinate, as confirmed by tetrazolium respiration tests. Seed mortality varied by species, with four weed species being inactivated within the first 7 days of composting at 39°C, whereas temperatures exceeding 60°C were required for two additional species. Uneven heat distribution in compost piles prolonged seed survival in cooler regions, underscoring the necessity of frequent turning to ensure uniform heating. Additionally, high moisture levels enhanced microbial and chemical degradation, further contributing to seed mortality.

Nishida et al. (1999) investigated the duration of heat exposure required for upland weed seed mortality under controlled temperature conditions [54]. Given that imported concentrate feed serves as a major pathway for invasive weed seeds into forage fields in Japan, understanding the effectiveness of composting in inactivating these seeds is essential for mitigating their spread. The study assessed seed viability following exposure to 55 and 60°C for varying durations, preceded by a 24-h water imbibition period at 15°C. The findings revealed that all tested species, except *Abutilon theophrasti* (velvetleaf), were completely inactivated at 55°C after 72 h and at 60°C after 24 h (Table 5). However, *A. theophrasti* exhibited greater heat resistance, requiring 120 h at 55°C and 30 h at 60°C for complete inactivation. Additionally, the study estimated the lethal duration (LD_{90}) for *A. theophrasti* using probit analysis, with confidence intervals of 42–58 h at 55°C and 10–17 h at 60°C. These results suggest that with proper composting management, including sufficient temperature and exposure duration, it is possible to effectively inactivate all weed seeds, including the most heat-resistant species.

Table 5 Effects of heat exposure duration on weed seed viability

		55°C					60°C			
		Thermal exposure time, hours								
	Untreated	24	48	72	96	120	3	6	24	30
Weed species	Viability rate, %									
Carolina horse nettle (*Solanum carolinense*)	99	72	7	0			67	9	0	
American black nightshade (*Solanum americanum*)	97	79	0				84	6	0	
Velvetleaf (*Abutilon theophrasti*)	93	23	12	9	2	0	39	23	7	0
American pokeweed (*Phytolacca americana*)	94	0					57	3	0	
Spiny amaranth (*Amaranthus spinosus*)	94	2	0				24	1	0	
Spleen amaranth (*Amaranthus patulus*)	97	38	0				74	0		
Pale smartweed (*Persicaria lapathifolia*)	83	0					0			
Fall panic grass (*Panicum dichotomiflorum*)	96	0					46	0		
Barnyard grass (*Echinochloa crus-galli*)	76	6	6	0			6	0		
Southern crabgrass (*Digitaria ciliaris*)	67	0					10	0		

Table partially modified based on Nishida et al. (1999) [54]

5.5 Pathogen Inactivation Through Composting Heat

The previous section discussed the effectiveness of composting heat in inactivating weed seeds. In addition to this function, composting heat also plays a crucial role in inactivating pathogens that may be present in compost materials.

Livestock manure often contains pathogenic bacteria, such as *Escherichia coli* (*E. coli*), which pose a risk of contaminating agricultural land and the surrounding environment. If not properly managed, these pathogens can have detrimental effects on human and animal health. Composting serves as an essential method to prevent such contamination, enabling the safe utilization of manure as an organic fertilizer.

The high temperatures attained during composting (55–60°C) are sufficient to inactivate a wide range of pathogenic microorganisms. However, in compost materials with high moisture content, such as cattle manure, achieving and sustaining these elevated temperatures can be challenging. To overcome this limitation, the following study proposes a straightforward and effective approach that involves incorporating organic matter rich in readily biodegradable compounds to enhance composting efficiency and facilitate the attainment of the thermal inactivation threshold.

Hanajima et al. (2006) investigated the effects of co-composting cattle feces with various organic wastes under high-moisture conditions on the reduction of *E. coli* [55]. Given that untreated livestock waste poses a hygiene risk due to potential contamination with pathogenic bacteria, including *E. coli* O157:H7, composting is widely regarded as an effective mitigation strategy. However, cattle feces typically have a high moisture content (80–85%) and lack easily biodegradable materials, making it difficult to reach the thermal death point necessary for pathogen inactivation.

The study demonstrated that incorporating certain organic wastes, such as tofu residue, rice bran, rapeseed meal, and food waste, significantly enhanced the thermophilic phase, allowing compost temperatures to exceed 55°C, a critical threshold for pathogen inactivation. The *E. coli* population declined from 10^6 CFU/g to below 10^2 CFU/g after 7 days of composting. The effectiveness of *E. coli* reduction was positively correlated with the presence of easily degradable carbon, measured as biochemical oxygen demand (BOD). A BOD value exceeding 166.2 mg O_2/g dry matter resulted in substantial *E. coli* reduction. These findings highlight that co-composting with suitable organic waste enhances microbial activity and heat generation, ensuring effective pathogen inactivation even under high-moisture conditions.

6 Conclusion and Future Perspectives

Composting technology represents a promising approach for addressing global challenges associated with agricultural residue management. As a microbially mediated process, its efficacy is governed by critical parameters, including aeration, temperature, moisture content, C/N ratio, pH, and microbial inoculation. Strategies including physical, chemical, and biological methods effectively enhance composting by increasing substrate biodegradability. Co-composting with complementary organic materials balances nutrient content and moisture levels, thus optimizing composting efficiency. Emerging techniques such as biochar amendments, nutrient adjustments, and membrane-covered composting have demonstrated significant advantages, notably reduced nitrogen losses, greenhouse gas emissions, odor control, and accelerated compost maturation. Furthermore, agricultural residue-derived compost has shown substantial benefits for soil fertility, improved soil structure, enhanced water retention, increased microbial biodiversity, and significant stimulation of plant growth and crop yields.

Several trends and unresolved challenges are shaping the future of composting research and practice:

Precision microbial regulation: Further developments are anticipated in precision microbial management through targeted inoculants and genetic insights into microbial communities, enabling tailored microbial interventions for specific waste streams [56].

Nutrient retention and optimization: Enhancing nutrient retention within composting systems, particularly nitrogen and phosphorus, will remain a critical research area, alongside exploring innovative amendments like biochar for improving compost effectiveness and environmental compatibility.

Emission control and sustainability: Regulatory pressures and increasing environmental awareness globally are likely to propel the adoption of enclosed or semipermeable membrane composting systems, reducing greenhouse gas emissions and odors significantly [57].

Internet of Things (IoT) and automation in composting: Advancements in real-time monitoring using IoT sensors and automated controls will further optimize composting efficiency and environmental compliance [43].

Addressing safety and contaminants: Several challenges remain, including developing economically viable solutions for the composting of lignocellulosic materials, accurately quantifying greenhouse gas mitigation potential, and addressing compost safety, particularly concerning pathogens, heavy metals, and microplastics. Future research must continue addressing these issues, refining compost production technologies, and strengthening compost quality standards to facilitate broader global adoption.

In conclusion, the composting of agricultural residue will continue to evolve, driven by technological advancements, environmental sustainability goals, and global agricultural needs. As composting becomes further integrated into sustainable agricultural practices worldwide, it will play a pivotal role in addressing global soil degradation, enhancing agricultural productivity, and contributing significantly toward achieving long-term environmental sustainability.

References

1. Kee SH, Chiongson JBV, Saludes JP et al (2021) Bioconversion of agro-industry sourced biowaste into biomaterials via microbial factories – a viable domain of circular economy. Environ Pollut 271:116311. https://doi.org/10.1016/j.envpol.2020.116311
2. Chopkova V, Petkova M, Shilev S (2023) Uncovering bacterial diversity during mesophilic and thermophilic phases of biowaste composting through next-generation sequencing. Appl Sci 13(5):3111. https://doi.org/10.3390/app13053111
3. Zhang J, Kong Y, Yang Y et al (2024) Effects of thermophilic bacteria inoculation on maturity, gaseous emission and bacterial community succession in hyperthermophilic composting. Sci Total Environ 927:172304. https://doi.org/10.1016/j.scitotenv.2024.172304
4. Watanabe K, Nagao N, Toda T et al (2009) The dominant bacteria shifted from the order "Lactobacillales" to Bacillales and Actinomycetales during a start-up period of large-scale, completely-mixed composting reactor using plastic bottle flakes as bulking agent. World J Microbiol Biotechnol 25(5):803–811. https://doi.org/10.1007/s11274-008-9952-7
5. Wang Y, Xu P, Wang Y et al (2024) Effects of aeration modes and rates on nitrogen conversion and bacterial community in composting of dehydrated sludge and corn straw. Front Microbiol 15:1372568. https://doi.org/10.3389/fmicb.2024.1372568

6. Cao Y, Wang J, Huang H et al (2019) Spectroscopic evidence for hyperthermophilic pretreatment intensifying humification during pig manure and rice straw composting. Bioresour Technol 294:122131. https://doi.org/10.1016/j.biortech.2019.122131
7. Cui P, Chen Z, Zhao Q et al (2019) Hyperthermophilic composting significantly decreases N_2O emissions by regulating N_2O-related functional genes. Bioresour Technol 272:433–441. https://doi.org/10.1016/j.biortech.2018.10.044
8. Shen B, Zheng L, Zheng X et al (2024) Insights from meta-analysis on carbon to nitrogen ratios in aerobic composting of agricultural residues. Bioresour Technol 413:131416. https://doi.org/10.1016/j.biortech.2024.131416
9. Azim K, Soudi B, Boukhari S et al (2018) Composting parameters and compost quality: a literature review. Org Agric 8(2):141–158. https://doi.org/10.1007/s13165-017-0180-z
10. Wang M, Wang X, Wu Y et al (2022) Effects of thermophiles inoculation on the efficiency and maturity of rice straw composting. Bioresour Technol 354:127195. https://doi.org/10.1016/j.biortech.2022.127195
11. Duan M, Zhang Y, Zhou B et al (2020) Effects of Bacillus subtilis on carbon components and microbial functional metabolism during cow manure-straw composting. Bioresour Technol 303:122868. https://doi.org/10.1016/j.biortech.2020.122868
12. Yi J, Wu HY, Wu J et al (2012) Molecular phylogenetic diversity of Bacillus community and its temporal-spatial distribution during the swine manure of composting. Appl Microbiol Biotechnol 93(1):411–421. https://doi.org/10.1007/s00253-011-3425-x
13. Agbagla-Dohnani A, NoziÈRe P, Gaillard-Martinie B et al (2003) Effect of silica content on rice straw ruminal degradation. J Agric Sci 140(2):183–192. https://doi.org/10.1017/S0021859603003034
14. Eusufzai MK, Deb SK, Maeda T et al (2013) Mass loss and C and N release from decomposing fresh and composted residues as affected by cold climate conditions. Environ Nat Resour Res 3(2):116. https://doi.org/10.5539/ENRR.V3N2P116
15. Zhang L, Sun X (2014) Effects of rhamnolipid and initial compost particle size on the two-stage composting of green waste. Bioresour Technol 163:112–122. https://doi.org/10.1016/j.biortech.2014.04.041
16. Alburquerque JA, McCartney D, Yu S et al (2008) Air space in composting research: a literature review. Compost Sci Util 16(3):159–170. https://doi.org/10.1080/1065657X.2008.10702374
17. Sun S, Guo C, Wang J et al (2024) Effect of initial moisture content, resulting from different ratios of vegetable waste to maize straw, on compost was mediated by composting temperatures and microbial communities at low temperatures. Chemosphere 357:141808. https://doi.org/10.1016/j.chemosphere.2024.141808
18. Cai R, Cao X, Jiang X et al (2024) The maturity, humus content, and microbial metabolic function of sheep manure compost on the Qinghai-Tibet Plateau can be significantly improved by reducing the moisture content. Environ Sci Pollut Res 31(14):21458–21470. https://doi.org/10.1007/s11356-024-32437-4
19. Wang Z, Meng CG, Zhe W et al (2013) Comparison of physicochemical parameters during the forced-aeration composting of sewage sludge and maize straw at different initial C/N ratios. J Air Waste Manage Assoc 63(10):1130–1236. https://doi.org/10.1080/10962247.2013.800616
20. Wang SP, Li XX, Sun ZY et al (2023) Evaluation of physicochemical properties, bacterial community, and product fertility during rice straw composting supplemented with different nitrogen-rich wastes. Bioresour Technol 369:128462. https://doi.org/10.1016/j.biortech.2022.128462
21. Li XX, Wang SP, Sun ZY et al (2021) Performance and microbial community dynamics during rice straw composting using urea or protein hydrolysate as a nitrogen source: a comparative study. Waste Manag 135:130–139. https://doi.org/10.1016/j.wasman.2021.08.026
22. Dong S, Li R, Zhou K et al (2024) Response of humification process to fungal inoculant in corn straw composting with two different kinds of nitrogen sources. Sci Total Environ 946:174461. https://doi.org/10.1016/j.scitotenv.2024.174461

23. Shaghaleh H, Zhu Y, Shi X et al (2023) Co-effects of nitrogen fertilizer and straw-decomposing microbial inoculant on decomposition and transformation of field composted wheat straw. Life 13(10):1993. https://doi.org/10.3390/life13101993
24. Cornell University (n.d.) The science and engineering of composting. Cornell Waste Management Institute. https://compost.css.cornell.edu/science.html. Accessed 23 Feb 2025
25. López-González JA, Suárez-Estrella F, Jurado MM et al (2024) Development of functional consortia for the pretreatment of compostable lignocellulosic waste: a simple and effective solution to a large-scale problem. J Environ Manag 356:120638. https://doi.org/10.1016/j.jenvman.2024.120638
26. Fujiwara S (1986) Composting basics and applications. In: Integrated agricultural technology system (soil and fertilizer edition). Nobunkyo (in Japanese), pp 51–63
27. Nguyen MK, Lin C, Hoang HG et al (2022) Evaluate the role of biochar during the organic waste composting process: a critical review. Chemosphere 299:134488. https://doi.org/10.1016/j.chemosphere.2022.134488
28. Tran HT, Lin C, Bui XT et al (2021) Bacterial community progression during food waste composting containing high dioctyl terephthalate (DOTP) concentration. Chemosphere 265:129064. https://doi.org/10.1016/j.chemosphere.2020.129064
29. Sebahire F, Faridullah F, Irshad M et al (2024) Effect of biochar on composting of cow manure and kitchen waste. Land 13(10):1545. https://doi.org/10.3390/land13101545
30. Liu Q, He X, Wang K et al (2023) Biochar drives humus formation during composting by regulating the specialized metabolic features of microbiome. Chem Eng J 458:141380. https://doi.org/10.1016/j.cej.2023.141380
31. Jia P, Huang Y, Zhang H et al (2024) Variation of microbial necromass carbon and its potential relationship with humification during composting of chicken manure with and without biochar addition. Bioresour Technol 409:131258. https://doi.org/10.1016/j.biortech.2024.131258
32. Xiao X, Chen B, Chen Z et al (2018) Insight into multiple and multilevel structures of biochars and their potential environmental applications: a critical review. Environ Sci Technol 52(9):5027–5047. https://doi.org/10.1021/acs.est.7b06487
33. Noor RS, Shah AN, Tahir MB et al (2024) Recent trends and advances in additive-mediated composting technology for agricultural waste resources: a comprehensive review. ACS Omega 9(8):8632–8653. https://doi.org/10.1021/acsomega.3c06516
34. Yuan N, Wang K, Liang M et al (2024) Exploring the influence of biochar-supported nano-iron oxide on phosphorus speciation transformation and bacterial community structure in aerobic pig manure composting processes. Microorganisms 12(12):2593. https://doi.org/10.3390/microorganisms12122593
35. Zhang S, Wang L, Zhou B et al (2024) Characteristics of humification, functional enzymes and bacterial community metabolism during manganese dioxide-added composting of municipal sludge. Environ Res 252:119151. https://doi.org/10.1016/j.envres.2024.119151
36. Li XX, Chu XL, Zhang WS et al (2025) In-depth comparison of sorghum straw composting process inoculated with mature compost or commercial microbial agent. Waste Biomass Valoriz 16(5):2469–2482. https://doi.org/10.1007/s12649-024-02832-6
37. Manu M, Kumar R, Garg A (2017) Performance assessment of improved composting system for food waste with varying aeration and use of microbial inoculum. Bioresour Technol 234:167–177. https://doi.org/10.1016/j.biortech.2017.03.023
38. Zhou L, Xin PY, Xiao WW et al (2023) Effects of bacterial inoculation on lignocellulose degradation and microbial properties during cow dung composting. Bioengineered 14(1):2185945. https://doi.org/10.1080/21655979.2023.2185945
39. Zhang Y, Lin B, Hao Y et al (2025) Two-stage inoculation with lignocellulose-degrading microorganisms in composting: enhanced humification efficiency and underlying mechanisms. Environ Res 271:120906. https://doi.org/10.1016/j.envres.2025.120906
40. Lu M, Hao Y, Lin B et al (2024) The bioaugmentation effect of microbial inoculants on humic acid formation during co-composting of bagasse and cow manure. Environ Res 252:118604. https://doi.org/10.1016/j.envres.2024.118604

41. Fu T, Shang GH, Shen C et al (2022) Moisture migration driven by the electric field causes the directional differentiation of compost maturity. Sci Total Environ 811:152415. https://doi.org/10.1016/j.scitotenv.2021.152415
42. Fu X, Zuo H, Wang Z et al (2025) Extreme thermophilic microbial inoculation for reducing NH_3 and N_2O emissions in hyperthermophilic aerobic composting of refinery waste activated sludge. J Environ Manag 380:124870. https://doi.org/10.1016/j.jenvman.2025.124870
43. Cao J, Qu H, Li R et al (2024) Effects of membrane covers and biochar on compost quality and greenhouse gas reduction in aerobic composting. Bioresources 19(2):3475. https://doi.org/10.15376/biores.19.2.3475-3488
44. Fu J, Fang Y, Chen Y et al (2024) Feasibility of vegetable residues compost used as the growth medium for Brassica chinensis L. seedlings: effects of membrane covering and superphosphate application. Agronomy 14(12):2836. https://doi.org/10.3390/agronomy14122836
45. Cao J, Li R, Qu H et al (2022) Effects of the membrane-covered technology and superphosphate on the compost quality and nitrogen containing gas emissions during aerobic composting. Bioresources 17(1):1781–1793. https://doi.org/10.15376/biores.17.1.1781-1793
46. Chen S, Sun X, Zhang H et al (2025) Influence of electric fields on the maturity and microbial communities during sludge and straw composting. Waste Biomass Valoriz 16(4):1941–1953. https://doi.org/10.1007/s12649-024-02780-1
47. Mi H, Shen C, Ding T et al (2023) Identifying the role of array electrodes in improving the compost quality of food waste during electric field-assisted aerobic composting. Bioresour Technol 388:129763. https://doi.org/10.1016/j.biortech.2023.129763
48. Zhang L, Wang X, Wang H et al (2024) Electric field as extracellular enzyme activator promotes conversion of lignocellulose to humic acid in composting process. Bioresour Technol 391:129948. https://doi.org/10.1016/j.biortech.2023.129948
49. Lin X, Wang N, Li F et al (2022) Evaluation of the synergistic effects of biochar and biogas residue on CO_2 and CH_4 emission, functional genes, and enzyme activity during straw composting. Bioresour Technol 360:127608. https://doi.org/10.1016/j.biortech.2022.127608
50. Medina J, Monreal CM, Antilén M et al (2020) Influence of inorganic additives on wheat straw composting: characterization and structural composition of organic matter derived from the process. J Environ Manag 260:110137. https://doi.org/10.1016/j.jenvman.2020.110137
51. Asai T, Tobina H, Maeda S et al (2016) Growth characteristics, yield and eating quality of paddy rice in different organic cultivation continued fifteen years Japanese. J Crop Sci 85(3): 274–281. https://doi.org/10.1626/jcs.85.274
52. Diacono M, Montemurro F (2011) Long-term effects of organic amendments on soil fertility. In: Sustainable agriculture, vol 2. Springer, Dordrecht, pp 761–786. https://doi.org/10.1007/978-94-007-0394-0_34
53. Larney FJ, Blackshaw RE (2003) Weed seed viability in composted beef cattle feedlot manure. J Environ Qual 32(3):1105–1113. https://doi.org/10.2134/jeq2003.1105
54. Nishida T, Kurokawa S, Shibata S et al (1999) Effect of duration of heat exposure on upland weed seed viability. J Weed Sci Technol 44:59–66. https://doi.org/10.3719/weed.44.59
55. Hanajima D, Kuroda K, Fukumoto Y et al (2006) Effect of addition of organic waste on reduction of Escherichia coli during cattle feces composting under high-moisture condition. Bioresour Technol 97(14):1626–1630. https://doi.org/10.1016/j.biortech.2005.07.034
56. Wang L, Li Y, Li X (2024) Microbe-aided thermophilic composting accelerates manure fermentation. Front Microbiol 15:1472922. https://doi.org/10.3389/fmicb.2024.1472922
57. Waqas M, Hashim S, Humphries UW et al (2023) Composting processes for agricultural waste management: a comprehensive review. Processes 11(3):731. https://doi.org/10.3390/pr11030731

Adv Biochem Eng Biotechnol (2026) 196: 283–300
https://doi.org/10.1007/10_2026_328

Published online: 30 June 2026

Valorization of Agricultural Residues Through Nutrient Enrichment for Animal Farming

Minmin Cai and Zhengzheng Zhao

Contents

Abstract The valorization of agricultural residues into high-value feed resources is essential for sustainable livestock production. Insects have emerged as efficient bioconversion agents due to their rapid growth, high feed conversion efficiency, and ability to utilize diverse organic wastes. This chapter reviews key insect species, including *Tenebrio molitor*, *Musca domestica*, *Acheta domesticus*, *Bombyx mori*, *Alphitobius diaperinus*, and *Hermetia illucens*, highlighting their roles in converting agricultural by-products into nutrient-rich biomass. Their nutritional profiles, including protein, lipids, and bioactive compounds, and their applications in poultry, aquaculture, and livestock feeds are summarized. Special attention is given to

M. Cai (✉) and Z. Zhao
National Key Laboratory of Agricultural Microbiology, College of Life Science and Technology, National Engineering Research Center of Microbial Pesticides, Huazhong Agricultural University, Wuhan, People's Republic of China
e-mail: cmm114@mail.hzau.edu.cn

black soldier fly larvae for their industrial potential and role in reducing antibiotic residues and antibiotic resistance genes. Overall, insect-based systems provide a sustainable strategy to recycle waste, enhance feed efficiency, and support circular agriculture.

Graphical Abstract

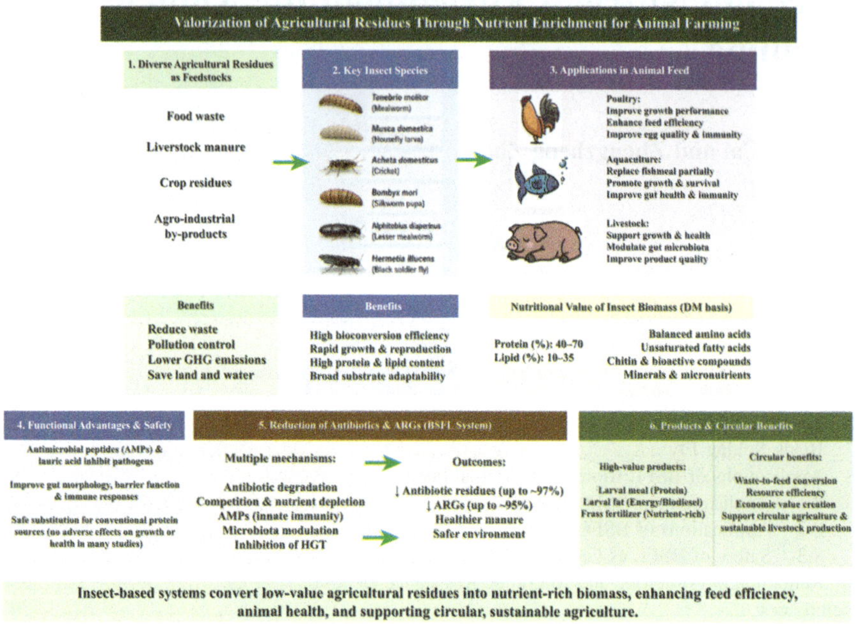

Keywords Antibiotic resistance genes, *Hermetia illucens*, Insect protein, Organic waste valorization, Sustainable feed

1 Introduction

Organic waste, which includes a diverse array of biodegradable materials derived from human activities, industrial operations, and daily living, often retains residual value despite being discarded. Common types include food waste, livestock manure, and agricultural residues. Although multiple treatment options exist, improper management of organic waste can result in severe environmental consequences. In recent years, biological treatment has gained prominence as a sustainable and eco-friendly alternative, employing microorganisms, earthworms (vermicomposting), or insect larvae for bioconversion. Among these, insects offer several notable advantages over microbes and earthworms, including higher bioconversion efficiency, rapid growth and reproduction, high protein yield, broad adaptability, and strong potential for industrial-scale implementation. As the most diverse and abundant animal populations on Earth, insects act as effective waste converters, generating fewer greenhouse gas emissions, consuming less water, and presenting a lower risk of zoonotic disease transmission.

Several other insect species have demonstrated promising potential in the valorization of organic residues and their subsequent use as animal feed ingredients. Among the most studied are mealworms (*Tenebrio molitor*), housefly larvae (*Musca domestica*), crickets (e.g., *Acheta domesticus*), and certain species of locusts and silkworms. These insects exhibit rapid growth, high feed conversion efficiency, and ability to utilize a wide range of organic substrates, including agricultural by-products, food waste, and agro-industrial residues. Edible insects generally show higher feed conversion efficiency than conventional livestock, converting a larger proportion of dietary protein into body mass.

2 Mealworms (*Tenebrio molitor*)

Mealworms are among the most widely studied insects for both waste bioconversion and feed production. They are capable of utilizing a broad range of low-value substrates, including wheat bran, cereal residues, vegetable trimmings, and food-processing by-products. This dietary flexibility makes them particularly suitable for circular bioeconomy systems, where agricultural side streams can be transformed into high-value biomass. Compared with many conventional livestock species, mealworms exhibit relatively efficient feed conversion, and studies have reported that they can convert a substantial proportion of dietary protein into edible body mass, often exceeding the efficiency observed in poultry and other traditional production systems [1]. In addition to their role in waste valorization, mealworms have gained considerable attention as an alternative protein ingredient in animal feed. Mealworm meal has been incorporated into poultry diets as a partial replacement for soybean meal or fishmeal, often without negative effects on growth or feed efficiency. In some trials, broilers fed diets containing mealworm meal demonstrated improved feed intake, better weight gain, and comparable carcass quality [2]. These results suggest that mealworms can provide both nutritional and functional benefits when used as a feed ingredient.

In aquaculture, mealworm meal has also been evaluated as a substitute for fishmeal. Species such as tilapia, carp, and trout have shown acceptable growth performance when part of the fishmeal fraction is replaced by mealworm-derived protein. This is particularly relevant given the increasing cost and environmental concerns associated with fishmeal production. The ability of mealworms to transform plant-based by-products into high-quality animal protein further enhances their value in sustainable aquafeed systems. Mealworms are also commonly used in small-scale or backyard poultry systems. Dried larvae are often provided as a protein-rich supplement to improve egg production, feather condition, and general bird health. Because they are easy to rear, store, and handle, mealworms are especially attractive for decentralized or smallholder production models.

From a nutritional perspective, mealworm larvae typically contain approximately 45–55% crude protein on a dry matter basis. Their amino acid profile is relatively balanced and includes high levels of essential amino acids such as lysine, leucine, and valine, which are critical for monogastric animals. The digestibility of mealworm protein is generally high, and studies have shown that replacing portions of soybean meal or fishmeal with mealworm protein can maintain or even improve feed conversion ratios and nutrient utilization. In addition to their protein fraction, mealworms contain substantial lipid levels, usually ranging from 25% to 35% of dry matter. Their fat is characterized by a high proportion of unsaturated fatty acids, particularly oleic and linoleic acids. These fatty acids are associated with beneficial effects on animal health and may improve the fatty acid composition of meat, eggs, or fish fillets. Moreover, the lipid fraction can serve as a valuable energy source in feed formulations, potentially reducing the need for added vegetable oils. Beyond basic nutrition, mealworms also contain bioactive compounds such as chitin, antimicrobial peptides, and micronutrients. These components may contribute to improved gut health, immune function, and overall performance in animals [3]. As research progresses, mealworms are increasingly viewed not only as a protein replacement but also as a functional feed ingredient with multiple benefits.

In summary, mealworms (*Tenebrio molitor*) combine high nutritional value with efficient biological conversion of diverse low-value substrates into protein- and lipid-dense biomass, making them a versatile ingredient in both poultry and aquaculture feeds. Their balanced amino acid profile, energy-rich lipids, and additional functional components support animal growth and health while contributing to more sustainable, circular production systems. As such, mealworms represent not only an effective protein alternative but also a multifunctional feed component with broad application potential.

3 Housefly Larvae (*Musca domestica*)

Housefly larvae, commonly referred to as maggots, are among the earliest insect species investigated for organic waste conversion. They are capable of feeding on a wide range of substrates, including livestock manure, food waste, and various organic residues [4]. Their rapid life cycle and high reproductive rate enable them to convert large quantities of waste into protein-rich biomass within a short period. One of the most notable advantages of housefly larvae is their ability to reduce the volume, moisture content, and pathogen load of manure. By consuming organic matter in waste streams, they contribute to waste stabilization and can significantly lower odor emissions and environmental pollution. This makes them particularly suitable for integration into livestock production systems, where manure management is a major concern.

From a feed perspective, dried housefly larvae meal has been widely evaluated as a substitute for conventional protein sources such as fishmeal and soybean meal. In poultry nutrition, inclusion of maggot meal has been associated with improved growth performance, better feed conversion ratios, and enhanced carcass characteristics. The high palatability of housefly larvae meal often stimulates feed intake, which can further contribute to improved growth rates. In aquaculture, housefly larvae meal has also shown promising results. Studies have demonstrated that it can replace a significant portion of fishmeal, often up to 25–50%, in diets for species such as tilapia, carp, and catfish without negatively affecting growth performance or feed utilization. This substitution potential is particularly important in regions where fishmeal is expensive or difficult to obtain.

Nutritionally, housefly larvae typically contain 40–60% crude protein on a dry matter basis. Their amino acid profile is comparable to that of fishmeal, especially in terms of essential amino acids such as lysine and methionine. This makes them a suitable protein source for monogastric animals, including poultry and fish. The lipid content of housefly larvae generally ranges from 20% to 30% of dry matter. Their fat fraction contains both saturated and unsaturated fatty acids, providing a readily available energy source for animal diets. In addition, the lipid composition can be influenced by the substrate used during rearing, offering opportunities to tailor the nutritional profile of the final product. Housefly larvae also contain micronutrients, enzymes, and bioactive compounds that may contribute to improved animal health [5]. Their rapid growth, high reproduction rate, and ability to utilize diverse waste streams make them a practical option for large-scale waste-to-feed systems, particularly in regions with high organic waste generation.

In summary, housefly larvae offer a highly nutritious, protein- and lipid-rich ingredient for animal feeds while simultaneously contributing to organic waste reduction. Their nutrient composition and balanced amino acid profile make them comparable to conventional protein sources, and their ability to convert diverse organic residues into valuable biomass underpins their practical application in integrated waste-to-feed systems. As a result, housefly larvae meal represents a promising sustainable alternative within circular agricultural production.

4 Crickets (*Acheta domesticus* and Related Species)

Crickets are increasingly recognized as one of the most promising insect species for sustainable protein production. They can be reared on a variety of plant-based by-products, including cereal bran, vegetable residues, and food processing waste. Their relatively high feed conversion efficiency, combined with low greenhouse gas emissions and reduced land requirements, makes them attractive for environmentally friendly production systems [6]. In terms of feed applications, cricket meal has been investigated as a protein source for poultry, aquaculture, and pet food. In aquaculture, partial replacement of fishmeal with cricket meal has generally maintained growth performance, feed conversion, and survival rates in species such as tilapia and trout. The high digestibility of cricket protein contributes to efficient nutrient utilization in these systems.

In poultry production, cricket meal has been used as a high-quality protein supplement. Feeding trials have shown improvements in growth rate, feed efficiency, and certain immune parameters in broilers. Crickets also appear to enhance gut health, possibly due to the presence of chitin and other functional components [7]. Nutritionally, crickets are known for their high protein content, typically ranging from 55% to 70% on a dry matter basis. Their amino acid profile is well balanced and comparable to conventional animal protein sources. This makes cricket meal a viable alternative to fishmeal or soybean meal in many feed formulations. The lipid content of crickets generally accounts for 10–20% of dry matter. Their fat contains significant levels of unsaturated fatty acids, including linoleic acid, which may positively influence the fatty acid composition of animal products. Additionally, crickets contain minerals such as zinc, iron, and phosphorus in relatively high concentrations, contributing to their nutritional value. Beyond their basic nutritional composition, crickets also offer practical advantages in farming. They require relatively simple housing, reproduce quickly, and can be reared at high densities. These characteristics make them suitable for both industrial-scale production and decentralized, small-scale farming systems.

In conclusion, the high protein content, strong digestibility, and well-rounded amino acid profile of crickets support their use as an effective substitute for traditional protein sources in both poultry and aquaculture diets. Their adaptability to different feeding substrates, simple husbandry requirements, and efficient nutrient utilization facilitate their incorporation into sustainable feed formulations and circular production frameworks, offering an environmentally friendly option that can help boost animal performance and resource efficiency.

5 Silkworm Pupae (*Bombyx mori*)

Silkworm pupae are a major by-product of the silk industry and represent a valuable source of protein and lipids. Each year, large quantities of pupae are generated during silk production, creating opportunities for their use in animal feed. Instead of being treated as waste, these pupae can be processed into a nutrient-rich meal for livestock and aquaculture [8]. Silkworm pupae meal has a long history of use in poultry and fish diets. It has been successfully incorporated into feed formulations for broilers, laying hens, carp, and tilapia. In many cases, it can partially replace fishmeal or soybean meal without compromising growth performance or feed efficiency. Some studies even report improved growth when silkworm pupae meal is included at appropriate levels.

On a dry matter basis, silkworm pupae typically contain about 50–65% crude protein, although values as high as 70% have been reported. Their amino acid profile is well balanced and particularly rich in lysine and methionine, which are often limiting in plant-based feed ingredients. This makes silkworm pupae an attractive protein source for monogastric animals. The lipid content of silkworm pupae generally ranges from 20% to 30% of dry matter. Their fat is characterized by high levels of unsaturated fatty acids, especially linolenic acid [9]. These fatty acids may improve the nutritional quality of animal products, such as eggs, meat, or fish fillets. In addition to protein and fat, silkworm pupae are rich in minerals such as calcium, phosphorus, and magnesium. They also contain bioactive compounds that may enhance immune function and overall animal health. From a sustainability perspective, the use of silkworm pupae aligns well with circular economy principles. Since the pupae are already produced as a by-product of silk manufacturing, their utilization in feed does not require additional land or major resource inputs. This makes them an efficient and environmentally friendly feed ingredient. Furthermore, silkworm pupae can be processed into various forms, including dried meal, defatted protein concentrates, or lipid fractions. These products can be tailored for specific feed applications, offering flexibility in feed formulation.

Overall, silkworm pupae serve as a rich and balanced nutrient source, supplying high levels of protein, essential amino acids, and health-promoting lipids that make them a practical substitute for traditional proteins in poultry and aquaculture diets. Their use valorizes a major agro-industrial by-product and aligns with sustainable production goals, while diverse processing options such as full-fat meal, defatted protein, or oil fractions enhance their applicability in varied feed formulations.

6 Lesser Mealworms (*Alphitobius diaperinus*)

The lesser mealworm, scientifically known as *Alphitobius diaperinus*, is another edible insect species showing increasing potential for agricultural waste valorization and animal feed applications. Although commonly considered a pest in poultry houses due to its scavenging behavior, recent research has highlighted the ability of lesser mealworm larvae to consume and bioconvert a variety of organic residual streams, including grain by-products, crop residues, and food processing waste, into nutrient-dense larval biomass [10]. As part of a broader trend in insect-based waste valorization, species like *A. diaperinus* complement more extensively studied insects such as yellow mealworms and black soldier flies in transforming low-value substrates into high-value feeds. Lesser mealworms are capable of feeding on agricultural by-products such as broken grains, damaged cereals, and residues from milling or brewing processes. Their relatively robust digestive physiology allows them to assimilate proteins, carbohydrates, and lipids from such substrates and convert them into their larval tissue more efficiently than many conventional livestock. This waste-to-biomass conversion holds promise for integrated circular systems, where the same feedstocks that are unsuitable for direct animal consumption can be transformed into high-quality insect-derived ingredients for poultry, swine, or aquaculture diets. In feed trials, lesser mealworm larvae have been included in diets for poultry and fish with encouraging results. When used as a partial replacement for conventional protein sources such as soybean meal or fishmeal, lesser mealworm meal has been shown to support satisfactory growth performance and feed efficiency in broiler chickens and various fish species. In broiler nutrition, inclusion of lesser mealworm meal has not only maintained body weight gain and feed conversion ratios, but also improved protein digestibility due to the high bioavailability of amino acids in insect protein. Similar positive responses have been documented in aquaculture species such as tilapia and carp, where diets partially incorporating lesser mealworm meal supported growth and health indicators comparable to traditional diets.

Nutritionally, lesser mealworm larvae are rich in crude protein, with contents often ranging between 45% and 60% on a dry matter basis – values that are competitive with other insect meals and conventional feed ingredients. Their amino acid profiles are balanced, providing essential amino acids such as lysine, methionine, and threonine at levels suitable for monogastric animals. This balanced amino acid pattern is critical for efficient protein synthesis and growth in poultry and fish, reducing the need for additional synthetic amino acid supplementation. Additionally, lesser mealworm larvae contain moderately high levels of lipids, often 20–30% of dry matter, with a composition that includes both saturated and unsaturated fatty acids. These lipids serve as concentrated energy sources in feed formulations and can positively influence the fatty acid profile of animal products when included at appropriate levels. Beyond protein and fat, lesser mealworm larvae also contain micronutrients such as iron, zinc, and B-vitamins, which contribute to the overall nutritional value of insect-derived feed ingredients [11]. Their chitin-rich exoskeleton, while indigestible to certain animals, has been associated with prebiotic effects, potentially promoting beneficial gut microbiota in poultry and fish. Research suggests that chitin and associated compounds in insect meals may support immune function and digestive health, although further investigations are needed to fully quantify these effects across diverse livestock systems.

From a practical perspective, lesser mealworms offer several advantages for integrated production systems. They can be reared at high densities on mixed organic substrates, require low water and land inputs compared with conventional livestock, and produce fewer greenhouse gas emissions. In addition, they are included in several regulatory frameworks addressing insect-derived feed ingredients; for example, species such as *Alphitobius diaperinus* are authorized alongside other insect meals for inclusion in aquaculture and poultry feeds in the European Union. Despite these advantages, there are challenges that accompany the use of lesser mealworm larvae as feed ingredients. Their rearing requires careful control of temperature, humidity, and diet composition to achieve consistent growth and nutrient profiles. Moreover, regulatory approval in some regions may still lag behind more established insect species due to safety assessments and feed hygiene standards. Ongoing research is addressing these hurdles, aiming to optimize rearing protocols, substrate formulations, and downstream processing to maximize both economic viability and nutritional quality. In summary, lesser mealworms (*Alphitobius diaperinus*) are emerging as a complementary insect species for waste bioconversion and animal feed applications. With nutrient profiles that support animal growth, flexibility in substrate utilization, and integration into regulatory frameworks for feed ingredients, they contribute to the diversification of insect-based resources in circular agricultural systems.

In summary, *Alphitobius diaperinus* shows strong potential as a sustainable feed ingredient because it can convert a wide range of organic residues into nutrient-rich biomass that supports performance in both poultry and aquaculture production. Its efficient use of nutrients, adaptability to mixed substrates, and growing inclusion in regulatory frameworks enhance its role as a complementary insect protein source that contributes to circular, resource-efficient agricultural systems.

7 Black Soldier Fly

Black soldier fly (BSF, *Hermetia illucens* L.) is widely distributed in tropical and subtropical climates. Its larvae (BSFL) can convert a broad range of organic waste, including livestock manure, decomposing fruits, vegetable waste, crop residues, and carrion. This makes BSFL particularly suited for the conversion of organic waste into valuable biomass while simultaneously reducing environmental burden. Studies have shown that BSFL reared on livestock manure can contain more than 40% crude protein and around 30% fat (on a dry matter basis), along with a rich profile of essential amino acids and minerals, making them an excellent source of animal feed protein [12]. In addition to being a high-quality protein supplement for livestock and poultry, BSFL-derived lipids can be used in biodiesel production, further enhancing their economic and environmental value (Fig. 1). Today, BSFL and their protein and fat derivatives are widely utilized in animal agriculture.

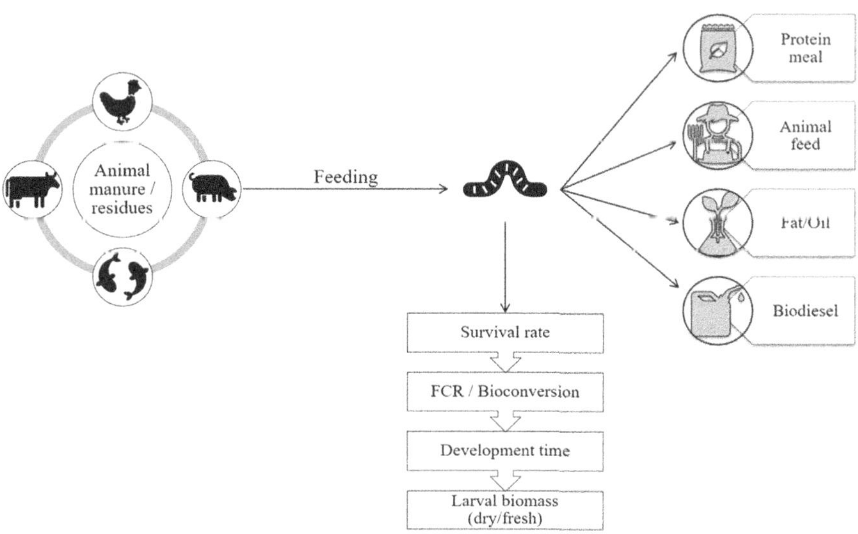

Fig. 1 Schematic summary of black soldier fly larvae role in organic waste bioconversion [13]

8 Applications of BSFL Biomass in Animal Feed

In 2017, the European Union authorized the use of BSFL and their derivatives in feed formulations for pets and aquaculture species, recognizing them as one of only four insect species legally approved for this purpose. Presently, nearly 80% of European insect farming enterprises are engaged in BSFL cultivation and processing. In the United States, the Food and Drug Administration (FDA) has approved dried BSFL for use in aquafeeds, particularly for farmed salmon in states like Idaho, Indiana, and Alaska, and in livestock and poultry feed formulations, including for broiler and layer chickens. As global market demand expands and rearing technologies advance, BSFL are increasingly being regarded as a mainstream candidate for commercial insect farming. Within the EU's legal framework, BSFL have even been classified as farm animals.

Several studies underscore the benefits of incorporating BSFL biomass into animal feed. For instance, replacing 25% of fishmeal with BSFL prepupae in rainbow trout diets significantly reduced dependence on marine-based feed ingredients. Likewise, Xiao et al. [14] reported that yellow catfish fed BSFL-supplemented diets exhibited a 29.12% increase in weight gain and an 11.18% improvement in specific growth rate, along with enhanced immunity and cost efficiency. In poultry and livestock farming, BSFL also show strong potential. Substituting soybean meal with BSFL meal in layer hen diets had no negative effects on bird health, egg production, or egg quality. Inclusion of 5%, 7.5%, or 10% BSFL powder in chick diets promoted growth without compromising meat quality. In swine, dietary supplementation with BSFL meal modulated the gut microbial composition and metabolic activity, improved mucosal immune balance, and enhanced pork quality.

Overall, the incorporation of BSFL biomass – either partially or entirely – as a replacement for conventional feed ingredients has been shown to improve animal growth performance, enhance immune function, and elevate the nutritional and sensory qualities of animal-derived products. These findings underscore the considerable potential of BSFL in animal feed applications.

8.1 Applications of BSFL Protein

BSFL are a rich source of protein, with crude protein content ranging from 38.0% to 46.2% during the larval stage, although a slight reduction occurs as larvae mature. Their protein profile includes 16 amino acids – comprising seven essential and nine nonessential amino acids – thereby meeting the dietary needs of both humans and animals [15]. Owing to this balanced composition, BSFL are widely regarded as a high-quality animal protein source. In recent years, food science research has proposed BSFL protein as a potential alternative protein source for human consumption [16].

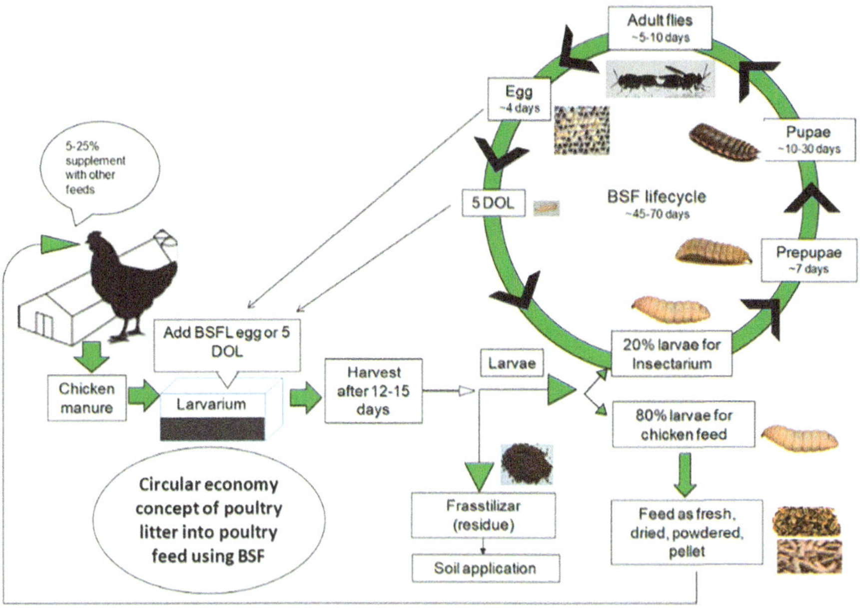

Fig. 2 The circular economy concept of bioconverting poultry litter into black soldier fly larvae meal, frasstilizer, and life cycle of black soldier fly [17]

At present, BSFL protein is primarily utilized in animal feed (Fig. 2). Substituting conventional protein ingredients in broiler diets with BSFL meal has been shown to reduce feed costs without negatively impacting feed intake, growth rate, feed conversion efficiency, or meat quality. Additionally, it supports gut microbiota regulation. In layer hens, BSFL inclusion has been associated with increased yolk mass, stable fatty acid profiles, enhanced yolk pigmentation, and reduced cholesterol content in eggs. In aquaculture, BSFL protein promotes growth in sturgeon, mitigates fat accumulation in the pyloric caeca of Atlantic salmon, and improves gut health and microbial diversity in freshwater crayfish. For ruminants, BSFL protein has been shown to reduce methane emissions in the rumen. In swine nutrition, partial replacement of fishmeal with BSFL protein in pig diets has proven effective, particularly in weaned piglets, where dietary inclusion of 4–8% BSFL protein increased lactic acid bacteria populations and strengthened intestinal barrier function.

8.2 *Applications of BSFL Fat*

Amid rising global energy demands and depletion of fossil fuel reserves, the environmental impacts of fossil fuel combustion – particularly greenhouse gas emissions – have become increasingly pressing. This has spurred interest in sustainable alternatives such as biodiesel, which is renewable, nontoxic, and biodegradable, making it one of the most promising green fuels. Due to their short life cycle and high lipid content, black soldier fly larvae (BSFL) have been identified as an excellent feedstock for biodiesel production.

BSFL are capable of accumulating substantial lipid reserves while converting organic waste. These lipids can be transesterified into biodiesel using catalysts such as lipase enzymes or bamboo biochar. The resulting biodiesel complies with international fuel standards, showing high oxidative stability and low viscosity due to its rich content of medium-chain saturated fatty acids, low levels of polyunsaturated fatty acids, and presence of natural antioxidants like phenolic compounds and sterols. Wang and Shelomi [18] applied microwave-assisted extraction and response surface optimization to recover lipids from BSFL, achieving a fatty acid profile of 22.54% oleic acid, 12.67% linoleic acid, and 6.45% palmitoleic acid – consistent with biodiesel quality standards.

Apart from energy production, BSFL fat serves as a valuable lipid source in animal feed. For example, replacing soybean oil with BSFL fat in Jian carp diets reduced abdominal fat deposition. In poultry, the inclusion of BSFL fat in broiler diets maintained optimal growth performance, meat quality, and carcass yield.

Notably, early-stage BSFL fat contains diverse unsaturated fatty acids, including rare odd-chain fatty acids like pentadecanoic acid (C15:0) and heptadecanoic acid (C17:0), as well as long-chain polyunsaturated fatty acids known for their cardioprotective and anticancer effects. These features underscore the potential of BSFL fat as a functional ingredient in the development of health-promoting foods and nutraceuticals.

8.3 Safety of BSFL as Feed

Numerous studies have demonstrated that replacing conventional protein sources such as soybean meal or fishmeal with BSFL meal in animal diets can significantly improve growth performance and digestive efficiency [19]. Upon exposure to exogenous microbes, BSFL can activate their innate immune system and produce antimicrobial peptides (AMPs) [20]. Additionally, BSFL are rich in bioactive compounds such as chitin, which can inhibit pathogenic bacterial growth and promote the proliferation and activity of beneficial intestinal microbiota [21]. When used as a protein source in swine diets, BSFL meal has been shown to markedly modulate gut microbial composition and colonic metabolite profiles in fattening pigs [22]. In meat-producing quails, partial substitution of dietary protein with BSFL meal resulted in increased saturated fatty acid content and reduced oxidative stress [23]. Moreover, the high lauric acid content in BSFL fat exhibits potent antimicrobial activity, particularly against Gram-positive bacteria.

Studies on weaning piglets have shown that complete replacement of fishmeal with BSFL meal in the diet maintains favorable growth performance while significantly enhancing the apparent digestibility and absorption of phosphorus (P) and calcium (Ca), likely due to the high mineral content in BSFL meal [24]. The presence of AMPs and chitosan in BSFL also contributes to improved disease resistance and antioxidant capacity in piglets. Under pathogenic challenges, BSFL supplementation has been associated with elevated levels of immunoglobulins IgG and IgA, indicating enhanced immune responsiveness. Histological analyses have further revealed that early dietary inclusion of BSFL meal improves intestinal morphology and confers resistance to *Escherichia coli* infections [25].

In aquaculture, BSFL meal has also showed significant promise. A diet in which 25% of the fishmeal was replaced with BSFL meal in yellow catfish led to improved immune indicators without compromising growth performance. Similarly, in zebrafish (*Danio rerio*), the amino acid and trace element profiles of BSFL meal were found to be well suited for growth requirements. The high lipid content (>30%) and the efficient lipid utilization by fish resulted in enhanced weight gain and overall better growth outcomes. For large-scale application, however, further long-term evaluations are essential to comprehensively assess its effects on growth and health indicators in both aquatic and terrestrial livestock, thereby establishing a solid safety profile for BSFL as an alternative feed ingredient.

A feeding trial in rainbow trout demonstrated that substituting 25% of the dietary fishmeal with BSFL protein reduced the total fishmeal content from 36% to 27%, and the fish oil content from 13% to 8% – without adverse effects on physiological parameters – highlighting the potential of insect-based proteins and lipids as sustainable replacements for traditional marine-derived feed components. Zarantoniello et al. [26] conducted a 6-month feeding study on zebrafish using diets in which 25% and 50% of fishmeal were replaced by BSFL meal. Utilizing a multidisciplinary approach including biometric, biochemical, histological, spectroscopic, and molecular analyses, they observed no significant negative effects. Despite a mild reduction in overall growth rate, there were no signs of intestinal inflammation or tissue damage, and stress and immune markers remained comparable across all groups, thus confirming the long-term safety of BSFL inclusion in fish diets.

Furthermore, BSFL-derived medium-chain fatty acids have been shown to possess strong antimicrobial and antiviral properties and play a crucial role in maintaining intestinal health [27]. They have also been reported to exert anti-inflammatory effects under pathological gut conditions [28]. However, the polysaccharide-rich chitin present in the exoskeletons of prepupal BSFL may interfere with the digestion and bioavailability of certain nutrients [29]. Excessive chitin levels in aquafeeds could potentially trigger intestinal inflammation and impair welfare in fish. Nevertheless, strategic supplementation with medium-chain fatty acids from BSFL may help mitigate intestinal mucosal inflammation in juvenile fish, further emphasizing the multifaceted benefits of insect-derived feed ingredients for animal health and performance.

9 Reduction of Antibiotic and Antibiotic Resistance Genes

In livestock production, sub-therapeutic applications of veterinary antibiotics are routinely added to animal feed to prevent diseases and enhance growth. National Ministry data show that this sector now accounts for approximately half of the country's total annual antibiotic usage, and this share is steadily rising. Because tetracyclines are poorly absorbed and slowly metabolized, roughly 70% are excreted unchanged in feces and urine, turning livestock manure into a serious reservoir of both active antibiotic residues and antibiotic resistance genes (ARGs). Once these enter soil, water, or air, they can infiltrate food chains and pose persistent environmental and human health risks. To address this, researchers have investigated the use of black soldier fly larvae (BSFL) and their gut microbiomes as a remediation strategy. Cai et al. [30] demonstrated that BSFL feeding on diets containing 20–80 mg/kg of tetracycline achieved nearly 97% degradation within 12 days – about 1.6 times faster than conventional composting – and in sterile larvae the degradation efficiency dropped by 20–50%, underscoring the essential role of the intestinal microbiota in facilitating antibiotic breakdown.

Subsequent studies confirmed that BSFL also degrade other antibiotic classes like enrofloxacin, sulfamethoxazole, and lincomycin. Researchers have isolated two functional groups of antibiotic-degrading bacteria from the larval gut. The first group shows high intrinsic activity in vitro. For example, *Proteus mirabilis* strain BSFL-3 degraded tylosin and enrofloxacin by 70.8 ± 0.8% and 65.9 ± 0.2%, respectively, across a wide pH (5.5–9.0) and temperature range (20–40°C). The second group – although weak alone – achieves strong synergy with larvae. For example, *Serratia marcescens* BSFL-6 removed only 28.5% of tetracycline in vitro, yet exceeded 60% when reintroduced into sterile larvae. When processing viscous pharmaceutical sludge, the larva–bacteria consortium boosted lincomycin removal from under 50% to over 80%. On the gene-level front, a 12-day BSFL treatment of chicken manure reduced key ARGs (e.g., *tetM*, *tetO*) by as much as 95%, outperforming both traditional composting and axenic larvae systems. Metagenomic analyses also revealed a shift from Proteobacteria and Bacteroidetes – often enriched in ARGs – toward more stable, low-resistance communities dominated by Firmicutes and Actinobacteria.

Moreover, although insects lack adaptive immunity, BSFL possess a strong innate immune system. They secrete antimicrobial peptides (AMPs) – such as cecropins, defensins, and attacins – when exposed to heavily contaminated manure, suppressing pathogens like *Escherichia coli* and *Salmonella enterica*. This, combined with nutrient depletion and falling antibiotic concentrations during larval growth, further curbs ARG-harboring bacteria and hinders horizontal gene transfer (Fig. 3). Overall, BSFL systems are user-friendly and low-cost and yield valuable by-products: frass can serve as nutrient-rich fertilizer, while larval biomass is convertible into animal feed or biodiesel. Through fine-tuning of larval gut microbiomes and processing parameters, this insect–microbe partnership holds great promise for scalable, antibiotic-residue mitigation in animal production systems – offering a viable strategy for greener livestock management.

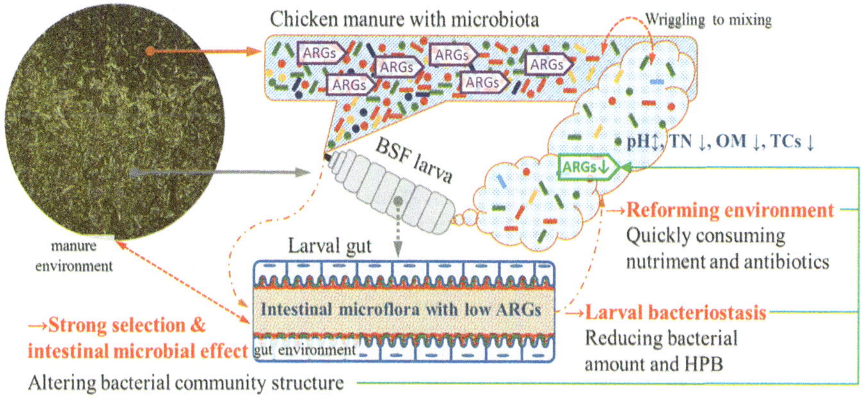

Fig. 3 Mechanisms by which the black soldier fly–microbiome system reduces antibiotic resistance genes in chicken manure

References

1. Hong J, Han T, Kim YY (2020) Mealworm (Tenebrio molitor larvae) as an alternative protein source for monogastric animal: a review. Animals (Basel) 10:2068. https://doi.org/10.3390/ani10112068
2. Parlar M, Ustundag AO (2024) Effects of adding mealworm (Tenebrio molitor L.) as a replacement for fish meal to broiler chicken diet. Trop Anim Health Prod 56:386. https://doi.org/10.1007/s11250-024-03986-5
3. Dalmoro YK, Franceschi CH, Stefanello C (2023) A systematic review and metanalysis on the use of Hermetia illucens and Tenebrio molitor in diets for poultry. Vet Sci 10:702. https://doi.org/10.3390/vetsci10120702
4. Hussein M, Pillai VV, Goddard JM, Park HG, Kothapalli KS, Ross DA et al (2017) Sustainable production of housefly (Musca domestica) larvae as a protein-rich feed ingredient by utilizing cattle manure. PLoS One 12:e0171708. https://doi.org/10.1371/journal.pone.0171708
5. Zhao Y, Wang W, Zhu F, Wang X, Wang X, Lei C (2017) Gut microbiota in housefly larvae. AMB Express 7:147. https://doi.org/10.1186/s13568-017-0446-7
6. Phesatcha B, Phesatcha K, Matra M, Wanapat M (2023) Cricket (Gryllus bimaculatus) meal pellets as a protein supplement. Anim Biosci 36:1384–1392. https://doi.org/10.5713/ab.22.0453
7. Fraijo-Valenzuela A, Arias-Moscoso JL, García-Pérez OD, Rodriguez-Anaya LZ, Gonzalez-Galaviz JR (2024) The biotechnological potential of crickets as a sustainable protein source for fishmeal replacement in aquafeed. BioTech (Basel) 13:51. https://doi.org/10.3390/biotech13040051
8. Rodríguez-Ortiz LM, Hincapié CA, Hincapié-Llanos GA, Osorio M (2024) Potential uses of silkworm pupae in food and feed: a systematic review. Front Insect Sci 4:1445636. https://doi.org/10.3389/finsc.2024.1445636
9. Tomotake H, Katagiri M, Yamato M (2010) Silkworm pupae as new sources of protein and lipid. J Nutr Sci Vitaminol 56:446–448. https://doi.org/10.3177/jnsv.56.446
10. Baliota GV, Rumbos CI, Gianotten N, Steeghs N, Athanassiou CG (2025) Review: The lesser mealworm as a nutrient pioneer: pathways to sustainable insect farming. Animal 19(Suppl 3): 101606. https://doi.org/10.1016/j.animal.2025.101606
11. Roncolini A, Milanović V, Aquilanti L et al (2020) Lesser mealworm powder as a novel baking ingredient. Food Res Int 131:109031. https://doi.org/10.1016/j.foodres.2020.109031

12. Mohan K, Rajan DK, Muralisankar T, Ganesan AR, Sathishkumar P, Revathi N (2022) Use of black soldier fly (Hermetia illucens L.) larvae meal in aquafeeds for a sustainable aquaculture industry: a review of past and future needs. Aquaculture 553:738095. https://doi.org/10.1016/j.aquaculture.2022.738095
13. Rehman KU, Hollah C, Wiesotzki K, Rehman RU, Rehman AU, Zhang J et al (2023) Black soldier fly as a tool for organic waste management: a mini-review. Waste Manag Res 41:81–97. https://doi.org/10.1177/0734242X221134528
14. Xiao XP, Jin P, Zheng LY, Cai M, Yu Z, Yu J, Zhang J (2018) Effects of black soldier fly larvae meal as fishmeal replacement. Aquac Res 49:1569–1577. https://doi.org/10.1111/are.13611
15. Liu X, Chen X, Wang H, Yang QQ, Rehman KU, Li W et al (2017) Dynamic changes of nutrient composition throughout the entire life cycle of black soldier fly. PLoS One 12: e0182601. https://doi.org/10.1371/journal.pone.0182601
16. Lu SY, Taethaisong N, Meethip W, Surakhunthod J, Sinpru B, Sroichak T et al (2022) Nutritional composition of black soldier fly larvae (Hermetia illucens L.) and its potential uses as alternative protein sources in animal diets: a review. Insects 13:831. https://doi.org/10.3390/insects13090831
17. Kullan ARK, Suresh A, Choi HL, Neumann EG, Hassan F (2024) Bioconversion of poultry litter into insect meal and organic frasstilizer using black soldier fly larvae as a circular economy model for the poultry industry: a review. Insects 16:12. https://doi.org/10.3390/insects16010012
18. Wang YS, Shelomi M (2017) Review of black soldier fly as animal feed and human food. Foods 6:91. https://doi.org/10.3390/foods6100091
19. Surendra KC, Tomberlin JK, van Huis A, Cammack JA, Heckmann LL, Khanal SK (2020) Rethinking organic wastes bioconversion using black soldier fly. Waste Manag 117:58–80. https://doi.org/10.1016/j.wasman.2020.07.050
20. Wu Q, Patocka J, Kuca K (2018) Insect antimicrobial peptides: a mini review. Toxins 10:461. https://doi.org/10.3390/toxins10110461
21. Wasko A, Bulak P, Polak-Berecka M et al (2016) Physicochemical structure of chitin from Hermetia illucens. Int J Biol Macromol 92:316–320. https://doi.org/10.1016/j.ijbiomac.2016.07.038
22. Yu M, Li Z, Chen W, Rong T, Wang G, Ma X (2019) Hermetia illucens larvae altered microbiota in pigs. J Anim Sci Biotechnol 10:50. https://doi.org/10.1186/s40104-019-0351-8
23. Cullere M, Tasoniero G, Giaccone V, Acuti G, Marangon A, Dalle Zotte A (2018) Black soldier fly as dietary protein source for broiler quails: meat proximate composition, fatty acid and amino acid profile, oxidative status and sensory traits. Animal 12:640–647. https://doi.org/10.1017/S1751731117001860
24. Biasato I, Renna M, Gai F, Dabbou S, Meneguz M, Perona G et al (2019) Partially defatted black soldier fly larva meal inclusion in piglet diets: effects on growth performance, nutrient digestibility, blood profile, gut morphology and histological features. J Anim Sci Biotechnol 10: 12. https://doi.org/10.1186/s40104-019-0325-x
25. Jin X, Yuan B, Liu M, Zhu M, Zhang X, Xie G et al (2021) Dietary Hermetia illucens larvae replacement alleviates diarrhea and improves intestinal barrier function in weaned piglets challenged with enterotoxigenic Escherichia coli K88. Front Vet Sci 8:647224. https://doi.org/10.3389/fvets.2021.647224
26. Zarantoniello M, Randazzo B, Truzzi C et al (2019) Six-month study on black soldier fly diets in zebrafish. Sci Rep 9:8598. https://doi.org/10.1038/s41598-019-45172-5
27. Liu Y (2015) Fatty acids, inflammation and intestinal health in pigs. J Anim Sci Biotechnol 6: 41. https://doi.org/10.1186/s40104-015-0041-5
28. Bertevello PL, De Nardi L, Torrinhas RS, Logullo AF, Waitzberg DL (2012) Partial replacement of x-6 fatty acids with medium-chain triglycerides, but not olive oil, improves colon cytokine response and damage in experimental colitis. J Parenter Enter Nutr 36:442–448. https://doi.org/10.1177/0148607111420153

29. Diener S, Zurbrugg C, Tockner K (2009) Conversion of organic material by black soldier fly larvae: establishing optimal feeding rates. Waste Manag Res 27:603–610. https://doi.org/10.1177/0734242X09103838
30. Cai M, Ma S, Hu R, Tomberlin JK, Thomashow LS, Zheng L, Li W, Yu Z, Zhang J (2018) Rapidly mitigating antibiotic resistant risks in chicken manure by Hermetia illucens bioconversion with intestinal microflora. Environ Microbiol 20:4051–4062. https://doi.org/10.1111/1462-2920.14354

Adv Biochem Eng Biotechnol (2026) 196: 301–336
https://doi.org/10.1007/10_2026_325

Published online: 26 May 2026

Anaerobic Digestion of Agricultural Residues for Biogas Production

Tian Yuan and Yue-Qin Tang

Contents

Abstract Anaerobic digestion (AD) is a promising method for treating organic waste while producing bioenergy and nutrient-rich fertilizer. Since the biogas generated from organic waste could replace fossil fuel use and the nutrients could be more efficiently recovered compared to composting, the AD process is essential for the development of a sustainable and circular economy. Agricultural residue is generated in large amounts worldwide and shows enormous potential for bioenergy

T. Yuan (✉)
Institute of Life and Environmental Sciences, University of Tsukuba, Tsukuba, Ibaraki, Japan
e-mail: yuan.tian.ga@u.tsukuba.ac.jp

Y.-Q. Tang (✉)
College of Architecture and Environment, Sichuan University, Chengdu, Sichuan, China
e-mail: tangyq@scu.edu.cn

generation. However, the high lignocellulose content restrains its biodegradability and limits its methane productivity during AD. Therefore, optimization of operating parameters and development of enhancement methods have been widely studied to enhance methane production from agricultural residue. In addition, the efficient utilization of anaerobic digestate has also been extensively studied to maximize nutrient recovery efficiency and prevent environmental and economic impacts of the AD process. This chapter summarizes the important factors, microbial mechanisms, and enhancement strategies for AD of agricultural residue and the potential utilization of anaerobic digestate, aiming to provide a comprehensive understanding and future perspectives of sustainable AD practices.

Graphical Abstract

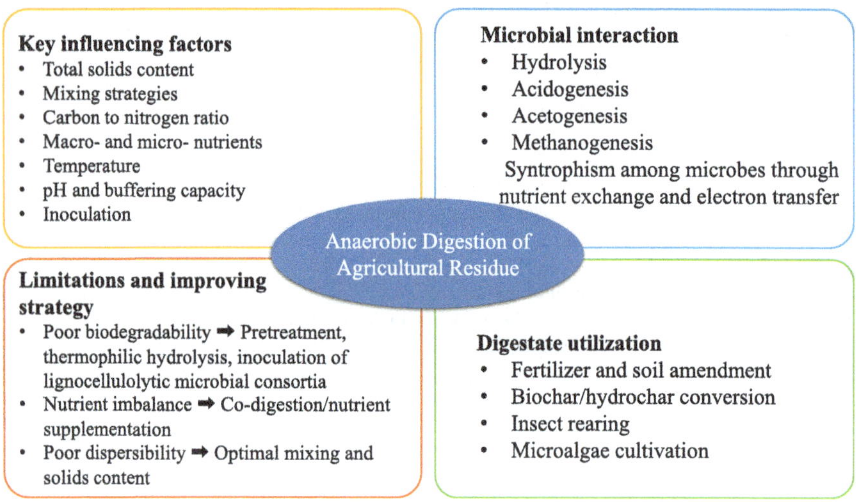

Keywords Anaerobic digestate, Bioenergy, Carbon neutrality, Co-digestion, Lignocellulose, Nanobubble, Nutrient recovery, Pretreatment

1 Introduction

Global energy consumption has rapidly grown in recent decades, reaching 183,230 TWh in 2023, with an annual growth rate of around 2% in the last 10 years [1]. Fossil fuels are still the major source (76.5%) of global energy consumption and are recognized as the largest contributor (92%) to global CO_2 emissions [2]. To cope with the climate changes caused by global warming, most countries have declared to achieve carbon neutrality by 2050. In this regard, the development of renewable energy has become increasingly prominent due to its important role in reducing CO_2 emissions [3]. Moreover, the depletion of natural resources, including fossil fuel and phosphorus, is also becoming the most pressing issue in the world, suggesting the urgent need to develop sustainable strategies and promote circular economy aiming to increase resource use efficiency and reduce waste generation [4].

As a result of the rapid increase in global population and life quality, vast amounts of solid waste are generated worldwide, causing serious impacts on the environment and economy. Due to the increased demand for food products, agricultural development is rapidly growing, resulting in generations of more than 5 billion tons of agricultural residue per year, with an estimated fuel value of 7.5×10^{15} kcal, equivalent to 0.8 billion kL of diesel [5, 6]. Besides, around 5 million tons of nitrogen, 1 million tons of phosphorus, and 12 million tons of potassium can be potentially recovered from agricultural residue [7]. However, up to 50% of them are discarded without proper treatment, and around 10% are directly burnt, resulting in vast amounts of nutrient loss and emissions of greenhouse gases (GHG) and atmospheric pollutants [8].

Anaerobic digestion (AD) is an alternative biological process for sustainable organic waste treatment and bioenergy and nutrient recovery. The biogas generated mainly contains around 50–75% methane (CH_4), with the rest being CO_2 and a small amount of other impurities, which could replace the use of fossil fuels. At the same time, the GHG emissions from organic waste could be avoided, thus contributing to carbon neutrality. In addition, the remaining residue after the AD process, called anaerobic digestate, contains rich mineralized and organic nutrients, exhibiting great potential as organic fertilizer. Compared to the traditional aerobic composting process, around half of carbon and nitrogen loss as CO_2, CH_4, and NH_3, the main contributors to global warming, acidification, and air pollution, could be mostly avoided in AD, improving resource recovery efficiency [9]. According to the report of the World Biogas Association [10], AD has the potential to generate 12,065 TWh of electricity from currently available major feedstocks, including livestock manure, food waste, sewage sludge, crop residues, and energy crops, which could meet 7% of global primary energy consumption and potentially reduce GHG emissions by 3,825 Mt CO_2 eq. In addition, anaerobic digestate is estimated to replace 5–7% of inorganic fertilizers currently in use. Among them, crop residues and energy crops contribute 29% and 34% of the total, exhibiting immense potential for bioenergy generation [10]. However, the recalcitrant nature of agricultural residue, primarily consisting of cellulose, hemicellulose, and lignin, usually results in low degradation rate during AD. Therefore, many improvement approaches have been developed and researched considering their remarkable potential for biogas production. Moreover, the complex interactions among various microorganisms involved in the AD process require careful operation and management to ensure the performance stability.

This chapter introduces the important factors affecting AD efficiency, degradation pathways, microbial interactions in the AD process, and enhancement strategies to improve biogas production from agricultural residue. Besides, the potential utilization of anaerobic digestate and its importance are also reviewed. The recent advancements are summarized to provide a comprehensive understanding and future perspectives of sustainable AD practices for agricultural residue utilization.

2 Anaerobic Digestion of Agricultural Residue

AD is a naturally occurring biological process that degrades organic matter into biogas by microorganisms in the absence of oxygen and can be divided into four stages, namely hydrolysis, acidogenesis, acetogenesis, and methanogenesis. The microbes involved in the first three stages belong to bacteria, while methanogens belong to the domain of archaea, having extremely slow growth rates of several days and strict requirements on the living environment compared to bacteria, only living in neutral or weak alkaline environments at suitable temperature ranges [11]. The complex interactions among different types of microbes require careful management of the AD operation to ensure process stability and efficient biogas production. In this section, the key factors affecting AD of agricultural residue, involved microbial mechanisms, and limitations are discussed in detail.

2.1 Key Factors for Efficient Biogas Production

2.1.1 Total Solid Content

According to the total solid content of the feedstock, AD processes can be categorized into three types: (1) wet (or liquid-state) AD at TS $<10\%$; (2) semisolid AD at TS of 10–15%; and (3) dry (or solid-state) AD at TS $\geq 15\%$ [12]. Some studies also considered TS >6–10% as high-solid AD based on the limit concerning diffusion behavior [13]. Dry AD or high-solid AD is designed to treat feedstocks with low moisture content, which reduces water consumption and achieves high volumetric biogas production in a compact reactor. However, the high solid content brings difficulty in efficient mass transfer and homogenous distribution of substrates and microorganisms, often resulting in low methane production rate and process instability. Lignocellulosic biomass mainly comprises polysaccharides, which could lead to increased production of volatile fatty acids (VFAs) at high solid conditions, causing the imbalance between acidogenesis and methanogenesis and the acidified pH condition, inhibiting methane production [14].

Currently, wet AD is the most widely applied system in the world; however, the treatment of large amounts of liquid digestate generated from wet AD systems is a major problem in most wet AD systems. Europe is the most advanced country in developing AD technology and has developed several commercial dry AD systems, like DRANCO, KOMPOGAS, and BEKON. The installation of dry AD systems has increased recently, making up around 35% of AD plants in Europe in 2015 [15].

2.1.2 Mixing Condition

Mixing is essential in AD systems to ensure the homogeneity of substrates and microorganisms and maintain a uniform temperature and pH environment. In the wet AD system, poor mixing can also lead to floating, sedimentation, and scum formation, resulting in clogging and restricted gas transfer [16]. Mechanical stirring, liquid/slurry recirculation, and gas recirculation or their combinations are commonly applied mixing methods. Mechanical stirring is the most applied mixing method in AD plants; however, the energy consumption is quite high, varying from 8% to 58% of the total energy demand in a full-scale anaerobic digester [16]. Liquid recirculation can be realized by pumping the mixture or leachate/percolate back into the AD reactor and is commonly used in dry AD systems [14]. In addition to improving reactor homogeneity, liquid recirculation also provides a washing effect of inhibitors such as VFAs and ammonia. Previous studies have reported remarkable enhancement (1–4 times) in methane production from dry AD systems with liquid recirculation [17, 18]. Gas recirculation can be conducted by pumping back biogas into the bottom of the AD reactor or introducing pressurized biogas into the reactor [19]. Gas recirculation not only provides a pneumatic mixing effect but also enhances biogas quality by accelerating the utilization of CO_2 by hydrogenotrophic methanogens and increasing the dissolution of H_2S [19, 20]. Recently, the combination of biogas and liquid recirculation in dry AD of agricultural wastes has been reported to greatly enhance methane production by 31% and 252% compared to the conditions with only liquid recirculation or no recirculation [18].

The mixing intensity and mixing mode can largely alter the performance of AD systems. Most studies reported that high mixing intensity and continuous mixing mode adversely affect biogas production while increasing energy consumption since the shear stress negatively affects microbial floc formation [12, 20]. Moreover, the long proliferation time of methanogens makes them sensitive to washout and inhibition, while the intensive mixing could disturb their growth, especially during the startup of AD or at high organic loading rates [19, 21].

2.1.3 Carbon-to-Nitrogen Ratios

Introducing feedstock with an optimal carbon-to-nitrogen (C/N) ratio is important for the AD process since carbon and nitrogen are two essential nutrients for microbial growth, and the appropriate C/N ratio could provide a dilution effect on inhibitors, like ammonia and VFAs, especially in the dry AD process. Lignocellulosic biomass usually has a high C/N ratio of 47–151, which could result in VFA accumulation and rapid depletion of nitrogen for protein synthesis, thus affecting the metabolism and energy transfer in microbial cells and resulting in low methane production [22, 23]. On the other hand, protein-rich substrates, like livestock manure, sewage sludge, and food waste, usually have a lower C/N ratio of 2–30, which could lead to ammonia accumulation, inhibiting methane production

[23]. Therefore, the co-digestion of substrates with high and low C/N ratios could improve methane production compared to mono-digestion. The optimal C/N ratio for efficient AD is reported in the range of 20–30 in most studies, while it also largely depends on the types of feedstocks [23].

2.1.4 Nutritional Elements

Except for carbon and nitrogen, macronutrients (P, Ca, Mg, K, etc.) and microelements (Fe, Ni, Co, Mo, Cu, Zn, etc.) also play vital roles in maintaining the stability and efficiency of AD since they are essential nutrients for cell growth and metabolism [24]. Lo et al. found that macronutrients, like Ca, K, Mg, and Na, show potential stimulation effects on AD of municipal solid waste [25]. Lei et al. revealed that phosphate addition at 475 mg/L accelerated biogas production in AD of rice straw [26]. Micronutrients are essential cofactors of various enzymes involved in the AD process. Among them, Fe, Ni, and Co are the most studied micronutrients to enhance AD performance as they are essential cofactors of important enzymes in methanogens, including acetyl-CoA, carbon monoxide dehydrogenase, hydrogenases, coenzyme F_{430}, methyl-coenzyme M reductase, CoM methylases, etc. [27]. In addition, Fe is also a cofactor of various enzymes in the hydrolysis and acidification stages and exhibits great reduction capacity, therefore providing a more favorable environment for AD [24]. Lignocellulosic biomass usually lacks micronutrients, such as Fe and Ni, which cause VFAs accumulation and affect the activities of methanogens and propionate oxidizing bacteria [14, 28]. The supplementation of micronutrients has been reported to largely enhance methane production by 30–59% in AD of lignocellulosic biomass [27]. However, excessive element concentrations also inhibit methanogenesis, requiring careful control during AD [24].

2.1.5 Temperature

According to the operating temperature, AD can be classified into psychrophilic (below 20°C), mesophilic (30–40°C), and thermophilic (50–60°C). Generally, higher temperatures accelerate hydrolysis and enhance the methane production rate, thus achieving higher treatment efficiency. However, acidification and ammonia inhibition could easily occur at thermophilic conditions, resulting in low or even halted methane production. The microbial structure is largely altered under different temperature ranges. Acetoclastic *Methanosaeta* was found to be predominant under mesophilic conditions, while hydrogenotrophic *Methanosarcina* and *Methanothermobacter* tend to be dominant in thermophilic conditions [29, 30]. Moreover, limited microbial populations could be enriched under thermophilic conditions, resulting in low microbial diversity and richness and thus more sensitivity to environmental changes [29, 30]. Thermophilic AD also requires high energy consumption and more precise process control to maintain system stability; therefore, mesophilic AD is currently more popular in applications. Nevertheless, considering the recalcitrant nature of lignocellulosic biomass, the thermophilic condition is more favorable to their degradation [14]. To combine the advantages of thermophilic and mesophilic conditions, temperature-phased anaerobic digestion (TPAD), including a short thermophilic phase (2–7 days) for enhanced hydrolysis, followed by a mesophilic phase (10–20 days) for methanogenesis, has been developed and proven to enhance methane production and shorten retention time in AD of agricultural wastes [31, 32]. Psychrophilic AD has been less studied and applied due to the extremely low biogas production efficiency at low temperatures, while it is more energy efficient, especially in cold regions. With appropriate adaptation and optimization of microorganisms, comparable methane production could be achieved under psychrophilic conditions [33–35].

2.1.6 pH and Buffering Capacity

The pH is a critical parameter affecting the stability of the AD process. The optimal pH range for methanogenesis is reported to be 6.8–7.2, while for hydrolysis and acidogenesis, the optimal pH is in the range of 5.5–6.5 [23]. The pH value is influenced by various factors in AD, including VFAs and ammonia concentrations, buffer capacity (alkalinity), etc. At the beginning stage of AD, rapid hydrolysis and acidification and slow growth rate of methanogens could lead to the accumulation of VFAs, resulting in pH reduction. When the pH is lower than 6.0, the activity of methanogens is inhibited, resulting in a long lag phase or even process failure [36]. With the degradation of protein, the increased ammonia concentration could partially recover the acidic pH. Subsequently, methanogens gradually consume VFAs, and the pH value increases and is stable at neutral or slightly alkaline conditions. When pH is higher than 8.0, high ammonia concentration inhibits methanogenesis, mainly due to the easy diffusion of free ammonia into the cell membrane, while microbial acclimatization was found to enhance the tolerance to high ammonia concentrations up to 5–10 g/L [37]. Some studies also reported similar methane production with improved methane content (>80%) at alkaline pH (8.0–9.0) compared to neutral conditions [19, 38]. Enhanced buffering capacity could prevent significant pH fluctuation in AD, which could be achieved by adjusting the organic loading rate or substrate-to-inoculum (S/I) ratio or adding bicarbonate or hydroxide salts [23]. The ratio of VFAs to alkalinity is often used as an indicator of process stability. A value below 0.4 is usually optimal for methane production [39].

2.1.7 Inoculation

Inoculation is generally necessary for the fast start-up of the AD process since inoculum could provide adapted anaerobic microbes, nutrients, and buffering capacity. Both inoculum type and inoculation size could largely affect AD efficiency. The degradation of lignocellulosic biomass is usually performed by lignocellulolytic bacteria that can secrete related enzymes, such as cellulases, hemicellulases, xylanase, peroxidase, and laccases. The bioaugmentation or addition of specific bacteria could largely enhance methane production from lignocellulosic biomass [40–42]. The inoculation size could be described as S/I ratio, feedstock-to-inoculum (F/I) ratio, or feedstock-to-microbe (F/M) ratio in batch AD systems or as organic loading rate (OLR) in continuous AD systems. They are usually calculated based on volatile solids (VS), total solids (TS), or chemical oxygen demand (COD). The greater inoculation size, described by a lower S/I ratio or OLR, could shorten the retention time and enhance weight-based methane yield but reduce volumetric productivity, while a lower inoculation size increases the risk of acidification but can improve the methane production rate, especially in continuous AD system [43]. The optimal S/I ratio or OLR largely depends on feedstock types and operating conditions. For less degradable substrates, like agricultural residue, the S/I ratio of 2–3 is usually optimal for methane production under mesophilic conditions, while for easily degradable substrates, like food waste, the S/I ratio should usually be lower than 1 to avoid over-acidification [44, 45]. Under thermophilic conditions, the optimal OLR or S/I ratio is usually higher due to the enhanced degradation efficiency [14, 46].

2.2 *Microbial and Biochemical Mechanisms*

2.2.1 Agricultural Residue Degradation Pathways and Related Microorganisms

Overview of the Degradation Process

Agricultural straw represents a primary form of agricultural residues. Regardless of the type, its main organic components are cellulose, hemicellulose, and lignin, along with minor amounts of other organic components such as proteins and fats. The structure of straw is robust, with cellulose tightly encapsulated by lignin and hemicellulose, posing challenges for microbial degradation. During AD, microbes primarily utilize the cellulose and hemicellulose components of straw; the degradation of lignin is relatively more challenging.

In AD microbial communities, certain microorganisms (mainly bacteria and possibly a few anaerobic fungi) can secrete cellulase or hemicellulase. Under the action of these hydrolases, cellulose and hemicellulose are hydrolyzed into various monosaccharides (mainly glucose and xylose, with small quantities of arabinose, galactose, mannose, etc.) (hydrolysis). The released monosaccharides are utilized by fermentation microbes to produce short-chain fatty acids, primarily acetate, propionate, (iso) butyrate, valerate, etc., while also producing H_2 and CO_2 (acidogenesis). Fermentation microbes include species capable of hydrolysis as well as those that are not, typically characterized by rich diversity and high metabolic variety. In the AD

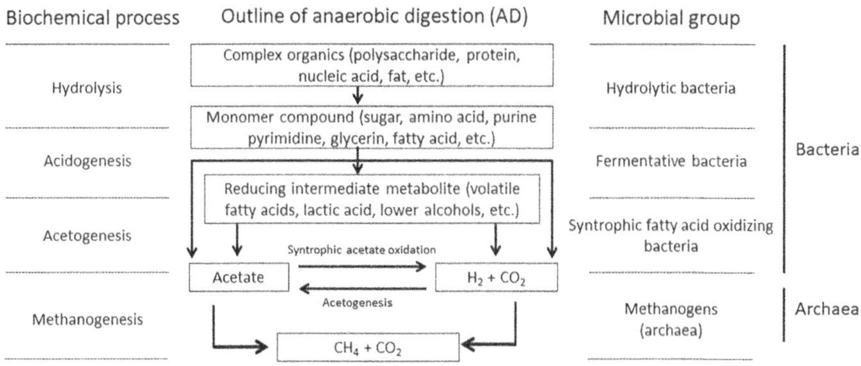

Fig. 1 Outline of anaerobic digestion of agricultural residues

system of relatively hard-to-degrade materials like straw, they hold a dominant position within the microbial community, influenced by the types of raw materials and AD process conditions.

Short-chain fatty acids above C2 are further oxidatively decomposed by various fatty acid-oxidizing bacteria to produce acetate, H_2, and CO_2 (acetogenesis) that can be utilized by methanogens. Fatty acid oxidation reactions are thermodynamically unfavorable under standard conditions; thus, a syntrophic relationship with methanogens, especially hydrogenotrophic methanogens (methanogens reducing CO_2 to methane with H_2), is necessary to reduce the partial pressure of H_2 and the concentration of acetate, making the actual free energy of the reaction negative. Enhancing the syntrophic relationship between fatty acid-oxidizing bacteria and methanogens in AD systems often effectively improves methane production efficiency and rate. Under certain conditions, such as high temperatures, ammonia stress, low loading rate, and long retention time, acetate can also be oxidized by acetate-oxidizing bacteria to generate H_2 and CO_2, though this reaction requires a syntrophic relationship with hydrogenotrophic methanogens [47–49]. Additionally, though H_2 and CO_2 produced during acidogenesis and acetogenesis can be directly utilized by hydrogenotrophic methanogens, under low temperatures and acidic and alkaline conditions where hydrogenotrophic methanogenesis is inhibited, homoacetogens can convert H_2 and CO_2 into acetate through the Wood-Ljungdahl pathway, which is then utilized by acetoclastic methanogens for methane production [50].

Acetate, H_2, and CO_2 produced during acidogenesis and acetogenesis can be utilized by acetoclastic and hydrogenotrophic methanogens to produce methane. Small amounts of methyl compounds generated during raw material degradation, such as methylamines, can be utilized by methylotrophic methanogens to produce methane. Most methanogens are monotrophic, but some species, such as those in the genus *Methanosarcina*, are multitrophic and can use acetate, H_2, and CO_2 and methyl compounds for methanogenesis [51].

Thus, under the action of various functional microorganisms, agricultural residues go through four main stages – hydrolysis, acidogenesis, acetogenesis, and methanogenesis – to complete the methanation process (Fig. 1).

Hydrolysis Process and Related Microorganisms

The hydrolysis of cellulose and hemicellulose relies on the synergistic action of various enzymes. Cellulose, a macromolecule composed of glucose monomers linked by β-1,4-glycosidic bonds, is randomly cleaved at noncrystalline regions of its chains by endoglucanases, releasing short-chain cellulose fragments and providing free ends for exoglucanases to act upon; exoglucanases then hydrolyze glycosidic bonds at the reducing or nonreducing ends of cellulose chains to release cellobiose. Cellobiose is subsequently hydrolyzed into glucose by β-glucosidase, which can also act on the glycosidic bonds of oligoglucosides to release glucose. Besides these three main types of enzymes, lytic polysaccharide monooxygenases (LPMOs), cellobiose dehydrogenase, cellodextrinase, and others also participate in the hydrolysis process, enhancing hydrolysis efficiency [52].

Hemicellulose, composed of multiple monomers such as xylose, arabinose, galactose, and mannose, linked by β-1,4-glycosidic bonds, β-1,3-glycosidic bonds, and others, includes various side chains besides the xylan backbone, thus requiring the cooperative action of multiple enzymes for effective degradation. Key hydrolases include endo-1,4-β-xylanase and β-xylosidase, along with α-glucuronidase, α-L-arabinofuranosidase, feruloyl esterase, acetyl xylan esterase, etc. [52].

In AD systems, bacteria primarily perform lignocellulosic hydrolysis, although anaerobic fungi have also been discovered in recent years [53]. Anaerobic cellulolytic bacteria mainly implement hydrolysis through the synthesis of multienzyme complexes known as cellulosomes [54]. These bacteria secrete various related enzymes and proteins (including the aforementioned hydrolases, anchoring proteins, scaffoldins, linker modules, etc.), assembling them into cellulosomes on the cell surface. Essentially, this complex is an efficient, highly organized cell surface enzyme system that not only exhibits high hydrolysis efficiency but also allows for the recycling of cellulose-hydrolyzing enzymes. Numerous species of anaerobic cellulolytic bacteria have been isolated and cultured, such as those from the genera *Clostridium*, *Ruminococcus*, *Butyrivibrio*, *Terrisporobacter*, *Bacteroides*, etc. [55, 56]. Additionally, several species of anaerobic cellulolytic fungi have been identified, encompassing genera such as *Neocallimastix*, *Caecomyces*, *Piromyces*, *Anaeromyces*, *Saccharomyces*, *Orpinomyces*, etc. [53, 57]. Previously, there was less attention paid to fungi in AD systems, but the discovery of these anaerobic fungi has opened up a bioaugmentation pathway for improving the AD efficiency of lignocellulosic residues.

Acidogenesis Process and Related Microorganisms

Once monosaccharides such as glucose are released, they can be immediately utilized by fermentative microorganisms. Although the species of bacteria with fermentation capabilities are diverse, and the fermentation products are varied, pyruvate serves as a critical node in the acidogenic pathway. The majority of acid-producing bacteria convert hexoses like glucose into pyruvate via the Embden-Meyerhof-Parnas (EMP) pathway, and pentoses like xylose into pyruvate through the coupling of the pentose phosphate pathway (PPP) and EMP. A few species can also convert hexoses to pyruvate via the Entner-Doudoroff (ED) pathway, such as *Zymomonas mobilis*. From pyruvate, various metabolic pathways diverge, leading to different types of fermentation products, including acetate, propionate, (iso)butyrate, lactate, ethanol, propanol, butanol, H_2, and CO_2.

Based on the proportion of fermentation products, acidogenic metabolic pathways can be classified into several types: acetate-ethanol type, propionate type, butyrate type, lactate type, and others.

The metabolic pathway where the main fermentation products are acetate and ethanol is referred to as the acetate-ethanol fermentation. During the production of acetate and ethanol, H_2 can be released, making this fermentation pathway suitable for H_2 production as well. Acetate can be produced through two primary routes: one is from pyruvate via the acetyl-CoA pathway; the other involves the oxidation of ethanol or acids such as propionate and butyrate. Ethanol is generated through the decarboxylation of pyruvate to acetaldehyde, followed by reduction to ethanol. Accumulation of ethanol in AD systems is toxic to microorganisms; hence, it is not usually a major fermentation product.

Fermentation pathways that predominantly produce propionate are called propionate-type fermentation. Theoretically, 1 mol of glucose can be converted into 2 mol of propionate. However, in practical scenarios, acetate is often co-produced during propionate formation by anaerobic microbes. There are mainly two pathways for propionate generation: one involves the reduction of pyruvate to lactate, which is then reduced to propionate; the other pathway involves genera such as *Propionibacterium*, *Corynebacterium*, and *Bifidobacterium* producing propionate through the transcarboxylase cycle.

Butyrate-type pathways primarily produce acetate and butyrate, along with H_2. Butyrate synthesis proceeds from pyruvate to acetyl-CoA, through a series of intermediates and reactions, ultimately branching into two pathways: one catalyzed by butyrate kinase (BK) to produce butyrate, and the other facilitated by butyryl-CoA:acetate CoA-transferase to generate butyrate. *Clostridium tyrobutyricum* is a typical representative of butyrate-producing fermentation.

Lactate fermentation occurs when pyruvate is converted to lactate under the action of lactate dehydrogenase (LDH). LDH exhibits stereospecificity, catalyzing the production of D-lactate, L-lactate, or racemic mixtures (DL). Based on the lactate product, fermentation can be categorized into homofermentative and heterofermentative lactic acid fermentation. Homofermentative lactic acid fermentation theoretically yields 2 mol of lactate per mole of glucose consumed, while heterofermentative lactic acid fermentation produces additional products such as CO_2, ethanol, and acetate.

In AD systems, acidogenic bacteria are diverse and widespread, encompassing numerous species across multiple phyla, including Firmicutes, Bacteroidetes, Spirochaetes, Actinobacteria, Proteobacteria, etc., such as *Acidipropionibacterium, Actinomyces, Acinetobacter, Bacteroides, Bacillus, Brooklawnia, Bifidobacterium, Butyrivibrio, Clostridium, Corynebacterium, Fermentimonas, Lactobacillus, Mesotoga, Olsenella, Propionibacterium, Propioniciclava, Petrimonas, Ruminococcus, Streptococcus*, and *Terrisporobacter* [58, 59]. Operational parameters of AD systems, such as temperature, pH, retention time, and loading rate, significantly affect the structure of the acidogenic microbial community, thus impacting the composition and distribution of fermentation products and consequently influencing subsequent H_2 production, acetogenesis, and methanogenesis processes.

Oxidation of Volatile Fatty Acids (VFAs) and Related Microorganisms

The intermediate metabolite VFAs produced during the acidogenesis process are further oxidized by VFA-oxidizing bacteria, yielding acetate, H_2, and CO_2 (acetogenesis) that can be utilized by methanogens. However, these oxidation processes are thermodynamically unfavorable under standard conditions (Table 1). Therefore, VFA-oxidizing bacteria typically form syntrophic relationships with hydrogenotrophic methanogens or sulfate-reducing bacteria to make the reaction's actual free energy negative. This metabolic characteristic makes it challenging to obtain pure cultures of these bacteria, with only a limited number of species successfully isolated so far. Acetate, propionate, (iso) butyrate, and (iso, neo) valerate are the primary intermediate metabolites. Here, we briefly describe the syntrophic degradation processes of these fatty acids, as illustrated in Fig. 2.

Table 1 Reactions of VFA degradation and methanogenesis and standard Gibbs free energy

Substrate	Reaction	Standard Gibbs free energy ($\Delta G^{0\prime}$)
Acetate	$Acetate^- + 4H_2O \rightarrow 2HCO_3^- + 4H_2 + H^+$	+104.6 kJ/mol
Propionate	$Propionate^- + 3H_2O \rightarrow acetate^- + 3H_2 + HCO_3^- + H^+$ $2\text{-}Propionate^- \rightarrow acetate^- + butyrate^-$	+71.7 kJ/mol +0.12 kJ/mol
Butyrate	$Butyrate^- + 2H_2O \rightarrow 2\text{-}acetate^- + 2H_2 + H^+$	+48.3 kJ/mol
Valerate	$Valerate^- + 2H_2O \rightarrow acetate^- + propionate^- + 2H_2$	+48.3 kJ/mol
Isovalerate	$Isovalerate^- + HCO_3^- + H_2O \rightarrow 3\text{-}acetate^- + H_2 + H^+$	+20.2 kJ/mol
H_2, CO_2	$4H_2 + HCO_3^- + H^+ \rightarrow CH_4 + 3H_2O$	−135.6 kJ/mol
Acetate	$CH_3COO^- + H_2O \rightarrow CH_4 + HCO_3^-$	−31.0 kJ/mol
CH_3OH	$4CH_3OH \rightarrow 3CH_4 + CO_2 + 2H_2O$	−105.0 kJ mol^{-1}

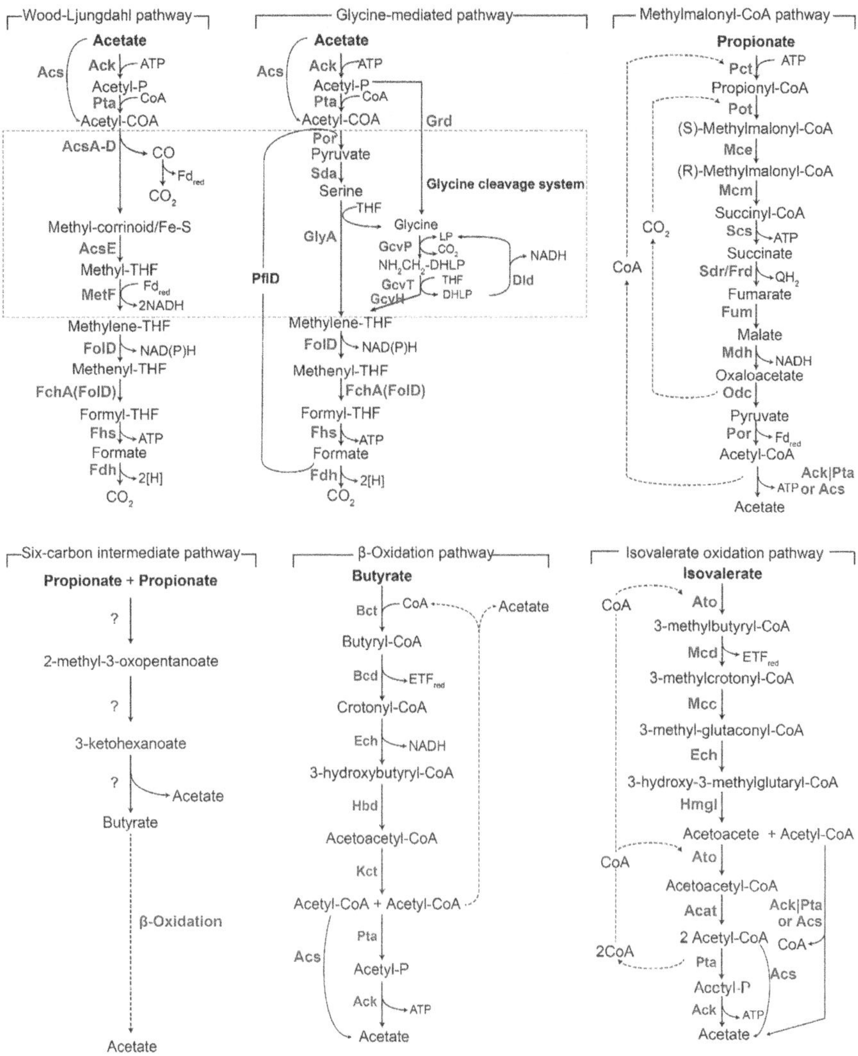

Fig. 2 Syntrophic oxidation pathways of VFAs

Syntrophic Oxidation of Acetate

Acetate can be directly utilized by acetoclastic methanogens to produce methane. Under conditions such as high temperatures, ammonia stress, or salt stress where acetoclastic methanogens are inhibited, acetate can be oxidized by acetate-oxidizing bacteria to produce H_2 and CO_2, which are then converted into methane by hydrogenotrophic methanogens. Two pathways for syntrophic acetate oxidation have been identified: Pathway 1 involves the reverse reaction of the Wood-Ljungdahl pathway, also known as the reverse acetyl-CoA pathway, with microorganisms like *Thermacetogenium phaeum* strain PB, *Clostridium ultunense* strain BS, and *Tepidanaerobacter acetatoxydans* strain Re1 capable of this pathway. Pathway 2 couples carboxylation steps of the Wood-Ljungdahl pathway with glycine cleavage, converting acetyl-CoA into methyltetrahydrofolate using enzymes from the glycine cleavage system, exemplified by *Thermotoga lettingae* strain TMO (Fig. 2). Recent studies using stable isotope probing (SIP) and metagenomics have identified potential novel acetate-oxidizing syntrophs in AD systems, including uncultured species within groups such as Synergistes group 4, Thermoanaerobacteraceae, Clostridia, Anaerolineae, and Gemmatimonadetes [47, 60].

Syntrophic Oxidation of Propionate

Propionate is oxidized by syntrophic propionate-oxidizing bacteria in conjunction with hydrogenotrophic methanogens to produce methane. Compared to other short-chain fatty acids (e.g., butyrate, valerate) or ethanol, the oxidation of propionate has the highest Gibbs free energy under standard conditions, making its oxidation particularly challenging. Thus, propionate and acetate accumulate more readily when AD is disrupted. Two known pathways for propionate oxidation include the methylmalonyl-CoA (MMC) pathway found in *Syntrophobacter*, *Pelotomaculum*, and *Desulfotomaculum* and a unique dismutation pathway found in *Smithella*, which yields butyrate as a by-product, which is further degraded by *Smithella* itself or butyrate oxidation bacteria to acetate and H_2 (Fig. 2). The fate of propionate oxidation in AD systems depends on operational conditions such as temperature and environmental factors like ammonia inhibition [61, 62].

Syntrophic Oxidation of Butyrate

Butyrate includes *n*-butyrate and isobutyrate. There are over ten strains of *n*-butyrate-oxidizing bacteria isolated, some of which can also oxidize isobutyrate. Most syntrophic butyrate-oxidizing bacteria belong to the genus *Syntrophomonas*, along with species from *Syntrophus*, *Themosyntropha*, *Syntrophothermus*, and *Algorimarina*. Novel syntrophic butyrate-oxidizing bacteria have been identified using SIP and metagenomic approaches [63]. The known pathway for *n*-butyrate oxidation is the β-oxidation pathway (Fig. 2), with some species capable of isomerizing isobutyrate into *n*-butyrate before oxidation.

Syntrophic Oxidation of Valerate

Valerate includes *n*-valerate, isovalerate, and 2-methylbutyrate (neo-valerate). Most *n*-valerate- and neo-valerate-oxidizing bacteria belong to *Syntrophomonas*. No pure culture capable of syntrophic isovalerate oxidation has been isolated yet, although a mixed culture utilizing isovalerate as the sole carbon source has been obtained

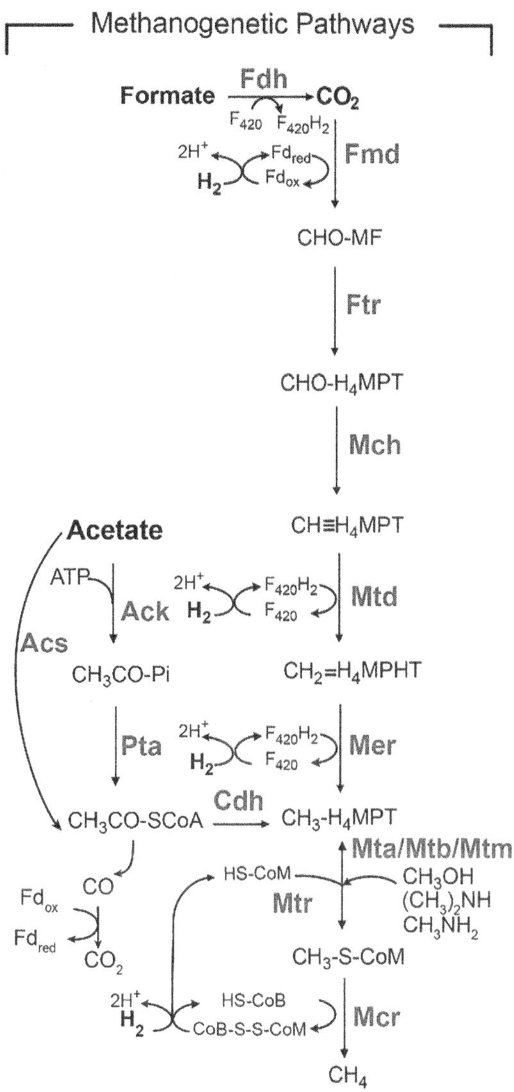

Fig. 3 Methanogenic pathways in AD

[64]. Using metagenomics, Chen et al. discovered novel isovalerate-oxidizing bacteria affiliated with *Clostridiales* and *Syntrophaceae*. Unlike other fatty acids, isovalerate oxidation does not proceed via the β-oxidation pathway but instead involves activation by coenzyme A transferase, followed by dehydrogenation, carboxylation, hydration, and cleavage to yield 3 mol of acetate and 1 mol of H_2 (Fig. 2) [65].

Methanogenesis Process and Related Microorganisms

Common methanogens in anaerobic digesters include acetoclastic methanogens like *Methanothrix*; multitrophic methanogens like *Methanosarcina*; hydrogenotrophic methanogens such as *Methanoculleus*, *Methanobacterium*, *Methanspirillum*, and *Methanothermobacter*; and methylotrophic methanogens like *Methanomassiliicoccus*. Methanogenic pathways are depicted in Fig. 3, with standard Gibbs free energies shown in Table 1.

The reduction of CO_2 for methane production can use electrons from H_2, formate, or direct electron transfer. Some electrogenic fatty acid-oxidizing bacteria release electrons that can be transferred extracellularly to methanogens capable of accepting external electrons, such as *Methanosarcina* and *Methanothrix* [66, 67], referred to as direct interspecies electron transfer. Utilization of H_2 and formate as electron carriers is termed indirect interspecies electron transfer. Operational conditions, substrate type, and fermentation pathways influence interspecies electron transfer patterns; for instance, Chen et al. observed that at high temperatures, propionate oxidation and methanogenesis primarily involve H_2, while at mesophilic temperatures, formate is the main carrier [61].

The distribution of methanogen species in anaerobic digesters is influenced by operational conditions, substrate composition, and environmental stress due to differences in substrate affinity, growth rates, and stress tolerance among different methanogens. For example, *Methanothrix* and *Methanosarcina* both utilize acetate for methane production, but *Methanothrix* exhibits higher acetate affinity and predominates under low loading conditions [68], whereas *Methanosarcina* shows greater stress tolerance [69, 70]. Some *Methanosarcina* species are multitrophic, capable of utilizing acetate, H_2, CO_2, and methyl compounds for methane production. However, whether all three methanogenic pathways function simultaneously in AD depends on reactor physicochemical parameters, substrate concentrations, and coexistence of different methanogens. Stress factors such as ammonia stress can inhibit specific methanogenic pathways, particularly acetoclastic methanogenesis [51].

Microbial Interactions in AD Systems

In AD microbial communities, interactions among members extend beyond occupying different positions in the "food chain." Despite the acidogenic fermentation process being energy yielding and not dependent on the consumption of its products such as H_2, a reduced H_2 partial pressure facilitates the acidogenesis process. Therefore, a mutualistic relationship exists between acidogenic fermenters and hydrogenotrophic methanogens. Hydrogen-producing acetogenic bacteria form strict syntrophic relationships with methanogens; enhancing interspecies electron transfer through methods like adding electron conductive materials or reducing intercellular spatial distances benefits the AD process [71]. Additionally, community members can form close relationships by exchanging amino acids, vitamins, siderophores, and other cofactors essential for cell growth. For instance, Yao et al. reported that most members of anaerobic microbial communities lack the ability to synthesize all essential amino acids, indicating that they must exchange these amino acids to establish interdependent survival patterns [72]. Further research by Yao et al. explored how the availability of amino acids influences interactions among community members and overall community function in AD [73].

Phages and Their Interactions with Prokaryotic Microbial Communities in AD Systems

Besides microorganisms, viruses also exist within AD systems. The microorganisms involved in AD are primarily prokaryotes; hence, the viruses are predominantly bacteriophages. Numerous studies have reported the significant role of bacteriophages in shaping prokaryotic microbial communities, affecting both their structure and function. To date, knowledge about bacteriophages targeting anaerobic microorganisms is limited, and research on the phageome in AD systems remains scarce. However, some initial studies have revealed the presence of diverse, largely unknown bacteriophages in AD systems [74]. Zeng et al. systematically analyzed the types, lifestyle characteristics, and infection features of bacteriophages targeting various functional groups of prokaryotic microbes in AD systems. They found that bacteriophages carry auxiliary metabolic genes involved in fatty acid oxidation, methanogenesis, and energy conservation [75]. This study provides insights into phage characteristics and their in situ interactions with prokaryotic hosts, highlighting their ecological role in AD systems. More work is needed to elucidate the dynamic interactions between the phageome and the prokaryotic microbiome in response to various environmental factors, providing guidance for developing phage-based regulation technologies.

2.2.2 Factors Influencing the Microbial Community Structure in Agricultural Residue AD Systems

Factors such as the type of agricultural residues, fermentation methods (single-stage or multistage, batch or continuous), reactor operational conditions (temperature, retention time, organic loading rate, etc.), pretreatment methods and conditions, and enhancement measures (microbial enhancement, conductive materials, etc.) all impact the structure of microbial communities within AD systems. These factors ultimately affect methane production efficiency and rates. For instance, Xu et al. discovered that the species richness and diversity of the microbial community in a two-stage AD system for methane production from rice straw were significantly higher than those in a single-stage AD system. The relative abundance of *Proteiniphilum*, capable of cellulose fermentation, along with methanogens *Methanothrix* and *Methanobacterium*, was markedly higher in the two-stage system. The differences in microbial community structure led to superior methane production efficiency of the two-stage system [76]. Liu et al. compared the microbial community structures in corn stover AD systems at four different temperatures ranging from 35 to 44°C. Results indicated significant variations in microbial community structures at different temperatures, with the relative abundance of the phylum Firmicutes increasing from 32.5% at 35°C to 66.6% at 44°C. Additionally, the relative abundances of *Methanoculleus* and *Methanomassiliicoccus* were significantly elevated at 44°C, which may contribute to enhanced methane production at this temperature [77]. Lu et al. found significant differences in microbial community structure between AD systems using low-temperature-urea-hydrothermal-pretreated corn stover and untreated corn stover. In the AD system treating pretreated corn stover, hydrolytic acidogenic bacteria (unknown species in Bacteroidales) accounted for 33.2%, while the multitrophic methanogen *Methanosarcina* comprised 72.3% [78].

As AD requires a complex microbial community, enhancing AD with a single microbe is generally ineffective, especially for recalcitrant substrates like straw. Li et al. utilized consortia of lignocellulolytic microbes from the rumen of ruminants and lactic acid bacteria to enhance the AD of rice straw, resulting in a fivefold increase in methane production efficiency [79]. Basak et al. employed an acclimatized microbiota rich in syntrophic bacteria and *Methanosarcina* for the AD of rice straw, achieving a methane yield 33.3% higher than that obtained with unacclimated AD sludge [80]. These findings underscore the critical importance of the microbial community structure in enhancing the efficiency of AD systems.

2.3 Limitations in Anaerobic Digestion of Agricultural Residues

2.3.1 Structural Recalcitrance and Poor Biodegradability

Compared to other feedstocks, the major challenge in AD of lignocellulosic biomass, such as straw, lies in their highly recalcitrant structure, which provides a natural barrier against microbial attack. The high degree of polymerization, crystalline structure of cellulose, and stable chemical bonds among cellulose, hemicellulose, and lignin further complicate enzymatic hydrolysis. Consequently, the biodegradation efficiency of lignocellulosic biomass during AD is relatively low, with volatile solid (VS) degradation ratios generally ranging from 30% to 50% [81, 82]. Although co-digestion with more degradable substrates can partially enhance the degradation rate, significantly improving the digestibility of straw remains challenging.

2.3.2 Nutrient Imbalance

Agricultural residues like straw are rich in carbon but deficient in essential nutrients such as nitrogen (N), phosphorus (P), and sulfur (S), which are crucial for microbial growth. In AD, an optimal C/N ratio of approximately 20 is generally recommended, whereas lignocellulosic feedstocks typically exhibit a C/N ratio of around 50. Moreover, critical trace elements such as Mn, Co, Zn, W, Se, and Ni are often insufficient in straw-based substrates to meet microbial requirements. Therefore, supplementation with these macro- and micronutrients or co-digestion with nutrient-rich materials is commonly employed to achieve balanced nutrition and improve anaerobic digestion efficiency [83, 84].

2.3.3 Poor Feedstock Dispersibility

Lignocellulosic biomass, such as straw, has low density and poor flowability, and some types also exhibit limited water absorption capacity. These characteristics hinder uniform dispersion within anaerobic digesters, leading to localized accumulation, such as crust formation on the upper region in digesters. This reduces contact between microorganisms and substrate and negatively affects mass and heat transfer, ultimately impairing degradation performance. Therefore, optimization of agitators and mixing strategies in digester systems is particularly important. Studies have demonstrated that computational fluid dynamics (CFD) simulations can be effectively used to optimize impeller configurations and mixing methods in rice straw AD systems, significantly enhancing methane production efficiency and shortening retention time [85].

3 Enhancement Strategies for Biogas Production from Agricultural Residues

3.1 Anaerobic Co-digestion

As mentioned above, lignocellulosic feedstocks such as straw are typically deficient in essential nutrients like N and P, making them unsuitable for standalone AD. Therefore, co-digestion with nutrient-rich substrates is commonly practiced. Common co-substrates include livestock manure [82], sewage sludge [86], and food waste [87]. These materials may pose challenges when digested alone – such as ammonia inhibition due to high nitrogen content; thus, co-digestion benefits both substrates. Wei et al. investigated co-digestion of corn stover and chicken manure at a TS ratio of 1:1, achieving a total solid (TS) removal rate of 50.5%, compared to 35.9% for corn stover and 59.4% for chicken manure alone, indicating improved straw degradation [82]. Tian et al. studied co-digestion of wheat straw and sewage sludge, reporting a VS removal rate of 57.1% for the mixture, higher than that of wheat straw (54.3%) or sludge (39.9%), demonstrating synergistic degradation effects [86]. However, regional availability of co-substrates must be considered, as long-distance transportation may increase operational costs.

3.2 *Pretreatment Methods*

As discussed above, lignocellulosic feedstocks possess rigid structures with poor microbial and enzymatic accessibility. To overcome this limitation, various pretreatment techniques have been developed in recent years aimed at disrupting structural barriers and exposing cellulose embedded within lignin, thereby enhancing biodegradability. Pretreatment methods include physical, chemical, biological, and combined approaches. While all these methods can effectively improve degradation efficiency and methane yield, economic feasibility and environmental impacts must be evaluated by comparing additional pretreatment costs with increased biogas revenues.

3.2.1 Physical Methods

Physical methods mainly include mechanical grinding, microwave irradiation, and extrusion. Mechanical size reduction lowers particle size and cellulose crystallinity, improves material dispersibility, and enhances heat and mass transfer efficiency, thus increasing biodegradability and methane production [88]. Microwave treatment alters cellulose ultrastructure, degrades part of the hemicellulose and lignin, and increases microbial and enzymatic accessibility, significantly boosting methane yield [89]. However, physical pretreatments are energy intensive, and microwave processing may generate inhibitory by-products such as furfural and phenolic compounds. Therefore, the input-output balance should be carefully assessed before implementation.

3.2.2 Chemical Methods

Chemical pretreatment involves the use of acids, alkalis, ionic liquids, organic solvents, calcium peroxide, ozone, etc., with acid/alkali treatments being the most widely studied due to their cost-effectiveness. Sulfuric acid is commonly used due to its low cost and effectiveness, though phosphoric acid, hydrochloric acid, and various organic acids have also been explored. Acid pretreatment dissolves part of the hemicellulose, exposes cellulose, and enhances enzyme accessibility, thereby promoting methane production [90]. Alkaline pretreatment using NaOH, KOH, $Ca(OH)_2$, ammonia, or urea targets lignin, disrupts polymeric structures, increases hemicellulose solubility, and enhances cellulose exposure [91]. Compared to NaOH, ammonia-based pretreatment not only adjusts the C/N ratio but also avoids introducing environmentally harmful sodium ions [92]. Although KOH shows excellent pretreatment performance, it is expensive. Combining KOH with ammonia or urea can reduce KOH dosage while adjusting C/N and enriching potassium in digestate without environmental risks, making it a recommended approach [93].

3.2.3 Biological Methods

Biological methods offer a nontoxic and potentially low-cost alternative to physical and chemical pretreatments. Microbial consortia and hydrolytic enzymes are primarily used. White-rot fungi were among the earliest biological agents tested for pretreatment due to their ability to specifically degrade lignin, expose cellulose, and shorten digestion times – methane yields have been reported to increase up to fivefold [94]. However, long pretreatment durations remain a major drawback. Pretreatment using prokaryotic microbial consortia, such as anaerobic digestion effluent containing abundant hydrolytic bacteria, has also shown promise in preliminary decomposition [95]. Cellulase enzymes hydrolyze cellulose and hemicellulose into monosaccharides, significantly accelerating digestion rates and enhancing methane yields [96], although enzyme cost should be taken into account.

3.2.4 Combined Pretreatment

Given the distinct mechanisms of different pretreatment methods, combining them can create synergistic effects, enhancing pretreatment outcomes while reducing overall input. Examples include urea-hydrothermal, ammonia-freezing-thawing, calcium peroxide-digestate, and calcium peroxide-cellulase combinations. Compared to single pretreatments, integrated strategies have demonstrated better AD performance [78, 97–99].

3.3 Novel Nanobubble Technology

Nanobubbles, also known as ultrafine bubbles, are defined as fine bubbles with a volume equivalent diameter of less than 1 μm according to the International Organization for Standardization (ISO 20480-1:2017). Nanobubbles exhibit excellent stability in water mainly due to their negatively charged surface, small size, low buoyancy, and high water mobility, thus showing efficient mass transfer capabilities [100]. These unique properties of nanobubbles as well as their ability to generate reactive oxygen species (ROS) upon collapse make them a promising green technology in diverse fields, such as agriculture for enhancing plant growth, biomedicine for efficient drug delivery and cancer therapy, and water and wastewater treatment [101, 102]. Since 2019, nanobubbles have been attempting to enhance methane production in AD of organic wastes. The nanobubble water generated from different gases (H_2, N_2, O_2, air, etc.) has all proven to enhance methane yield by 10–43% when treating different organic wastes, including waste-activated sludge, food waste, pig manure, corn straw, etc. [100]. It was shown that nanobubbles can enhance enzymatic activities and enrich hydrolytic/acidifying bacteria and methanogenic archaea, thus promoting organic degradation and methane production

Table 2 Enhanced AD of lignocellulosic biomass by nanobubble technology

Feedstock	Operation conditions of AD	Type of nanobubble water (NBW)	Increment in methane yield	Effects on substrate degradation and enzyme activities	References
Erigeron canadensis	35 °C 30 days	O_2	16%	• 2–90% enhanced activities of dehydrogenase, α-glucosidase and coenzyme F_{420}	[104]
Rice straw	37 °C 45 days	O_2 N_2	18% 7%	• 9% and 18% enhanced cellulose and hemicellulose reduction with O2 NBW • Enhanced activities of electron transport system (ETS) and coenzyme F_{420}	[105]
Rice straw	37 °C 57 days	CO_2 N_2 H_2	14% 9% 5%	• 6–11% increased soluble carbohydrates and 4–25% increased VFAs during hydrolysis and acidification stages	[106]
Corn straw	35 °C 26 days	CO_2	17%	• 11–23% enhanced activities of ETS, β-glucosidase, dehydrogenase, and coenzyme F_{420}	[107]
Corn straw	35 °C 22 days Total solids (TS) of 4%	H_2	25%	• 33% and 26% enhanced cellulose and hemicellulose reduction	[108]
Cellulose	35 °C 50 days	Air N_2	18% 10%	• 21–29% enhanced cellulose reduction	[109]
Cellulose	35 °C 18 days	O_2 N_2 Air	30% 9% 8%	• 8–14% enhanced cellulose reduction • 10–38% enhanced cellulase activity, and 1.1–1.7 times higher ETS activity	[110]
Acetic acid and lignin	35 °C 10 days	N_2 CO_2	22% 17%	• 11–13% higher lignin reduction	[111]
Waste-activated sludge and alkaline lignin	37 °C 20 days	N_2	17%	• 10% higher lignin reduction	[112]

NBW used as pretreatment					
Rice straw	37 °C 5 days soaking and 45 days AD	N_2 CO_2 O_2 (soaking pretreatment)	No increment compared to deionized water (DW) soaking group	–	[113]
Corn straw	35 °C 75 days TS of 15%	O_2 N_2 (pre-augmentation of inoculum)	47% 18%	• 2–55% enhanced activities of cellulase, xyanases, and lignin peroxidase in pre-augmented sludge	[114]
Corn straw		O_2 N_2 (pre-augmentation of inoculum)	22% 10%	• 60–88% enhanced cellulose reduction and 22–23% enhanced hemicellulose reduction	[115]

[103]. As shown in Table 2, nanobubbles also show great potential to enhance degradation of lignocellulosic biomass. Previous studies speculated that ROS produced from nanobubble collapse may assist the cleavage of C–C or C–O bonds in lignin molecules and attack the active sites on the benzene ring for opening, thus leading to lignin depolymerization and degradation [112]. In addition, nanobubble water derived from O_2 or air could provide a micro-oxygen environment, thus facilitating the decomposition of refractory organic compounds [103]. Although nanobubble technology shows great potential to enhance methane production during AD of organic waste, previous research mainly focused on evaluating the effects on AD performance, enzymatic activities, and microbial metabolism. Moreover, nanobubble water is generally produced from clean water and added in AD process, which may result in waste of water resources and increased generation of liquid digestate. Therefore, more research is necessary to clarify the in-depth mechanisms and develop in situ nanobubble generation that is feasible for practical application.

4 Potential Utilization of Anaerobic Digestate from Agricultural Residues

Although biogas is the main product of the AD process, large amounts of digestate containing mineralized nutrients (N, P, K, etc.) and stabilized organics (humic substances, etc.) remain after AD and require further processing to prevent environmental pollution and nutrient loss. It was estimated that around 50–100 L digestate could be generated when producing 1 m^3 biogas [116]. In most AD plants, the digestate is first separated into liquid and solid phases by mechanical dewatering. Then, solid digestate is processed to produce organic fertilizer, while liquid digestate is treated as wastewater or concentrated into liquid fertilizer products. Previous studies estimated that solid-liquid separation and liquid digestate treatment consume 8–26% of the total energy produced in AD of food waste that has high energy content; if feedstock with low methane potential were used, digestate treatment would consume all the produced energy and even results in negative energy balance in the AD plant [117]. Agricultural residue usually contains lower moisture content and can be treated by dry AD systems, which could largely save the cost of digestate treatment. To fully recover nutrients and energy and minimize the environmental impact of the AD process, several sustainable approaches have been studied and developed to utilize anaerobic digestate as summarized in the following subsections.

4.1 Fertilizer and Soil Amendment

Digestate could provide essential nutrients for plant growth and enhance soil fertility by increasing organic matter, elevating soil pH, and increasing exchangeable bases.

Although the direct use of digestate as fertilizer has been investigated in several previous studies [118, 119], potential pollutants like heavy metals and pathogenic microorganisms pose risks to soil and plant health, necessitating the appropriate pretreatment before land application. Composting and pasteurization treatment are commonly used to inactive viable pathogens, converting solid digestate into more stable and safe fertilizer products [120]. AD operated at thermophilic conditions could largely reduce pathogen numbers in the digestate [121]. In addition, high VFAs and ammonia concentrations during AD are also conducive to eliminating pathogens, and dry mesophilic AD showed excellent inactivation performance for almost all pathogenic bacteria [121].

The nutrient composition of digestate largely depends on feedstock types. Solid digestate is usually rich in stabilized organic matter, exhibiting slow releasing properties, and is preferable as soil amendments, while liquid digestate contains more mineralized nutrients, more efficiently providing nutrients for plant growth. Ammonia is the main inorganic nitrogen form in anaerobic digestate, which is phytotoxic at high concentrations and easily leads to N_2O and NH_3 emissions during long-term storage and application, requiring careful management during utilization. However, enhanced plant growth could be achieved with proper application strategies primarily due to the biostimulatory compounds contained in the digestate [122]. Despite the high nitrogen emissions from anaerobic digestate, replacing mineral fertilizer with digestate could significantly decrease the carbon footprint by eliminating CO_2 emissions during the production of mineral fertilizers and direct emissions after application [123].

4.2 *Thermochemical Conversion*

Although digestate shows high fertilizer potential, the excess application and long-term storage could result in nutrient leaching and environmental pollution. Thermochemical process provides a rapid and efficient approach to reducing the digestate volume and converting it into biochar or hydrochar products that could be used as biofuel or soil conditioners. Digestate contains large amounts of water, which could be a suitable substrate for hydrothermal treatment. Hydrochar, bio-oil, and syngas could be produced at different treatment temperatures, that is, via hydrothermal carbonization (180–250°C), liquefaction (200–400°C), and gasification (300–500°C) [124]. Hydrothermal carbonization is most studied due to the relatively low energy input. The produced hydrochar could retain water and nutrients and provide habitat to microbes in the soil, enhancing soil fertility and plant growth and preventing nutrient leaching [125]. The aqueous phase after hydrothermal treatment contains rich soluble carbon and nitrogen, which has been recently recognized as a potential source of fertilizer [126]. Studies have also revealed the high heating value of hydrochars derived from maize silage digestate and garden waste ranging from 25 to 36 MJ/kg, which is comparable to the heating value of coal (27 MJ/kg) [124]. Pyrolysis (300–1,000°C) could also convert solid digestate into biochar;

however, the low moisture content of <30% is required for the feedstock; thus, solid-liquid separation and further drying process are necessary to pretreat the digestate, making the process energy intensive [124].

4.3 Insect Conversion

In recent years, insect conversion of organic waste has been increasingly concerned for the sustainable production of animal feed or biofuel. During the AD process, around 40–50% of organic matter in the raw feedstock is converted into biogas, while the remaining residue still contains relatively high protein content due to the relatively slow degradation of proteins with complex structures. Therefore, it could be potentially used for rearing insects, which can be more safely used as protein-rich animal feed or feedstock for biofuel production with less impurities or potential pollutants compared to anaerobic digestate. Several studies investigated the valorization of anaerobic digestate using black soldier fly larvae, suggesting the high potential of insect conversion for protein-rich feed production or biofuel production [127–130]. However, heavy metal concentration and pathogenic infection are potential risks during insect conversion of digestate requiring further investigation [131].

4.4 Microalgae Cultivation

Microalgae cultivation is also a promising solution to convert the nutrients in liquid digestate into value-added products, like biofuel, animal feed, or other biochemicals. In addition, biogas purification could be realized at the same time due to the excellent CO_2 absorption capacity of microalgae [132]. However, microalgae are also sensitive to various environmental conditions, the growth of which could be easily inhibited by high-strength nutrient concentrations and other toxic compounds in digestate, such as ammonia and metals [133]. High turbidity, imbalance of nutrient compositions, and presence of competing microbes could also limit microalgae growth in the digestate [134, 135]. To address the above issues, many studies have focused on various pretreatment methods of liquid digestate for enhanced microalgae cultivation, including dilution, filtration, oxidation, nitrification, and so on [136]. The selection of suitable microalgal strains that have a high tolerance to environmental stress is also essential for successful cultivation in the digestate [137].

5 Conclusion

Anaerobic digestion of agricultural residue exhibits great potential for bioenergy generation and carbon neutrality. However, the recalcitrant nature of agricultural residue generally requires suitable operating conditions and enhancement strategies to maximize methane production. Various factors, such as feedstock types, operational conditions, reactor configuration, and enhancement methods, impact the microbial community structure within AD systems, eventually affecting the methane production efficiency. The complex microbial community involved in AD processes, including the presence of fungi and phages, is not yet fully understood and requires further investigation to advance fundamental knowledge and enable the development of new regulatory technologies.

Dry AD at high solid conditions ($\geq$15% of TS) with mild mixing, such as intermittent liquid recirculation, is economically friendly for treating agricultural residue with high solid contents. Thermophilic hydrolysis and inoculation with acclimated inoculum can efficiently enhance biodegradability and shorten the digestion time. Co-digestion with nitrogen-rich substrates is a commonly applied method to enhance methane production by balancing the C/N ratio and nutritional element contents. Various pretreatment methods, including mechanical, chemical, and biological treatments, have been shown to improve AD efficiency; however, their economic feasibility requires careful assessment. Nanobubble technology has recently emerged as a green approach to enhance lignocellulose degradation and subsequent methane production, and ongoing technological advancements are expected to improve its practical applicability.

The anaerobic digestate remaining after AD contains rich mineralized nutrients and stabilized organics, indicating high potential for use as fertilizers and soil amendments. However, appropriate strategies to reduce nitrogen emissions during digestate storage and application need to be developed to further enhance the environmental benefits of the AD process. In addition, thermochemical processes, insect conversion, and microalgae cultivation have been demonstrated as efficient approaches for converting digestate into value-added products, such as biofuels, soil conditioners, and animal feed. Further efforts are required to optimize AD processes and the posttreatment of anaerobic digestate to improve overall environmental performance and economic efficiency.

References

1. Ritchie H, Rosado P, Roser M (2024) Energy production and consumption. Our World in Data. https://ourworldindata.org/energy-production-consumption. Accessed 9 Feb 2026
2. Ge M, Friedrich J, Vigna L (2024) Where do emissions come from? 4 charts explain greenhouse gas emissions by sector. World Resources Institute. https://www.wri.org/insights/4-charts-explain-greenhouse-gas-emissions-countries-and-sectors#:~:text=The%20energy%20sector%20produces%20the,%25)%20and%20buildings%20(6.6%25). Accessed 9 Feb 2026
3. Yuan X, Su CW, Umar M et al (2022) The race to zero emissions: can renewable energy be the path to carbon neutrality? J Environ Manag 308:114648. https://doi.org/10.1016/j.jenvman.2022.114648
4. Wang J, Azam W (2024) Natural resource scarcity, fossil fuel energy consumption, and total greenhouse gas emissions in top emitting countries. Geosci Front 15(2):101757. https://doi.org/10.1016/j.gsf.2023.101757

5. Lal R (2005) World crop residues production and implications of its use as a biofuel. Environ Int 31(4):575–584. https://doi.org/10.1016/j.envint.2004.09.005
6. Krishna VV, Mkondiwa M (2023) Economics of crop residue management. Ann Rev Resour Econ 15(1):19–39. https://doi.org/10.1146/annurev-resource-101422-090019
7. Singh A, Choudhary S, Meena RP et al (2022) Nutrient supplying potential of crop residues in Indian agriculture. In: Ahmad F, Sultan M (eds) Agricultural waste-new insights. IntechOpen, London. https://doi.org/10.5772/intechopen.108970
8. Smerald A, Rahimi J, Scheer C (2023) A global dataset for the production and usage of cereal residues in the period 1997–2021. Sci Data 10(1):685. https://doi.org/10.1038/s41597-023-02587-0
9. Lin L, Xu F, Ge X et al (2018) Improving the sustainability of organic waste management practices in the food-energy-water nexus: a comparative review of anaerobic digestion and composting. Renew Sustain Energy Rev 89:151–167. https://doi.org/10.1016/j.rser.2018.03.025
10. Jain S, Newman D, Nzihou A et al (2019) Global potential of biogas. World Biogas Association, London. https://www.worldbiogasassociation.org/wp-content/uploads/2019/07/WBA-globalreport-56ppa4_digital.pdf. Accessed 9 Feb 2026
11. Liu X, Yan Z, Yue ZB (2011) Biogas. In: Moo-Young M (ed) Comprehensive biotechnology, vol 3. 2nd edn, pp 99–114. https://doi.org/10.1016/B978-0-08-088504-9.00165-3
12. Karrabi M, Ranjbar FM, Shahnavaz B et al (2023) A comprehensive review on biogas production from lignocellulosic wastes through anaerobic digestion: an insight into performance improvement strategies. Fuel 340:127239. https://doi.org/10.1016/j.fuel.2022.127239
13. Ellacuriaga M, Cascallana JG, González R et al (2021) High-solid anaerobic digestion: reviewing strategies for increasing reactor performance. Environments 8(8):80. https://doi.org/10.3390/environments8080080
14. Yang L, Xu F, Ge X et al (2015) Challenges and strategies for solid-state anaerobic digestion of lignocellulosic biomass. Renew Sustain Energy Rev 44:824–834. https://doi.org/10.1016/j.rser.2015.01.002
15. Kariyama ID, Zhai X, Wu B (2018) Influence of mixing on anaerobic digestion efficiency in stirred tank digesters: a review. Water Res 143:503–517. https://doi.org/10.1016/j.watres.2018.06.065
16. Degueurce A, Tomas N, Le Roux S et al (2016) Biotic and abiotic roles of leachate recirculation in batch mode solid-state anaerobic digestion of cattle manure. Bioresour Technol 200:388–395. https://doi.org/10.1016/j.biortech.2015.10.060
17. Xu Z, Gu W, Shi Y et al (2025) Integration of biogas and liquid recirculation enhance methane production in high-solid anaerobic digestion of pig manure and rice husks. Bioresour Technol 416:131807. https://doi.org/10.1016/j.biortech.2024.131807
18. Yuan T, Zhang Z, Lei Z et al (2022) A review on biogas upgrading in anaerobic digestion systems treating organic solids and wastewaters via biogas recirculation. Bioresour Technol 344:126412. https://doi.org/10.1016/j.biortech.2021.126412
19. Zhao J, Hou T, Lei Z et al (2020) Effect of biogas recirculation strategy on biogas upgrading and process stability of anaerobic digestion of sewage sludge under slightly alkaline condition. Bioresour Technol 308:123293. https://doi.org/10.1016/j.biortech.2020.123293
20. Lindmark J, Thorin E, Fdhila RB et al (2014) Effects of mixing on the result of anaerobic digestion. Renew Sustain Energy Rev 40:1030–1047. https://doi.org/10.1016/j.rser.2014.07.182
21. Myszograj S, Stadnik A, Płuciennik-Koropczuk E (2018) The influence of trace elements on anaerobic digestion process. Civil Environ Eng Rep 28(4):105–115. https://doi.org/10.2478/ceer-2018-0054
22. Siddique MNI, Wahid ZA (2018) Achievements and perspectives of anaerobic co-digestion: a review. J Clean Prod 194:359–371. https://doi.org/10.1016/j.jclepro.2018.05.155

23. Wang Z, Hu Y, Wang S et al (2023) A critical review on dry anaerobic digestion of organic waste: characteristics, operational conditions, and improvement strategies. Renew Sustain Energy Rev 176:113208. https://doi.org/10.1016/j.rser.2023.113208
24. Romero-Güiza MS, Vila J, Mata-Alvarez J et al (2016) The role of additives on anaerobic digestion: a review. Renew Sustain Energy Rev 58:1486–1499. https://doi.org/10.1016/j.rser.2015.12.094
25. Lo HM, Chiu HY, Lo SW et al (2012) Effects of different SRT on anaerobic digestion of MSW dosed with various MSWI ashes. Bioresour Technol 125:233–238. https://doi.org/10.1016/j.biortech.2012.08.084
26. Lei Z, Chen J, Zhang Z et al (2010) Methane production from rice straw with acclimated anaerobic sludge: effect of phosphate supplementation. Bioresour Technol 101(12): 4343–4348. https://doi.org/10.1016/j.biortech.2010.01.083
27. Bardi MJ, Vinardell S, Astals S et al (2023) Opportunities and challenges of micronutrients supplementation and its bioavailability in anaerobic digestion: a critical review. Renew Sustain Energy Rev 186:113689. https://doi.org/10.1016/j.rser.2023.113689
28. Schmidt T, Nelles M, Scholwin F et al (2014) Trace element supplementation in the biogas production from wheat stillage–optimization of metal dosing. Bioresour Technol 168:80–85. https://doi.org/10.1016/j.biortech.2014.02.124
29. Tian Z, Zhang Y, Li Y et al (2015) Rapid establishment of thermophilic anaerobic microbial community during the one-step startup of thermophilic anaerobic digestion from a mesophilic digester. Water Res 69:9–19. https://doi.org/10.1016/j.watres.2014.11.001
30. Ao T, Xie Z, Zhou P et al (2021) Comparison of microbial community structures between mesophilic and thermophilic anaerobic digestion of vegetable waste. Bioprocess Biosyst Eng 44:1201–1214. https://doi.org/10.1007/s00449-021-02519-5
31. Dooms M, Benbelkacem H, Buffière P (2018) High solid temperature phased anaerobic digestion from agricultural wastes: putting several reactors in sequence. Biochem Eng J 130: 21–28. https://doi.org/10.1016/j.bej.2017.11.011
32. de Diego-Díaz B, Peñas FJ, Fernández-Rodríguez J (2021) Sustainable management of lignocellulosic wastes: temperature strategies for anaerobic digestion of artichoke. J Clean Prod 280:124479. https://doi.org/10.1016/j.jclepro.2020.124479
33. Saady NMC, Massé DI (2013) Psychrophilic anaerobic digestion of lignocellulosic biomass: a characterization study. Bioresour Technol 142:663–671. https://doi.org/10.1016/j.biortech.2013.05.089
34. Rusín J, Chamrádová K, Basinas P (2021) Two-stage psychrophilic anaerobic digestion of food waste: comparison to conventional single-stage mesophilic process. Waste Manag 119: 172–182. https://doi.org/10.1016/j.wasman.2020.09.039
35. Akindolire MA, Rama H, Roopnarain A (2022) Psychrophilic anaerobic digestion: a critical evaluation of microorganisms and enzymes to drive the process. Renew Sustain Energy Rev 161:112394. https://doi.org/10.1016/j.rser.2022.112394
36. Zhou J, Zhang R, Liu F et al (2016) Biogas production and microbial community shift through neutral pH control during the anaerobic digestion of pig manure. Bioresour Technol 217:44–49. https://doi.org/10.1016/j.biortech.2016.02.077
37. Jiang Y, McAdam E, Zhang Y et al (2019) Ammonia inhibition and toxicity in anaerobic digestion: a critical review. J Water Process Eng 32:100899. https://doi.org/10.1016/j.jwpe.2019.100899
38. Rincón-Pérez J, Celis LB, Morales M et al (2021) Improvement of methane production at alkaline and neutral pH from anaerobic co-digestion of microalgal biomass and cheese whey. Biochem Eng J 169:107972. https://doi.org/10.1016/j.bej.2021.107972
39. Elazhara M, Bouchabchoubb A, Elazhara F et al (2022) Influence of volatile fatty acids/ alkalinity ratio on methane production during mesophilic anaerobic digestion: stability, efficiency and optimization. Desalin Water Treat 257:142–149. https://doi.org/10.5004/dwt.2022.28580

40. Hu Y, Hao X, Wang J et al (2016) Enhancing anaerobic digestion of lignocellulosic materials in excess sludge by bioaugmentation and pre-treatment. Waste Manag 49:55–63. https://doi.org/10.1016/j.wasman.2015.12.006
41. Ozbayram EG, Kleinsteuber S, Nikolausz M et al (2018) Bioaugmentation of anaerobic digesters treating lignocellulosic feedstock by enriched microbial consortia. Eng Life Sci 18(7):440–446. https://doi.org/10.1002/elsc.201700199
42. Ferraro A, Massini G, Miritana VM et al (2020) A novel enrichment approach for anaerobic digestion of lignocellulosic biomass: process performance enhancement through an inoculum habitat selection. Bioresour Technol 313:123703. https://doi.org/10.1016/j.biortech.2020.123703
43. Xu F, Wang F, Lin L et al (2016) Comparison of digestate from solid anaerobic digesters and dewatered effluent from liquid anaerobic digesters as inocula for solid state anaerobic digestion of yard trimmings. Bioresour Technol 200:753–760. https://doi.org/10.1016/j.biortech.2015.10.103
44. Shahbaz M, Ammar M, Zou D et al (2019) An insight into the anaerobic co-digestion of municipal solid waste and food waste: influence of co-substrate mixture ratio and substrate to inoculum ratio on biogas production. Appl Biochem Biotechnol 187:1356–1370. https://doi.org/10.1007/s12010-018-2891-3
45. Li Y, Chen Z, Peng Y et al (2022) Deeper insights into the effects of substrate to inoculum ratio selection on the relationship of kinetic parameters, microbial communities, and key metabolic pathways during the anaerobic digestion of food waste. Water Res 217:118440. https://doi.org/10.1016/j.watres.2022.118440
46. Liu C, Wang W, Anwar N et al (2017) Effect of organic loading rate on anaerobic digestion of food waste under mesophilic and thermophilic conditions. Energy Fuel 31(3):2976–2984. https://doi.org/10.1021/acs.energyfuels.7b00018
47. Zeng Y, Zheng D, Li LP et al (2024) Metabolism of novel potential syntrophic acetate-oxidizing bacteria in thermophilic methanogenic chemostats. Appl Environ Microbiol 90:e01090-23. https://doi.org/10.1128/aem.01090-23
48. Werner JJ, Garcia ML, Perkins SD et al (2014) Microbial community dynamics and stability during an ammonia-induced shift to syntrophic acetate oxidation. Appl Environ Microbiol 80:3375–3383. https://doi.org/10.1128/AEM.00166-14
49. Shigematsu T, Tang YQ, Kawaguchi H et al (2003) Effect of dilution rate on structure of a mesophilic acetate-degrading methanogenic community during continuous cultivation. J Biosci Bioeng 96:547–558. https://doi.org/10.1016/S1389-1723(04)70148-6
50. Pan X, Zhao L, Li C et al (2021) Deep insights into the network of acetate metabolism in anaerobic digestion: focusing on syntrophic acetate oxidation and homoacetogenesis. Water Res 190:116774. https://doi.org/10.1016/j.watres.2020.116774
51. Yi Y, Dolfing J, Jin G et al (2023) Thermodynamic restrictions determine ammonia tolerance of methanogenic pathways. Water Res 232:119664. https://doi.org/10.1016/j.watres.2023.119664
52. Li Y, Song W, Han X et al (2022) Recent progress in key lignocellulosic enzymes: enzyme discovery, molecular modifications, production, and enzymatic biomass saccharification. Bioresour Technol 363:127986. https://doi.org/10.1016/j.biortech.2022.127986
53. Liang J, Liu S, Du Z et al (2024) Recent advances in methane and hydrogen production from lignocellulosic degradation with anaerobic fungi. Bioresour Technol 413:131544. https://doi.org/10.1016/j.biortech.2024.131544
54. Yoav S, Barak Y, Shamshoum M et al (2017) How does cellulosome composition influence deconstruction of lignocellulosic substrates in *Clostridium (Ruminiclostridium) thermocellum* DSM 1313? Biotechnol Biofuels 10:222. https://doi.org/10.1186/s13068-017-0909-7
55. Bayer EA, Kenig R, Lamed R (1983) Adherence of *Clostridium thermocellum* to cellulose. J Bacteriol 156:818–827. https://doi.org/10.1128/jb.156.2.818-827.1983

56. Artzi L, Bayer EA, Moraïs S (2017) Cellulosomes: bacterial nanomachines for dismantling plant polysaccharides. Nat Rev Microbiol 15:83–95. https://doi.org/10.1038/nrmicro.2016.164
57. Shariat Panahi HK, Dehhaghi M, Guillemin GJ (2022) A comprehensive review on anaerobic fungi applications in biofuels production. Sci Total Environ 829:154521. https://doi.org/10.1016/j.scitotenv.2022.154521
58. Li BY, Xia ZY, Gou M et al (2022) Production of volatile fatty acid from fruit waste by anaerobic digestion at high organic loading rates: performance and microbial community characteristics. Bioresour Technol 346:126648. https://doi.org/10.1016/j.biortech.2021.126648
59. Chen YW, Feng G, Hong X et al (2024) Effects of high solid content and straw proportion on volatile fatty acids production from straw, sludge and food wastes: performance and microbial community characteristics. Appl Biol Chem 67:83. https://doi.org/10.1186/s13765-024-00935-1
60. Zheng D, Wang HZ, Gou M et al (2019) Identification of novel potential acetate-oxidizing bacteria in thermophilic methanogenic chemostats by DNA stable isotope probing. Appl Microbiol Biotechnol 103:8631–8645. https://doi.org/10.1007/s00253-019-10078-9
61. Chen YT, Zeng Y, Wang HZ et al (2020) Different interspecies electron transfer patterns during mesophilic and thermophilic syntrophic propionate degradation in chemostats. Microb Ecol 80:120–132. https://doi.org/10.1007/s00248-020-01485-x
62. Wang HZ, Yan YC, Gou M et al (2019) Response of propionate-degrading methanogenic microbial communities to inhibitory conditions. Appl Biochem Biotechnol 189:233–248. https://doi.org/10.1007/s12010-019-03005-1
63. Yi Y, Wang HZ, Chen YT et al (2020) Identification of novel butyrate- and acetate-oxidizing bacteria in butyrate-fed mesophilic anaerobic chemostats by DNA-based stable isotope probing. Microb Ecol 79:285–298. https://doi.org/10.1007/s00248-019-01400-z
64. Stieb M, Schink B (1986) Anaerobic degradation of isovalerate by a defined methanogenic coculture. Arch Microbiol 144:291–295. https://doi.org/10.1007/BF00410965
65. Chen YT, Zeng Y, Li J et al (2020) Novel syntrophic isovalerate-oxidizing bacteria and their energetic cooperation with methanogens in methanogenic chemostats. Environ Sci Technol 54(15):9618–9628. https://doi.org/10.1021/acs.est.0c01840
66. Perego C, König R, Cuomo M et al (2025) *Shewanella oneidensis* and *Methanosarcina barkeri* augmentation and conductive material effects on long-term anaerobic digestion performance. Biotechnol Biofuels Bioprod 18:10. https://doi.org/10.1186/s13068-025-02609-6
67. He H, Zeng Y, Dong H et al (2025) Enrichment of *Methanothrix* species via riboflavin-loaded granular activated carbon in anaerobic digestion of high-concentration brewery wastewater amidst continuous inoculation of *Methanosarcina barkeri*. Water Res 268:122739. https://doi.org/10.1016/j.watres.2024.122739
68. Tang YQ, Shigematsu T, Morimura S et al (2015) Dynamics of the microbial community during continuous methane fermentation in continuously stirred tank reactors. J Biosci Bioeng 119:375–383. https://doi.org/10.1016/j.jbiosc.2014.09.014
69. Zhang H, Yuan W, Dong Q et al (2022) Integrated multi-omics analyses reveal the key microbial phylotypes affecting anaerobic digestion performance under ammonia stress. Water Res 213:118152. https://doi.org/10.1016/j.watres.2022.118152
70. Feng G, Zeng Y, Wang HZ et al (2023) *Proteiniphilum* and *Methanothrix harundinacea* became dominant acetate utilizers in a methanogenic reactor operated under strong ammonia stress. Front Microbiol 13:1098814. https://doi.org/10.3389/fmicb.2022.1098814
71. Xu P, Liu H, Liu C et al (2024) Syntrophic methane production from volatile fatty acids: focus on interspecies electron transfer. Sci Total Environ 946:174410. https://doi.org/10.1016/j.scitotenv.2024.174410

72. Yao J, Zeng Y, Wang MX et al (2021) Energy availability determines strategy of microbial amino-acid synthesis in volatile fatty acids-fed anaerobic methanogenic chemostats. Front Microbiol 12:744834. https://doi.org/10.3389/fmicb.2021.744834
73. Yao J, Zhang Q, Gou G et al (2024) High synthetic cost-amino acids reduce member interactions of acetate-degrading methanogenic microbial community. Front Microbiol 15: 1368215. https://doi.org/10.3389/fmicb.2024.1368215
74. Bhattarai B, Bhattacharjee AS, Coutinho FH et al (2024) Investigating the viral ecology and contribution to the microbial ecology in full-scale mesophilic anaerobic digesters. Chemosphere 349:140743. https://doi.org/10.1016/j.chemosphere.2023.140743
75. Zeng Y, Zhong XZ, Chen YT et al (2025) Characteristics of phages and their interactions with hosts in anaerobic reactors. Environ Microbiol 27:e70040. https://doi.org/10.1111/1462-2920.70040
76. Xu Z, Yuan H, Li X (2021) Anaerobic bioconversion efficiency of rice straw in continuously stirred tank reactor systems applying longer hydraulic retention time and higher load: one-stage vs. two-stage. Bioresour Technol 321:124206. https://doi.org/10.1016/j.biortech.2020.124206
77. Liu C, Wachemo AC, Tong H et al (2018) Biogas production and microbial community properties during anaerobic digestion of corn stover at different temperatures. Bioresour Technol 261:93–103. https://doi.org/10.1016/j.biortech.2017.12.076
78. Lu Y, Sun Y, Zhang L et al (2023) Substance bioconversion, hydrolases activity, and metagenomic analysis to unravel the enhanced biomethanation of corn stover with urea-hydrothermal pretreatment. J Environ Manag 333:117466. https://doi.org/10.1016/j.jenvman.2023.117466
79. Li J, Wu Y, Zhao J et al (2022) Bioaugmented degradation of rice straw combining two novel microbial consortia and lactic acid bacteria for enhancing the methane production. Bioresour Technol 344:126148. https://doi.org/10.1016/j.biortech.2021.126148
80. Basak B, Patil SM, Kumar R et al (2022) Syntrophic bacteria- and *Methanosarcina*-rich acclimatized microbiota with better carbohydrate metabolism enhances biomethanation of fractionated lignocellulosic biocomponents. Bioresour Technol 360:127602. https://doi.org/10.1016/j.biortech.2022.127602
81. Lu Y, Yuan H, Zuo X et al (2021) Biomethane yield, physicochemical structures, and microbial community characteristics of corn stover pretreated by urea combined with mild temperature hydrotherm. Polymers 13:2207. https://doi.org/10.3390/polym13132207
82. Wei Y, Gao Y, Yuan H et al (2022) Effects of organic loading rate and pretreatments on digestion performance of corn stover and chicken manure in completely stirred tank reactor (CSTR). Sci Total Environ 815:152499. https://doi.org/10.1016/j.scitotenv.2021.152499
83. Liu CM, Yuan HR, Zou DX et al (2015) Improving biomethane production and mass bioconversion of corn stover anaerobic digestion by adding NaOH retreatment and trace elements. Biomed Res Int 125241. https://doi.org/10.1155/2015/125241
84. Janke L, Weinrich S, Leite AF et al (2017) Optimization of semi-continuous anaerobic digestion of sugarcane straw codigested with filter cake: effects of macronutrients supplementation on conversion kinetics. Bioresour Technol 245:35–43. https://doi.org/10.1016/j.biortech.2017.08.084
85. Shen F, Tian L, Yuan H et al (2013) Improving the mixing performances of rice straw anaerobic digestion for higher biogas production by computational fluid dynamics (CFD) simulation. Appl Biochem Biotechnol 171:626–642. https://doi.org/10.1007/s12010-013-0375-z
86. Tian W, Chen Y, Shen Y et al (2020) Effects of hydrothermal pretreatment on the mono- and co-digestion of waste activated sludge and wheat straw. Sci Total Environ 732:139312. https://doi.org/10.1016/j.scitotenv.2020.139312
87. Chen X, Yuan H, Zou D et al (2015) Improving biomethane yield by controlling fermentation type of acidogenic phase in two-phase anaerobic co-digestion of food waste and rice straw. Chem Eng J 273:254–260. https://doi.org/10.1016/j.cej.2015.03.067

88. Witaszek K, Pilarski K, Niedbała G et al (2020) Energy efficiency of comminution and extrusion of maize substrates subjected to methane fermentation. Energies 13(8):1887. https://doi.org/10.3390/en13081887
89. Fang Y, Si B, Qiu J et al (2020) Bioconversion of bamboo shoot shell to methane assisted by microwave irradiation and fungus metabolism. Sci Total Environ 724:138268. https://doi.org/10.1016/j.scitotenv.2020.138268
90. Domanski J, Marchut-Mikolajczyk O, Cieciura-Wloch W et al (2020) Production of methane, hydrogen and ethanol from *Secale cereale* L. straw pretreated with sulfuric acid. Molecules 25(4):1013. https://doi.org/10.3390/molecules25041013
91. Bianco F, Senol H, Papirio S (2021) Enhanced lignocellulosic component removal and biomethane potential from chestnut shell by a combined hydrothermal-alkaline pretreatment. Sci Total Environ 762:144178. https://doi.org/10.1016/j.scitotenv.2020.144178
92. Li J, Wachemo AC, Yu GQ et al (2020) Enhanced anaerobic digestion performance of corn stalk pretreated with freezing-thawing and ammonia: an experimental and theoretical study. J Clean Prod 247:119112. https://doi.org/10.1016/j.jclepro.2019.119112
93. Luo J, Li J, Zhang L, Li N et al (2020) Effects of different potassium and nitrogen pretreatment strategies on anaerobic digestion performance of rice straw. RSC Adv 10:25547–25556. https://doi.org/10.1039/D0RA02136A
94. Shah TA, Ullah R (2019) Pretreatment of wheat straw with ligninolytic fungi for increased biogas productivity. Int J Environ Sci Technol 16:7497–7508. https://doi.org/10.1007/s13762-019-02277-8
95. Wei Y, Lan Y, Li X et al (2022) Effect of wheat straw pretreated with liquid fraction of digestate from different substrates on anaerobic digestion performance and microbial community characteristics. Sci Total Environ 818:151764. https://doi.org/10.1016/j.scitotenv.2021.151764
96. Zhao X, Zheng Z, Cai Y et al (2020) Accelerated biomethane production from lignocellulosic biomass: pretreated by mixed enzymes secreted by *Trichoderma viride* and *Aspergillus* sp. Bioresour Technol 309:123378. https://doi.org/10.1016/j.biortech.2020.123378
97. Huang WB, Yuan HR, Li XJ (2020) Multi-perspective analyses of rice straw modification by *Pleurotus ostreatus* and effects on biomethane production. Bioresour Technol 296:122365. https://doi.org/10.1016/j.biortech.2019.122365
98. Yuan H, Guan R, Cao C et al (2022) Combined modifications of CaO and liquid fraction of digestate for augmenting volatile fatty acids production from rice straw: microbial and proteomics insights. Bioresour Technol 364:128089. https://doi.org/10.1016/j.biortech.2022.128089
99. Li Z, Zhang J, Kong C et al (2021) Methane production from wheat straw pretreated with CaO_2/cellulase. RSC Adv 11:20541. https://doi.org/10.1039/D1RA02437J
100. Hou T, Song H, Cui Z et al (2024) Nanobubble technology to enhance energy recovery from anaerobic digestion of organic solid wastes: potential mechanisms and recent advancements. Sci Total Environ 931:172885. https://doi.org/10.1016/j.scitotenv.2024.172885
101. Dehariya D, Eswar K, Tarafdar A et al (2023) Recent advances of nanobubble-based systems in cancer therapeutics: a review. Biomed Eng Adv 5:100080. https://doi.org/10.1016/j.bea.2023.100080
102. Marcelino KR, Ling L, Wongkiew S et al (2023) Nanobubble technology applications in environmental and agricultural systems: opportunities and challenges. Crit Rev Environ Sci Technol 53(14):1378–1403. https://doi.org/10.1080/10643389.2022.2136931
103. Wang X, Lei Z, Shimizu K et al (2021) Recent advancements in nanobubble water technology and its application in energy recovery from organic solid wastes towards a greater environmental friendliness of anaerobic digestion system. Renew Sustain Energy Rev 145:111074. https://doi.org/10.1016/j.rser.2021.111074
104. Pei L, Song Y, Chen G et al (2024) Enhancement of methane production from anaerobic digestion of Erigeron canadensis via O_2-nanobubble water supplementation. Chemosphere 354:141732. https://doi.org/10.1016/j.chemosphere.2024.141732

105. Wang E, Sun H, Xing F et al (2024) Metagenomic analysis reveals metabolic mechanism of enhancing lignocellulosic anaerobic digestion mediated by CO_2/O_2-nanobubble water. Bioresour Technol 414:131622. https://doi.org/10.1016/j.biortech.2024.131622
106. Zhu Y, Lyu T, Li D et al (2024) Process mechanisms of nanobubble technology enhanced hydrolytic acidification of anaerobic digestion of lignocellulosic biomass. Chem Eng J 480: 147956. https://doi.org/10.1016/j.cej.2023.147956
107. Song H, Hou T, Jiao Y et al (2023) Supplementation of CO_2-nanobubble water to enhance the methane production from anaerobic digestion of corn straw. Chemosphere 313:137613. https://doi.org/10.1016/j.chemosphere.2022.137613
108. He C, Song H, Liu L et al (2022) Enhancement of methane production by anaerobic digestion of corn straw with hydrogen-nanobubble water. Bioresour Technol 344:126220. https://doi.org/10.1016/j.biortech.2021.126220
109. Wang X, Yuan T, Guo Z et al (2020) Enhanced hydrolysis and acidification of cellulose at high loading for methane production via anaerobic digestion supplemented with high mobility nanobubble water. Bioresour Technol 297:122499. https://doi.org/10.1016/j.biortech.2019.122499
110. Wang X, Yuan T, Lei Z et al (2020) Supplementation of O_2-containing gas nanobubble water to enhance methane production from anaerobic digestion of cellulose. Chem Eng J 398: 125652. https://doi.org/10.1016/j.cej.2020.125652
111. Ho TH, Yang X, Nie J et al (2020) Effect of nanobubble water on anaerobic methane production from lignin. Res Chem Intermed 46:4767–4780. https://doi.org/10.1007/s11164-020-04250-4
112. Yang X, Nie J, Wei Y et al (2020) Simultaneous enhancement on lignin degradation and methane production from anaerobic co-digestion of waste activated sludge and alkaline lignin supplemented with N_2-nanobubble water. Bioresour Technol Rep 11:100470. https://doi.org/10.1016/j.biteb.2020.100470
113. Wang E, Sun H, Chen P et al (2023) Two-step anaerobic digestion of rice straw with nanobubble water. Bioresour Technol 376:128928. https://doi.org/10.1016/j.biortech.2023.128928
114. Wang X, Lei Z, Zhang Z et al (2023) Use of nanobubble water bioaugmented anaerobically digested sludge for high-efficacy energy production from high-solids anaerobic digestion of corn straw. Sci Total Environ 863:160825. https://doi.org/10.1016/j.scitotenv.2022.160825
115. Wang X, Lei Z, Shimizu K et al (2020) Improved methane production from corn straw using anaerobically digested sludge pre-augmented by nanobubble water. Bioresour Technol 311: 123479. https://doi.org/10.1016/j.biortech.2020.123479
116. Elsayed M, Abomohra AEF, Ai P et al (2019) Acetogenesis and methanogenesis liquid digestates for pretreatment of rice straw: a holistic approach for efficient biomethane production and nutrient recycling. Energy Convers Manag 195:447–456. https://doi.org/10.1016/j.enconman.2019.05.011
117. Tampio E, Marttinen S, Rintala J (2016) Liquid fertilizer products from anaerobic digestion of food waste: mass, nutrient and energy balance of four digestate liquid treatment systems. J Clean Prod 125:22–32. https://doi.org/10.1016/j.jclepro.2016.03.127
118. Horta C, Carneiro JP (2021) Use of digestate as organic amendment and source of nitrogen to vegetable crops. Appl Sci 12(1):248. https://doi.org/10.3390/app12010248
119. Asp H, Bergstrand KJ (2024) Nutrient utilization and growth of tomato crops fertilized with solid anaerobic digestate. Sci Hortic 329:112986. https://doi.org/10.1016/j.scienta.2024.112986
120. Tiong YW, Sharma P, Xu S et al (2024) Enhancing sustainable crop cultivation: the impact of renewable soil amendments and digestate fertilizer on crop growth and nutrient composition. Environ Pollut 342:123132. https://doi.org/10.1016/j.envpol.2023.123132
121. Jiang Y, Xie SH, Dennehy C et al (2020) Inactivation of pathogens in anaerobic digestion systems for converting biowastes to bioenergy: a review. Renew Sustain Energy Rev 120: 109654. https://doi.org/10.1016/j.rser.2019.109654

122. Sani MNH, Amin M, Bergstrand KJ et al (2025) Harnessing biostimulants from biogas digestates for high-value resource recovery: a review. Environ Chem Lett 23:139–164. https://doi.org/10.1007/s10311-024-01801-8
123. de Oliveira Fernandes MA, Baêta BEL, Adarme OFH et al (2025) LCA-based carbon footprint analysis of anaerobic digestion of coffee husk waste. Renew Sustain Energy Rev 207:114993. https://doi.org/10.1016/j.rser.2024.114993
124. Wang Z, Li J, Yan B et al (2024) Thermochemical processing of digestate derived from anaerobic digestion of lignocellulosic biomass: a review. Renew Sustain Energy Rev 199: 114518. https://doi.org/10.1016/j.rser.2024.114518
125. Bona D, Bertoldi D, Borgonovo G et al (2023) Evaluating the potential of hydrochar as a soil amendment. Waste Manag 159:75–83. https://doi.org/10.1016/j.wasman.2023.01.024
126. Shan G, Li W, Bao S et al (2023) Evaluating the aqueous phase obtained from hydrothermal carbonization of municipal sludge as possible liquid fertilizer for plant growth: an analysis of heavy metals and their molecular composition. J Clean Prod 404:136989. https://doi.org/10.1016/j.jclepro.2023.136989
127. Elsayed M, Ran Y, Ai P et al (2020) Innovative integrated approach of biofuel production from agricultural wastes by anaerobic digestion and black soldier fly larvae. J Clean Prod 263: 121495. https://doi.org/10.1016/j.jclepro.2020.121495
128. Fu SF, Wang DH, Xie Z et al (2022) Producing insect protein from food waste digestate via black soldier fly larvae cultivation: a promising choice for digestate disposal. Sci Total Environ 830:154654. https://doi.org/10.1016/j.scitotenv.2022.154654
129. Kofroňová J, Melliti A, Vurm R (2024) Biogas digestate and sewage sludge as suitable feeds for black soldier fly (*Hermetia illucens*) larvae. Toxics 12(6):414. https://doi.org/10.3390/toxics12060414
130. Brodeur D, Deschamps MH, Vandenberg G et al (2024) Integration of pretreated crop residues to improve the valorization of biogas digestate by the black soldier fly (*Hermetia illucens* L.; Diptera: Stratiomyidae) larvae. Waste Biomass Valoriz 15(5):2671–2685. https://doi.org/10.1007/s12649-023-02340-z
131. Guan D, Zhao J, Wang Y et al (2024) A critical review on sustainable management and resource utilization of digestate. Process Saf Environ Prot 183:339–354. https://doi.org/10.1016/j.psep.2024.01.029
132. Ho SH, Nagarajan D, Ren NQ et al (2018) Waste biorefineries – integrating anaerobic digestion and microalgae cultivation for bioenergy production. Curr Opin Biotechnol 50: 101–110. https://doi.org/10.1016/j.copbio.2017.11.017
133. Uggetti E, Sialve B, Latrille E et al (2014) Anaerobic digestate as substrate for microalgae culture: the role of ammonium concentration on the microalgae productivity. Bioresour Technol 152:437–443. https://doi.org/10.1016/j.biortech.2013.11.036
134. Xia A, Murphy JD (2016) Microalgal cultivation in treating liquid digestate from biogas systems. Trends Biotechnol 34(4):264–275. https://doi.org/10.1016/j.tibtech.2015.12.010
135. Chuka-ogwude D, Ogbonna JC, Moheimani NR (2021) Depth optimization of inclined thin layer photobioreactor for efficient microalgae cultivation in high turbidity digestate. Algal Res 60:102509. https://doi.org/10.1016/j.algal.2021.102509
136. Al-Mallahi J, Ishii K (2022) Attempts to alleviate inhibitory factors of anaerobic digestate for enhanced microalgae cultivation and nutrients removal: a review. J Environ Manag 304: 114266. https://doi.org/10.1016/j.jenvman.2021.114266
137. Chong CC, Cheng YW, Ishak S et al (2022) Anaerobic digestate as a low-cost nutrient source for sustainable microalgae cultivation: a way forward through waste valorization approach. Sci Total Environ 803:150070. https://doi.org/10.1016/j.scitotenv.2021.150070

Adv Biochem Eng Biotechnol (2026) 196: 337–362
https://doi.org/10.1007/10_2026_326

Published online: 24 June 2026

Valorization of Agricultural Residues for Biohydrogen Production via Dark Fermentation

Cao Guangli, Wu Jiwen, Liu Bingfeng, Fan Shengqiang, Xie Guojun, Sun Jiazheng, Shanshan Yang, and Guan Yujing

Contents

C. Guangli (✉), W. Jiwen, L. Bingfeng, F. Shengqiang, X. Guojun, S. Jiazheng, S. Yang, and G. Yujing
State Key Laboratory of Urban-Rural Water Resources and Environment, School of Environment, Harbin Institute of Technology, Harbin, China
e-mail: caogl@hit.edu.cn; shanshanyang@hit.edu.cn

Abstract Biohydrogen has emerged as a promising alternative energy carrier due to its renewability and environmental friendliness. In recent decades, significant progress has been made in biohydrogen production from organic waste, particularly agricultural residues, which represent one of the most abundant renewable resources globally. This chapter provides a comprehensive overview of the latest advances in the biotechnological conversion of agricultural biomass into biohydrogen, with a particular focus on dark fermentation as a sustainable and efficient approach. It begins by addressing the structural complexity of agricultural waste and the associated challenges in microbial conversion. The core mechanisms of dark fermentation are discussed, along with key microbial communities involved and the primary factors affecting biohydrogen yield. The chapter further evaluates pretreatment methods, fermentation processes, and emerging strategies for improving performance. In addition, it highlights the importance of machine learning-based modeling, process integration, techno-economic assessment, and life cycle analysis in advancing the practical application of dark fermentation for biohydrogen production from agricultural biomass. The purpose is to provide the theoretical and technical basis for the valorization of agricultural residue into biohydrogen.

Graphical Abstract

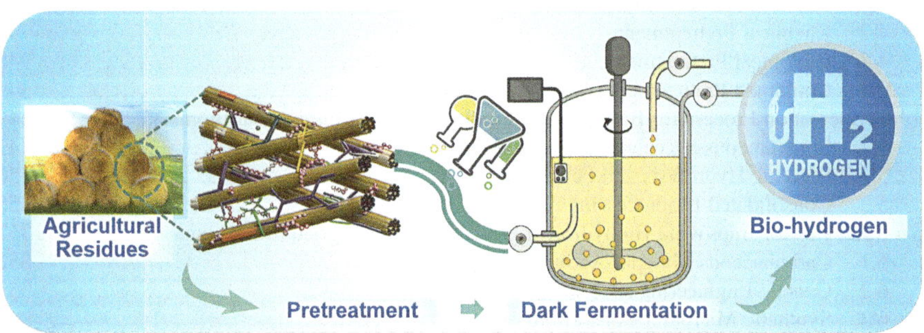

Keywords Agricultural residue, Bioconversion, Biohydrogen, Dark fermentation, Pretreatment

1 Introduction

The global transition toward sustainable energy systems has intensified the demand for clean, renewable energy sources. Hydrogen, as a carbon-neutral energy carrier, holds significant promise due to its high energy density and its potential to decarbonize sectors such as transportation, industry, and power generation. Traditionally, hydrogen production has relied heavily on fossil fuel-based processes such as steam methane reforming and coal gasification, which are energy intensive and contribute substantially to greenhouse gas emissions. Therefore, developing sustainable and environmentally friendly hydrogen production technologies is crucial for achieving global climate goals and promoting a low-carbon economy.

Biohydrogen production refers to microbial processes that utilize renewable biomass to generate and release molecular hydrogen, and it encompasses several distinct production pathways. Based on the microorganisms involved and their metabolic mechanisms, biohydrogen production can be classified into biophotolysis, photofermentation, and dark fermentation [1–3] (as shown in Fig. 1). Among the technologies, biophotolysis has a relatively low hydrogen yield. As a result, current research and applications primarily focus on photofermentation and dark fermentation. Compared to photofermentation, dark fermentation offers several advantages, including independence from light conditions, faster microbial growth and metabolic rates, higher and more stable hydrogen production rates, a wide range of available substrates, and lower costs. Consequently, dark fermentation has emerged as a major research focus in the field of biohydrogen production and a critical direction for industrial application breakthroughs. It has rapidly developed into one of the most promising and environmentally friendly methods for sustainable hydrogen production [4–7].

Agricultural residues, primarily composed of cellulose, hemicellulose, and lignin, represent the abundant renewable organic resource on the earth. It is generated in vast quantities through agricultural activities and agro-industrial processes. The valorization of agricultural residues into value-added products not only contributes to waste management and environmental protection but also aligns with the

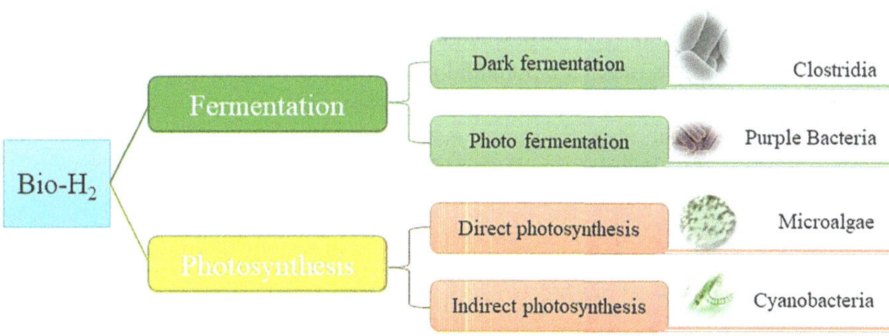

Fig. 1 Biohydrogen production technology

principles of a circular bioeconomy. Utilizing agricultural residues for biohydrogen production addresses critical issues related to feedstock availability, sustainability, and cost-effectiveness. The conversion of underutilized biomass resources into biohydrogen adds economic value to the agricultural sector, promoting rural development and sustainable resource utilization. However, the recalcitrant structure of agricultural biomass, primarily due to the presence of lignin and the crystalline characteristics of cellulose, poses significant challenges to efficient microbial degradation and biohydrogen production, necessitating effective pretreatment strategies.

This chapter aims to systematically investigate the sustainable conversion of agricultural biomass into biohydrogen via dark fermentation. Firstly, the physicochemical properties of agricultural biomass and its key obstacles in microbial transformation are analyzed. Subsequently, the basic principles and metabolic mechanisms of dark fermentation are elaborated, with a focus on analyzing key microbial metabolic pathways and their influencing factors on biohydrogen production. The chapter also focuses on various pretreatment strategies used to enhance the degradability of lignocellulosic substrates and introduces various emerging technologies currently being promoted to improve biohydrogen yield. Finally, by integrating the latest research progress and looking forward to future development directions, this chapter aims to provide a theoretical basis and technical support for building a sustainable biohydrogen production system and promote the high-value application of agricultural biomass in renewable energy systems.

2 Characteristics of Agricultural Biomass

2.1 Sources of Agricultural Biomass

Globally, agricultural biomass is primarily derived from crop residues such as rice straw, wheat stalks, corn husks, and sugarcane leaves, as well as agro-industrial by-products including fruit peels, oilseed cakes, molasses, and bagasse [8, 9]. Other important sources include animal manure, which is often underutilized but abundant in quantity. These biomass types vary in composition, biodegradability, and energy potential, requiring tailored pretreatment and conversion strategies to maximize biohydrogen production. The widespread availability, renewability, and low cost of agricultural biomass make it a potential raw material for sustainable biohydrogen production. The effective utilization of these resources not only provides a clean energy alternative but also addresses issues related to waste management and environmental sustainability in rural and agricultural sectors.

2.2 Composition of Agricultural Biomass

Agricultural biomass is primarily composed of three major polymeric constituents: cellulose, hemicellulose, and lignin [10]. Cellulose, accounting for approximately

30–50% of the biomass structure, is a linear homopolymer of β-1,4-linked glucose units, forming highly ordered crystalline microfibrils that provide mechanical strength and rigidity to plant cell walls. Hemicellulose, which constitutes about 20–35%, is a heterogeneous group of polysaccharides composed of various sugar monomers such as xylose, mannose, galactose, and arabinose. Unlike cellulose, hemicellulose has an amorphous structure and is more easily hydrolyzed. Lignin, making up 15–30% of lignocellulosic biomass, is a complex aromatic polymer formed through the radical coupling of phenylpropanoid units. It acts as a binding agent, embedding cellulose and hemicellulose in a three-dimensional matrix that enhances plant structural integrity and provides resistance against microbial attack and environmental stresses. The complex and tightly cross-linked nature of these components, particularly the intimate association between lignin and polysaccharides, poses significant challenges for efficient biomass deconstruction and processes of biohydrogen conversion.

2.3 *Challenges in Agricultural Biomass Utilization*

Despite its abundance and promising characteristics, the efficient utilization of agricultural biomass presents several fundamental challenges. One of the primary obstacles is its structural recalcitrance, which stems from the robust lignin-carbohydrate complex that limits microbial and enzymatic access to fermentable sugars. The crystalline structure of cellulose further impedes enzymatic hydrolysis, reducing the efficiency of biomass conversion processes. Although hemicellulose is comparatively easier to degrade, it contains a high proportion of pentose sugars, which are not readily metabolized by most conventional hydrogen-producing microorganisms, thus limiting the synthesis rate and conversion yield of biohydrogen. Additionally, agricultural biomass exhibits considerable heterogeneity in chemical composition and physical properties across different feedstock types, seasons, and geographic regions, making standardization and process optimization difficult. Such variability significantly impacts pretreatment effectiveness and consistency of downstream fermentation outcomes. Furthermore, several pretreatment techniques intended to enhance biomass digestibility can generate inhibitory by-products, including furfural, hydroxymethylfurfural (HMF), and phenolic compounds, all of which can severely hamper microbial performance during dark fermentation. Overall, overcoming the physicochemical barriers associated with agricultural biomass is critical to unlocking its full potential as a sustainable feedstock for biohydrogen production.

3 Fundamentals of Dark Fermentation for Biohydrogen Production

3.1 Mechanism and Pathways of Dark Fermentation

Dark fermentative hydrogen production refers to the process by which hydrogen-producing microorganisms decompose and metabolize organic biomass to generate biohydrogen under anaerobic and non-illuminated conditions [11, 12]. This process involves two key oxygen-sensitive enzymes: hydrogenase and formate hydrogen lyase. Hydrogenases (HYDs) are further categorized into [FeFe]-hydrogenases and [NiFe]-hydrogenases, where [FeFe]-hydrogenases primarily catalyze hydrogen production, while [NiFe]-hydrogenases mainly catalyze hydrogen oxidation reactions [13, 14]. Due to the absence of a complete electron transport chain in microbial cells, excess electrons generated during dehydrogenation in anaerobic dark fermentation must be rapidly neutralized to maintain intracellular redox balance and ensure normal metabolic activity. The reduction of protons to molecular hydrogen by hydrogenases serves as the primary regulatory mechanism for disposing of surplus electrons during anaerobic fermentation [15]. Formate hydrogen lyase (FHL) is a multienzyme complex widely distributed in facultative anaerobic bacteria, notably *Escherichia coli*. It catalyzes the disproportionation of formate into carbon dioxide and molecular hydrogen, thereby serving as a crucial mechanism for redox balancing under anaerobic conditions. The complex is composed of formate dehydrogenase, which oxidizes formate to CO_2 and releases electrons, and a [NiFe]-hydrogenase, which couples these electrons with protons to generate H_2. Through this coordinated activity, FHL not only disposes of excess reducing equivalents but also contributes to microbial hydrogen metabolism and has attracted interest for biohydrogen production research [16].

The biochemical mechanisms of dark fermentative biohydrogen production can be summarized into two pathways: the pyruvate decarboxylation hydrogen production pathway and the NADH oxidation hydrogen production pathway, as shown in Fig. 2. Pyruvate decarboxylation represents the main direct hydrogen-producing

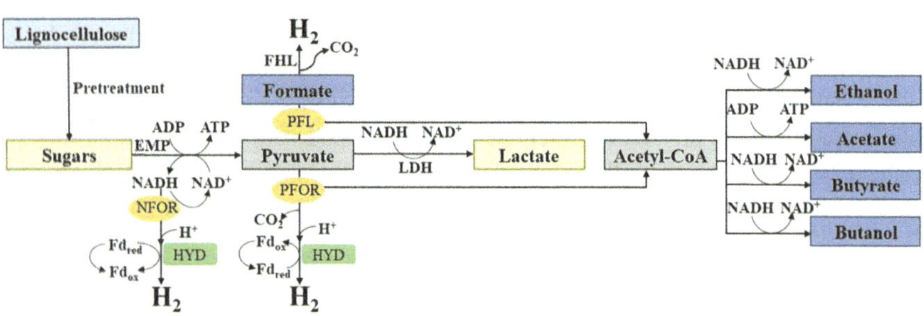

Fig. 2 Typical metabolic pathways for converting biomass substrate into hydrogen during dark fermentation (*FHL* formate hydrogenlyase, *LDH* lactate dehydrogenase, *[Fe–Fe]-Hyd* [Fe–Fe]-hydrogenase) [17, 18]

route within cells and is divided into two sub-pathways: the pyruvate-formate lyase (PFL) pathway and the pyruvate-ferredoxin oxidoreductase (PFOR) pathway. Initially, glucose is metabolized into pyruvate via the Embden-Meyerhof-Parnas (EMP) glycolysis pathway, generating NADH as reducing power. In facultative anaerobic hydrogen-producing bacteria, the PFL pathway predominates, where pyruvate-formate lyase catalyzes the cleavage of pyruvate into formate. Subsequently, FHL, comprising formate dehydrogenase (FDH) and hydrogenase, disproportionates formate into CO_2 and H_2 by direct electron transfer, independent of NAD^+. In strict anaerobic hydrogen-producing bacteria such as *Clostridium*, pyruvate is decarboxylated to acetyl-CoA via PFOR, releasing electrons that are transferred to hydrogenase through ferredoxin. The resulting acetyl-CoA is further metabolized into products such as acetate, butyrate, and ethanol. In addition, pyruvate metabolism can also generate competing by-products, including lactate and propionate, which consume reducing equivalents required for hydrogen generation [15, 17, 18].

Theoretically, each mole of glucose can yield up to 12 mol of biohydrogen in an ideal dark fermentation system (Eq. 1). However, in practice, the formation of volatile fatty acids (VFAs) – including acetate, propionate, butyrate, ethanol, butanol, and acetone – consumes a significant portion of the available reducing power, resulting in a hydrogen yield much lower than the theoretical maximum. For instance, when acetate is the sole fermentation product, 1 mol of glucose produces 4 mol of hydrogen (Eq. 2), whereas if butyrate is the end product, only 2 mol of hydrogen are produced per mole of glucose (Eq. 3) [19]. In actual dark fermentation systems, factors such as the metabolic inhibition of hydrogen-producing bacteria and acidification caused by VFAs generally limit the practical hydrogen yield to less than 2.5 mol H_2/mol glucose [17]. Currently, it is considered economically viable for dark fermentation systems if 60–80% of the energy contained in the substrate is converted into hydrogen [20]:

$$C_6H_{12}O_6 + 6H_2O \rightarrow 6CO_2 + 12H_2 \tag{1}$$

$$C_6H_{12}O_6 + 2H_2O \rightarrow 2CO_2 + 2CH_3COOH + 4H_2 \tag{2}$$

$$C_6H_{12}O_6 \rightarrow 2CO_2 + CH_3CH_2CH_2COOH + 2H_2 \tag{3}$$

3.2 Key Microorganisms Involved

The dark fermentation of agricultural biomass for hydrogen production first requires microorganisms with cellulose degradation ability. *Clostridium thermocellum* has been extensively studied in this context. Bayer et al. introduced the concept of the cellulosome and investigated its composition and spatial organization in *C. thermocellum* [21]. In addition to *C. thermocellum*, other well-characterized anaerobic cellulolytic bacteria include *Clostridium josui*, *Ruminococcus albus*, *Clostridium cellulovorans*, and *Ruminococcus flavefaciens* [22].

Given that microbial hydrogen-producing capacity remains a key bottleneck in biohydrogen development, isolating high-performance strains from natural environments has become a central research focus. For instance, Xing et al. isolated *Ethanoligenens harbinense* YUAN-3 from sludge, achieving a hydrogen yield of 2.81 mol H_2/mol glucose [23]. *Enterobacter cloacae* IIT-BT, purified by Kumar, yielded 2.7 mol H_2/mol glucose [24]. Kadar et al. [25] evaluated *Caldicellulosiruptor saccharolyticus* on various sugars: it produced 2.24 mol H_2/mol xylose (maximum rate: 11.3 mmol/L h), 2.5 mol H_2/mol glucose (10.7 mmol/L h), and 2.32 mol H_2/mol of glucose-xylose mixtures (9.2 mmol/L h). Similarly, Cao et al. [26] isolated a thermophile, *Thermoanaerobacterium thermosaccharolyticum* W16, which efficiently fermented both glucose and xylose, yielding up to 2.42 mol H_2/mol glucose and 2.19 mol H_2/mol xylose, with hydrogen production rates reaching 12.9 and 10.7 mmol/L/h, respectively. With mixed sugars, yields ranged from 2.23 to 2.37 mol H_2/mol substrate, with rates of 11.2–12.7 mmol/L/h. Despite these advances, pentose-fermenting strains remain limited. Currently, only a few thermophilic species are known to co-ferment hexoses and pentoses effectively. Enhancing pentose utilization is crucial for expanding the substrate spectrum and improving the efficiency of lignocellulosic hydrolysate conversion.

Screening strains that can directly degrade lignocellulosic biomass and have efficient hydrogen production capacity is still a hot research topic at present. It is difficult for a single wild-type strain to efficiently complete the three main processes of cellulase secretion, hydrolysis, and hydrogen production without affecting efficiency. However, fermentation strategies such as engineering modification or mixed culture may achieve better hydrogen production than pure cultivation of wild-type strains. The co-culture system constructed by Liu et al. using cellulose-degrading bacterium JN4 and efficient hydrogen-producing bacterium GD17 significantly increased hydrogen production by two times compared to the pure culture system of cellulose-degrading bacterium [27]. Cheng et al. developed a co-culture hydrogen production system using sugarcane bagasse as the carbon source, employing *thermophilic Clostridium* ATCC 27405 in combination with a thermophilic anaerobic bacterium SCUT27. They found that the hydrogen production in the co-culture system increased by 13.7% and 25.05%, respectively, compared to the pure culture of the two bacteria [28]. Therefore, isolating microbial strains with efficient cellulose degradation and hydrogen production capabilities from natural habitats is of great significance for constructing feasible consolidated bioprocessing (CBP) systems and promoting the development of biohydrogen production.

3.3 Main Parameters Affecting Biohydrogen Production

The production of biohydrogen through dark fermentation is influenced by a series of physical, chemical, and biological parameters, which determine the efficiency, yield, and stability of the process. The most critical factors include substrate type and concentration, pH value, temperature, hydraulic retention time (HRT), inoculum source, and inhibitory compounds.

3.3.1 Inoculum Selection

The choice of inoculum significantly impacts the efficiency of dark fermentative hydrogen production. Strict anaerobes, particularly species from the genus *Clostridium*, are well known for their high hydrogen-producing capabilities [29]. These bacteria utilize substrate-level phosphorylation for energy generation and lack cytochrome systems, channeling excess electrons toward hydrogen production. Facultative anaerobes such as *Enterobacter* and *Klebsiella* offer operational robustness by tolerating occasional oxygen exposure, which can be advantageous in nonideal industrial settings [30]. However, they typically produce lower hydrogen yields compared to strict anaerobes. Thermophilic bacteria, such as *Thermoanaerobacterium thermosaccharolyticum*, thrive at elevated temperatures (50–60°C), which accelerate metabolic rates, reduce contamination risks, and potentially enhance hydrogen production [31]. In engineering practice, stable operation is typically achieved by employing mixed microbial consortia or inocula acclimated from waste-activated sludge or anaerobic digestate [32]. The selection of appropriate strains, or microbial consortia that are suitable for specific substrates and operating conditions, is essential for maximizing biohydrogen yields.

3.3.2 Substrate Characteristics

The characteristics of fermentation substrate make a critical difference in dark fermentative biohydrogen production [33, 34]. Simple carbohydrates such as glucose and sucrose are readily fermentable and generally yield higher hydrogen outputs. In contrast, pentose sugars like xylose and arabinose, abundant in lignocellulosic biomass hydrolysates, often result in lower hydrogen yields due to less favorable metabolic pathways [35]. Complex biomass such as corn stover and wheat straw requires pretreatment to break down their recalcitrant structure and release fermentable sugars.

Optimal substrate concentration is critical for maintaining high biohydrogen production rates. Excessive substrate levels or residual hydrolysis inhibitors such as furfural and phenolic compounds in the substrate can lead to substrate inhibition, osmotic stress, and metabolic shifts toward solventogenesis, resulting in decreased biohydrogen yields [36]. Conversely, insufficient substrate concentrations can limit microbial activity and hydrogen production. Balancing substrate availability with microbial metabolic capacity is essential for process optimization.

3.3.3 pH Value

Maintaining an optimal pH range is vital for sustaining microbial viability and enzymatic activity [37]. Dark fermentation generally favors slightly acidic conditions, with optimal pH values between 5.5 and 6.5. Deviations from this range can inhibit hydrogenase activity and shift metabolic pathways toward non-hydrogen-producing end products. Buffering capacity, achieved through the use of bicarbonate systems or controlled alkali addition, helps stabilize pH fluctuations caused by the accumulation of organic acids during fermentation [38].

3.3.4 Temperature

Temperature profoundly influences microbial growth rates, enzymatic activities, and metabolic pathways. Mesophilic conditions (35–40°C) are commonly used and suitable for most *Clostridium* and *Enterobacter* species, whereas thermophilic conditions (50–60°C) favor species like those in the genus *Thermoanaerobacterium.* Operating at thermophilic temperatures can enhance reaction kinetics, reduce contamination, and improve hydrogen yields, but requires more energy input for reactor heating [39].

3.3.5 Hydraulic Retention Time

HRT impacts the balance between microbial growth and substrate utilization. Short HRTs can lead to biomass washout and reduced hydrogen productivity, while excessively long HRTs may promote the accumulation of inhibitory by-products. Optimizing HRT is essential for maintaining a stable and productive fermentation system.

3.3.6 Inhibitory Compounds

During dark fermentation, on the one hand, inhibitory compounds generated during the hydrolysis and saccharification of lignocellulosic substrates can suppress the activity of hydrogen-producing bacteria; on the other hand, by-products such as volatile fatty acids (VFAs), alcohols, and solvents can accumulate and further inhibit microbial activity [40]. Compounds like lactate, ethanol, butanol, and formate consume reducing equivalents (e.g., NADH), diverting electron flow away from hydrogen production. Strategies to mitigate inhibitor formation include optimizing operational parameters, employing microbial consortia capable of synergistic substrate utilization, and integrating product recovery processes.

3.3.7 Hydrogen Partial Pressure

High hydrogen partial pressure in the reactor can thermodynamically inhibit further hydrogen evolution by shifting the redox balance unfavorably. Continuous gas removal via sparging, vacuum extraction, or membrane-based systems helps maintain low hydrogen concentrations and sustain high production rates.

In conclusion, the temperature, pH, and other conditions of the biohydrogen production system have a significant impact on hydrogen-producing bacteria. Therefore, it is necessary to select the optimal parameters and control them accurately according to the reaction conditions, in order to improve the stability and efficiency of biohydrogen production.

4 Pretreatment of Agricultural Biomass

The pretreatment of agricultural biomass has been extensively investigated as a crucial step to facilitate enzymatic hydrolysis and microbial fermentation by improving the accessibility of cellulose and hemicellulose while removing or altering lignin structures. Particularly for agricultural residues such as straws, various physical, chemical, and combined pretreatment methods have been developed to maximize the release of reducing sugars and enhance biohydrogen production efficiency.

4.1 Physical Pretreatment

Physical pretreatment methods, including mechanical milling, grinding, and steam explosion, are among the most straightforward approaches. Advanced physical methods, including high-energy radiation, microwave, ultrasonic, and gamma-ray treatments, have also been explored to disrupt the lignocellulosic structure [41]. These methods aim to reduce particle size, disrupt the fibrous structure, and increase the specific surface area of biomass. However, these methods often require significant energy inputs, leading to high operational costs. For instance, mechanical milling to achieve particle sizes below 400 mesh can enhance enzymatic hydrolysis efficiency but may consume up to 50% of the total energy input of the process [42].

Furthermore, the issue of toxicity inhibitor production also needs to be taken into consideration. Steam explosion pretreatment leads to the release of high concentrations of toxic compounds, including furfural, 5-HMF, acetic acid, formic acid, and various phenolic derivatives. These degradation products can severely inhibit microbial fermentation if not properly managed. Consequently, physical pretreatments are typically employed in conjunction with other methods to improve overall efficiency.

4.2 Chemical Pretreatment

Chemical pretreatment involves the application of chemical agents such as acids, alkalis, organic solvents, ionic liquids, metal chlorides, and plasma treatments to disrupt the lignocellulosic matrix of agricultural residues [43]. Among these, acid pretreatment using sulfuric acid (H_2SO_4) and hydrochloric acid (HCl) is the most widely studied [44]. Research conducted by Wang and Saratale explored the effects of dilute acid treatments on cornstalks and sorghum husks, respectively. They found that treating cornstalks with 0.6% HCl and sorghum husks with 0.2% H_2SO_4 resulted in the highest hydrogen yields [45, 46]. However, acid pretreatment also leads to the generation of inhibitory compounds, such as furfural, 5-hydroxymethylfurfural (5-HMF), and acetic acid, which can negatively impact subsequent fermentation processes [47].

Alkaline pretreatment, using agents such as sodium hydroxide (NaOH), calcium hydroxide ($Ca(OH)_2$), and sodium bicarbonate ($NaHCO_3$), targets the lignin and hemicellulose components by breaking ester and ether bonds, thereby reducing biomass crystallinity and increasing porosity. Among these, sodium hydroxide (NaOH) is the most commonly used for its efficacy. For instance, treating rice straw with 1% NaOH at 170°C enhanced biohydrogen production by 82.05% [48]. Calcium hydroxide offers a milder alternative, with benefits including low cost, safety, and environmental compatibility. Pretreatment with $Ca(OH)_2$ has been shown to disrupt lignocellulosic structures effectively, enhancing biodegradability and methane yield [49]. In recent years, methods for conducting alkali pretreatment of straw at low temperatures have been reported. At −15°C, alkali pretreatment effectively removes over 70% of lignin content from straw, significantly enhancing biomass digestibility. A biohydrogen yield as high as 22.08 mmol/L has been achieved following NaOH pretreatment under optimized conditions [50].

4.3 Biological Pretreatment

Biological pretreatment mainly relies on the metabolic activities of microorganisms, particularly fungi such as white-rot, soft-rot, and brown-rot species, to degrade lignin selectively. White-rot fungi are especially effective due to their production of ligninolytic enzymes like laccase, manganese peroxidase, and lignin peroxidase. These enzymes synergistically break down lignin, enhancing the accessibility of cellulose and hemicellulose. For example, *Phanerochaete chrysosporium* has been used to pretreat corn stover, achieving a 34.3% reduction in lignin content and a 20.3% increase in enzymatic hydrolysis efficiency over 15 days [51]. Similarly, *Coriolus versicolor* MTCC 138 achieved a 27.1% lignin removal in rice straw after 20 days of treatment [52]. Despite their environmental friendliness and low energy requirements, biological methods are limited by slow processing times and are not yet viable for large-scale industrial applications.

4.4 Combined Pretreatment

To overcome the limitations of individual pretreatment methods, combined pretreatment strategies are frequently employed. These approaches integrate physical, chemical, and sometimes biological methods to achieve synergistic effects. Dai et al. proposed a combination of bacterial treatment and an NaOH/urea solution. This integrated approach significantly enhanced the reducing sugar yield, achieving increases of 1.4-fold and 3.92-fold compared to NaOH/urea pretreatment alone and untreated rice straw, respectively [53]. An approach of microwave-assisted acid and alkali pretreatment techniques was used on lignocellulosic biomass pretreatment, achieving a reducing sugar yield of up to 35.6 g/L, indicating its effectiveness in enhancing enzymatic hydrolysis [54]. Teramoto et al. explored a combined pretreatment method involving ethanol cooking followed by ball milling on lignocellulosic materials [55]. This approach effectively reduced the particle size and crystallinity of the biomass, thereby improving its enzymatic digestibility.

Overall, pretreatment is an indispensable prerequisite for the efficient valorization of lignocellulosic biomass. By selectively removing or modifying lignin and enhancing the accessibility of cellulose and hemicellulose, pretreatment enables more efficient enzymatic hydrolysis and microbial fermentation processes. Continued innovation in pretreatment technologies, particularly in developing cost-effective, environmentally friendly methods, will be critical to advancing the sustainable conversion of lignocellulosic residues into biofuels and bioproducts.

5 Fermentation Process for Biohydrogen Production from Agricultural Biomass

In the context of lignocellulosic biohydrogen production, three principal fermentation strategies have been established: separate hydrolysis and fermentation (SHF), simultaneous saccharification and fermentation (SSF), and consolidated bioprocessing (CBP) [56].

5.1 Simultaneous Saccharification and Fermentation

Simultaneous saccharification and fermentation (SSF) integrates the enzymatic hydrolysis of complex polysaccharides and the microbial fermentation of resultant sugars into a single-step process [57]. The advantage of SSF lies in its process simplification, reduced operational time, and minimized contamination risk, as sugar release and consumption occur concurrently. However, a major challenge in SSF arises from the differing optimal conditions required for enzymatic hydrolysis and microbial hydrogen production. While cellulolytic enzymes generally exhibit maximal activity at moderately high temperatures and slightly acidic pH, fermentative hydrogen-producing microorganisms often prefer neutral pH and specific temperature ranges, depending on whether mesophilic or thermophilic strains are used. Thermophilic bacteria, such as *Clostridium thermocellum* and *Thermoanaerobacterium*, are particularly suitable for SSF systems because they can simultaneously hydrolyze cellulose and produce hydrogen at elevated temperatures (50–60°C). As reported, simultaneous fermentation of various cellulosic waste materials by *Clostridium thermocellum* ATCC 27405 demonstrated effective biohydrogen production, with yields up to 1.27 mmol H_2/mmol glucose equivalent utilized [58]. Despite these advantages, cellulose hydrolysis often remains the rate-limiting step in SSF, and incomplete saccharification can limit the availability of fermentable sugars, thereby constraining hydrogen yield [59].

5.2 Separate Hydrolysis and Fermentation

In the process of separate hydrolysis and fermentation (SHF), enzymatic or microbial hydrolysis of complex carbohydrates is conducted independently before the fermentation stage [60]. By decoupling these processes, it is possible to optimize hydrolysis and fermentation conditions separately, achieving higher efficiency in each step. During the hydrolysis phase, polysaccharides are broken down into mono reducing sugars. These hydrolysates are then fermented by hydrogen-producing bacteria under strictly anaerobic conditions. For example, hydrogen production from agricultural residues via SHF has been reported to reach cumulative yields of 287.46 mL/g TS with wheat straw and up to 410.43 mL/g TS with corncob, demonstrating the feasibility of this strategy for biohydrogen generation [61]. Overall, SHF allows for better control over substrate concentration, pH, and temperature, which is particularly advantageous when dealing with substrates rich in cellulose and hemicellulose. Overall, SHF provides higher flexibility and control but requires additional operational steps compared to SSF.

5.3 Consolidated Bioprocessing

Consolidated bioprocessing (CBP) represents a promising and integrative strategy for the efficient conversion of lignocellulosic biomass into biohydrogen. Unlike traditional multistep approaches such as SHF and SSF, CBP combines enzyme production, biomass hydrolysis, and microbial fermentation into a single bioreactor system. This streamlining of the process eliminates the need for externally added hydrolytic enzymes and separate reactors, thereby reducing operational complexity, capital investment, and overall production costs.

Numerous studies have demonstrated the potential of cellulolytic bacteria such as *Clostridium thermocellum* and *Thermoanaerobacterium thermosaccharolyticum* in direct hydrogen production from lignocellulosic materials. Table 1 summarizes key

Table 1 Hydrogen production reported in cellulosic substrate fermentation studies

Microorganism	Cellulosic substrate	Temperature (°C)	Y_{H2}	References
Clostridium cellulolyticum	Cellulose MN301	37	1.7 mol H_2/mol glucose	[62]
	Microcrystalline cellulose		1.6 mol H_2/mol glucose	
Clostridium populeti	Cellulose MN301	37	1.6 mol H_2/mol glucose	
	Microcrystalline cellulose		1.4 mol H_2/mol glucose	
Clostridium termitidis strain CT1112	α-Cellulose	37	7.7 mmol H_2/L culture	[63]
Clostridium acetobutylicum X9	Microcrystalline cellulose	37	33.7 mmol H_2/L culture	[58]
	Steam-exploded corn straw		8.4 mmol H_2/L culture	[64]
Ruminococcus albus	Sweet sorghum	37	2.7 mmol H_2/g-substrate	[65]
C. thermocellum 27405	Microcrystalline cellulose	60	15 mmol H_2/L culture	[63]
	Cornstalk powder		9.1 mmol H_2/L culture	
	Corncob powder		9.4 mmol H_2/L culture	
Caldicellulosiruptor saccharolyticus	Sweet sorghum	70	30.17 mL/g raw material	[66]
	Sugarcane bagasse		19.21 mL/g raw material	
	Wheat straw		44.86 mL/g raw material	
	Maize leaves		38.14 mL/g raw material	
	Silphium		8.14 mL/g raw material	
Thermoanaerobacterium thermosaccharolyticum M18	Microcrystalline cellulose	60	10.86 mmol H_2/g Avicel	[67]
	Corncob		3.28 mmol H_2/g-substrate	
	Cornstalk		3.47 mmol H_2/g-substrate	
	wheat straw		3.53 mmol H_2/g-substrate	

Y_{H2}: Yield of hydrogen production based on per g/mol substrate consumed or per liter volume microbial strains that have demonstrated the capacity to directly ferment cellulose into biohydrogen. Among them, thermophilic strains consistently outperform mesophilic ones, in terms of not only substrate degradation efficiency but also overall biohydrogen productivity. Their enhanced performance under high-temperature conditions underscores their value in the development of viable CBP systems.

6 Strategies for Improving Biohydrogen Production from Agricultural Biomass

Biohydrogen production from agricultural biomass during dark fermentation is often constrained by several biochemical and process-related factors. These include accumulation of organic solvents (lactate, ethanol, butyrate), unfavorable oxidation-reduction conditions, electron competition by alternative pathways, and limited electron transfer efficiency within microbial communities. Till now, extensive research has focused on developing exogenous regulation strategies to enhance the biohydrogen production process during dark fermentation.

6.1 Carbonaceous Material Addition

Carbonaceous materials, such as biochar, carbon nanotubes, and graphene derivatives, have been widely investigated in fermentative hydrogen production systems [68, 69]. Owing to their high specific surface area and abundant porous structures, these materials provide favorable sites for microbial attachment and can synergistically adsorb toxic intermediates (e.g., 5-hydroxymethyl furfural and phenols), thereby mitigating microbial inhibition. In addition, when prepared at elevated pyrolysis temperatures, such carbon materials often exhibit the ability of electrical conductivity, which promotes extracellular electron transfer (EET) among microbial cells [70]. These conductive networks also facilitate direct interspecies electron transfer (DIET), enabling a larger fraction of electrons to be channeled toward hydrogen production rather than diverted to competing metabolic pathways [71].

Metal oxides such as Fe_2O_3, Fe_3O_4, MnO_2, and NiO nanoparticles function as redox mediators by accepting and donating electrons during microbial metabolism [72]. Consequently, metal oxides are often employed as functional modifiers of carbonaceous materials to further enhance hydrogen-producing capacity. The inclusion of these metal oxides helps modulate intracellular redox balance and redirect electron flux toward the biohydrogen-generating way. Furthermore, iron plays a central role in the active site of hydrogenase; thus, supplementation with iron-containing metal oxides has been reported to enhance hydrogenase activity. Incorporation of such metal oxides in biochar or other carbon-based carriers has been shown to significantly improve hydrogen yields through synergistic promotion of

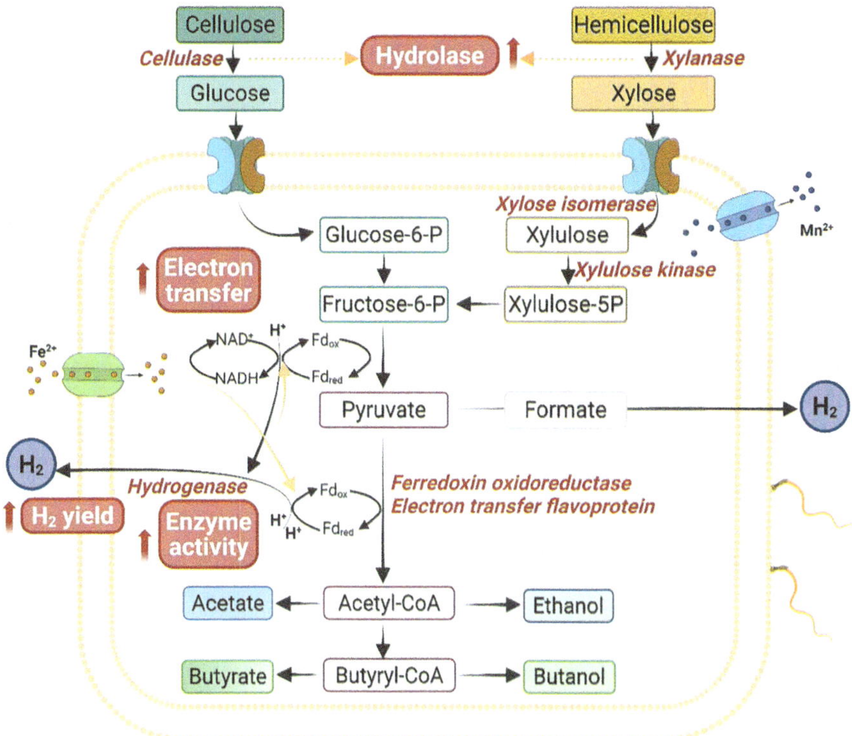

Fig. 3 The mechanism of exogenous carbon materials enhancing dark fermentative hydrogen production

both electron transfer and enzymatic function [73]. The mechanism of exogenous carbon materials enhancing dark fermentative hydrogen production is shown in Fig. 3.

6.2 *Genetic Engineering*

Genetic engineering of hydrogen-producing microorganisms offers a powerful way to overcome intrinsic metabolic limitations [74]. Strategies include (1) overexpression of hydrogenase genes; (2) knockout of competing pathways such as lactate dehydrogenase (LDH) and alcohol dehydrogenase (ADH) genes to minimize electron diversion; and (3) introduction of heterologous hydrogenases with superior kinetics or oxygen tolerance. For instance, engineered strains of *Clostridium acetobutylicum* lacking solventogenic pathways demonstrated a substantial increase in hydrogen yield compared to wild-type strains [75]. Nevertheless, the substantial investment required for the development of genetically engineered strains poses a barrier to their widespread application in industrial settings.

6.3 Synthetic Microbial Consortium

Constructing synthetic consortia with complementary metabolic capabilities can optimize overall substrate utilization. Co-cultures of hydrogen-producing bacteria and auxiliary species (e.g., those consuming inhibitory by-products) can stabilize fermentation systems and prolong hydrogen evolution phases. A study demonstrated that co-culturing *Bacillus cereus* A1 and *Brevundimonas naejangsanensis* B1 enhanced hydrogen production by redirecting metabolic fluxes and facilitating mutualistic interactions, such as formate exchange, which stabilized the fermentation process and increased hydrogen yields [76]. It was reported that co-cultivation of *Bradyrhizobium japonicum* with a transgenic *Chlamydomonas reinhardtii* strain expressing soybean leghemoglobin significantly enhanced hydrogen production and growth by accelerating oxygen consumption and respiration rates [77].

6.4 Fermentation Conditions' Adjustment

During the process of hydrogen production, the accumulation of VFAs leads to a rapid decline in pH during the exponential fermentation phase [78]. When the pH drops to a certain threshold, and assuming that substrate availability remains sufficient, a metabolic shift from acidogenesis to solventogenesis occurs, resulting in a subsequent increase in pH. Therefore, controlling the pH during the fermentation process helps to mitigate the toxic inhibitory effects of VFA accumulation on microbial activity and enhances the biohydrogen production potential of the fermentation system. Dynamic pH control systems that adjust buffering capacity or automatically add acids/bases can significantly stabilize hydrogen yields over prolonged fermentations. The most commonly employed approach is the addition of buffering agents, such as calcium carbonate, which minimizes sharp pH fluctuations arising from acid accumulation and maintains a favorable microenvironment for hydrogen-producing bacteria [79, 80]. Moreover, certain buffering materials, such as calcium carbonate, can also supply essential ions that promote microbial growth and enzymatic activity, further improving the overall stability and efficiency of the hydrogen fermentation process.

In addition, ORP is also an important parameter in the fermentative hydrogen production system [81]. Lower (more negative) ORP values favor reductive pathways and enhance hydrogen evolution. Controlling ORP through gas sparging (e.g., N_2 or CO_2 stripping), redox-active additives, or electrochemical methods provides a way to direct microbial metabolism toward efficient hydrogen production.

Enhancing biohydrogen production requires a multifaceted approach grounded in a deep understanding of microbial metabolism, electron flow dynamics, and community ecology. Strategies that promote efficient electron transfer, optimize intracellular redox balance, and stabilize microbial consortia show particular promise. Future research should prioritize integrating multiple enhancement techniques synergistically, applying systems biology tools for predictive modeling, and designing scalable, resilient biohydrogen production platforms suitable for industrial deployment.

7 Conclusion and Future Perspectives

The production of biohydrogen from lignocellulosic biomass provides a promising pathway for sustainable and renewable energy generation. However, there are still challenges throughout the entire process chain, from efficient biomass deconstruction to microbial conversion, system integration, and industrial scalability. The recent progress has laid the foundation for further breakthroughs, with a focus on pretreatment optimization, microbial genetic engineering, process enhancement, and industrial demonstration. To fully realize the potential of agricultural residues, future research should prioritize the following five areas:

7.1 Development of Advanced Pretreatment Technologies

Due to the complex and highly integrated structure of cellulose, hemicellulose, and lignin, pretreatment remains one of the primary bottlenecks in the bioconversion of agricultural residues. Future efforts should focus on developing environmentally sustainable and cost-effective pretreatment systems that synergistically combine mechanical disruption, chemical solubilization, and biological delignification. The goal is to maximize fermentable sugar release while minimizing the generation of inhibitory by-products. Incorporating intelligent control systems capable of real-time monitoring of key parameters such as pH, temperature, viscosity, solid loading, residence time, and inhibitor concentration can further improve process efficiency, stability, and reproducibility. Additionally, the conversion of lignin into high-value products, such as bio-based carbon materials or aromatic compounds, holds great promise for offsetting pretreatment costs and enhancing the overall economic viability of biohydrogen production. The establishment of closed-loop, modular pretreatment platforms will be essential to support the transition toward sustainable and scalable lignocellulosic biorefinery operations.

7.2 *Construction of Efficient Microbial Strains or Synthetic Microbiomes*

In the process of biohydrogen production from lignocellulose, natural microbial strains often exhibit limitations in substrate adaptability, hydrogen production, and tolerance to inhibitory compounds. Future research should focus on developing high-performance engineered strains through the integration of synthetic biology and systems biology. The strategy may include metabolic pathway reconstruction, such as enhancing the expression of [FeFe]-hydrogenase, blocking competitive pathways, and optimizing electron transfer chains to improve power utilization efficiency and increase hydrogen productivity. In addition, the ability to introduce degradation inhibitory compounds (such as phenolic compounds) can enhance the robustness of the strain.

The synthetic microbiome method provides an effective solution for collaborative biomass conversion, in which cellulose-degrading and hydrogen-producing strains are assembled into a stable co-culture system. These alliances can be regulated through group sensing and signaling mechanisms to maintain functional balance and improve system efficiency. The future direction should also explore bioreactors with spatial structures and self-regulated population control modules, enabling microbial communities to operate efficiently under complex organic conditions and fully utilize all lignocellulosic components.

7.3 *Process Integration and Cascade Valorization*

Biohydrogen production via single-step anaerobic fermentation still faces bottlenecks in conversion efficiency and economic feasibility. In the future, resource utilization and system energy efficiency should be improved through process integration and cascade utilization strategies. A two-stage process of "solid-state enzymatic hydrolysis and liquid hydrogen production" can be adopted, in which the first stage efficiently hydrolyzes lignocellulose through fungal solid-state fermentation, and the second stage uses hydrogen-producing anaerobic bacteria for liquid-phase anaerobic fermentation, achieving spatiotemporal decoupling of saccharification and biohydrogen production processes. In addition, the coupled photosynthetic hydrogen production system can further convert intermediate products such as acetic acid into hydrogen gas, with a theoretical yield increase from 4 mol H_2/mol glucose to 12 mol H_2/mol glucose. Microbial electrolysis cells (MECs) offer another promising pathway, where electroactive bacteria oxidize organic matter at the anode and drive hydrogen evolution at the cathode. Looking forward, the development of modular, multifunctional bioreactor systems that integrate pretreatment, fermentation, and electrochemical conversion supported by AI-driven real-time process control will be essential.

7.4 Application of Machine Learning and Data-Driven Modeling

As experimental data and process parameters accumulate in different raw materials, pretreatment methods, and fermentation systems, the integration of machine learning (ML) and data-driven modeling becomes increasingly valuable. These tools can reveal complex nonlinear relationships between operational variables and hydrogen production, which are difficult to capture through traditional mechanical models. By training algorithms on multiple datasets, including microbial dynamics, substrate composition, environmental conditions, and reactor performance, machine learning-based models can be used to predict process outcomes, optimize experimental design, and identify key control points. In addition, the prediction framework can support the expansion of biohydrogen production by providing scenario analysis under various input conditions. Future research should focus on developing interpretable, robust, and transferable models to guide the intelligent design and automation of laboratory-scale, pilot-scale, and full-scale hydrogen production systems.

7.5 Techno-Economic Assessment and Life Cycle Analysis

To ensure the economic feasibility and environmental sustainability of agricultural residues' biohydrogen conversion, a comprehensive technical and economic assessment (TEA) and life cycle analysis (LCA) are essential. TEA can provide insights into capital investment, operating costs, energy input-output ratio, and potential cost reduction leverage for different process configurations. On the other hand, life cycle assessment evaluates the environmental footprint of biohydrogen systems throughout the entire production chain, including greenhouse gas emissions, water and energy consumption, and waste generation. TEA and LCA can jointly determine the trade-off between efficiency, cost, and environmental performance, supporting research and policy level decision-making. Future research should emphasize scenario-based assessments and integrate local biomass availability, renewable energy inputs, and policy incentives, ultimately promoting the implementation of biohydrogen production within the framework of a circular bioeconomy.

In conclusion, sustainable biohydrogen production from agricultural residues via dark fermentation has evolved from a theoretical concept to a multidisciplinary research frontier spanning microbial ecology, materials science, and process engineering. By harnessing cross-disciplinary innovation and embracing integrated optimization strategies, dark fermentation has the potential to emerge as a viable, scalable, and environmentally friendly pathway for clean hydrogen generation in the context of a circular bioeconomy.

References

1. Saravanan A, Senthil Kumar P, Khoo KS et al (2021) Biohydrogen from organic wastes as a clean and environment-friendly energy source: production pathways, feedstock types, and future prospects. Bioresour Technol 342:126021. https://doi.org/10.1016/j.biortech.2021.126021
2. Cao Y, Liu H, Liu W et al (2022) Debottlenecking the biological hydrogen production pathway of dark fermentation: insight into the impact of strain improvement. Microb Cell Factories 21(1):166. https://doi.org/10.1186/s12934-022-01893-3
3. Dzulkarnain ELN, Audu JTW, Dagang WRZW et al (2022) Microbiomes of biohydrogen production from dark fermentation of industrial wastes: current trends, advanced tools and future outlook. Bioresour Bioprocess 9(1):16. https://doi.org/10.1186/s40643-022-00504-8
4. Zheng Y, Zhang Q, Zhang Z et al (2022) A review on biological recycling in agricultural waste-based biohydrogen production: Recent developments. Bioresour Technol 347:126595. https://doi.org/10.1016/j.biortech.2021.126595
5. Singh T, Alhazmi A, Mohammad A et al (2021) Integrated biohydrogen production via lignocellulosic waste: opportunity, challenges & future prospects. Bioresour Technol 338:125511. https://doi.org/10.1016/j.biortech.2021.125511
6. Harirchi S, Wainaina S, Sar T et al (2022) Microbiological insights into anaerobic digestion for biogas, hydrogen or volatile fatty acids (VFAs): a review. Bioengineered 13(3):6521–6557. https://doi.org/10.1080/21655979.2022.2035986
7. Huang J, Chen X, Hu B et al (2022) Bioaugmentation combined with biochar to enhance thermophilic hydrogen production from sugarcane bagasse. Bioresour Technol 348:126790. https://doi.org/10.1016/j.biortech.2022.126790
8. Alan H, Köker AR (2023) Analyzing and mapping agricultural waste recycling research: an integrative review for conceptual framework and future directions. Resour Policy 85:103987. https://doi.org/10.1016/j.resourpol.2023.103987
9. Guo JH, Zhang Y, Fang JJ et al (2024) Reduction and reuse of forestry and agricultural bio-waste through innovative green utilization approaches: a review. Forests 15(8):1372. https://doi.org/10.3390/f15081372
10. Bajpai P (2016) Structure of lignocellulosic biomass. In: Pretreatment of lignocellulosic biomass for biofuel production. Springerbriefs in molecular science. Springer, Singapore. https://doi.org/10.1007/978-981-10-0687-6_2
11. Wang S, Zhang T, Bao M et al (2020) Microbial production of hydrogen by mixed culture technologies: a review. Biotechnol J 15(1):1900297. https://doi.org/10.1002/biot.201900297
12. Turon V, Trably E, Fouilland E et al (2016) Potentialities of dark fermentation effluents as substrates for microalgae growth: a review. Process Biochem 51(11):1843–1854. https://doi.org/10.1016/j.procbio.2016.03.018
13. Argun H, Kargi F (2011) Bio-hydrogen production by different operational modes of dark and photo-fermentation: an overview. Int J Hydrog Energy 36(13):7443–7459. https://doi.org/10.1016/j.ijhydene.2011.03.116
14. Hallenbeck PC (2009) Fermentative hydrogen production: principles, progress, and prognosis. Int J Hydrog Energy 34(17):7379–7389. https://doi.org/10.1016/j.ijhydene.2008.12.080
15. Nandi R, Sengupta S (1998) Microbial production of hydrogen: an overview. Crit Rev Microbiol 24(1):61–84. https://doi.org/10.1080/10408419891294181
16. Sinha P, Roy S, Das D (2015) Role of formate hydrogen lyase complex in hydrogen production in facultative anaerobes. Int J Hydrog Energy 40(29):8806–8815. https://doi.org/10.1016/j.ijhydene.2015.05.076
17. Lukajtis R, Kucharska K, Glinka M et al (2018) Hydrogen production from biomass using dark fermentation. Renew Sustain Energy Rev 91:665–694. https://doi.org/10.1016/j.rser.2018.04.043

18. Hallenbeck PC, Benemann JR (2002) Biological hydrogen production; fundamentals and limiting processes. Int J Hydrog Energy 27(11–12):1185–1193. https://doi.org/10.1016/S0360-3199(02)00131-3
19. Hallenbeck PC (2012) Microbial technologies in advanced biofuels production, vol 1. Springer
20. Logan BE (2004) Peer reviewed: extracting hydrogen and electricity from renewable resources. Environ Sci Technol 38(9):160–167. https://doi.org/10.1021/es040468s
21. Bayer E, Setter E, Lamed R (1985) Organization and distribution of the cellulosome in *Clostridium thermocellum*. J Bacteriol 163(2):552–559. https://doi.org/10.1128/jb.163.2.552-559.1985
22. Talia P, Sede SM, Campos E et al (2012) Biodiversity characterization of cellulolytic bacteria present on native Chaco soil by comparison of ribosomal RNA genes. Res Microbiol 163(3):221–232. https://doi.org/10.1016/j.resmic.2011.12.001
23. Xing D, Ren N, Li Q et al (2006) *Ethanoligenens harbinense* gen. nov., sp. nov., isolated from molasses wastewater. Int J Syst Evol Microbiol 56(4):755–760. https://doi.org/10.1099/ijs.0.63926-0
24. Kumar N, Das D (2000) Enhancement of hydrogen production by *Enterobacter cloacae* IIT-BT 08. Process Biochem 35(6):589–593. https://doi.org/10.1016/S0032-9592(99)00109-0
25. Kádár Z, De Vrijek T, van Noorden GE et al (2004) Yields from glucose, xylose, and paper sludge hydrolysate during hydrogen production by the extreme thermophile *Caldicellulosiruptor saccharolyticus*. Appl Biochem Biotechnol 114(1):497–508. https://doi.org/10.1385/abab:114:1-3:497
26. Ren N, Cao G, Wang A et al (2008) Dark fermentation of xylose and glucose mix using isolated *Thermoanaerobacterium thermosaccharolyticum* W16. Int J Hydrog Energy 33(21):6124–6132. https://doi.org/10.1016/j.ijhydene.2008.07.107
27. Liu Y, Yu P, Song X et al (2008) Hydrogen production from cellulose by co-culture of *Clostridium thermocellum* JN4 and *Thermoanaerobacterium thermosaccharolyticum* GD17. Int J Hydrog Energy 33(12):2927–2933. https://doi.org/10.1016/j.ijhydene.2008.04.004
28. Cheng J, Zhu M (2013) A novel anaerobic co-culture system for bio-hydrogen production from sugarcane bagasse. Bioresour Technol 144:623–631. https://doi.org/10.1016/j.biortech.2013.07.018
29. Wang J, Yin Y (2021) *Clostridium* species for fermentative hydrogen production: an overview. Int J Hydrog Energy 46(70):34599–34625. https://doi.org/10.1016/j.ijhydene.2021.08
30. Liu W, Pang J, Wu D et al (2023) Hydrogen production by a novel *Klebsiella pneumoniae* strain from sheep rumen uses corn straw as substrate. Energy 282:128210. https://doi.org/10.1016/j.energy.2023.128210
31. Zhang K, Cao G, Ren N (2019) Bioaugmentation with *Thermoanaerobacterium thermosaccharolyticum* W16 to enhance thermophilic hydrogen production using corn stover hydrolysate. Int J Hydrog Energy 44(12):5821–5829. https://doi.org/10.1016/j.ijhydene.2019.01.045
32. Amekan Y, Wangi DSAP, Cahyanto MN et al (2018) Effect of different inoculum combination on biohydrogen production from melon fruit waste. Int J Renew Energy Dev 7(2):101–109. https://doi.org/10.14710/ijred.7.2.101-109
33. Choi J, Ahn Y (2014) Characteristics of biohydrogen fermentation from various substrates. Int J Hydrog Energy 39(7):3152–3159. https://doi.org/10.1016/j.ijhydene.2013.12.050
34. Akutsu Y, Li YY, Harada H et al (2009) Effects of temperature and substrate concentration on biological hydrogen production from starch. Int J Hydrog Energy 34(6):2558–2566. https://doi.org/10.1016/j.ijhydene.2009.01.048
35. Wang J, Wan W (2009) Factors influencing fermentative hydrogen production: a review. Int J Hydrog Energy 34(2):799–811. https://doi.org/10.1016/j.ijhydene.2008.11.015
36. Reed MC, Lieb A, Nijhout HF (2010) The biological significance of substrate inhibition: a mechanism with diverse functions. Bioessays 32(5):422–429. https://doi.org/10.1002/bies.200900167

37. Wang H, Xu J, Sheng L et al (2018) A review on bio-hydrogen production technology. Int J Energy Res 42(11):3442–3453. https://doi.org/10.1002/er.4044
38. Guo S, Lu C, Wang K et al (2020) Enhancement of pH values stability and photo-fermentation biohydrogen production by phosphate buffer. Bioengineered 11(1):291–300. https://doi.org/10.1080/21655979.2020.1736239
39. Jain R, Panwar NL, Jain SK et al (2024) Bio-hydrogen production through dark fermentation: an overview. Biomass Convers Biorefinery 14(12):12699–12724. https://doi.org/10.1007/s13399-022-03282-7
40. Bundhoo MAZ, Mohee R (2016) Inhibition of dark fermentative bio-hydrogen production: a review. Int J Hydrog Energy 41(16):6713–6733. https://doi.org/10.1016/j.ijhydene.2016.03.057
41. Xu H, Li B, Mu X (2016) Review of alkali-based pretreatment to enhance enzymatic saccharification for lignocellulosic biomass conversion. Ind Eng Chem Res 55(32):8691–8705. https://doi.org/10.1021/acs.iecr.6b01907
42. Licari A, Monlau F, Solhy A et al (2016) Comparison of various milling modes combined to the enzymatic hydrolysis of lignocellulosic biomass for bioenergy production: glucose yield and energy efficiency. Energy 102:335–342. https://doi.org/10.1016/j.energy.2016.02.083
43. Ren N, Zhao L, Chen C et al (2016) A review on bioconversion of lignocellulosic biomass to H_2: key challenges and new insights. Bioresour Technol 215:92–99. https://doi.org/10.1016/j.biortech.2016.03.124
44. Guerra-Rodríguez E, Portilla-Rivera OM, Jarquín-Enríquez L et al (2012) Acid hydrolysis of wheat straw: a kinetic study. Biomass Bioenergy 36:346–355. https://doi.org/10.1016/j.biombioe.2011.11.005
45. Saratale GD, Kshirsagar SD, Saratale RG et al (2015) Fermentative hydrogen production using sorghum husk as a biomass feedstock and process optimization. Biotechnol Bioprocess Eng 20:733–743. https://doi.org/10.1007/s12257-015-0172-3
46. Brodeur G, Yau E, Badal K et al (2011) Chemical and physicochemical pretreatment of lignocellulosic biomass: a review. Enzyme Research 1:787532. https://doi.org/10.4061/2011/787532
47. Palmqvist E, Hahn-Hägerdal B (2000) Fermentation of lignocellulosic hydrolysates. II: inhibitors and mechanisms of inhibition. Bioresour Technol 74(1):25–33. https://doi.org/10.1016/S0960-8524(99)00161-3
48. Kim DH, Jo IS, Kang BJ et al (2022) Evaluation of bio-hydrogen production using rice straw hydrolysate extracted by acid and alkali hydrolysis. Int J Hydrog Energy 47(88):37385–37393. https://doi.org/10.1016/j.ijhydene.2021.12.177
49. Song Z, Yang G, Han X et al (2013) Optimization of the alkaline pretreatment of rice straw for enhanced methane yield. Biomed Res Int 1:968692. https://doi.org/10.1155/2013/968692
50. Dong L, Cao GL, Zhao L et al (2018) Alkali/urea pretreatment of rice straw at low temperature for enhanced biological hydrogen production. Bioresour Technol 267:71–76. https://doi.org/10.1016/j.biortech.2018.05.055
51. Zhao L, Cao G, Wang A et al (2012) Fungal pretreatment of cornstalk with *Phanerochaete chrysosporium* for enhancing enzymatic saccharification and hydrogen production. Bioresour Technol 114:365–369. https://doi.org/10.1016/j.biortech.2012.03.076
52. Sahni N, Phutela UG (2013) Comparative profile of paddy straw pretreated with standard and isolated lignocellulolytic fungal cultures. J Yeast Fungal Res 4(7):92–99. https://doi.org/10.5897/JYFR2013.0118
53. Dai Y, Si M, Zhang N et al (2015) Combination of biological pretreatment with NaOH/Urea pretreatment at cold temperature to enhance enzymatic hydrolysis of rice straw. Bioresour Technol 198:725–731. https://doi.org/10.1016/j.biortech.2015.09.091
54. Zhu S, Wu Y, Zi N et al (2006) Comparison of three microwave/chemical pretreatment processes for enzymatic hydrolysis of rice straw. Biosyst Eng 93(3):279–283. https://doi.org/10.1016/j.biosystemseng.2005.11.013

55. Teramoto Y, Tanaka N, Lee SH et al (2008) Pretreatment of eucalyptus wood chips for enzymatic saccharification using combined sulfuric acid-free ethanol cooking and ball milling. Biotechnol Bioeng 99(1):75–85. https://doi.org/10.1002/bit.21522
56. Li J, Zhang Y, Shi S et al (2020) Effect of residual extractable lignin on acetone-butanol-ethanol production in SHF and SSF processes. Biotechnol Biofuels 13:67. https://doi.org/10.1186/s13068-020-01710-2
57. Valles A, Álvarez-Hornos FJ, Martínez-Soria V et al (2020) Comparison of simultaneous saccharification and fermentation and separate hydrolysis and fermentation processes for butanol production from rice straw. Fuel 282:118831. https://doi.org/10.1016/j.fuel.2020.118831
58. Magnusson L, Islam R, Sparling R et al (2008) Direct hydrogen production from cellulosic waste materials with a single-step dark fermentation process. Int J Hydrog Energy 33(20): 5398–5403. https://doi.org/10.1016/j.ijhydene.2008.06.018
59. Sitthikitpanya S, Reungsang A, Prasertsan P et al (2018) Two-stage thermophilic bio-hydrogen and methane production from lime-pretreated oil palm trunk by simultaneous saccharification and fermentation. Int J Hydrog Energy 43(9):4284–4293. https://doi.org/10.1016/j.ijhydene.2018.01.063
60. Moreno-Dávila I, Herrera-Ramírez EB, Rodríguez-Garza MM et al (2017) Bio-hydrogen production by SSF of paper industry wastes using anaerobic biofilms: a comparison of the use of wastes with/without pretreatment. Int J Hydrog Energy 42(51):30267–30273. https://doi.org/10.1016/j.ijhydene.2017.09.054
61. Li Y, Zhang Z, Zhu S et al (2018) Comparison of bio-hydrogen production yield capacity between asynchronous and simultaneous saccharification and fermentation processes from agricultural residue by mixed anaerobic cultures. Bioresour Technol 247:1210–1214. https://doi.org/10.1016/j.biortech.2017.09.053
62. Li D, Chen H (2007) Biological hydrogen production from steam-exploded straw by simultaneous saccharification and fermentation. Int J Hydrog Energy 32(12):1742–1748. https://doi.org/10.1016/j.ijhydene.2006.12.011
63. Levin DB, Islam R, Cicek N et al (2006) Hydrogen production by *Clostridium thermocellum* 27405 from cellulosic biomass substrates. Int J Hydrog Energy 31(11):1496–1503. https://doi.org/10.1016/j.ijhydene.2006.06.015
64. Ren N, Wang A, Gao L et al (2008) Bioaugmented hydrogen production from carboxymethyl cellulose and partially delignified corn stalks using isolated cultures. Int J Hydrog Energy 33(19):5250–5255. https://doi.org/10.1016/j.ijhydene.2008.05.020
65. Ntaikou I, Gavala HN, Kornaros M et al (2008) Hydrogen production from sugars and sweet sorghum biomass using *Ruminococcus albus*. Int J Hydrog Energy 33(4):1153–1163. https://doi.org/10.1016/j.ijhydene.2007.10.053
66. Ivanova G, Rákhely G, Kovács KL (2009) Thermophilic biohydrogen production from energy plants by *Caldicellulosiruptor saccharolyticus* and comparison with related studies. Int J Hydrog Energy 34(9):3659–3670. https://doi.org/10.1016/j.ijhydene
67. Cao G, Zhao L, Wang A et al (2014) Single-step bioconversion of lignocellulose to hydrogen using novel moderately thermophilic bacteria. Biotechnol Biofuels 7:1–13. https://doi.org/10.1186/1754-6834-7-82
68. Igalavithana AD, You S, Zhang L et al (2022) Progress, barriers, and prospects for achieving a "Hydrogen Society" and opportunities for biochar technology. ACS ES&T Eng 2(11): 1987–2001. https://doi.org/10.1021/acsestengg.1c00510
69. Lou T, Yin Y, Wang J (2024) Recent advances in effect of biochar on fermentative hydrogen production: performance and mechanisms. Int J Hydrog Energy 57:315–327. https://doi.org/10.1016/j.ijhydene.2024.01.039
70. Wu J, Dong L, Zhou C et al (2019) Enhanced butanol-hydrogen coproduction by *Clostridium beijerinckii* with biochar as cell's carrier. Bioresour Technol 294:122141. https://doi.org/10.1016/j.biortech.2019.122141

71. Zhang Y, Ye Y, Fang M et al (2025) Biogas production enhancement from anaerobic digestion with magnetic biochar: insights into the functional microbes and DIET. J Environ Manag 381: 125347. https://doi.org/10.1016/j.jenvman.2025.125347
72. Liu J, Jiang J, Meng Y et al (2020) Preparation, environmental application and prospect of biochar-supported metal nanoparticles: a review. J Hazard Mater 388:122026. https://doi.org/10.1016/j.jhazmat.2020.122026
73. Wu J, You S, Cao G et al (2025) Unlocking bio-H_2 production potential from straw–sludge co-substrates: the role of Fe/Mn-biochar in alleviating carbon catabolite repression. Chem Eng J 519:165592. https://doi.org/10.1016/j.cej.2025.165592
74. Srirangan K, Pyne ME, Chou CP (2011) Biochemical and genetic engineering strategies to enhance hydrogen production in photosynthetic algae and cyanobacteria. Bioresour Technol 102(18):8589–8604. https://doi.org/10.1016/j.biortech.2011.03.087
75. Oh SE, Zuo Y, Zhang H et al (2009) Hydrogen production by *Clostridium acetobutylicum* ATCC 824 and megaplasmid-deficient mutant M5 evaluated using a large headspace volume technique. Int J Hydrog Energy 34(23):9347–9353. https://doi.org/10.1016/j.ijhydene.2009.09.084
76. Wang S, Tang HZ, Peng F et al (2019) Metabolite-based mutualism enhances hydrogen production in a two-species microbial consortium. Commun Biol 2(1):82. https://doi.org/10.1038/s42003-019-0331-8
77. Wu S, Li X, Yu J et al (2012) Increased hydrogen production in co-culture of *Chlamydomonas reinhardtii* and *Bradyrhizobium japonicum*. Bioresour Technol 123:184–188. https://doi.org/10.1016/j.biortech.2012.07.055
78. Moon C, Jang SZ, Yun Y-M et al (2015) Effect of the accuracy of pH control on hydrogen fermentation. Bioresour Technol 179:595–601. https://doi.org/10.1016/j.biortech.2014.10.128
79. Kim DH, Kim SH, Jung KW et al (2011) Effect of initial pH independent of operational pH on hydrogen fermentation of food waste. Bioresour Technol 102(18):8646–8652. https://doi.org/10.1016/j.biortech.2011.03.030
80. Wang D, Zhang D, Xu Q et al (2019) Calcium peroxide promotes hydrogen production from dark fermentation of waste activated sludge. Chem Eng J 355:22–32. https://doi.org/10.1016/j.cej.2018.07.192
81. Song J, An D, Ren N et al (2011) Effects of pH and ORP on microbial ecology and kinetics for hydrogen production in continuously dark fermentation. Bioresour Technol 102(23): 10875–10880. https://doi.org/10.1016/j.biortech.2011.09.024

Adv Biochem Eng Biotechnol (2026) 196: 363–386
https://doi.org/10.1007/10_2026_327

Published online: 7 May 2026

Life Cycle Analysis for Agricultural Residues and Related Products

Penjit Srinophakun, Anusith Thanapimmetha, Nutchapon Chiarasumran, Khemmathin Lueangwattanapong, and Maythee Saisriyoot

Contents

Abstract Agricultural residues represent an abundant and underutilized resource that can be redirected from disposal pathways toward value-added applications such as composting, mulching, bioenergy, land amendment, and bioplastics. This chapter examines these pathways through the perspective of life cycle assessment (LCA), highlighting how agricultural residues can contribute to waste reduction, soil improvement, greenhouse gas mitigation, and circular resource use. Across the reviewed case studies, environmental outcomes are shown to depend not only on the residue itself, but also on the selected conversion pathway, management practice, system boundary, and local implementation context. Composting and mulching can

P. Srinophakun, A. Thanapimmetha, N. Chiarasumran, K. Lueangwattanapong, and M. Saisriyoot (✉)
Department of Chemical Engineering, Faculty of Engineering, Kasetsart University, Bangkok, Thailand
e-mail: fengmts@ku.ac.th

reduce environmental burdens, although benefits remain sensitive to process modifications, additive use, and field conditions. Energy recovery from residues can offer substantial gains when carbon-intensive reference systems are displaced, but performance varies with logistics and conversion efficiency. Land application may improve soil quality and carbon retention while also shifting greenhouse gas fluxes depending on amendment type and site conditions. Likewise, residue-derived bioplastics can lower impacts relative to fossil-based plastics under favorable feedstock and end-of-life scenarios, but these benefits are not universal. Overall, the chapter shows that LCA is a valuable tool for comparing residue management options while also emphasizing the need for integrated assessment frameworks that account for environmental, technical, and implementation-related factors when selecting sustainable valorization strategies.

Graphical Abstract

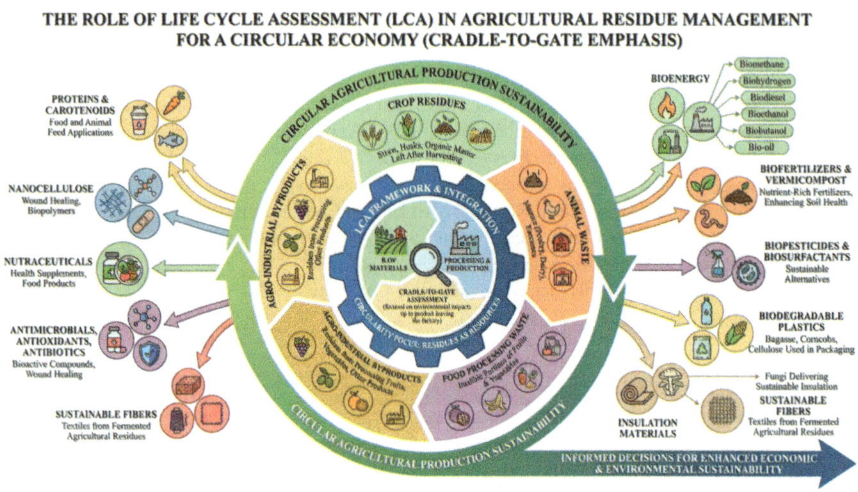

Keywords Agricultural residues, Bioenergy, Bioplastics, Circular economy, Composting, Environmental sustainability, Land application, Life cycle assessment, Mulching, Residue valorization

1 Introduction

Agricultural crop residues are byproducts of crop production, including materials such as straw, husks, and other organic matter left after harvesting, encompassing a wide range of organic materials generated from various sources. These residues are significant for their potential use in multiple applications, including bioenergy, such as biomethane, biohydrogen, biodiesel, bioethanol, biobutanol, and bio-oil produced through anaerobic digestion or fermentation; biofertilizers and vermicompost yielded by converting organic waste into nutrient-rich fertilizers, enhancing soil health and fertility; biopesticides and biosurfactants derived from agricultural waste offering sustainable alternatives to chemical pesticides and surfactants; biodegradable plastics made from agricultural residues such as bagasse, corncobs, and cellulose used in packaging and other applications; insulation materials created from fungi delivering sustainable insulation for construction; sustainable fibers for textiles produced from fermented agricultural residues; antimicrobials, antioxidants, and antibiotics developed from bioactive compounds, which are substances that have a biological effect when introduced into the body, extracted from agricultural residues; nutraceuticals used in health supplements and food products; nanocellulose for wound healing and biopolymers; and proteins and carotenoids, used in food and animal feed applications.

Animal waste, including manure from poultry, dairy, and slaughterhouses, is another significant source of agricultural residues. This waste is often used in composting and as raw material for biogas production, contributing to renewable energy generation. Manure and animal carcasses are significant components of agricultural waste. Food processing waste comprises inedible portions of fruits and vegetables, such as citrus peels and banana skins. These residues are increasingly being explored for the production of nanomaterials with applications in various industries, including pharmaceuticals and environmental remediation. Agro-industrial byproducts include residues from the processing of fruits, vegetables, and other food products, such as grape, banana, and olive processing waste. These byproducts can be transformed into bioactive compounds in pharmaceuticals and cosmetics.

Life cycle assessment (LCA) plays a crucial role in the management of agricultural residues by providing a comprehensive framework for evaluating the environmental impacts of various management strategies. LCA can be approached in different ways, including *cradle-to-grave*, which examines a product's entire life cycle, from raw material extraction ("cradle") to disposal ("grave"), including transportation and use phases; *cradle-to-gate*, which focuses on the environmental impact of a product up to a point it leaves the factory ("gate"), excluding the use and disposal stages; and *cradle-to-cradle*, which is aligned with circular economy principles. These principles aim to design products for recycling or reuse, creating a closed-loop system that minimizes waste, and *gate-to-gate* focuses on the environmental impact of a product's production process within a specific part of the value chain without considering upstream or downstream effects. Each scope focuses on a specific stage of a product's life cycle to assess its environmental impact.

LCA supports the transition from linear to circular agricultural production systems, in which residues are not seen as waste but as valuable resources. By integrating LCA into agricultural residue management, stakeholders can make informed decisions that align with bioeconomic principles, thereby enhancing the economic and environmental sustainability of agricultural systems. This approach helps us see the potential of what was once considered waste and how it can be transformed into valuable resources through key processes, such as composting, energy production, land application, or even bioplastic production.

2 Composting and Mulching of Agricultural Residues and Their Environmental Impact

Composting of agricultural residues is a powerful tool for environmental stewardship. This strategy not only reduces waste by transforming organic materials into valuable compost but also enriches soil fertility. Recycling organic materials back into the production cycle reduces the need for raw materials and lowers residue management costs, thereby supporting a circular economy [1, 2]. Composting agricultural residues enhances soil structure, fertility, and microbial activity, thereby facilitating nutrient cycling and promoting soil vitality, which are crucial for sustainable crop production. The addition of compost enhances the physical, chemical, and biological properties of soil, thereby increasing agricultural productivity [2, 3].

Composting is a powerful tool for reducing greenhouse gas (GHG) emissions and water pollution. Compared with other waste management practices, such as burning or landfilling, composting significantly reduces methane and nitrous oxide (N_2O) emissions, which are major contributors to global warming. It also helps to reduce water pollution caused by nutrient runoff from chemical fertilizers. When combined with advanced technologies, composting can further reduce the environmental impact by lowering the emissions associated with waste decomposition [1, 4].

On the other hand, composting has some environmental impacts, such as emissions from transportation and energy use during the composting process. For instance, while organic manure compost made from maize residual roots can enhance crop yield and soil health, its application may increase carbon dioxide (CO_2) emissions from the decomposition of organic matter. Additionally, composting effectiveness can vary with the type of waste, composting method, and local conditions [1]. Therefore, optimizing composting practices and integrating them with other waste management strategies can enhance environmental benefits.

Mulching, another effective strategy, involves using agricultural residues to cover the soil, which helps with moisture retention, temperature regulation, and weed control. Mulching can effectively manage residues with minimal environmental impacts [5]. These benefits are primarily associated with improved soil health, reduced GHG emissions, and enhanced resource efficiency. Agricultural practices can become more sustainable and environmentally friendly by utilizing biodegradable and organic mulch materials.

Mulching helps conserve soil moisture by reducing evaporation, which is crucial in arid and semiarid regions where water scarcity is a significant concern [6, 7]. It enhances soil fertility by improving physicochemical properties, such as nutrient cycling and soil structure, thereby fostering microbial activity and plant growth [6, 8]. Organic mulches enrich soil fauna and improve nutrient availability by lowering the soil pH [7].

Mulching with crop residues is a key strategy for reducing indirect GHG emissions. The carbon footprint of agricultural systems can be significantly lowered by decreasing the need for agronomic inputs and farm operations. Biodegradable mulch films, for instance, have been shown to reduce N_2O emissions, a potent GHG, thereby contributing to a more sustainable agricultural system [9].

Mulching, by reducing the need for irrigation and fertilizers, helps conserve water and reduce resource consumption. It minimizes the use of pesticides by suppressing weed growth, thereby reducing potential health impacts and environmental toxicity. Biodegradable mulches have a lower net environmental impact on aquatic pollution and toxicity indicators than conventional polyethylene films. The combination of straw return with biodegradable film mulching has been identified as a synergistic cultivation measure that offers both environmental and economic benefits, such as reduced global warming potential (GWP, a measure of how much a given mass of greenhouse gas is estimated to contribute to global warming over a given time horizon) and improved crop yields [10].

Mulching practices can enhance the aesthetic and economic values of agricultural landscapes, contributing to sustainable food production and ecological balance [11]. However, the production and disposal of biodegradable mulches can still have environmental impacts, and mulching effectiveness can vary with crop type, management practices, and climatic conditions [11, 12]. Consequently, ongoing research and optimization of mulch materials and methods are essential to maximize their environmental benefits and address any associated concerns.

2.1 *Case Studies on Composting and Mulching of Agricultural Residues*

- Composting agricultural residues, specifically those derived from olive, elderberry, and grape waste, has demonstrated significant potential in reducing greenhouse gas (GHG) emissions. Notably, the treatment involving olive waste resulted in a 46% reduction in N_2O emissions, a 32% reduction in CO_2 emissions, and a 34% reduction in GWP [13]. The application of alkaline organic residues, such as wood ash and paper sludge, as soil amendments has shown promising outcomes in mitigating GHG emissions. For example, pellets composed of deinking paper sludge (DPS) and wood ash (WA) were found to decrease CO_2 emissions by 10% and N_2O emissions by 40% [14, 15]. Furthermore, the composting of tree leaf litter with effective microorganisms and green sunn hemp plants has the potential to significantly reduce GHG emissions, achieving a reduction of approximately 310 kg CO_2-eq (a metric used to compare the emissions of different greenhouse gases by converting them to the equivalent amount of carbon dioxide based on their global warming potential) [16].

- Thermophilic composting of fresh chicken manure and rice husk supplemented with iron minerals (imTC) increased the production of reactive oxygen species (ROS), which were highly reactive molecules containing oxygen, resulting in a significant 45.12% reduction in GHG emissions [17].
- Modified lignin (ML), derived from straw, demonstrated the ability to mitigate NH_3 and GHG emissions during composting. The ML treatment could reduce NH_3 emissions by 36.3% and decrease the GWP by 31.0–64.6% compared to other additives, such as biochar and phosphogypsum. This advantage is attributed to ML's enhanced adsorption capacity for ammonium and nitrate [18].
- Biochars obtained from mature composts and certified compost originating from a composting plant effectively mitigated CO_2 emissions during the intensive phase of composting, with even low doses (3–6% per dry matter) capable of reducing total measured gas emissions by up to 50%. Biochar also enhanced the nutrient content of the compost, making it a beneficial additive for reducing emissions and improving compost quality [19].
- The gate-to-gate life cycle of biochar production from corn residue, rice straw, and sugarcane leaves significantly reduced the GWP impact, achieving reductions of 2.46–3.22 times compared to open burning, with the lowest GWP recorded at −1,833.19 to −1,473.21 kg CO_2-eq [20].
- Rape residue mulching has the potential to elevate greenhouse gas (GHG) emissions of CO_2 and N_2O while concurrently reducing methane (CH_4) emissions. This results in a 9–30% increase in GWP, yet a notable reduction in net GWP of 33–71% and in GHG intensity of 35–72% compared to scenarios without residue treatment [21].
- Mulching has been shown to enhance maize yield by 18.68–41.80%, increase total fixed carbon in straw by 23.64%, grain by 28.87%, and root biomass by 46.31%. However, it also raises GHG emissions (CO_2 by 10.78% and N_2O by 3.41%) and decreases soil organic carbon sequestration (SOCS) by 6.57% while reducing GHG intensity by 13.61% [22].
- Straw mulching (SR-M) exhibits a lesser impact on CH_4 emissions (14.86% reduction) compared to straw incorporation (SR-I), which demonstrates a more substantial effect (129.39% reduction). Nevertheless, SR-M has a greater impact on N_2O emissions (18.47% reduction) than SR-I (7.76% reduction) [23].
- Straw mulching (SM) may increase CO_2 and N_2O emissions relative to no mulching; however, it functions as a sink for net ecosystem GHG emissions, resulting in a reduction of GWP by 14% in maize and 2% in wheat [24, 25].
- Soil mulching (SM) significantly enhances crop yield by 21.84% compared to no soil mulching (NSM) while increasing GWP by 11.38%. For specific crops, such as maize and wheat, in arid, semiarid, and semi-humid zones, it is recommended to employ complete mulching with straw or gravel, combined with drip irrigation, to mitigate the adverse effects of SM on GHG emissions. Conversely, the application of SM can substantially reduce CH_4 emissions for rice in humid regions [26].

The composting studies point to a fairly clear direction: the environmental burden is not determined solely by the residue, but can be shifted by how the composting system is managed. Additives such as iron minerals, modified lignin, wood ash, paper sludge, and biochar do not all act in the same way, yet they repeatedly affect the same core processes – gas formation, nitrogen retention, and carbon stabilization. This is why the reported reductions in emissions are substantial in some systems but not automatically transferable to others. What matters is whether the treatment changes the biochemical and physicochemical conditions that control loss pathways during composting.

The mulching studies are less about straightforward improvement than about balance. Several cases show better yield, lower net GWP, or reduced GHG intensity, but they also show that some gas components can increase concurrently. In other words, mulching should not be treated as uniformly beneficial or detrimental. Its environmental value depends on crop type, climate zone, irrigation regime, and whether the main objective is yield improvement, carbon retention, methane reduction, or overall GHG mitigation.

3 Energy Production from Agricultural Residues: An Environmental Perspective

The key environmental benefits of energy production from agricultural residues are both substantial and multifaceted. These residues are rich in lignocellulosic biomass, which consists of cellulose, lignin, and hemicellulose, making them suitable for biofuel production and other value-added applications. The benefits of biofuels primarily focus on reducing GHG emissions, thereby substantially reducing the carbon footprint, improving waste management, and promoting sustainable agricultural practices. Importantly, the energy produced creates additional income streams for farmers, thereby enhancing their economic stability [27, 28]. Agricultural residues can enhance energy security and support rural development by providing a local and renewable energy source that is critical in regions heavily reliant on traditional biomass, where agricultural residues can offer a sustainable alternative and contribute to energy security [29–31].

3.1 Economic Considerations in the Utilization of Agricultural Residues

The economic feasibility of biofuel production is closely tied to the availability and cost of agricultural residues, efficiency of conversion technologies, scale of operations, and supportive policies and incentives, all of which vary widely by region and policy framework. These elements collectively determine the competitiveness of biofuels in the energy market compared with conventional fossil fuels.

Agricultural residues such as straw, husks, and bagasse are abundant and relatively low-cost feedstocks for biofuel production. Their availability is influenced by regional agricultural practices and crop types, which vary significantly worldwide. For example, residues are more readily available in regions with high agricultural output, thereby reducing transportation and procurement costs [32, 33].

The choice between biochemical and thermochemical conversion technologies affects the cost-effectiveness of biofuel production. These technological advancements and process optimizations can enhance conversion efficiency, reduce production costs, and improve financial sustainability. However, the efficiency of converting agricultural residues into biofuels is often limited by the complex composition of lignocellulosic biomass, which requires advanced pretreatment technologies to break down effectively. These advanced pretreatment technologies are urgently needed to enhance conversion rates and reduce energy consumption. The conversion of agricultural residues into biofuels and biochar requires advanced technologies that are not always economically viable. The need for pretreatment and variability in feedstock composition can affect conversion efficiency and economic feasibility.

The infrastructure required for biofuel production, including processing facilities and supply-chain logistics, is a significant cost factor. The seasonal and regional availability of agricultural residues complicates logistics and transportation, impacting the consistent supply of feedstock for bioenergy production. Moreover, the collection, transportation, and storage of agricultural residues pose significant logistical challenges. These issues are exacerbated by the seasonal and regional availability of residues, resulting in inconsistent supply chains [34, 35]. Regions with established agricultural industries may have better infrastructure, thereby reducing initial investment costs [33, 36]. Efficient logistics for feedstock collection and transportation is crucial for minimizing costs and ensuring a steady supply of raw materials.

Developing supportive policy frameworks and incentives is crucial for promoting biofuel production and adoption of bioenergy technologies. Subsidies, tax incentives, and renewable energy mandates can enhance the economic attractiveness of biofuels [36, 37]. Policies should promote sustainable farming practices and foster stakeholder collaborations [37, 38]. However, in many regions, such supportive policies and incentives are lacking, which can hinder the development of the necessary infrastructure and technology. Consequently, it is crucial to educate agricultural communities on the benefits and practices of utilizing residues for energy production. This action can facilitate better management and utilization of these resources, thereby reducing reliance on traditional residue disposal methods such as burning [35, 39].

3.2 Environmental Impacts of Diverse Feedstocks

- LCA of biofuels derived from agricultural residue feedstock is significantly influenced by the types of feedstock used and the conversion processes and technologies employed. Different agricultural residues present unique environmental impacts and benefits when used as feedstocks for biofuel production. The choice of feedstock significantly impacts the overall sustainability, GHG emissions, and energy efficiency of the biofuel production process.
- Banana peduncle has lower emissions than first-generation bioethanol; however, it still presents higher impacts in categories such as freshwater ecotoxicity and human toxicity. Its production process, particularly in the saccharification and fermentation stages, is an energy-intensive contributor to environmental impacts [40].
- Corncobs have been identified as offering the best overall benefits among various biomass feedstocks, with a favorable net energy balance and a lower GWP than other residues such as corn and wheat straw. Bioethanol production from corncobs yields significant GHG reductions and energy efficiency improvements [41].
- Rice and wheat straw are commonly used for biofuel production; however, they present challenges, including high resource consumption and emissions during production. Rice straw, for instance, significantly impacts respiratory inorganics and resource depletion, whereas wheat straw contributes to land release issues, affecting total GHG emissions [42, 43].
- Algae and waste cooking oil (WCO) stand out as promising feedstocks for biodiesel production, offering significant reductions in GHG emissions compared with fossil fuels. Algae, in particular, have a high potential for carbon capture and can be cultivated on nonarable land, thereby reducing their impact on land use. The production processes for algae-based biodiesel can be energy intensive, but the environmental benefits depend on the efficiency of both cultivation and conversion processes. WCO, on the other hand, utilizes waste materials, which can further enhance its environmental benefits by reducing waste and associated emissions [44].
- Noncellulosic fermentable carbohydrates (FCs) for biofuels, sourced from energy cane, sweet sorghum, sugar beet, maize, and Brazilian sugarcane, have varying environmental impacts. This diversity underscores the complexity of the decision-making process for biofuel production. Energy cane exhibited the lowest overall environmental impacts, while sugar beet had high GHG emissions and fossil fuel use but low eutrophication potential, highlighting trade-offs among feedstocks [45].

For all feedstocks, using agricultural residues for energy production can affect soil health and nutrient cycling, potentially leading to soil degradation and underscoring the urgency of sustainable practices such as incorporating residues into soil management strategies, which are essential for maintaining soil productivity [30, 39]. Although converting residues into energy products can reduce GHG emissions, improper management can lead to pollution. For instance, burning residues releases harmful pollutants, negating potential environmental benefits [35].

The energy case studies are useful because they make clear that "energy recovery" is not one single pathway. Electricity generation, liquid biofuels, gaseous fuels, co-firing, and other advanced routes operate at different efficiencies and displace different fossil fuels. For that reason, the spread in reported environmental benefits is not surprising. A residue-to-energy system looks much stronger when it replaces a carbon-intensive benchmark and when transport, drying, pretreatment, and conversion losses are kept under control.

What emerges from these studies is not a single best technology, but a recurring pattern: the result is shaped by the match between the residue, the conversion route, and the system being displaced. That point is important for interpretation. A favorable result in one study cannot simply be generalized to all residue-based energy systems, because the outcome also depends on logistics, local energy mix, and life cycle boundary selected for the analysis. The case studies, therefore, add value when read comparatively, not as isolated success stories.

3.3 *Energy Production from Agricultural Residues: A Review of Case Studies*

- In Mexico, with the optimization of the supply chain, the integration of solid biofuels from agricultural residues obtained from diverse Mexican crops, which did not have any use in the market, into the electric power grid, compared to conventional fuels as energy sources, could lead to a significant reduction of 20% in CO_2 emissions. This reduction was a substantial step toward mitigating climate change and improving air quality [46].
- In Ghana, hydrogen-based electricity from agricultural residues from 12 significant crops could reduce GHG emissions by 10.26% annually [28].
- In regions like Piedmont, Italy, replacing diesel fuel in the regional tractors' fleet with biomethane derived from agricultural waste, including crop residues and livestock manure, could lead to a substantial reduction in CO_2 emissions of around 16.8% annually [47].
- In Greece, blending lignite with biomass residues, such as sunflower husk, corn husk, and other agricultural wastes, significantly reduced nonrenewable CO_2 emissions. Specifically, blends containing 70, 50, and 30 wt% biomass resulted in approximately 75%, 57%, and 34% reductions in nonrenewable CO_2 emissions per produced megajoule (MJ), respectively [48].
- The torrefaction of pequi seeds, a type of agricultural residue from an exotic tropical fruit native to the Brazilian savanna, demonstrated substantial CO_2 emission reduction when used as biofuel. The optimized torrefaction process could lead to a considerable CO_2-eq reduction of 188.65 kg by substituting diesel fuel, providing a reassuring sign of the solution's effectiveness [49].

- In Europe, the potential CO_2 savings from advanced biofuels derived from agricultural residues, such as rapeseed straw, rice straw, forestry slash, and sawdust, ranged from 60% to 85% in most cases [50].
- Compared to fossil fuels, global net lignocellulosic bioethanol produced from crop residues could yield relative emission savings ranging from 70% to 79%. Regarding total transport emissions, the savings ranged from 4% to 23% when comparing realistic, sustainable, and theoretical potentials. Their averaged footprints (MJ^{-1}) ranged 0.14–0.24 m^2 of land, 74–120 L of water, and 28–44 g of CO_2-eq [51]
- From an attributional LCA, straw-derived biodiesel in Sweden using oleaginous yeast significantly reduced GWP by 33–80% compared to fossil fuels, with primary fossil energy used for biodiesel production ranging from 0.33 to 0.80 MJ_{prim}/MJ. When all straw fractions were converted to energy carriers, and no lignin was returned to the soil, the highest avoided climate impact was achieved [52].
- Using biodiesel (BD100) from sunflower for agricultural practices at a local scale in Italy could lead to a significant reduction in CO_2 emissions, with a potential decrease of up to 57% due to the renewability of CO_2 from its combustion, which was fixed by the growing biomass (sunflower) [53].
- The reduction in GHG emissions in the USA states (Alabama, Florida, and Georgia) from substituting gasoline with ethanol produced from winter wheat grain and straw was 187 g CO_2-eq/km. The reduction increased to 208 g CO_2-eq/km when ethanol was produced from grain and wheat straw [54].
- In the USA, GHG savings from displacing gasoline with cellulosic ethanol from agricultural residues, such as corn stover and wheat straw, ranged from 0.61 to 0.82 billion metric tons of CO_2-eq, depending on whether the forest biomass was included or not over the 2015–2035 period [55].
- If 20% of the world's rice straw is utilized for ethanol production, it could generate approximately 40×10^9 L of ethanol annually, replacing about 25×10^9 L of fossil fuel-based gasoline and potentially reducing net GHG emissions by up to 70×10^6 tons of CO_2-eq annually [56].
- In Alberta, Canada, from cradle-to-gate LCA evaluation, when used for electricity generation, biochars derived from pine woodchips, wheat, and grape pomace could reduce emissions, compared to coal-derived electricity, by 82%, 80%, and 75%, respectively [57].
- Integrating biomass (woody type) with low-carbon hydrogen production, such as through natural gas pyrolysis, could further enhance the carbon efficiency of biomass-to-liquid fuel processes. Moreover, a cradle-to-grave LCA investigation revealed that the integration achieves up to an 80% reduction in GHG emissions compared to natural gas-based methanol [58].
- In scope of well-to-gate LCA (emissions through the point of production), the GWP of the generation of the yeast lipid and its conversion into various products by the fermentation of pentose sugar extracted from sugarcane bagasse was found to be 260.03, 572.16, 27.83, and 85.19 kg CO_2-eq for nonpolar lipid, lube base

oil, biodiesel, and BTEX (benzene, toluene, ethyl benzene, and xylene) processes, respectively [59].
- From a cradle-to-grave LCA study, electricity generation in the Ivory Coast from cocoa cortex, oil palm fiber, and sugarcane bagasse impacted global warming on the scales of 0.12, 0.16, and 0.19 kg CO_2-eq/kWh, respectively. Its eutrophication impacts were 0.38, 0.4, and 0.91 g PO_4^{3-}/kWh, respectively [31].
- Based on six types of typical aircraft widely used in China, the LCA approach within the boundaries of feedstock, fuel, and flight stages indicated that the GHG emissions of cellulose jet fuel from agricultural residues were 17.9 g CO_2-eq/MJ. In contrast, those of petroleum-based jet fuel were 90.2 g CO_2-eq/MJ. Moreover, the order of GHGs in the well-to-wheel process was as follows: agriculture residue < corn stover < beanstalk < wheat straw < rice straw [60].

The feedstock comparison studies show why residue choice cannot be reduced to a simple ranking. Different materials enter the system with very different moisture contents, compositions, collection demands, pretreatment needs, and conversion compatibilities. As a result, a residue that looks attractive from a greenhouse gas perspective may perform less well in water use, toxicity, or broader resource demand, while another may be easier to process or more energetically efficient, even if its climate advantage is smaller.

This matters because feedstock selection is often discussed as though one residue should emerge as the "best" option. The case studies suggest otherwise. What they really show is that performance depends on the conversion pathway's demands and on which impact categories are prioritized. In practice, feedstock choice should be treated as a matching problem between material properties, process requirements, and environmental priorities, rather than as a universal hierarchy.

4 Land Application of Agricultural Residues: An Environmental Perspective

The application of agricultural residues to land has varying environmental impacts, depending on the type of residue and the application method. However, it is important to note that this practice has significant benefits. Environmental outcomes are influenced by factors such as residue composition, management practices, and local environmental conditions. The primary environmental concerns include GHG emissions, eutrophication, and soil contamination. But on the flip side, potential benefits include improved soil fertility and reduced reliance on chemical fertilizers, offering a ray of hope for sustainable agriculture practices.

The open burning of residues such as corn and rice straw releases substantial amounts of GHG and pollutants. Conversely, converting residues into biochar or using them in composting can reduce GWP by sequestering carbon and improving soil health [20]. Additionally, biochar produced from agricultural residues can enhance soil health by improving nutrient retention and water-holding capacity. It also acts as a carbon sink, sequestering carbon and mitigating the impacts of climate change [29, 61].

Incorporating crop residues into soil can enhance soil fertility and organic carbon sequestration. However, it is crucial to understand that the balance between these effects depends on site-specific conditions such as soil type and climate. This dependency verifies the importance of considering the unique environmental context of each agricultural site. Residues, such as wheat straw and maize husks, contribute to soil fertility by providing organic matter and essential nutrients, including nitrogen, phosphorus, and potassium. The decomposition of residues is influenced by their carbon-to-nitrogen (C:N) ratio, with residues such as maize and wheat decomposing more slowly, thereby providing long-term benefits for soil moisture conservation and carbon sequestration.

In addition to composting and mulching, alternative management practices such as conventional tillage (CT) and no-tillage (NT) also impact environmental outcomes. In NT systems, retaining residues can reduce marine aquatic ecotoxicity and GWP compared with CT systems. However, it is important to note that the effectiveness of these practices depends largely on the site-specific management strategies employed. Incorporating residues into the soil can increase N_2O emissions but reduce NO_3^- leaching, highlighting the need for tailored approaches to balance these effects. Agricultural residues can be repurposed into valuable resources, such as biogas, biochar, and bio-adsorbents, thereby mitigating environmental impacts and providing economic benefits.

The application of agricultural residues presents several challenges, including the potential for soil contamination by heavy metals and other pollutants. Organic residues, such as pruning waste and compost, can contain heavy metals like copper (Cu), zinc (Zn), cadmium (Cd), chromium (Cr), nickel (Ni), and lead (Pb). However, studies have shown that most pruning residues meet the legal thresholds for heavy metal content, suggesting minimal soil toxicity when applied appropriately. However, commercial peat substrates may exceed these thresholds, particularly for chromium, indicating a potential risk when used as a soil amendment [62]. However, it also offers opportunities for sustainable agriculture by recycling nutrients and improving soil health. The key is to manage these residues effectively and maximize their benefits while minimizing environmental risks.

Although the land application of agricultural residues can offer environmental benefits, it also presents challenges. The effectiveness of these practices largely depends on the management strategies employed and the specific environmental context. The land application of agricultural residues, such as food processing, wash water, and solid residuals, can provide nutrients to the soil, thereby reducing the need for synthetic fertilizers. However, potential environmental impacts, such as eutrophication and carcinogenic emissions, must be managed through appropriate techniques. Substituting mineral fertilizers with organic residues in agricultural practices can yield significant environmental benefits. This substitution reduces the environmental footprint associated with the production and application of synthetic fertilizers.

4.1 Utilization of Agricultural Residues in Land Application: Case Studies

- Compared to no residue amendment, incorporating soybean and maize-soybean mixed residues into the soil significantly increased grain yield by 24% and 21%, respectively. This finding was significant as it demonstrated the potential of residue incorporation to boost crop productivity. Additionally, soybean and mixed residues reduced GHG intensity by 25% compared to maize residue, indicating the role of residue utilization in mitigating climate change [63].
- The incorporation of straw-derived biochar in nitrogen fertilizer application (BN) in maize farmland in northwest China significantly increased the net ecosystem carbon budget (NECB, the balance between carbon uptake and release in an ecosystem) by 295.8% (1,996.79 kg soil carbon/ha/year) compared to total straw incorporation. At the same time, it also reduced heterotrophic respiration (Rh) and N_2O emissions by 8.2% and 20.9%, respectively, thereby decreasing the net GWP and GHG intensity of the farmland ecosystems by 56.4% (1,609.49 kg CO_2-eq/ha/year) and 46.5%, respectively. Integrated straw-derived biochar utilization led to a reduction in the average carbon footprint (CF) and nitrogen footprint (NF) of the maize agroecosystem by 350.2% (4,004.85 kg CO_2-eq/ha/year) and 16.5% (12.68 kg N-eq/ha/year), respectively, while improving net ecosystem economic benefits (NEEBs) by $624.7/ha/year [64].
- In Eastern India, soil incorporation with agricultural residues, particularly rice crop residues, showed a significant reduction in emissions relative to those from burning. The reduction was particularly notable in the wet hydrologic category, with a change of 3.1×10^3 kg CO_2-eq. Moreover, biochar has been found to reduce further emissions relative to burning, offering a promising solution for reducing GHG emissions in agricultural practices [65].
- The type of bio-based residue amendments from agricultural soils (sandy loam and clay, representing vast production areas in northwestern Europe) significantly impacted GHG emissions. Residues such as sewage sludge, aquatic plant material, and compressed beet leaves yielded the highest GWP across all soil types. This finding indicates that residue type was a stronger determinant of GHG emissions than soil microbial community composition [66].
- Using residual waste materials (RWMs) as soil amendments affected the magnitude, direction, and timing of GHG fluxes in agricultural soil. Different amendments showed varying effects on CO_2, CH_4, and N_2O fluxes, with dehydrated food waste (DFW) producing the most significant fluxes of CO_2 at 25°C, N_2O at 20°C saturation, and CH_4 at all moisture contents and temperatures [67].
- It was found that cover crop (CC) residues, especially those from legumes, could have a significant influence on GHG emissions. The addition of CC residues could enhance the abundance of N-fixing bacteria, indirectly affecting N_2O production. This finding opened up new possibilities for reducing GHG emissions and offered a more favorable long-term GWP among various organic amendments [68].

The land application studies are helpful because they bring together several mechanisms that are often discussed separately. In some cases, the benefit comes mainly from retaining carbon in the soil or improving soil properties. In others, the more important effect is the avoidance of open burning or the reduction of emissions relative to alternative residue management. There are also cases in which the amendment form changes the timing or magnitude of greenhouse gas fluxes rather than simply decreasing across all indicators.

That is why residue addition alone does not tell the whole story. The environmental effect of land application depends on the amendment itself, the receiving soil, and the management objective. A practice that is beneficial for yield or carbon retention in one setting may not behave the same way under different soil or hydrological conditions. Read together, these studies suggest that land application is often promising, but its performance is strongly conditioned by site context and by the reason the amendment is being used in the first place.

5 Bioplastics: A Sustainable Alternative

Bioplastics produced from agricultural residues offer a sustainable alternative to petroleum-based plastics with significantly lower environmental impacts across various metrics when evaluated through an LCA framework. Bioplastics are typically derived from agricultural waste materials, such as potato peels, cassava, sugarcane, corn stover, wheat straw, and rice husks, through biochemical processes. These materials are abundant and inexpensive and do not compete with food production, making them an attractive option for sustainable bioplastic production. Bioplastics from agricultural residues often have lower GHG emissions than petroleum-based plastics, as the carbon in bioplastics is derived from atmospheric CO_2 captured by plants. Bioplastics, such as polylactic acid (PLA) and polyhydroxyalkanoates (PHA), are derived from biomass, reducing reliance on fossil fuels and mitigating GHG emissions. Additionally, their ability to decompose naturally minimizes long-term pollution and supports the health of both soil and marine ecosystems. Key steps in bioplastic production include starch extraction; enzymatic hydrolysis, which converts biomass into fermentable sugars, fermentation, in which microorganisms metabolize sugars into bioplastic precursors such as lactic acid (LA); and polymerization, in which chemical processes transform monomers into polymers.

PLA and polyhydroxybutyrate (PHB) are derived from renewable resources, including lignocellulosic and food wastes. Their production involves energy-intensive stages, such as biomass processing, pretreatment, and fermentation, as well as water-intensive agricultural processes. Therefore, their production contributes to high CO_2 emissions [69, 70] and to a large water footprint [71]. Compared with polystyrene (PS), which was found to have the highest impact on climate change, and polyethylene terephthalate (PET), which exhibits the most significant effect on ozone-layer depletion, PLA has higher environmental impacts in the categories of marine eutrophication, freshwater eutrophication, and human toxicity, primarily due to the agricultural phase involved in growing the raw materials for PLA production [72].

Bioplastics are designed to decompose naturally and reduce waste accumulation. However, their biodegradation in natural environments can lead to higher GHG emissions than engineered end-of-life processes, such as anaerobic digestion, as the methane produced can be captured and utilized, thereby reducing the overall carbon footprint of the process. This scenario highlights the need for effective disposal systems, which are crucial for minimizing the environmental impact of bioplastics. Recycling initiatives and robust disposal systems for biodegradable materials can reduce the need for new resources and mitigate environmental impacts, making them a shared responsibility for all stakeholders [73, 74].

Despite their environmental benefits, bioplastics face production costs and scalability challenges. Factors, such as feedstock selection and production efficiency, influence the economic feasibility of bioplastics. However, advancements in synthesis techniques and material properties, such as blending and incorporation of nanomaterials, are necessary to enhance the performance of bioplastics and make them competitive with conventional plastics. The potential of bioplastics in sustainable applications, particularly in packaging, is a promising development that is gaining increasing regulatory support and technological advancements [75].

5.1 Case Studies on Bioplastics: An Analytical Overview

- The GWP in the cradle-to-gate analysis for 1 kg of PHB, using methane as its raw material, and PLA, using corn stover as its raw material, were quantified as 3.09 kg and 5.79 kg CO_2-eq, respectively, which were higher than those of fossil-based plastics. Nevertheless, the gate-to-grave analysis revealed that, under well-controlled waste management scenarios, the superior end-of-life (EOL) options for biodegradable plastics, such as anaerobic digestion and incineration, yielded an overall reduction in GWP [76].
- The GWPs for producing 1 ton of fermented lactic acid from various feedstocks – corn, wheat straw, sugarcane bagasse, and food waste – were 3,056.5, 4,224.99, 2.26×10^4, and 2,446.71 kg CO_2-eq, respectively. Their water usage (WU) accounted for 5.47×10^5, 8.29×10^4, 1×10^6, and 2.76×10^4 kg, respectively. Their primary energy demands (PEDs) were 4.26×10^4, 3.88×10^4, 3.1×10^5, and 3.52×10^4 MJ, respectively [77].
- Bioplastic production from potato peels released GHG emissions ranging between 0.354 and 0.623 kg CO_2-eq per kg, significantly lower than the 2.37 kg CO_2-eq per kg of polypropylene (PP) [78].
- PLA and PHB production from biomass waste collected from Romania generated 512.7 and 345 kg of CO_2-eq per kg of polymer [69].
- A cradle-to-grave LCA analysis indicated that for 1 kg of PLA derived from vine shoot waste, 4.3 kg CO_2-eq was required for composting PLA with natural gas,

3.7 kg CO_2-eq for incinerating PLA with natural gas, and 1.9 kg CO_2-eq for incinerating PLA with wood pellets for electricity generation [79].
- The fermentation process for producing lactic acid from wheat straw contributed significantly to global warming, ranging from 1.38 to 0.44 kg CO_2-eq [80].
- A Chinese cradle-to-gate LCA analysis revealed that bio-based PBAT derived from forestry waste outperforms fossil-based PBAT in 13 out of 18 environmental impact categories, with reductions ranging from 15% to 93%. The GWP of bio-based PBAT was 3.72 kg CO_2-eq, 63% lower than that of fossil-based PBAT [81].
- The mean cradle-to-grave GWPs of 1 kg PLA derived from corn stover combusted for the biorefinery were 4.3 kg CO_2-eq for composting PLA with natural gas, 3.7 kg CO_2-eq for incinerating PLA for electricity with natural gas, and 1.9 kg CO_2-eq for incinerating PLA for electricity with wood pellets [82].
- The cradle-to-grave life cycle of bioplastic from various feedstocks: lignocellulosic biomass from forestry, maize biomass, food waste digestate, and food waste accounted for −16.3 to +23.5, 0.3 to 1.0, 1.0 to 4.8, and −0.1 to +0.4 kg CO_2-eq per kg bioplastic, respectively. The anaerobic digestion process yielded 0.23 kg CH_4 per kg bioplastic, contributing to low-carbon energy solutions. Long-term carbon sequestration from digestate application was estimated to be 13.2% of the carbon applied. Methane fugitive emissions from the system were modeled at various rates, impacting overall emissions. Incineration emissions were calculated with an energy conversion efficiency of 22% and process emissions of 0.053 kg CO_2-eq/kg bioplastic [83].

The bioplastic case studies are especially sensitive to system definition. Some studies show clear reductions relative to fossil-based plastics, while others report higher burdens at the production stage or large variation across feedstocks and disposal routes. This is not really a contradiction. It reflects the fact that bioplastic performance is shaped by several linked decisions: where the carbon comes from, how intensively the material is processed, how much water and energy are consumed upstream, and what happens after use.

These studies suggest that residue-derived bioplastics should not be presented as automatically superior alternatives. Their advantage becomes more convincing when low-burden residues replace dedicated feedstocks and when end-of-life management avoids uncontrolled degradation or recovers value through energy recovery or controlled treatment. Conversely, upstream process intensity and unfavorable disposal assumptions can erode much of that benefit. The broader lesson from these cases is that the environmental promise of bioplastics depends as much on system design and waste management as on the use of biomass itself.

6 Conclusion

The use of agricultural residues offers numerous environmental benefits and presents a promising avenue for enhancing agricultural sustainability while mitigating environmental impacts. These practices reduce waste, improve soil health, lower greenhouse gas emissions, and support sustainable agriculture. However, it is crucial to consider the potential risks and challenges associated with such practices. Addressing issues such as soil contamination and developing effective management strategies are essential for maximizing the benefits of agricultural residue utilization. Moreover, increased awareness and education among farmers, particularly in developing countries, can empower them to adopt sustainable practices and contribute to global efforts to combat climate change and resource depletion.

By analyzing various impact categories, LCA plays a crucial role in identifying the most sustainable options for converting agricultural residues into valuable products. This approach supports circular economic practices and helps optimize the environmental benefits of these strategies. However, it is important to consider the broader context of agricultural residue valorization, including economic feasibility, social acceptance, and technological advancements. The choice of feedstock and the specific conditions under which valorization processes are implemented can also significantly influence the environmental outcomes. Therefore, a holistic approach that integrates LCA with other sustainability assessment tools is essential for optimizing agricultural residue valorization strategies.

Future work in agricultural residue valorization should move beyond generalized claims of sustainability and focus more directly on the conditions that determine environmental performance in practice. In particular, greater attention is needed to feedstock-specific logistics, regional management conditions, and end-of-life scenarios, since these factors repeatedly shift life cycle outcomes across composting, energy production, land application, and bioplastic systems. More consistent use of harmonized system boundaries, functional units, and impact categories would also improve comparability across studies and make the evidence base more useful for decision-making.

At the same time, future assessments should better link environmental outcomes to technical feasibility and implementation constraints. This includes the integration of LCA with techno-economic analysis, policy evaluation, and site-specific agricultural data so that residue management pathways can be judged not only by theoretical environmental benefit, but also by practical scalability and local relevance. Such an approach would help identify residue utilization strategies that are robust under real operating conditions and would support the development of circular agricultural systems that are both environmentally sound and practically deployable.

References

1. Serafini LF, Feliciano M, Rodrigues MA, Gonçalves A (2023) Systematic review and meta-analysis on the use of LCA to assess the environmental impacts of the composting process. Sustainability 15:1394. https://doi.org/10.3390/su15021394
2. Doughmi A, Cherkaoui E, Khamar M, Nounah A (2024) Organic waste valorization through composting as part of a circular economy. Technol Sci 29:210–218. https://doi.org/10.55549/epstem.1566173

3. Sathiyapriya S, Prabhaharan J, Sheeba S et al (2024) Nutrient recycling through composting: harnessing agricultural wastes for sustainable crop production. Plant Sci Today 11:1–13. https://doi.org/10.14719/pst.5627
4. Liu Z, Wang X, Li S et al (2022) Advanced composting technologies promotes environmental benefits and eco-efficiency: a life cycle assessment. Bioresour Technol 346:126576. https://doi.org/10.1016/j.biortech.2021.126576
5. Brás I, Silva E, Raimondo R et al (2024) Valorisation of forest and agriculture residual biomass – the application of life cycle assessment to analyse composting, mulching, and energetic valorisation strategies. Sustainability 16:630. https://doi.org/10.3390/su16020630
6. Ravichandran M, Samiappan S, Pandiyan R, Rajesh Kannan V (2022) Improvement of crop and soil management practices through mulching for enhancement of soil fertility and environmental sustainability: a review. J Exp Biol Agric Sci 10:697–712. https://doi.org/10.18006/2022.10(4).697.712
7. El-Beltagi HS, Basit A, Mohamed HI et al (2022) Mulching as a sustainable water and soil saving practice in agriculture: a review. Agronomy 12:1881. https://doi.org/10.3390/agronomy12081881
8. Kumar P, Usadadiya VP (2023) Mulching: an efficient technology for sustainable agriculture production. Int J Plant Soil Sci 35:887–896. https://doi.org/10.9734/ijpss/2023/v35i203880
9. Long B, Li F, Wang K et al (2024) Biodegradable mulch film reduces carbon footprint in maize production. Authorea. https://doi.org/10.22541/au.173453823.34953089/v1
10. Chen Q, Zhang X, Zhao Y et al (2024) Straw return with biodegradable film mulch: a synergistic cultivation measure with environmental and economic benefits based on life cycle assessment. Agronomy 14:2993. https://doi.org/10.3390/agronomy14122993
11. Iqbal R, Raza MAS, Valipour M et al (2020) Potential agricultural and environmental benefits of mulches – a review. Bull Natl Res Cent 44:75. https://doi.org/10.1186/s42269-020-00290-3
12. Tan Q, Yang L, Wei F et al (2023) Comparative life cycle assessment of polyethylene agricultural mulching film and alternative options including different end-of-life routes. Renew Sustain Energy Rev 178:113239. https://doi.org/10.1016/j.rser.2023.113239
13. Pereira JLS, Costa T, Figueiredo V et al (2024) Influence of different agro-food waste on ammonia and greenhouse gas emissions during composting. Agronomy 14:220. https://doi.org/10.3390/agronomy14010220
14. Marouani E, Ziadi N, Lévesque V et al (2021) Mitigation of CO_2, CH_4 and N_2O from acidic clayey soil amended with fertilizer pellets based on alkaline organic residues. Waste Biomass Valoriz 12:3813–3827. https://doi.org/10.1007/s12649-020-01276-y
15. Marouani E, Ziadi N, Lévesque V et al (2024) Granulation of paper residues with wood ash and evaluation of greenhouse gas emissions. In: Proceedings of the 3rd international EUROMAGH conference 2020, pp 270–281. https://doi.org/10.1007/978-981-97-2000-2_31
16. Singh AK, Goswami T, Majumdar B, Kar G (2024) Mitigation of CO_2 emissions in agro-ecosystems by composting of tree leaf litter with effective microorganisms and green sunnhemp plants. Commun Soil Sci Plant Anal 55:411–424. https://doi.org/10.1080/00103624.2023.2269187
17. Du X, Xing R, Lin Y et al (2024) Reduced greenhouse gas emission by reactive oxygen species during composting. Bioresour Technol 404:130910. https://doi.org/10.1016/j.biortech.2024.130910
18. Wang W, Wang X, Zhang X et al (2024) Modified lignin can achieve mitigation of ammonia and greenhouse gas emissions simultaneously in composting. Bioresour Technol 402:130840. https://doi.org/10.1016/j.biortech.2024.130840
19. Stegenta-Dąbrowska S, Syguła E, Bednik M, Rosik J (2024) Effective carbon dioxide mitigation and improvement of compost nutrients with the use of composts' biochar. Materials 17:563. https://doi.org/10.3390/ma17030563
20. Prateep Na Talang R, Na Sorn W, Polruang S, Sirivithayapakorn S (2024) Alternative crop residue management practices to mitigate the environmental and economic impacts of open

burning of agricultural residues. Sci Rep 14:14372. https://doi.org/10.1038/s41598-024-65389-3
21. Zhisheng Z, Cao C-G, Guo L-J, Li C-F (2014) The effects of rape residue mulching on net global warming potential and greenhouse gas intensity from no-tillage paddy fields. Sci World J 2014:198231. https://doi.org/10.1155/2014/198231
22. Pang J, Tian Z, Zhang M et al (2025) Enhancing carbon sequestration and greenhouse gas mitigation in semiarid farmland: the promising role of biochar application with biodegradable film mulching. J Integr Agric 24:517–535. https://doi.org/10.1016/j.jia.2023.12.011
23. Huang T, Wen S, Zhang M et al (2024) Effect on greenhouse gas emissions (CH_4 and N_2O) of straw mulching or its incorporation in farmland ecosystems in China. Sustain Prod Consum 46: 223–232. https://doi.org/10.1016/j.spc.2024.02.024
24. Wei H, Zhang K, Chai N et al (2024) Exploring low-carbon mulching strategies for maize and wheat on-farm: spatial responses, factors and mitigation potential. Sci Total Environ 906: 167441. https://doi.org/10.1016/j.scitotenv.2023.167441
25. Song Q, Yang B, Xu M et al (2024) A three-year record of CO_2, CH_4 and N_2O emissions in maize fields influenced by mulching methods on the Loess Plateau, China. Agric Ecosyst Environ 367:108963. https://doi.org/10.1016/j.agee.2024.108963
26. Guo C, Liu X (2022) Effect of soil mulching on agricultural greenhouse gas emissions in China: a meta-analysis. PLoS One 17:e0262120. https://doi.org/10.1371/journal.pone.0262120
27. Santolini E, Bovo M, Barbaresi A et al (2024) Valorization of corn wastes: assess the environmental impacts in a life-cycle prospective. In: Grajales LM, Serra JCV, Collicchio E (eds) Agroenergy. Woodhead Publishing, pp 131–149
28. Seglah PA, Neglo KAW (2024) Valorization of agricultural residues for hydrogen-based electricity generation towards circular bioeconomy. Open J Energy Efficiency 13:57–78. https://doi.org/10.4236/ojee.2024.132004
29. Phadtare P, Kalbande S (2024) Assessment of agro-residues as biochar feedstock: cotton stalk, soybean straw, and pigeon pea stalk. Int J Res Agron 7:36–40. https://doi.org/10.33545/2618060x.2024.v7.i6sa.809
30. Mekunye F, Makinde P (2024) Sustainable biofuel production from agricultural waste: advances in biochemical and thermochemical conversion pathways. Curr Res Progress Agric Sci 6(6):117–139. https://doi.org/10.9734/bpi/crpas/v6/3285
31. Modeste BY, Bouich A, Oladapo OS et al (2023) Life cycle assessment (LCA) of agricultural residues (cocoa cortex, sugarcane bagasse and oil palm fiber) for power generation in boilers through fuel combustion. SHS Web Conf 175:01050. https://doi.org/10.1051/shsconf/202317501050
32. Samanta P, Senapati T, Dey S, Ghosh AR (2023) An overview of biomass conversion from agricultural waste. In: Agriculture waste management and bioresource, pp 46–77
33. Sarkar JD, Sarkar AK, Mondal P (2023) Sustainable manipulation of agricultural residues in bioenergy production. In: Rakshit A, Biswas A, Sarkar D et al (eds) Handbook of energy management in agriculture. Springer Nature, Singapore, pp 713–737
34. Assaye Y (2024) Agricultural waste for energy storage, conversion and agricultural applications. Am J Mod Energy 10:38–41. https://doi.org/10.11648/j.ajme.20241003.11
35. Kumar M, Patle GT, Lal M et al (2024) An overview of crop residue management challenges, opportunities, and the way forward for sustainable food-energy security in India. Ecol Environ Conserv 30:S486–S502. https://doi.org/10.53550/eec.2024.v30i06s.071
36. Kichonge B, Kivevele T (2023) Viability of non-edible oilseed plants and agricultural wastes as feedstock for biofuels production: a techno-economic review from an African perspective. Biofuels Bioprod Biorefin 17:1382–1410. https://doi.org/10.1002/bbb.2489
37. Guddaraddi A, Singh A, Amrutha G et al (2023) Sustainable biofuel production from agricultural residues an eco-friendly approach: a review. Int J Environ Climate Chang 13:2905–2914. https://doi.org/10.9734/ijecc/2023/v13i102956

38. Zhang Z (2024) Outlook for sustainable utilization of agricultural waste and biomass fuel and materials production. Adv Econ Manag Polit Sci 125:1–8. https://doi.org/10.54254/2754-1169/125/2024.17768
39. Sarkar A (2024) Strategies for effective management of crop residues. Ecol Environ Conserv 30:1509–1515. https://doi.org/10.53550/eec.2024.v30i04.006
40. Daniel KM, Escaño JV, Palada JL, Aberilla JM (2024) Life cycle assessment of bioethanol produced from banana peduncle. IOP Conf Ser Mater Sci Eng 1318:012012. https://doi.org/10.1088/1757-899X/1318/1/012012
41. Yin T, Huhe T, Li X et al (2024) Research on life cycle assessment and performance comparison of bioethanol production from various biomass feedstocks. Sustainability 16:1788. https://doi.org/10.3390/su16051788
42. Liu L, Ouyang Z, Hu C, Li J (2024) Quantifying direct CO_2 emissions from organic manure fertilizer and maize residual roots using 13C labeling technique: a field study. Sci Total Environ 906:167603. https://doi.org/10.1016/j.scitotenv.2023.167603
43. Patel K, Vashist M, Goyal D et al (2025) From waste to resource: a life cycle assessment of biochar from agricultural residue. Environ Prog Sustain Energ 44:e14558. https://doi.org/10.1002/ep.14558
44. Cerón-Ferrusca M, Romero RR, Alanis C, Natividad R (2024) Algae-based and other emerging neat/modified feedstocks. In: Balakrishna RG, Mohan S, Zaki Sharara T (eds) Developments in biodiesel: feedstock, production, and properties. Royal Society of Chemistry, pp 113–119
45. Ortiz-Reyes E, Anex RP (2022) Economic and environmental performance of non-cellulosic fermentable carbohydrates production for biofuels and chemicals. J Clean Prod 353:131526. https://doi.org/10.1016/j.jclepro.2022.131526
46. Martínez-Guido SI, García-Trejo JF, Gutiérrez-Antonio C, Gómez-Castro FI (2022) Optimal agriculture residues revalorization as a biofuel alternative in electric power grids. In: Yamashita Y, Kano M (eds) Computer aided chemical engineering. Elsevier, pp 595–600
47. Assandri D, Bagagiolo G, Cavallo E, Pampuro N (2022) Replacing agricultural diesel fuel with biomethane from agricultural waste: assessment of biomass availability and potential energy supply in Piedmont (north-west Italy). Agronomy 12:2996. https://doi.org/10.3390/agronomy12122996
48. Vasileiadou A (2024) Reduction of CO_2 emissions through the co-combustion of lignite with biomass residues: renewable and non-renewable CO_2 per produced megajoule and fuel characterization. Therm Sci Eng Progress 50:102566. https://doi.org/10.1016/j.tsep.2024.102566
49. Silveira EA, Barcelo R, Cruz Lamas G et al (2024) Biofuel from agro-industrial residues as sustainable strategy for CO_2 mitigation: statistical optimization of pequi seeds torrefaction. Energy Convers Manag 304:118222. https://doi.org/10.1016/j.enconman.2024.118222
50. Malins C, Searle S, Baral A et al (2014) Wasted: Europe's untapped resource: an assessment of advanced biofuels from wastes and residues
51. Holmatov B, Schyns JF, Krol MS et al (2021) Can crop residues provide fuel for future transport? Limited global residue bioethanol potentials and large associated land, water and carbon footprints. Renew Sustain Energy Rev 149:111417. https://doi.org/10.1016/j.rser.2021.111417
52. Karlsson H, Ahlgren S, Sandgren M et al (2017) Greenhouse gas performance of biochemical biodiesel production from straw: soil organic carbon changes and time-dependent climate impact. Biotechnol Biofuels 10:217. https://doi.org/10.1186/s13068-017-0907-9
53. Coppola F, Bravi M, Ridolfi R, Tiezzi E (2007) Decreasing greenhouse effect in agriculture using biodiesel – when green may be enough, pp 213–221
54. Persson T, Garcia AGY, Paz JO et al (2010) Reduction in greenhouse gas emissions due to the use of bio-ethanol from wheat grain and straw produced in the south-eastern USA. J Agric Sci 148:511–527. https://doi.org/10.1017/S0021859610000316
55. Wang W (2022) Agricultural and forestry biomass for meeting the renewable fuel standard: implications for land use and GHG emissions. Energies (Basel) 15. https://doi.org/10.3390/en15238796

56. Bhattacharyya P, Barman D (2018) Crop residue management and greenhouse gases emissions in tropical rice lands. In: Muñoz MÁ, Zornoza R (eds) Soil management and climate change. Academic Press, pp 323–335
57. Akbari M, Oyedun AO, Gemechu E, Kumar A (2021) Comparative life cycle energy and greenhouse gas footprints of dry and wet torrefaction processes of various biomass feedstocks. J Environ Chem Eng 9:105415. https://doi.org/10.1016/j.jece.2021.105415
58. Ostadi M, Zang G, Bromberg L et al (2024) Enhancing biomass-to-liquid conversion through synergistic integration of natural gas pyrolysis: process options and environmental implications. Energy Convers Manag 302:118142. https://doi.org/10.1016/j.enconman.2024.118142
59. Sharma T, Dasgupta D, Singh J et al (2020) Yeast lipid-based biofuels and oleochemicals from lignocellulosic biomass: life cycle impact assessment. Sustain Energy Fuels 4:387–398. https://doi.org/10.1039/C9SE00540D
60. Liu Z, Liu H, Yang X (2023) Life cycle assessment of the cellulosic jet fuel derived from agriculture residue. Aerospace 10:129. https://doi.org/10.3390/aerospace10020129
61. Shah J, Valaki J (2023) A comprehensive review on feasibility of different agro residues for production of bio-oil, bio-char and pyro-gas. J Kejuruteraan 35:77–93. https://doi.org/10.17576/jkukm-2023-35(1)-08
62. Rodríguez-Espinosa T, Navarro-Pedreño J, Gómez Lucas I et al (2023) Environmental risk from organic residues. Sustainability 15:192. https://doi.org/10.3390/su15010192
63. Raseduzzaman M, Gaudel G, Ali MR et al (2024) Cereal-legume mixed residue addition increases yield and reduces soil greenhouse gas emissions from fertilized winter wheat in the North China Plain. Agronomy 14:6. https://doi.org/10.3390/agronomy14061167
64. Guo R, Qian R, Naseer MA et al (2024) Integrated straw-derived biochar utilization to increase net ecosystem carbon budget and economic benefit and reduce the environmental footprint. Field Crop Res 307:109247. https://doi.org/10.1016/j.fcr.2023.109247
65. Urban Cordeiro E, Arenas-Calle L, Woolf D et al (2024) The fate of rice crop residues and context-dependent greenhouse gas emissions: model-based insights from Eastern India. J Clean Prod 435:140240. https://doi.org/10.1016/j.jclepro.2023.140240
66. Ho A, Ijaz UZ, Janssens TKS et al (2017) Effects of bio-based residue amendments on greenhouse gas emission from agricultural soil are stronger than effects of soil type with different microbial community composition. GCB Bioenergy 9:1707–1720. https://doi.org/10.1111/gcbb.12457
67. Waggoner A (2016) Effects of residual waste material as agricultural soil amendments on soil greenhouse gas fluxes. Open Access Master's Theses
68. Brenzinger K, Drost SM, Korthals G, Bodelier PLE (2018) Organic residue amendments to modulate greenhouse gas emissions from agricultural soils. Front Microbiol 9:3035. https://doi.org/10.3389/fmicb.2018.03035
69. Senila L, Kovacs E, Resz MA et al (2024) Life cycle assessment (LCA) of bioplastics production from lignocellulosic waste (study case: PLA and PHB). Polymers (Basel) 16:3330. https://doi.org/10.3390/polym16233330
70. Synani K, Abeliotis K, Velonia K et al (2024) Environmental impact and sustainability of bioplastic production from food waste. Sustainability 16:5529. https://doi.org/10.3390/su16135529
71. Aliyeva S (2024) Life cycle assessment of polymer matrix production of nano-ZnO antimicrobial packaging materials. In: PAHTEI-proceedings of Azerbaijan High Technical Educational Institutions, pp 127–135
72. Fonseca A, Ramalho E, Gouveia A et al (2023) Life cycle assessment of PLA products: a systematic literature review. Sustainability 15:12470
73. Stagner J (2016) Methane generation from anaerobic digestion of biodegradable plastics – a review. Int J Environ Stud 73:462–468. https://doi.org/10.1080/00207233.2015.1108607
74. Gadaleta G, De Gisi S, Picuno C et al (2024) Assessment of methane production, disintegration, and biodegradation potential of bioplastic waste in anaerobic digestion systems. J Environ Chem Eng 12:111658. https://doi.org/10.1016/j.jece.2023.111658

75. Mafe AN, Edo GI, Akpoghelie PO et al (2025) Comparative analysis of the environmental impact of biopolymer-based and conventional plastic packaging in food engineering applications. Al-Mustaqbal J Sustain Eng Sci 2. https://doi.org/10.62723/2959-5932.1017
76. Park D, Lee H, Won W (2024) Unveiling the environmental gains of biodegradable plastics in the waste treatment phase: a cradle-to-crave life cycle assessment. Chem Eng J 487:150540. https://doi.org/10.1016/j.cej.2024.150540
77. Qiu J, Tao K, Yang Y et al (2025) Environmental impacts of polylactic acid synthesis via diverse biorefinery strategies: progressive analysis based on dynamic lifecycle assessment. ACS Sustain Chem Eng 13:898–910. https://doi.org/10.1021/acssuschemeng.4c07568
78. Samer M, Hijazi O, Mohamed BA et al (2022) Environmental impact assessment of bioplastics production from agricultural crop residues. Clean Techn Environ Policy 24:815–827. https://doi.org/10.1007/s10098-021-02145-5
79. Kovacs E, Hoaghia MA, Senila L et al (2022) Life cycle assessment of biofuels production processes in viticulture in the context of circular economy. Agronomy 12:1320. https://doi.org/10.3390/agronomy12061320
80. Rebolledo-Leiva R, Ladakis D, Ioannidou S-M et al (2023) Attributional and consequential life cycle perspectives of second-generation polylactic acid: the benefits of integrating a recycling strategy. J Clean Prod 420:138354. https://doi.org/10.1016/j.jclepro.2023.138354
81. Luo C, Zhou Y, Chen Z et al (2024) Comparative life cycle assessment of PBAT from fossil-based and second-generation generation bio-based feedstocks. Sci Total Environ 954:176421. https://doi.org/10.1016/j.scitotenv.2024.176421
82. Li J, Wang Y, Xu C et al (2024) Bioplastic derived from corn stover: life cycle assessment and artificial intelligence-based analysis of uncertainty and variability. Sci Total Environ 946: 174349. https://doi.org/10.1016/j.scitotenv.2024.174349
83. Bishop G, Styles D, Lens PNL (2022) Land-use change and valorisation of feedstock side-streams determine the climate mitigation potential of bioplastics. Resour Conserv Recycl 180: 106185. https://doi.org/10.1016/j.resconrec.2022.106185

The manufacturer's authorised representative in the EU is Springer Nature Customer Service Centre GmbH, Europaplatz 3, 69115 Heidelberg, Germany. If you have any concerns regarding our products, please contact ProductSafety@springernature.com

Printed and bound by CPI Group (UK) Ltd, Croydon, CR0 4YY
16/09/2026
02217815-0011